Gold–Polymetallic Deposits in Convergent Margins

Gold–Polymetallic Deposits in Convergent Margins

Guest Editors

Haocheng Yu
Hao Song
Mingyang Wang

MDPI

Basel • Beijing • Wuhan • Barcelona • Belgrade • Novi Sad • Cluj • Manchester

Guest Editors
Haocheng Yu
School of Earth Sciences and Resources
China University of Geosciences
Beijing
China

Hao Song
Applied Nuclear Technology in Geoscience Key Laboratory of Sichuan Province
Chengdu University of Technology
Chengdu
China

Mingyang Wang
State Key Laboratory of Hydro-science and Engineering
Tsinghua University
Beijing
China

Editorial Office
MDPI AG
Grosspeteranlage 5
4052 Basel, Switzerland

This is a reprint of the Special Issue, published open access by the journal *Minerals* (ISSN 2075-163X), freely accessible at: https://www.mdpi.com/journal/minerals/special_issues/6EYHX37285.

For citation purposes, cite each article independently as indicated on the article page online and as indicated below:

Lastname, A.A.; Lastname, B.B. Article Title. *Journal Name* **Year**, *Volume Number*, Page Range.

ISBN 978-3-7258-8047-8 (Hbk)
ISBN 978-3-7258-8048-5 (PDF)
https://doi.org/10.3390/books978-3-7258-8048-5

Contents

Xian-Fa Xue, Sheng-Xiang Lu, Shou-Xu Wang, Da-Hu Yuan, Zheng-Wang Zeng, Jin-Hong Qiu and Jie Wang
Geochronological Constraints on the Genesis of the Changshitougounao Gold Deposit, Qinling Orogen
Reprinted from: *Minerals* **2025**, *15*, 903, https://doi.org/10.3390/min15090903 1

Yangxin Lu, Weiming Jiang, Molei Zhao, Yuanzhi Zhou, Jie Yang, Kunfeng Qiu and Qiuming Cheng
Anomaly Detection in Mineral Micro-X-Ray Fluorescence Spectroscopy Based on a Multi-Scale Feature Aggregation Network
Reprinted from: *Minerals* **2025**, *15*, 970, https://doi.org/10.3390/min15090970 16

Fei Ren, Zheng-Jiang Ding, Zhong-Yi Bao, Jun-Wei Wang, Shun-Xi Ma, Tao Niu, et al.
Source and Precipitation Process of Gold in the Linglong Gold Deposit, Jiaodong Peninsula: Constraints from Trace Elements of Pyrite and S-Pb Isotopes
Reprinted from: *Minerals* **2025**, *15*, 1220, https://doi.org/10.3390/min15111220 35

Ning-Han Sun, Tao Cui, Rong-Zhi Zhou, Yu-Ying Li, Qiang Fu, Lian-Yuan Qin, et al.
Genesis of High-Grade Gold Mineralization at the Guocheng Deposit, Jiaodong Peninsula: Constraints from Magnetite Geochemistry
Reprinted from: *Minerals* **2025**, *15*, 1267, https://doi.org/10.3390/min15121267 58

Yi-Xue Gao, Shan-Shan Li, Chuan-Peng Liu, Ming-Qian Wu, Zhen Shang, Ze-Yu Yang, et al.
Fluorite Composition Constraints on the Genesis of the Weishan REE Deposit, Luxi Terrane
Reprinted from: *Minerals* **2026**, *16*, 69, https://doi.org/10.3390/min16010069 78

Yi-Xue Gao, Shan-Shan Li, Chuan-Peng Liu, Ming-Qian Wu, Zhen Shang, Yi-Zhan Sun, et al.
Apatite Geochemical Signatures of REE Ore-Forming Processes in Carbonatite System: A Case Study of the Weishan REE Deposit, Luxi Terrane
Reprinted from: *Minerals* **2026**, *16*, 112, https://doi.org/10.3390/min16010112 101

Chang-Qing Dong, Zhao-Hua Bai, Ke Yang, Meng-Zhen Hao, Jia-Yi Wang, Hao Zhou and Jia-Nan Fu
Genesis of the Yawan Gold Deposit, West Qinling Orogen: Insights from Calcite U-Pb Geochronology and Geochemistry of Sulfides
Reprinted from: *Minerals* **2026**, *16*, 114, https://doi.org/10.3390/min16010114 124

Jia-Hui Zhou, Nan Ju, Qun-Feng Miao, Zhuo-Er Teng, Xiao-Dong Wang, Xiao-Xia Li, et al.
Petrogenesis of Rhyolitic Porphyry Hosting the Newly Discovered Dengshang Mo Deposit, Northern Hebei Province
Reprinted from: *Minerals* **2026**, *16*, 249, https://doi.org/10.3390/min16030249 142

Jiabao Jia, Yueliang Hu, Lin Gao, Yulu Lv, Junjie Wang, Xiaomei Yang, et al.
Whole-Rock Element Analyses Constraining the Magmatic Evolution and Metallogenesis of the Jiaojia Fault Zone, Jiaodong Gold Province
Reprinted from: *Minerals* **2026**, *16*, 350, https://doi.org/10.3390/min16040350 159

ElSayed A. Saber, H. M. Hamouda, A. S. Hamid and Ahmed A. El-Sheikh
Trace-Element Systematics and Multivariate Insights into Gold Fertility of Arsenopyrite from the Um Rus Orogenic Gold Deposit, Central Eastern Desert, Egypt
Reprinted from: *Minerals* **2026**, *16*, 439, https://doi.org/10.3390/min16050439 180

Jin-Shuai Zhang, Hao-Cheng Yu, Guo-Long Yan, Ming Ma, Tao Cui, Ya-Peng Li, et al.
Genesis of the Longkou Gold Deposit in the Northeastern Jiaolai Basin: Constraints from Sericite Rb-Sr Geochronology and Pyrite Geochemistry
Reprinted from: *Minerals* **2026**, *16*, 485, https://doi.org/10.3390/min16050485 . . . . **206**

Zhe-Han Li, Hong-Jie Tan, Ding-Fan Xing, Tao Yang, Tong-Tong He, Jia-Nan Fu and Kun-Feng Qiu
Hyperspectral Drill Core Spectroscopy for Alteration Zoning in Orogenic Gold Deposits: Insights from the Liba Deposit, West Qinling, China
Reprinted from: *Minerals* **2026**, *16*, 526, https://doi.org/10.3390/min16050526 . . . . **222**

minerals

MDPI

Article

Geochronological Constraints on the Genesis of the Changshitougounao Gold Deposit, Qinling Orogen

Xian-Fa Xue [1], Sheng-Xiang Lu [2,*], Shou-Xu Wang [2], Da-Hu Yuan [3], Zheng-Wang Zeng [4], Jin-Hong Qiu [1] and Jie Wang [1,5,*]

1 State Key Laboratory of Geological Processes and Mineral Resources, Frontiers Science Center for Deep-Time Digital Earth, School of Earth Sciences and Resources, China University of Geosciences, Beijing 100083, China; xianfaxue@email.cugb.edu.cn (X.-F.X.); 2001240055@email.cugb.edu.cn (J.-H.Q.)

2 Shandong Gold Geology and Mineral Exploration Co., Ltd., Laizhou 261400, China; shiuhuwanggold@163.com

3 The Third Geological and Mineral Exploration Institute of Gansu Provincial Bureau of Geology and Mineral Resources, Lanzhou 730000, China; yuandahugsth@163.com

4 Northwest Gold Co., Ltd., Lanzhou 730050, China; zzwanggold@163.com

5 Centre for Exploration Targeting, School of Earth Sciences, The University of Western Australia, 35 Stirling Hwy, Crawley, WA 6009, Australia

* Correspondence: lushx@sd-gold.com (S.-X.L.); jiewang@email.cugb.edu.cn (J.W.); Tel.: +86-15253507131 (S.-X.L.); +86-15689940717 (J.W.)

Abstract

The Western Qinling Orogenic Belt, China's second-largest Au-metallogenic province, hosts numerous polymetallic deposits, with gold resources particularly concentrated in the northwestern Xiahe–Hezuo area. The Changshitougounao gold deposit, located south of the Xiahe Fault, comprises disseminated ores controlled by near E–W-trending faults and is primarily hosted in quartz diorite and the Lower Triassic Longwuhe Formation. Zircon LA–ICP–MS U–Pb dating of fresh quartz diorite yields an age of 241.8 ± 2.6 Ma. Two generations of monazite were identified: type I magmatic monazite and type II hydrothermal monazite. Type I monazite is intergrown with feldspar, quartz, and biotite, and in situ LA–ICP–MS U–Pb analysis gives an age of 239.2 ± 2.2 Ma. Type II monazite occurs as irregular granular aggregates associated with Au-bearing sulfides and hydrothermal sericite, with an in situ U–Pb age of 230 ± 3.5 Ma. Apatite, also coeval with Au-bearing sulfides and type II monazite, yields an LA–ICP–MS U–Pb age of 230.9 ± 2.5 Ma and 230.7 ± 3.0 Ma. Zircon and type I monazite thus constrain the emplacement of the ore-bearing quartz diorite to ca. 240 Ma, whereas hydrothermal type II monazite and apatite constrain the timing of mineralization to ca. 230 Ma. The ~10 Ma interval between magmatism and mineralization indicates that goldmineralization in the Changshitougounao deposit is decoupled from Early Triassic magmatic activity. Integrating previous studies of the West Qinling geodynamic evolution, we infer that the Changshitougounao deposit formed during collisional orogenesis, in response to the closure of the Paleo-Tethys Ocean. Consequently, the Changshitougounao gold deposit is best classified as an orogenic gold system. Pyrite–arsenopyrite and sericite alteration serve as effective exploration vectors, and the contact zone between quartz diorite veins and slate represents a favorable structural setting for ore prospecting.

Keywords: Zircon U–Pb; apatite U–Pb; monazite U–Pb; Changshitougounao orogenic gold deposit; West Qinling

1. Introduction

Accurately constraining the timing of orogenic gold mineralization is fundamental to understanding both deposit genesis and geodynamic evolution [1–3]. The West Qinling Orogenic Belt, situated between the North China and South China Blocks, hosts over 2280 t of gold, making it China's second-largest gold province (Figure 1) [4,5]. The Xiahe–Hezuo area represents the largest gold-bearing metallogenic cluster in West Qinling, with more than 665 t of gold and numerous polymetallic deposits [6,7]. The host rocks comprise widespread Early–Middle Triassic granitoids and Devonian–Permian sedimentary rocks [8,9]. Zircon U–Pb geochronology indicates regional magmatic activity at ca. 250–235 Ma [10–14]. There are two predominant deposit types in the Xiahe–Hezuo area: orogenic gold (-antimony) deposits and magmatic–hydrothermal deposits [15–17]. Hydrothermal sericite, apatite, and monazite ages suggest that orogenic gold mineralization was concentrated between ca. 235 Ma and 210 Ma, exemplified by Zaozigou (106 t), Jiagantan (153 t), Yidinan (34 t), and Nanban (2.6 t) deposits [13,14,18–20]. In contrast, intrusion-related gold systems, such as the Ludousou and the Dewulu Cu–Au skarn, record emplacement ages of ca. 250–235 Ma [15,21,22]. The Changshitougounao deposit, whose orebodies are entirely concealed and located south of the Xiahe–Hezuo fault, makes precise determination of its mineralization timing crucial for understanding its genesis. This also provides a framework for constraining its origin and guiding future exploration in the Gannan area. In this study, we describe the geology of the Changshitougounao gold deposit and innovatively apply coeval U–Pb geochronology of magmatic zircon, hydrothermal monazite, and apatite to precisely constrain the timing of magmatic activity and gold mineralization.

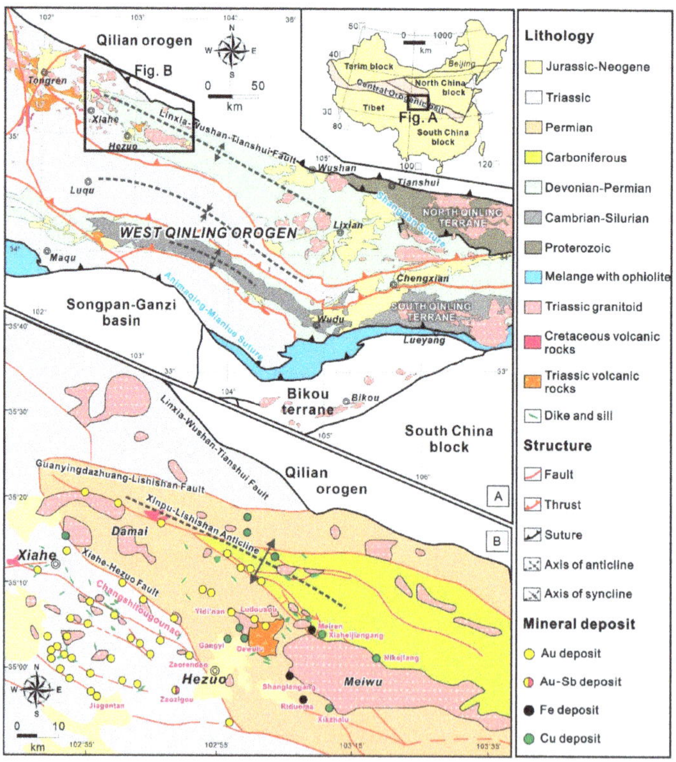

Figure 1. Simplified geological map of the West Qinling Orogenic Belt (**A**) and simplified geological map and mineral deposit distribution of the Xiahe–Hezuo district (**B**) (modified after [21]).

2. Regional Geology

The Qinling Orogenic Belt, a principal segment of the Central China Orogenic Belt, extends > 1500 km E–W across central China (Figure 1) [23–25]. Its northern boundary is delineated by the Lingbao–Lushan–Wuyang Fault and Qilian Orogen, demarcating the North China Block [26,27]. The southern margin is defined by the Mianlue–Bashan–Xiangguang and Songpan–Ganzi Faults, separating it from the South China Block [2,28]. Three major sutures, Kuangping, Shandan, and Mianlue, subdivide the belt into the southern North China Block, the North Qinling Block, the South Qinling Block, and the northern South China Block [29]. The Kuangping Suture, located between the South China Block and the North Qinling Block, is characterized by extensive Paleo- to Mesoproterozoic ophiolitic mélanges, dominated by greenschist- and amphibolite-facies volcanic-arc rocks dated at ca. 1.45–0.95 Ga [27]. The Shangdan Suture to the south contains Paleozoic ophiolitic assemblages with subduction-related volcanic-sedimentary suites [30]. The Mianlue Suture, located between the South Qinling Block and the South China Block, comprises ophiolitic mélanges, ocean–island basalts, and island-arc volcanic sequences [31–33].

Bounded by the Huicheng Basin and Foping Anticline, the Qinling Belt is conventionally divided into the East and West Qinling Orogenic Belts [34,35]. The West Qinling Orogen preserves a record of polyphase tectonism associated with the Paleo-Tethys Ocean evolution, culminating in the Late Triassic North China–South China block collision, constituting the eastern Paleo-Tethys Oceans [36,37]. Devonian–Triassic sedimentary cover is widespread, whereas exposure of Precambrian basement is sparse [31]. Mesozoic magmatism was vigorous, especially during the Early to Middle Triassic, and gave rise to numerous orogenic and magmatic–hydrothermal gold deposits, making it China's second largest gold-bearing district [5,17,24,38–42]. Situated at the northwestern margin of the West Qinling Orogen, the Xiahe–Hezuo area comprises Carboniferous–Triassic marine sedimentary sequences intruded by Early–Middle Triassic granitoids, with subsidiary Permian volcanics and Cretaceous volcaniclastic units [37]. NWW-trending structures, notably the Xiahe–Hezuo Fault and the Xinbao–Lishishan anticline, dominate the regional structural fabric. South of the Xiahe–Hezuo Fault, the area is subdivided into northeast and southwest domains, marked by porphyry–skarn gold deposits genetically linked to Triassic granitoids, and by vein–disseminated gold–antimony deposits, respectively [43,44].

3. Deposit Geology

The Changshitougounao gold deposit (102°34′15″–102°41′00″E, 35°04′30″–35°10′00″N) is situated near Damai Village, east of Xiahe County in Gansu Province. Systematic exploration, initiated in 2014 by the Gansu Provincial Bureau of Geology and Minerals Exploration and Development has defined resources > 1 t Au at 2.53 g/t. The Changshitougounao is a concealed deposit with no surface expression, and it remains at an early exploration stage without any underground workings [45,46]. Mineralization is hosted principally within Triassic quartz diorite intrusions and the Longwuhe Formation metasedimentary sequence (Figure 2). The Longwuhe Formation comprises predominantly silty and argillaceous slates interlayered with minor clastic limestone and fine sandstone. Structural control is exerted by both regionally extensive E–W-trending and secondary E–W-trending faults. The former serve as the principal ore conduits and dilatant zones, while the latter are subordinate fractures that localize mineralization. Quartz diorite dikes intrude along the E–W fracture corridors. Alteration assemblages are characteristic of mid- to low-temperature fluids and include silicification, pyritization, carbonatization, and arsenopyrite mineralization. Pyrite and arsenopyrite are the principal sulfide phases, with subordinate stibnite locally present. Gold occurs exclusively as invisible inclusions or lattice-bound forms within pyrite and arsenopyrite. And no native gold was observed in our petrographic analysis [45].

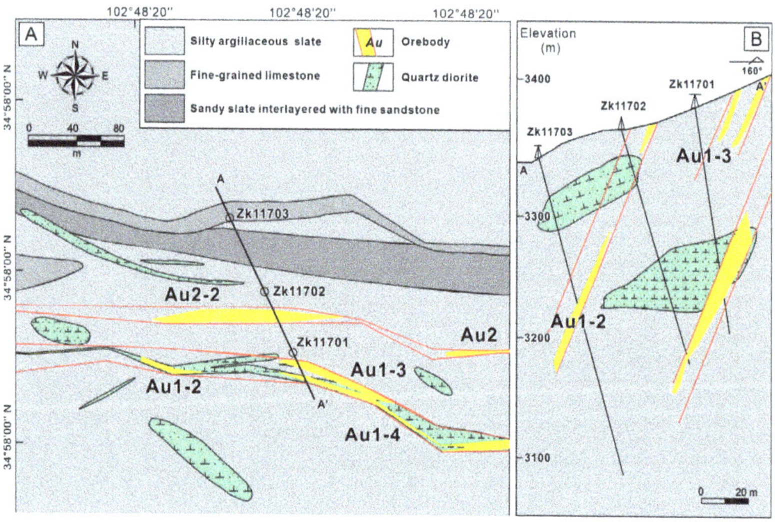

Figure 2. Simplified geological map of the Changshitougounao gold deposit (**A**) and schematic cross-section along exploration profile A–A′ (**B**).

4. Samples and Analytical Methods

4.1. Sample Descriptions

Three quartz diorite samples were selected from drill holes Zk11701 and Zk11702 along exploration line 117: one fresh sample (24CSTGN01) and two mineralized samples (24CSTGN02 and 24CSTGN03). Zircon U–Pb dating was performed on 24CSTGN01 to constrain the crystallization age of the intrusive body. Hydrothermal monazite and apatite U–Pb dating was conducted on 24CSTGN02 and 24CSTGN03 to determine the timing of mineralization.

The fresh quartz diorite (sample 24CSTGN01) is gray-black and massive, with discernible quartz grains and minor coarse-grained dark minerals, including biotite and hornblende (Figure 3). Phenocrysts constitute ~50% of the rock volume, comprising plagioclase (~25%), quartz (~15%), biotite (~5%), and hornblende (~5%). Plagioclase and quartz phenocrysts typically range from 0.2 to 1.0 mm. Biotite and hornblende phenocrysts are generally 0.2–0.3 mm. Plagioclase and quartz occur as tabular grains, biotite as euhedral flakes, and hornblende as prismatic crystals. The fine-grained groundmass is cryptocrystalline, comprising plagioclase (15%–30%), quartz (5%–25%), and accessory zircon and apatite.

The two mineralized quartz diorite samples (24CSTGN02 and 24CSTGN03) are also gray-black but display pervasive alteration (Figure 4). Primary minerals include quartz, feldspar, sericite, pyrite, arsenopyrite, and minor stibnite. Plagioclase and biotite phenocrysts exhibit varying degrees of sericitization to fine-grained sericite. The groundmass consists of plagioclase, quartz, and subordinate K-feldspar, with accessory apatite, monazite, and sericite (Figure 5). Pyrite occurs both as euhedral grains and as disseminated aggregates, locally accompanied by intense sericitization. Detailed petrographic observations of monazite and apatite were carried out using both optical microscopy (Leica DM2700P, Leica Microsystems, Wetzlar, Germany) and scanning electron microscopy (SEM) (Thermo Fisher Scientific, Waltham, MA, USA) at China University of Geosciences, Beijing, to ensure that the dated grains accurately reflect the age of the gold deposit.

Figure 3. Hand specimen of quartz diorite (**A**) and photomicrographs showing petrographic features (**B–D**). Pl = plagioclase, Qz = quartz, Bt = biotite, and Hbl = hornblende.

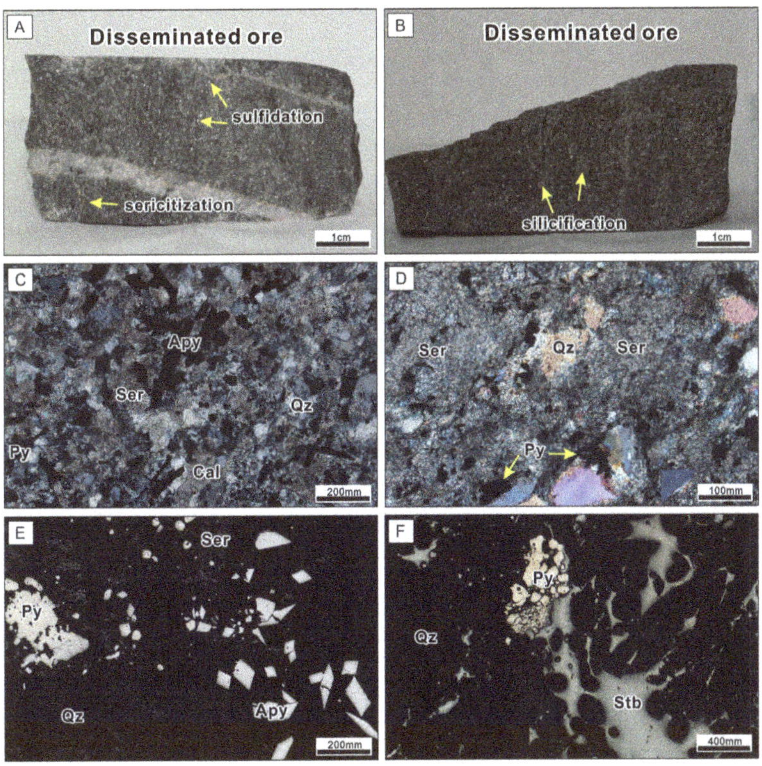

Figure 4. Hand specimens and petrographic characteristics of mineralized quartz diorite. Hand specimens showing silicification, sulfidation, and sericitization (**A**,**B**); photomicrographs exhibiting sericitization and carbonatization (**C**,**D**); and pyrite–arsenopyrite–stibnite assemblage (**E**,**F**). Ser = sericite, Cal = calcite, Py = pyrite, Apy = arsenopyrite, and Stb = stibnite.

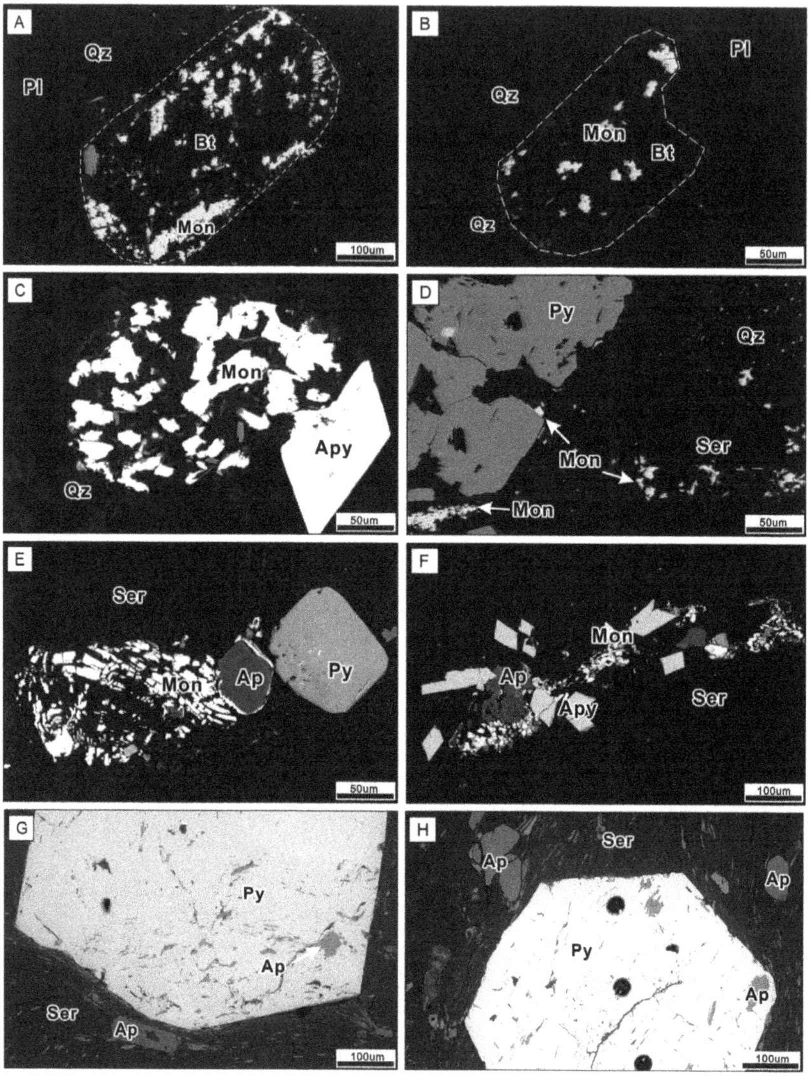

Figure 5. Photomicrographs showing the texture and mode of occurrence of type I (**A**,**B**) and type II (**C**–**F**) monazite, and apatite grains (**G**,**H**). The yellow circles highlight biotite grain boundaries. Mon = monazite), Ap = apatite.

4.2. Analytical Methods

4.2.1. Zircon and Apatite LA–ICP–MS U–Pb Dating

The fresh quartz diorite and mineralized samples were first crushed to 40–60 mesh. Zircon and apatite crystals were separated using conventional heavy-liquid and magnetic techniques, then handpicked under a Leica M205C binocular microscope (Leica Microsystems, Wetzlar, Germany) at the Langfang Regional Geological Survey, Hebei Province. Selected grains were mounted in epoxy, polished to ~50% thickness to expose internal features, and cleaned in a 5% HNO_3 ultrasonic bath. Before U–Pb analysis, mounts were carbon-coated for cathodoluminescence (CL) imaging. Zircon CL micrographs were acquired with a JXA-880 electron microscope (JEOL Ltd., Tokyo, Japan) (20 kV, 20 nA) at the

Langfang Regional Geological Survey, Hebei Province. Apatite CL images were captured on a Tescan GAIA 3 SEM (TESCAN ORSAY HOLDING, a.s., Brno, Czech Republic) at the Beijing Research Institute of Uranium Geology. All grains were meticulously examined to document internal zoning, inclusions, and fractures.

Zircon LA–ICP–MS U–Pb age dating was performed using a Thermo Fisher Neptune multicollector ICP-MS (Thermo Fisher Scientific, Waltham, MA, USA) coupled with an ESI New Wave 193 nm FX ArF excimer laser ablation system (Elemental Scientific Lasers, Bozeman, MT, USA) at Tianjin Center, China Geological Survey. The laser beam spot diameter was set to 30 μm with an 8 Hz repetition rate. The zircon standard GJ-1 was used for U–Pb age calibration, and NIST SRM 610 glass was used as an external standard for determining U, Th, and Pb concentrations in zircon. Apatite LA–ICP–MS U–Pb age determinations were carried out using a laser ablationmulticollector ICP-MS system consisting of a New Wave 193-FX ArF excimer laser coupled to a Thermo Fisher Neptune MC–ICP–MS at Tianjin Center, China Geological Survey. The laser spot size was 40 μm. NIST SRM 610 glass was used as the external standard for elemental concentrations, and Otter Lake monazite was used as the apatite U–Pb age standard. Isotopic ratios of zircon and apatite were calculated using ICPMSDataCal 11.8 [47], and data processing methods were used with Isoplot 3.0 [48].

4.2.2. Monazite LA–ICP–MS U–Pb Dating

Monazite grains were first identified in doubly polished thin sections using optical microscopy and SEM at China University of Geosciences, Beijing, then targeted for in situ U–Pb analysis. Geochronological measurements were conducted at Yanduzhongshi Geological Analysis Laboratories Ltd. Analyses were performed on an Analytik Jena PQMS quadrupole ICP-MS coupled to a NWR193 193 nm ArF excimer laser, employing 30 μm spot sizes at ~4 J/cm^2 laser fluence. Common-Pb corrections followed the ^{207}Pb-based are detailed in the references [49,50]. NIST SRM 610 glass was used as an external standard for the determination of U, Th, and Pb concentrations. Monazite 44069 was used as the external standard for monazite U–Pb dating analyses. Concordia diagrams U–Pb ages were processed using Isoplot 3.0.

5. Results

5.1. Zircon U–Pb Geochronology

LA–ICP–MS U–Pb analyses of the fresh quartz diorite (sample 24CSTGN01) is shown in Supplementary Table S1. Cathodoluminescence (CL) imaging (Figure 6A) reveals that most grains are euhedral with well-developed oscillatory zoning. Grain lengths range from 100 to 200 μm, with aspect ratios of 2:1 to 3:1 and Th/U ratios above 0.28, consistent with magmatic zircon [51]. Nineteen grains were analyzed, of which seventeen yielded concordant ages. The weighted mean $^{206}Pb/^{238}U$ age is 241.8 ± 2.6 Ma (MSWD = 1.6; n = 17; Supplementary Table S1, Figure 6B).

5.2. Monazite U–Pb Geochronology

Two distinct types of monazites (Type I and Type II) have been identified based on textural and compositional criteria. Type I monazite predominantly occurs within biotite (Figure 5A,B). It typically forms euhedral to subhedral single crystals. LA–ICP–MS analyses reveal uniformly high Th abundances, ranging from 4340 to 73,000 ppm (mean: 21,027 ppm), with Th/U ratios constrained between 24.1 and 968.2. In situ U–Pb dating of this monazite type, summarized in Supplementary Table S2 and displayed on a Tera–Wasserburg concordia diagram, yielded a lower intercept age of 239.2. ± 2.2 Ma (MSWD = 0.34, n = 18) (Figure 7A).

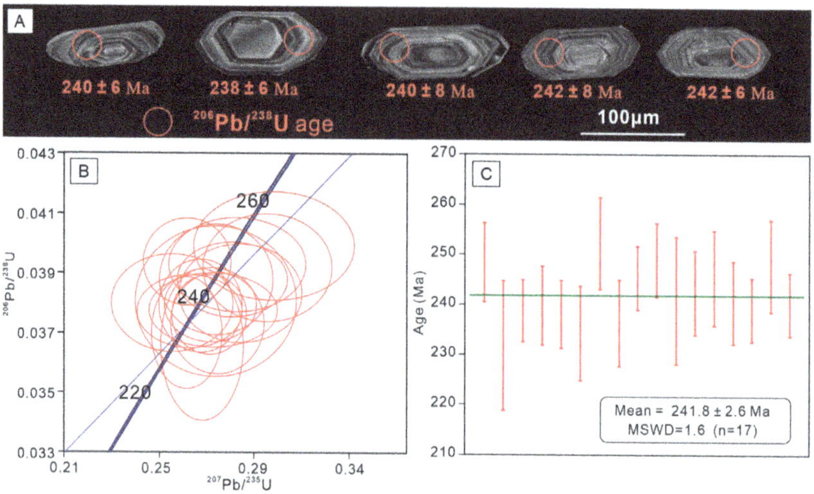

Figure 6. Representative zircon cathodoluminescence images (**A**) and zircon LA–ICP–MS U–Pb concordia diagrams of fresh quartz diorite (**B**,**C**).

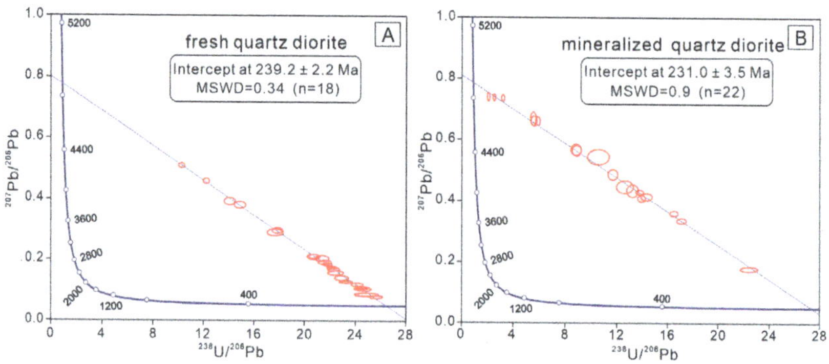

Figure 7. In situ LA–ICP–MS U–Pb Tera–Wasserburg concordia diagrams of type I (**A**) and type II (**B**) monazite grains.

Type II monazite is characteristically anhedral, with abundant dissolution cavities, and commonly forms irregular aggregates of fine-grained anhedral crystals (Figure 5C–F). This type exhibits highly variable Th concentrations (884–42,700 ppm; mean = 18,229 ppm) and a broad range of Th/U ratios (0.51–159.3). In situ U–Pb dating defines a lower intercept age of 231.0 ± 3.5 Ma (MSWD = 0.9, n = 22; Supplementary Table S2; Figure 7B).

5.3. Apatite U–Pb Geochronology

Apatite separates from samples 24CSTGN02 and 24CSTGN03 occur as semi-euhedral to anhedral grains within or adjacent to pyrite and arsenopyrite. CL images reveal two types of luminescence: one appears black with greenish rims, and the other is uniformly green. Despite these differences, both types yield consistent LA–ICP–MS lower intercept ages of 230.9 ± 2.5 Ma (MSWD = 1.6, n = 28; Supplementary Table S3; Figure 8) and 230.7 ± 3.0 Ma (MSWD = 3.0, n = 42; Supplementary Table S3; Figure 8), indicating consistent ages.

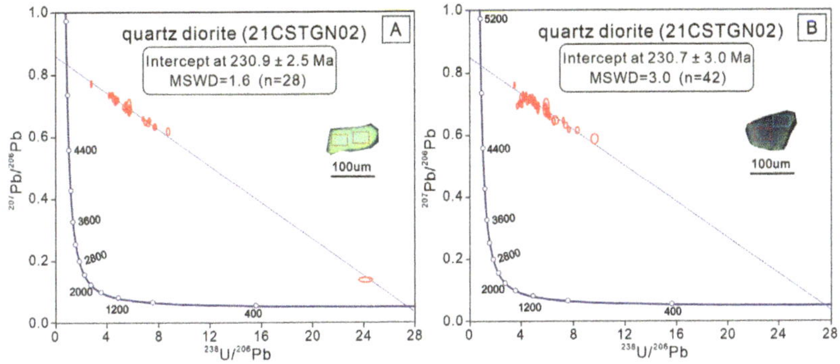

Figure 8. LA–ICP–MS U–Pb Tera–Wasserburg concordia diagrams of apatite grains. Representative cathodoluminescence images with identified analytical spot and U–Pb age (Ma) value are also shown (**A**,**B**). Representative cathodoluminescence images of apatite grains with identified analytical spot are also shown. Red box circle: U-Pb beam.

6. Discussion

6.1. Origin of Monazite and Apatite

Monazite and apatite are ubiquitous accessory silicate minerals that can form via magmatic, metamorphic, and hydrothermal processes [52–54]. Type I monazite occurs as euhedral grains intergrown with quartz and magmatic biotite, consistent with a magmatic origin (Figure 5A,B) [55]. In contrast, Type II monazite forms irregular aggregates of multiple anhedral to subhedral micro-grains, a habit distinct from magmatic monazite (Figure 5C–F). Petrographic observations show that Type II monazite is intimately associated with pyrite, arsenopyrite, and sericite, indicating coeval crystallization with sulfide mineralization from a common hydrothermal fluid [21,41,42,56–58]. Monazite has a relatively high closure temperature of approximately 700–750 °C, making it a robust geochronometer that is generally resistant to resetting during low- to medium-grade metamorphism [59]. Therefore, the low- to medium-grade metamorphism that occurred in the Xiahe–Hezuo district during north-directed subduction and collision of the Paleo-Tethys Ocean (~250–230 Ma) would not have reset the U–Pb isotopic system in monazite [42,55,60,61]. Additionally, apatite occurs mainly at the margins of or within sulfide phases, and cathodoluminescence images show evidence of hydrothermal dissolution and reprecipitation, indicating alteration by mineralizing hydrothermal fluids (Figure 5E–H).

6.2. Magmatism Timing

The ore-hosting intrusion of the Changshitougounao gold deposit is a quartz diorite. LA–ICP–MS U–Pb dating of magmatic zircon from the fresh quartz diorite yields an age of 241.8 ± 2.6 Ma, while magmatic monazite yields 239.2 ± 2.2 Ma. These ages overlap within analytical uncertainty and indicate a crystallization age of ca. 240 Ma for the host intrusion.

Regional studies document widespread Early–Middle Triassic magmatism in the Xiahe–Hezuo area (ca. 250–235 Ma; Figure 9) [17,19,38,62]. Specifically, the Ayishan intrusion crystallized at 242–238 Ma, the Dewulu complex at 247–238 Ma, the Tongren intrusion at 242–238 Ma, the Meiwu intrusion at 251–243 Ma, and the host rock at Zaozigou, Ludousou, Zaorendao, and Yidinan similarly crystallized at ca. 235–250 Ma [3,22,36,63–66]. In close agreement with these regional data, zircon and magmatic monazite from the fresh quartz diorite of the Changshitougounao gold deposit define ages of ca. 240 Ma, confirming that the Changshitougounao gold deposit formed during the Early Triassic magmatic episode in the West Qinling orogen.

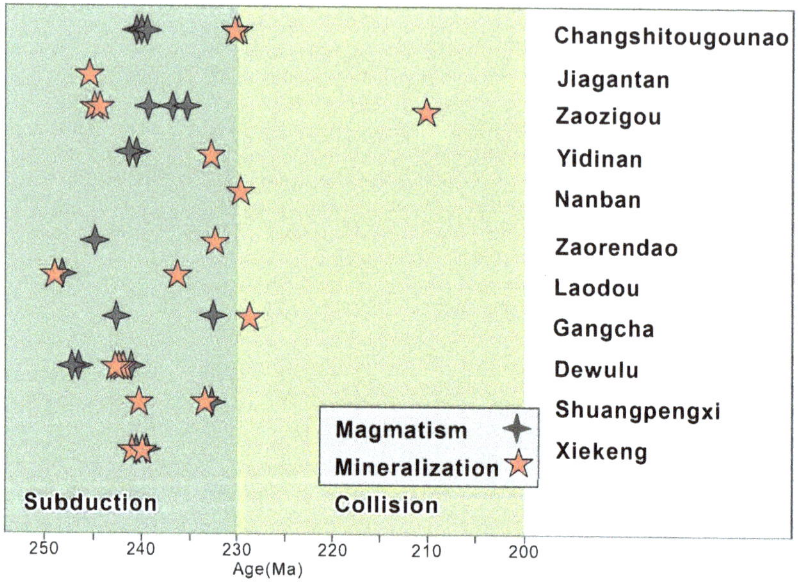

Figure 9. Magmatic activity and mineralization ages in the Xiahe–Hezuo district. Previous data are from [11,13–15,19,21,22,41,42,65,67–73].

6.3. Mineralization Timing

The in situ LA–ICP–MS U–Pb ages of hydrothermal monazite (231.5 ± 3.5 Ma) and apatite (230.9 ± 2.5 Ma, 230.7 ± 3.0 Ma) constrain gold mineralization at the Changshitougounao gold deposit to ca. 230 Ma. This age is consistent with regional orogenic gold mineralization in the Xiahe–Hezuo district, which is concentrated between 235 Ma and 210 Ma, as exemplified by the Zaozigou deposit (ca. 211 Ma, hydrothermal monazite in situ LA–ICP–MS U–Pb), and the Yidinan deposit (232 Ma, hydrothermal monazite in situ LA–ICP–MS U–Pb) [21,67]. In contrast, regional magmatic–hydrothermal deposits formed earlier, between 250 Ma and 235 Ma, including the Laodou deposit (249.1 Ma and 235 Ma, hydrothermal sericite $^{40}Ar/^{39}Ar$), Au–Cu skarn deposits adjacent to the Dewulu intrusion (ca. 240 Ma, hydrothermal sericite $^{40}Ar/^{39}Ar$), and the Xiekeng–Jiangligou–Shuangpengxi polymetallic deposit (ca. 240 Ma, molybdenite Re–Os) [2,15,19,22,65,68,74,75]. Thus, the ca. 230 Ma mineralization age at Changshitougounao aligns with the regional orogenic gold event. These data bracket mineralization in the district coeval with, or slightly younger than, Early Triassic magmatism.

6.4. Genetic Classification of the Changshitougounao Gold Deposit

The Xiahe–Hezuo polymetallic belt, located at the western margin of the West Qinling Orogenic Belt, contains the region's largest polymetallic mineral resources. Deposits here are chiefly classified as either orogenic or magmatic–hydrothermal. Orogenic gold deposits, such as Zaozigou (106 t), Yidinan (34 t), Nanban (2.6 t), and Wanken (4 t), are typically hosted in Triassic metasedimentary successions and Early–Middle Triassic granitoids and are controlled by secondary regional fault sets [7,76]. Host-rock alteration is dominated by intense silicification, sericitization, pyritization, arsenopyritization, and carbonatization. Orebodies exhibit disseminated and vein-hosted textures, with pyrite, arsenopyrite, stibnite, and sphalerite as the principal sulfides [19,21,76,77]. Hydrothermal minerals dating of these orogenic systems yields mineralization ages of 235–210 Ma (Figure 9). By contrast, magmatic–hydrothermal deposits, for example, the Xiekeng–Jiangligou–Shuangpengxi

Cu–Au–Fe–Mo skarn deposit, the Laodou intrusion-related gold deposit, Gangcha gold deposit and the Dewulu Au–Cu skarns deposits, are hosted entirely within granitoid intrusions and formed synchronously with magmatism [19,22,68,69,74,78,79]. Both the quartz diorite host and the gold mineralization date of the Laodou deposit, to ~249 Ma [22]. Likewise, the Dewulu intrusion and its surrounding skarn-hosted Au–Cu deposits formed at ca. 239 Ma [15,75]. These systems feature mineral assemblages of magnetite–siderite, pyrite, chalcopyrite, bornite, magnetite, chalcocite, arsenopyrite, and native gold, with supergene goethite, azurite, and malachite. Host-rock alteration displays potassic alteration, chloritization, silicification, sericitization, and carbonatization [68,74].

The Changshitougounao gold deposit differs markedly from magmatic–hydrothermal deposits (Supplementary Table S4). Mineralization is concentrated at the contact between quartz diorite and Triassic slate. Alteration is dominated by intense pyritization, arsenopyritization, silicification, sericitization, and carbonatization. Ore minerals include pyrite, arsenopyrite, apatite, and monazite, with minor stibnite. The orebodies display strong structural control by shear zones and lack alteration characteristics typical of magmatic–hydrothermal deposits. Hydrothermal monazite and apatite yield in situ LA–ICP–MS U–Pb ages that are ~10 Ma younger than the host quartz diorite emplacement, indicating that gold mineralization is not directly linked to magmatic intrusion. These characteristics, such as alteration assemblage, mineralogy, structural control, and timing, align with the orogenic gold model of Goldfarb et al. [5]. Previous geochemical and U–Pb zircon data indicate three major Indosinian plutonic pulses in the West Qinling Orogen at ca. 250–235 Ma, 228–215 Ma, and 215–185 Ma, corresponding to subduction-related magmatism, syn-collision granite emplacement, and post-collision magmatic activity, respectively [17,35,40,60]. By coupling these regional emplacement phases with the ~230 Ma mineralization age at Changshitougounao, we interpret the Changshitougounao gold deposit as having formed during collisional orogenesis associated with closure of the Paleo-Tethys Ocean. The deposit is best classified as an orogenic gold system. Pyritization, arsenopyritization, and sericitization are key hydrothermal alteration vectors for exploration. The most favorable mineralization site is the contact between granitoid intrusions and Triassic slate.

7. Conclusions

(1) Zircon U–Pb ages from fresh quartz diorite and coeval Type I magmatic monazite, constrain the emplacement of the ore-hosting intrusion to ca. 240 Ma. In contrast, hydrothermal apatite and Type II monazite yield U–Pb ages of ca. 230 Ma, indicating gold mineralization occurred 10 Ma after intrusion and suggesting a limited genetic link between magmatism and mineralization.

(2) Mineralization is structurally controlled by secondary regional faults at the contact between the quartz diorite and Triassic slate. Hydrothermal alteration is dominated by silicification, pyritization, arsenopyritization, and sericitization. These alteration assemblages, combined with shear-zone control, distinguish the Changshitougounao gold deposit from nearby magmatic–hydrothermal systems and align it with classical orogenic gold deposits.

(3) The Changshitougounao gold deposit formed during collisional orogenesis, associated with closure of the Paleo-Tethys Ocean. Sericitization, pyritization, and arsenopyritization are identified as key exploration vectors. The contact between Early–Middle Triassic granitoids and underlying slate represents the most prospective horizon.

Supplementary Materials: The following supporting information can be downloaded at: https://www.mdpi.com/article/10.3390/min15090903/s1, Table S1. LA-ICP-MS zircon U-Pb age of the Changshitougounao gold deposit (24CSTGN01); Table S2. In-situ LA-ICP-MS U-Pb data of two types (magmatic and hydrothermal) monazite grains (24CSTGN01 and 24CSTGN02); Table S3. LA-ICP-MS U-Pb apatite data for mineralized quartz diorite (24CSTGN02 and 24CSTGN03); Table S4. Comparison of typical Orogenic and Magmatic Deposits in the Western Qinling.

Author Contributions: Conceived the ideas, X.-F.X., J.W. and S.-X.L.; map compilation, J.-H.Q. and S.-X.W.; data curation, formal analysis, and investigation, X.-F.X., J.-H.Q., Z.-W.Z. and D.-H.Y.; writing—original draft, X.-F.X. and S.-X.L. All authors have read and agreed to the published version of the manuscript.

Funding: This study was financially supported by the National Natural Science Foundation of China (42303067), the Shandong Gold Group Collaborative Research Grant (23B1014), projects of Ministry of Natural Resources, a new round of Mineral Exploration Breakthrough Strategic Action Science and Technology Support Project (ZKKJ202410), and the Frontiers Science Center for Deep-time Digital Earth (2652023001), and the Fundamental Research Funds for the Central Universities, China (2-9-2024-008).

Data Availability Statement: The data set is presented directly in the present study.

Acknowledgments: We are grateful to Hao-Cheng Yu for their field guidance and constructive discussions. We are grateful to Jia-Hui Zhou for the assistance with the figures. We also deeply thank the anonymous reviewers and editors for their helpful comments and suggestions.

Conflicts of Interest: Authors Sheng-Xiang Lu and Shou-Xu Wang was employed by the Shandong Gold Geology and Mineral Exploration Co., Ltd. Author Zheng-Wang Zeng was employed by the Northwest Gold Co., Ltd. The remaining authors declare that the research was conducted in the absence of any commercial or financial relationships that could be construed as a potential conflict of interest.

References

1. Goldfarb, R.J.; Groves, D.I. Orogenic gold: Common or evolving fluid and metal sources through time. *Lithos* **2015**, *233*, 2–26. [CrossRef]
2. Qiu, K.F.; Taylor, R.D.; Song, Y.H.; Yu, H.C.; Song, K.R.; Li, N. Geologic and geochemical insights into the formation of the Taiyangshan porphyry copper–molybdenum deposit, Western Qinling Orogenic Belt, China. *Gondwana Res.* **2016**, *35*, 40–58. [CrossRef]
3. Yang, L.Q.; Deng, J.; Goldfarb, R.J.; Zhang, J.; Gao, B.F.; Wang, Z.L. 40Ar/39Ar geochronological constraints on the formation of the Dayingezhuang gold deposit: New implications for timing and duration of hydrothermal activity in the Jiaodong gold province, China. *Gondwana Res.* **2014**, *25*, 1469–1483. [CrossRef]
4. Deng, J.; Wang, Q.F.; Li, G.J. Tectonic evolution, superimposed orogeny, and composite metallogenic system in China. *Gondwana Res.* **2017**, *50*, 216–266. [CrossRef]
5. Goldfarb, R.; Qiu, K.; Deng, J.; Chen, Y.; Yang, L. *Orogenic Gold Deposits of China*; Society of Economic Geologists: Littleton, CO, USA, 2019; pp. 236–324.
6. Qiu, K.F.; Deng, J.; He, D.Y.; Rosenbaum, G.; Zheng, X.; Williams-Jones, A.E.; Yu, H.C.; Balen, D. Evidence of Vertical Slab Tearing in the Late Triassic Qinling Orogen (Central China) From Multiproxy Geochemical and Isotopic Imaging. *J. Geophys. Res. Solid Earth* **2023**, *128*, e2022JB025514. [CrossRef]
7. Yu, H.C.; Qiu, K.F.; Deng, J.; Zhu, R.; Mathieu, L.; Sai, S.X.; Sha, W.J. Exhuming and Preserving Epizonal Orogenic Au-Sb Deposits in Rapidly Uplifting Orogenic Settings. *Tectonics* **2022**, *41*, e2021TC007165. [CrossRef]
8. Qiu, K.F.; Yu, H.C.; Gou, Z.Y.; Liang, Z.L.; Zhang, J.L.; Zhu, R. Nature and origin of Triassic igneous activity in the Western Qinling Orogen: The Wenquan composite pluton example. *Int. Geol. Rev.* **2017**, *60*, 242–266. [CrossRef]
9. Qiu, K.F.; Yu, H.C.; Hetherington, C.J.; Huang, Y.Q.; Yang, T.; Deng, J. Tourmaline composition and boron isotope signature as a tracer of magmatic-hydrothermal processes. *Am. Mineral.* **2021**, *106*, 1033–1044. [CrossRef]
10. Zhu, L.M.; Zhang, G.W.; Yang, T. Geochronology, petrogenesis and tectonic implications of the Zhongchuan granitic pluton in the Western Qinling metallogenic belt. *Geol. J.* **2013**, *48*, 310–334. [CrossRef]
11. Luo, B.; Zhang, H.; Xiao, Z. Petrogenesis and tectonic implications of the Early Indosinian Meiwu pluton in West Qinling, central China. *Earth Sci. Front.* **2012**, *19*, 199–213.
12. Wei, P.; Mo, X.X.; Yu, X.H.; Huang, X.F.; Ding, Y.; Li, X.W. Geochemistry, chronology and geological significance of the granitoids in Xiahe, West Qinling. *Acta Petrol. Sin.* **2013**, *29*, 3981–3992.

13. Huang, Y.Q.; Qiu, K.F.; Yu, H.C.; Jin, D.G.; He, D.Y.; Xiao, C.H.; Wang, Y.X. Petrogenesis of ore-hosting porphyry in the Gelouang gold deposit, West Qinling and its geological implications. *Acta Petrol. Sin.* **2020**, *36*, 1567–1585.
14. Xue, X.F.; Qiu, K.F.; He, D.Y.; Huang, Y.Q.; Zhu, R.; Yu, H.C.; Jiang, G.P. Petrogenesis of the quartz diorite porphyry in the Zaozigou, West Qinling and its geological implications. *Acta Petrol. Sin.* **2023**, *39*, 463–483. [CrossRef]
15. Qiu, K.F.; Deng, J. Petrogenesis of granitoids in the Dewulu skarn copper deposit: Implications for the evolution of the Paleotethys ocean and mineralization in Western Qinling, China. *Ore Geol. Rev.* **2017**, *90*, 1078–1098. [CrossRef]
16. Qiu, K.F.; Marsh, E.; Yu, H.C.; Pfaff, K.; Gulbransen, C.; Gou, Z.Y.; Li, N. Fluid and metal sources of the Wenquan porphyry molybdenum deposit, Western Qinling, NW China. *Ore Geol. Rev.* **2017**, *86*, 459–473. [CrossRef]
17. He, D.Y.; Qiu, K.F.; Yu, H.C.; Jowitt, S.M.; Zheng, X.; Mazumder, R.; Deng, J. Lithospheric architecture and evolution of the Qinling Orogen of Central China and associated controls on metallogeny. *Earth-Sci. Rev.* **2025**, *264*, 105092. [CrossRef]
18. He, D.Y.; Qiu, K.F.; Santosh, M.; Yu, H.C.; Long, Z.Y.; Wang, J.Y. Inhomogeneous crust-mantle interaction and Triassic tectonic escape of a Proterozoic microplate: A tale of the Bikou Terrane. *Lithos* **2021**, *396–397*, 106227. [CrossRef]
19. Yu, H.C.; Qiu, K.F.; Nassif, M.T.; Geng, J.Z.; Sai, S.X.; Duo, D.W.; Huang, Y.Q.; Wang, J. Early orogenic gold mineralization event in the West Qinling related to closure of the Paleo-Tethys Ocean–Constraints from the Ludousou gold deposit, central China. *Ore Geol. Rev.* **2020**, *117*, 103217. [CrossRef]
20. Pang, Y.; Yan, L.J.; Liu, Y.; Tang, L.; Zhu, R.; Liu, G.F. Seismic Wave Finite-Difference Forward Modeling for Orogenic Gold Deposits. *Minerals* **2022**, *12*, 1465. [CrossRef]
21. Qiu, K.F.; Yu, H.C.; Deng, J.; McIntire, D.; Gou, Z.Y.; Geng, J.Z.; Chang, Z.S.; Zhu, R.; Li, K.N.; Goldfarb, R. The giant Zaozigou Au-Sb deposit in West Qinling, China: Magmatic- or metamorphic-hydrothermal origin? *Miner. Depos.* **2020**, *55*, 345–362. [CrossRef]
22. Jin, X.Y.; Li, J.W.; Hofstra, A.H.; Sui, J.X. Magmatic-hydrothermal origin of the early Triassic Laodou lode gold deposit in the Xiahe-Hezuo district, West Qinling orogen, China: Implications for gold metallogeny. *Miner. Depos.* **2016**, *52*, 883–902. [CrossRef]
23. Deng, J.; Wang, Q.F. Gold mineralization in China: Metallogenic provinces, deposit types and tectonic framework. *Gondwana Res.* **2016**, *36*, 219–274. [CrossRef]
24. Deng, J.; Wang, C.M.; Bagas, L.; Santosh, M.; Yao, E.Y. Crustal architecture and metallogenesis in the south-eastern North China Craton. *Earth-Sci. Rev.* **2018**, *182*, 251–272. [CrossRef]
25. Dong, Y.; Santosh, M. Tectonic architecture and multiple orogeny of the Qinling Orogenic Belt, Central China. *Gondwana Res.* **2016**, *29*, 1–40. [CrossRef]
26. Yang, L.Q.; Deng, J.; Qiu, K.F.; Ji, X.Z.; Santosh, M.; Song, K.R.; Song, Y.H.; Geng, J.Z.; Zhang, C.; Hua, B. Magma mixing and crust–mantle interaction in the Triassic monzogranites of Bikou Terrane, central China: Constraints from petrology, geochemistry, and zircon U–Pb–Hf isotopic systematics. *J. Asian Earth Sci.* **2015**, *98*, 320–341. [CrossRef]
27. Dong, Y.P.; Zhou, M.F.; Zhang, G.W.; Zhou, D.W.; Liu, L.; Zhang, Q. The Grenvillian Songshugou ophiolite in the Qinling Mountains, Central China: Implications for the tectonic evolution of the Qinling orogenic belt. *J. Asian Earth Sci.* **2008**, *32*, 325–335. [CrossRef]
28. Yang, L.Q.; Deng, J.; Dilek, Y.; Qiu, K.F.; Ji, X.Z.; Li, N.; Ryan, D.T.; Yu, J.Y. Structure, geochronology, and petrogenesis of the Late Triassic Puziba granitoid dikes in the Mianlue suture zone, Qinling Orogen, China. *Geol. Soc. Am. Bull.* **2015**, *127*, 1831–1854. [CrossRef]
29. Yang, L.; Song, S.; Allen, M.B.; Su, L.; Dong, J.; Wang, C. Oceanic accretionary belt in the West Qinling Orogen: Links between the Qinling and Qilian orogens, China. *Gondwana Res.* **2018**, *64*, 137–162. [CrossRef]
30. Liu, L.; Liao, X.; Wang, Y.; Wang, C.; Santosh, M.; Yang, M.; Zhang, C.; Chen, D. Early Paleozoic tectonic evolution of the North Qinling Orogenic Belt in Central China: Insights on continental deep subduction and multiphase exhumation. *Earth-Sci. Rev.* **2016**, *159*, 58–81. [CrossRef]
31. Yang, L.Q.; Deng, J.; Li, N.; Zhang, C.; Ji, X.Z.; Yu, J.Y. Isotopic characteristics of gold deposits in the Yangshan Gold Belt, West Qinling, central China: Implications for fluid and metal sources and ore genesis. *J. Geochem. Explor.* **2016**, *168*, 103–118. [CrossRef]
32. Deng, J.; Wang, Q.; Li, G.; Santosh, M. Cenozoic tectono-magmatic and metallogenic processes in the Sanjiang region, southwestern China. *Earth-Sci. Rev.* **2014**, *138*, 268–299. [CrossRef]
33. Li, S.Z.; Jahn, B.; Zhao, S.J.; Dai, L.M.; Li, X.Y.; Suo, Y.H.; Guo, L.L.; Wang, Y.M.; Liu, X.C.; Lan, H.Y. Triassic southeastward subduction of North China Block to South China Block: Insights from new geological, geophysical and geochemical data. *Earth-Sci. Rev.* **2017**, *166*, 270–285. [CrossRef]
34. Mao, J.; Goldfarb, R.J.; Zhang, Z.; Xu, W.; Qiu, Y.; Deng, J. Gold deposits in the Xiaoqinling–Xiong'ershan region, Qinling Mountains, central China. *Miner. Depos.* **2002**, *37*, 306–325. [CrossRef]
35. Dong, Y.; Zhang, G.; Neubauer, F.; Liu, X.; Genser, J.; Hauzenberger, C. Tectonic evolution of the Qinling orogen, China: Review and synthesis. *J. Asian Earth Sci.* **2011**, *41*, 213–237. [CrossRef]

36. Yu, H.C.; Guo, C.A.; Qiu, K.F.; McIntire, D.; Jiang, G.P.; Gou, Z.Y.; Geng, J.Z.; Pang, Y.; Zhu, R.; Li, N.B. Geochronological and geochemical constraints on the formation of the giant Zaozigou Au-Sb deposit, West Qinling, China. *Minerals* **2019**, *9*, 37. [CrossRef]
37. Xiao, C.H.; Chen, Z.L.; Han, F.B.; Wei, L.X.; Ma, L.T. Provenance and tectonic significance of Lower Triassic clastic sedimentary rocks of the West Qinling Orogenic Belt, Central China. *J. Asian Earth Sci.* **2022**, *224*, 105026. [CrossRef]
38. Yu, H.C.; Qiu, K.F.; Sai, S.X.; McIntire, D.; Pirajno, F.; Duo, D.W.; Miggins, D.; Wang, J.; Jia, R.Y.; Wu, M.Q. Paleo-tethys late triassic orogenic gold mineralization recorded by the Yidi'nan gold deposit, West Qinling, China. *Ore Geol. Rev.* **2020**, *116*, 103211. [CrossRef]
39. Yu, H.C.; Qiu, K.F.; Hetherington, C.J.; Chew, D.; Huang, Y.Q.; He, D.Y.; Geng, J.Z.; Xian, H.Y. Apatite as an alternative petrochronometer to trace the evolution of magmatic systems containing metamict zircon. *Contrib. Mineral. Petrol.* **2021**, *176*, 68. [CrossRef]
40. Qiu, K.F.; Deng, J.; Yu, H.C.; Rasbury, T.; Tang, Y.W.; Zhu, R.; Zhang, P.C.; Goldfarb, R. The Zaozigou orogenic gold-antimony deposit, West Qinling Orogen, China: Structural controls on multiple mineralization events. *Geol. Soc. Am. Bull.* **2024**, *136*, 4218–4232. [CrossRef]
41. Fan, C.; Yu, H.C.; Tamer, M.T.; Zhang, L.; Wang, J.; Liu, P.X.; Xue, X.F.; Li, C.; Wang, Y.X. Ore genesis of the Nanban gold deposit, West Qinling Orogen: Insights from monazite geochronology and geochemistry of sulfides. *Ore Geol. Rev.* **2025**, *179*, 106531. [CrossRef]
42. Wang, J.; Yu, H.C.; Petrella, L.; Duo, D.W.; Wang, L.; Yang, F.; Zhou, H.Y.; Li, C.; Qiu, K.F. Fluid evolution and genesis of the Yidinan granitoid-hosted orogenic gold deposit (China). *Geol. Soc. Am. Bull.* **2025**, *137*, 1945–1963. [CrossRef]
43. Zhang, L.; Qiu, K.; Hou, Z.; Pirajno, F.; Shivute, E.; Cai, Y. Fluid-rock reactions of the Triassic Taiyangshan porphyry Cu-Mo deposit (West Qinling, China) constrained by QEMSCAN and iron isotope. *Ore Geol. Rev.* **2021**, *132*, 104068. [CrossRef]
44. Wang, Y.; Li, B.; Chen, A.; Duan, X.; Jiang, W.Q. Numerical simulation of copper–gold mineralization in the Xiahe-Hezuo area of the western Qinling orogenic belt: Significance for prospecting and factors influencing similarities and differences in favorable mineralization areas. *Ore Geol. Rev.* **2025**, *182*, 106598. [CrossRef]
45. Deng, X.T. *Exploration Report on the Changshitougounao–Zhayouliang Gold Prospect, Xiahe County, Gannan Tibetan Autonomous Prefecture, Gansu Province*; The Third Geological and Mineral Exploration Institute of Gansu Provincial: Lanzhou, China, 2016.
46. Cao, X.P.; Deng, X.T. Geological and Mineralization Characteristics and Exploration Indicators of the Changshitougounao Gold Deposit, Xiahe. *Gansu Sci. Technol.* **2016**, *32*, 27–30.
47. Liu, Y.; Gao, S.; Hu, Z.; Gao, C.; Zong, K.; Wang, D. Continental and Oceanic Crust Recycling-induced Melt-Peridotite Interactions in the Trans-North China Orogen: U-Pb Dating, Hf Isotopes and Trace Elements in Zircons from Mantle Xenoliths. *J. Petrol.* **2009**, *51*, 537–571. [CrossRef]
48. Ludwig, K.R. *ISOPLOT 3.0: A Geochronological Toolkit for Microsoft Excel*; Berkeley Geochronology Center Special Publication; Berkeley Geochronology Center: Berkeley, CA, USA, 2003.
49. Andersen, T. Correction of common lead in U–Pb analyses that do not report 204P. *Chem. Geol.* **2002**, *192*, 59–79. [CrossRef]
50. Ludwig, K.R. *User's Manual for Isoplot 3.75: A Geochronological Toolkit for Microsoft Excel*; Berkeley Geochronology Center Special Publication; Berkeley Geochronology Center: Berkeley, CA, USA, 2012; Volume 5, p. 75.
51. Fu, J.N.; Pirajno, F.; Yang, F.; Shivute, E.; Sun, Y.Z.; Ai, N.; Qiu, K.F. Integration of zircon and apatite U–Pb geochronology and geochemical mapping of the Wude basalts (Emeishan large igneous province): A tool for a better understanding of the tectonothermal and geodynamic evolution of the Emeishan LIP. *Geosci. Front.* **2021**, *12*, 573–585. [CrossRef]
52. Zou, Y.Z.; Li, S.S.; Zhang, H.F.; Tamer, M.T.; Bishikwabo, G.K.; Shivute, E.T.; Dong, W.M.; Vatuva, A.; Zhang, C.H. Petrogenesis, magmatic and metamorphic evolution of Salem granite: Implications for the Damara Orogeny. *Ore Geol. Rev.* **2025**, *179*, 106512. [CrossRef]
53. Tamer, M.T.; Ketcham, R.A. How many vs which: On confined track selection criteria for apatite fission track analysis. *Chem. Geol.* **2023**, *634*, 121584. [CrossRef]
54. Tamer, M.T.; Chung, L.; Ketcham, R.A.; Gleadow, A.J. The need for fission-track data transparency and sharing. *Geochronology* **2025**, *7*, 45–58. [CrossRef]
55. Qiu, K.F.; Yang, L.Q. Genetic feature of monazite and its U-Th-Pb dating: Critical considerations on the tectonic evolution of Sanjiang Tethys. *Acta Petrol. Sin.* **2011**, *27*, 2721–2732.
56. Schandl, E.S.; Gorton, M.P. A textural and geochemical guide to the identification of hydrothermal monazite: Criteria for selection of samples for dating epigenetic hydrothermal ore deposits. *Econ. Geol.* **2004**, *99*, 1027–1035. [CrossRef]
57. Boatner, L.A. Synthesis, structure, and properties of monazite, pretulite, and xenotime. *Rev. Mineral. Geochem.* **2002**, *48*, 87–121. [CrossRef]
58. Xiong, X.; Zhu, L.; Yang, P.; Uysal, I.T.; Zhao, J.-x.; Zhang, S.; Ding, L.; Li, B.; Ma, Y.; Zhang, G. Texture, geochemistry and U–Pb geochronology of monazite of different origins: Implications for the magmatic–hydrothermal evolution of the Guilingou porphyry Mo-W deposit, South Qinling Orogen, central China. *Ore Geol. Rev.* **2025**, *182*, 106648. [CrossRef]

59. Poitrasson, F.; Chenery, S.; Shepherd, T.J. Electron microprobe and LA-ICP-MS study of monazite hydrothermal alteration: Implications for U-Th-Pb geochronology and nuclear ceramics. *Geochim. Cosmochim. Acta* **2000**, *64*, 3283–3297. [CrossRef]
60. Hu, F.; Liu, S.; Ducea, M.N.; Chapman, J.B.; Wu, F.; Kusky, T. Early Mesozoic magmatism and tectonic evolution of the Qinling Orogen: Implications for oblique continental collision. *Gondwana Res.* **2020**, *88*, 296–332. [CrossRef]
61. Dong, Y.; Yang, Z.; Liu, X.; Sun, S.; Li, W.; Cheng, B.; Zhang, F.; Zhang, X.; He, D.; Zhang, G. Mesozoic intracontinental orogeny in the Qinling Mountains, central China. *Gondwana Res.* **2016**, *30*, 144–158. [CrossRef]
62. Jin, W.; Zhang, Q.; He, D.; Jia, X. SHRIMP dating of adakites in western Qinling and their implications. *Acta Petrol. Sin.* **2005**, *21*, 959–966.
63. Xu, X.; Chen, J.; Gao, T.; Li, P.; Li, T. Granitoid magmatism and tectonic evolution in northern edge of the Western Qinling terrane, NW China. *Acta Petrol. Sin.* **2014**, *30*, 371–389.
64. Li, N.; Chen, Y.J.; Santosh, M.; Pirajno, F. Compositional polarity of Triassic granitoids in the Qinling Orogen, China: Implication for termination of the northernmost paleo-Tethys. *Gondwana Res.* **2015**, *27*, 244–257. [CrossRef]
65. Chen, B.; Zhang, Y.; Li, K. Genesis of the Jiagantan gold deposit in West Qinling, China: Constraints from Re-Os dating, in-situ trace elements and S isotope of gold bearing minerals. *Acta Petrol. Sin.* **2024**, *40*, 1767–1783. [CrossRef]
66. Gou, Z.Y.; Yu, H.C.; Qiu, K.F.; Geng, J.Z.; Wu, M.Q.; Wang, Y.G.; Yu, M.H.; Li, J. Petrogenesis of Ore-Hosting Diorite in the Zaorendao Gold Deposit at the Tongren-Xiahe-Hezuo Polymetallic District, West Qinling, China. *Minerals* **2019**, *9*, 76. [CrossRef]
67. Sui, J.X.; Li, J.W.; Jin, X.Y.; Vasconcelos, P.; Zhu, R. 40Ar/39Ar and U-Pb constraints on the age of the Zaozigou gold deposit, Xiahe-Hezuo district, West Qinling orogen, China: Relation to early Triassic reduced intrusions emplaced during slab rollback. *Ore Geol. Rev.* **2018**, *101*, 885–899. [CrossRef]
68. Zhang, T. The Geochemistry Characteristics and Metallogenic Series of Gangcha Pluton in Tongren, Qinghai Province. Ph.D. Thesis, China University of Geosciences, Beijing, China, 2015.
69. Du, B.S.; Shen, J.F.; Santosh, M.; Liu, H.M.; Liu, J.J.; Wang, S.H.; Xu, K.X. Genesis of the Gangcha gold deposit, West Qinling Orogen, China: Constraints from Rb-Sr geochronology, in-situ sulfur isotopes and trace element geochemistry of pyrite. *Ore Geol. Rev.* **2021**, *138*, 104350. [CrossRef]
70. Geng, J.Z.; Qiu, K.F.; Gou, Z.Y.; Yu, H.C. Tectonic regime switchover of Triassic Western Qinling Orogen: Constraints from LA-ICP-MS zircon U–Pb geochronology and Lu–Hf isotope of Dangchuan intrusive complex in Gansu, China. *Geochemistry* **2017**, *77*, 637–651. [CrossRef]
71. An, W.T.; Chen, J.P.; Chen, X.W.; Wei, J.; Wei, Z.X. Ideal element distribution pattern and characteristics of primary halo in the fault-controlled ore zone of the Yidinan gold deposit, Gansu Province, China. *Nat. Resour. Res.* **2020**, *29*, 2867–2880. [CrossRef]
72. Luo, B.J.; Zhang, H.F.; Xu, W.C.; Yang, H.; Zhao, J.H.; Guo, L.; Zhang, L.Q.; Tao, L.; Pan, F.B.; Gao, Z.Y. The magmatic plumbing system for Mesozoic high-Mg andesites, garnet-bearing dacites and porphyries, rhyolites and leucogranites from West Qinling, central China. *J. Petrol.* **2018**, *59*, 447–482. [CrossRef]
73. Lu, Y. Skarn Copper (Gold) Metallogeny and Metallogenic Regularities in the West Section of the Western Qinling Orogen. Ph.D. Thesis, China University of Geosciences, Beijing, China, 2016.
74. Jiang, Z. Metallogenic Model and Exploration Potential of the Gangcga Gold Deposit in Gansu. Ph.D. Thesis, China University of Geosciences, Beijing, China, 2017.
75. Sui, J.X.; Li, J.W.; Wen, G.; Jin, X.Y. The Dewulu reduced Au-Cu skarn deposit in the Xiahe-Hezuo district, West Qinling orogen, China: Implications for an intrusion-related gold system. *Ore Geol. Rev.* **2017**, *80*, 1230–1244. [CrossRef]
76. Yang, F.; Qiu, K.F.; He, D.Y.; Huang, Y.Q.; Wang, Y.X.; Fu, N.; Yu, H.C.; Xue, X.F. Mineralogy and geochemistry of gold-bearing sulfides in the Wanken gold deposit, West Qinling Orogen. *Earth Sci. Front.* **2023**, *30*, 371.
77. Hu, X.; Ding, Z.; Gong, Y.; Li, K.; He, M. Sediment-hosted disseminated gold deposits in orogenic belts: An example from the giant Jiagantan gold deposit in the West Qinling orogen, China. *Ore Geol. Rev.* **2022**, *146*, 104950. [CrossRef]
78. Guo, X.; Yan, Z.; Wang, Z.; Wang, T.; Wu, F.; Fu, C.; Hou, K.; Li, J. Geological characteristics and associated magmatic ages of the Xiekeng skarn-type Cu-Au deposit in the West Qinling terrane. *Acta Petrol. Sin.* **2011**, *27*, 3811–3822.
79. Kong, C.S.; Shen, J.; Santosh, M.; Li, S.; Cao, W.; Peng, Z.; Liu, H.; Li, J.; Li, J.; Bao, L.; et al. Age and genesis of the Gangcha gold deposit, western Qinling orogen, China. *Geol. J.* **2017**, *53*, 1871–1885. [CrossRef]

minerals

MDPI

Article

Anomaly Detection in Mineral Micro-X-Ray Fluorescence Spectroscopy Based on a Multi-Scale Feature Aggregation Network

Yangxin Lu [1,2,†], Weiming Jiang [3,†], Molei Zhao [1,2,*], Yuanzhi Zhou [3], Jie Yang [1,4], Kunfeng Qiu [1,3] and Qiuming Cheng [1,3]

1 Frontiers Science Center for Deep-Time Digital Earth, State Key Lab of Geological Processes and Mineral Resources, China University of Geosciences, Beijing 100083, China; yangxinlu@email.cugb.edu.cn (Y.L.)
2 School of Artificial Intelligence, China University of Geosciences, Beijing 100083, China
3 School of Earth Sciences and Resources, China University of Geosciences, Beijing 100083, China
4 Institute of Earth Sciences, China University of Geosciences, Beijing 100083, China
* Correspondence: molei.zhao@cugb.edu.cn
† These authors contributed equally to this work and should be considered as co-first authors.

Abstract

Micro-X-ray fluorescence spectroscopy (micro-XRF) integrates spatial and spectral information and is widely employed for multi-elemental analyses of rock-forming minerals. However, its inherent limitation in spatial resolution gives rise to significant pixel mixing, thereby hindering the accurate identification of fine-scale or anomalous mineral phases. Furthermore, most existing methods heavily rely on manually labeled data or predefined spectral libraries, rendering them poorly adaptable to complex and variable mineral systems. To address these challenges, this paper presents an unsupervised deep aggregation network (MSFA-Net) for micro-XRF imagery, aiming to eliminate the reliance of traditional methods on prior knowledge and enhance the recognition capability of rare mineral anomalies. Built on an autoencoder architecture, MSFA-Net incorporates a multi-scale orthogonal attention module to strengthen spectral–spatial feature fusion and employs density-based adaptive clustering to guide semantically aware reconstruction, thus achieving high-precision responses to potential anomalous regions. Experiments on real-world micro-XRF datasets demonstrate that MSFA-Net not only outperforms mainstream anomaly detection methods but also transcends the physical resolution limits of the instrument, successfully identifying subtle mineral anomalies that traditional approaches fail to detect. This method presents a novel paradigm for high-throughput and weakly supervised interpretation of complex geological images.

Keywords: micro-XRF imaging; anomaly detection; deep autoencoder; multi-scale attention; unsupervised mineral recognition

1. Introduction

Minerals are vital carriers that record geological evolution and preserve critical information about the Earth's interior. Their crystal structures retain key physicochemical parameters such as temperature, pressure, chemical composition, and elemental occurrence states during mineral formation, making them essential for deciphering rock-forming processes and reconstructing geological histories [1–5]. In recent years, micro-X-ray fluorescence spectroscopy (micro-XRF) has gained widespread application in mineral recognition, geological exploration [6], and related fields due to its ability to perform large-scale, non-destructive analysis at a relatively low cost. This technique offers simple sample preparation and rapid scanning, with micro-XRF images combining high-dimensional spectral data with two-dimensional spatial structure, providing critical data support for interpreting rock microstructures [7].

However, the application of micro-XRF still faces several challenges. On the one hand, its limited spatial resolution often leads to the "mixed-pixel" problem, where a single pixel may contain signals from multiple mineral phases. This not only blurs mineral boundaries but also makes it difficult to detect rare or trace minerals. On the other hand, the large volume of micro-XRF images and inherent noise redundancy impose high demands on mineral recognition algorithms [8]. To address these issues, supervised learning methods have been introduced to enhance the automatic analysis of micro-XRF for image classification and segmentation tasks, including spectral angle mapper (SAM) and convolutional neural networks (CNNs) [9,10]. Nevertheless, these approaches generally rely on pre-constructed spectral libraries or manually annotated labels, limiting their ability to handle pixel-mixing errors or recognize unknown minerals, resulting in poor model generalization. While certain approaches have sought to integrate traditional SAM methods with linear programming for unsupervised estimation of mineral abundances in fine-grained mixtures [11], they remain dependent on prior spectral libraries and annotations, thus falling short of truly unsupervised analysis. With the rapid development of deep learning, a range of mineral recognition studies has emerged, demonstrating superior performance over traditional methods [12–15]. For instance, Kim et al. [16] applied ANN models to build mineral classifiers, exploring correlations between elemental abundance data from XRF and mineral recognition data from μ-XRD.

Nonetheless, these deep learning-based approaches still face two fundamental limitations: First, label-dependent models lack generalization abilities when dealing with unknown minerals or complex compositions. Second, even deep networks tend to focus on global high-level semantic features when optimized for classification tasks, making them less sensitive to pixel-level, subtle, rare anomalies and fine-grained variations.

In this study, we treat sparse, irregularly distributed, and compositionally anomalous mineral grains in micro-XRF images as "anomalous" pixels—features often overlooked by conventional methods. Anomaly detection algorithms aim to identify patterns deviating from the dominant distribution and are naturally suited for capturing low-frequency pixel-level features such as fine-grained minerals and subtle compositional shifts [17]. In studies of hyperspectral images, autoencoder-based reconstruction error frameworks are widely adopted for unsupervised detection due to their ability to learn unknown patterns without requiring labels [18]. However, mineral anomalies in micro-XRF images often exhibit both spatial sparsity and spectral perturbation. Existing methods, while capable of representing features, often ignore the joint modeling of spatial–spectral information across multiple scales. Without mechanisms for dynamic attention aggregation, they fail to enhance response signals from anomalies embedded within mixed pixels, thus limiting the model's sensitivity to subtle mineral variations.

To address the challenges of spatial resolution and label dependency in anomaly detection for micro-XRF images, we propose an unsupervised deep learning model—multi-scale feature aggregation network (MSFA-Net)—that integrates orthogonal attention mechanisms with an anomaly detection framework.

The contributions of this work are as follows:

1 We design an unsupervised MSFA-Net based on an autoencoder architecture, enabling end-to-end anomaly detection and reconstruction while jointly modeling spatial features and reconstruction errors.

2 We introduce a multi-scale orthogonal attention (MOA) module to effectively integrate and disentangle spatial–spectral features at different scales, enhancing the model's ability to discriminate subtle mineral anomalies within mixed pixels.

3 We incorporate a DBSCAN-based spatial clustering strategy to generate pseudo-labels, guiding the reconstruction loss to assign higher semantic weights to anomalous regions. This improves the anomaly response capability while reducing dependence on labels or spectral libraries.

This study not only improves anomaly detection performance but also shifts micro-XRF analysis from a rule-driven to a data-driven paradigm, presenting an unsupervised framework for intelligent interpretation of complex mineral imagery. It offers both theoretical and practical support for geological analysis under high-throughput, weak-label conditions.

2. Methodology

2.1. MSFA-Net Framework for Micro-XRF Anomaly Detection

Owing to its ability to capture multi-scale contextual information without a corresponding increase in parameters through grouping strategies and inexpensive features, MSFA-Net has been successfully applied in several other domain analysis tasks [19,20]. However, such approaches have not yet been extended to micro-XRF anomaly detection. Unlike label-dependent models, in micro-XRF anomaly detection, MSFA-Net does not require prior annotations or spectral libraries, thereby improving its generalization ability to unknown minerals. By employing a multi-scale orthogonal attention (MOA) module, the network dynamically aggregates features at different resolutions while disentangling spatial–spectral redundancies. This enables the model to not only preserve global semantic consistency but also highlight subtle, fine-grained anomalies embedded within mixed pixels. Moreover, the integration of multi-scale aggregation ensures that both coarse-grained contextual cues and fine-grained local variations are captured, directly addressing the difficulty of detecting sparse anomalies. In this way, MSFA-Net effectively mitigates the dual limitations of label dependency and sensitivity loss, providing a principled solution to anomaly detection in micro-XRF images.

MSFA-Net adopts an autoencoder-based architecture and integrates a multi-scale attention module to learn latent image features in an unsupervised manner [21]. The framework has three parts, namely, the encoder, MOA, and decoder. The overall architecture is illustrated in Figure 1. Prior to model input, the raw micro-XRF image data, $X_{\text{recon}} \in \mathbb{R}^{B \times C_{\text{in}} \times H \times W}$, undergoes preprocessing. For each pixel at location (h, w), its spectral profile is represented as a C_{in}-dimensional vector. The collection of these vectors forms the input dataset, where $N = H \times W$ denotes the total number of pixels, and C_{in} is the number of spectral bands in the input data.

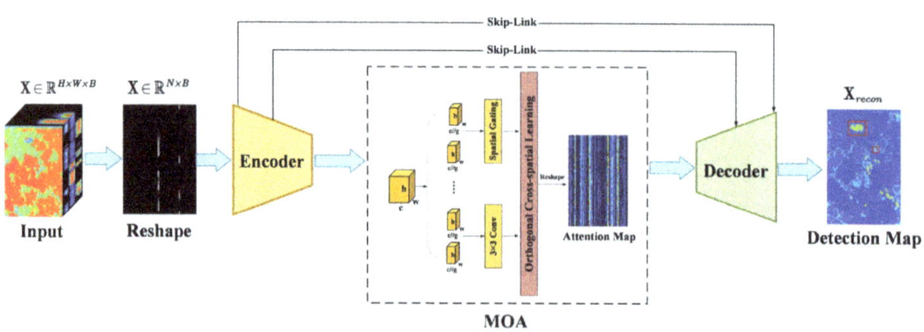

Figure 1. MSFA-Net model structure.

2.1.1. Encoder

As the feature extraction pathway of MSFA-Net, the encoder is designed to progressively reduce the dimensionality of the high-dimensional spectral vector associated with each pixel while abstracting semantic features at multiple levels. This process increases the depth of feature channels while compressing the original spectral information, $X_{h,w}$, into a more compact and informative latent representation. The encoder consists of three hierarchically stacked modules. Each module includes a linear layer, a normalization layer, and a ReLU activation function, which together enable nonlinear feature transformation and contribute to the stability of the training process. The encoding procedure can be formally expressed as follows:

$$M = \mathrm{En}(\mathrm{X}, Out_{\mathrm{En}}). \tag{1}$$

Here, M denotes the encoded spectral–spatial representation, $\mathrm{E}n(\cdot)$ is the encoder function, X represents the original spectral data of the micro-XRF image, and Out_{En} indicates the output dimension of the encoder.

The latent features, F_{latent}, generated by the encoder contain the image's global high-level semantic information, serving as the input to the subsequent multi-scale attention module. In addition, the feature maps e_1 and e_2, produced at different stages of the encoder, are transmitted to the corresponding upsampling layers in the decoder through skip connections, allowing the network to retain spatial details that may be lost during the downsampling process.

2.1.2. Multi-Scale Orthogonal Attention Module

In previous studies, the use of micro-XRF images for mineral recognition and classification has often been hindered by the fact that the average grain size of the sample is smaller than the spatial resolution of the instrument. As a result, a single pixel may contain multiple mineral components, leading to mixed pixels, blurred mineral boundaries, and reduced classification accuracy [22]. Similarly, in geological exploration, multi-scale data analysis and multi-source information fusion have been proven to improve the accuracy of mineral boundary identification and the extraction of anomalous features [23,24]. To address this issue, this study designs and incorporates the MOA module. Inspired by EMA and OrthoNets [25,26], the MOA module adopts a multi-branch structure to simultaneously capture both global and local spatial contextual information. During feature fusion, it integrates Schmidt orthogonalization to reduce spectral redundancy and enhance inter-channel independence. The overall structure is illustrated in Figure 2.

Firstly, the module divides the input feature channels into G groups for multi-branch parallel processing. Subsequently, global context features are extracted using 1D average pooling along the horizontal (X Avg Pool) and vertical (Y Avg Pool) directions, generating preliminary spatial attention maps. Based on this, the module performs Schmidt orthogonalization on the spectral dimension to ensure that learned spectral features are mutually independent, thereby suppressing redundancy.

The orthogonalized spectral features are then fused with the spatial attention maps. Furthermore, the module introduces a bidirectional attention interaction mechanism, enabling deep interaction between the fused spatial–spectral features and another branch that captures local spatial information. This design allows the network to adaptively aggregate multi-scale spatial contexts and dynamically assign attention weights to each pixel. Through this architecture, the MOA module significantly enhances feature representation, suppresses noise and redundancy, and improves the discriminability of mixed-pixel regions.

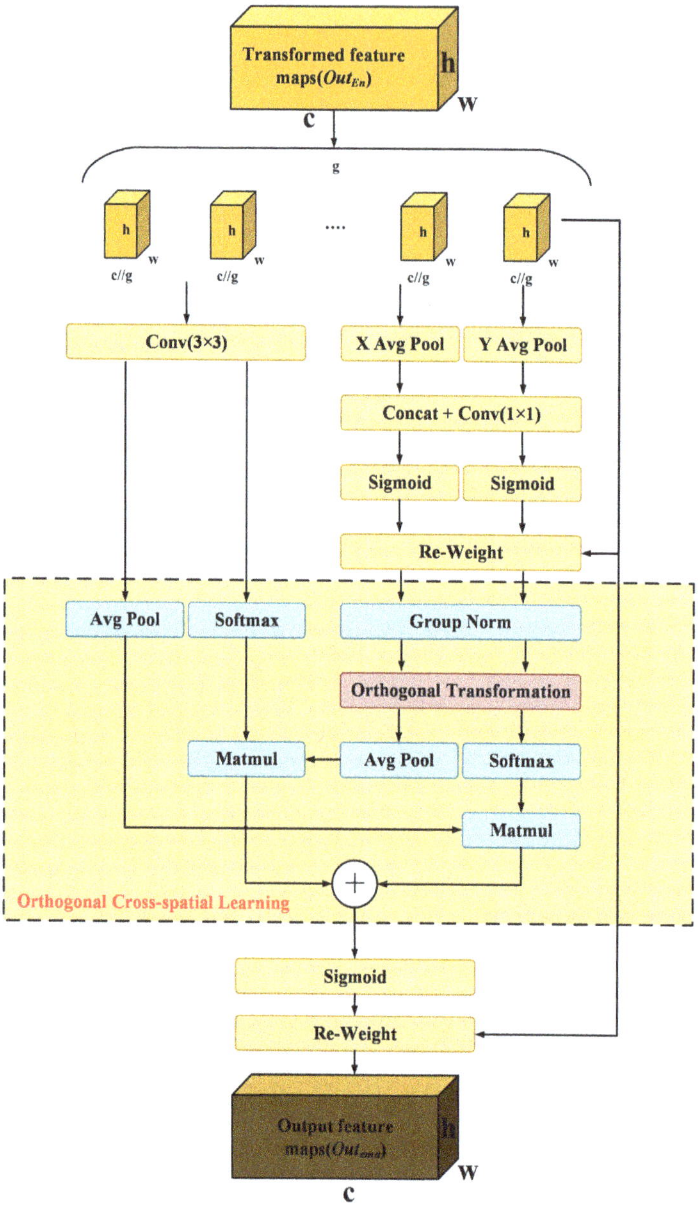

Figure 2. The structures of our MOA module. G represents the divided groups. X represents features from pooling along the height (h) direction. Y represents features from pooling along the width (w) direction.

An input feature map, $x_{\text{in}} \in \mathbb{R}^{B\times C_{\text{latent}}\times H_{\text{latent}}\times W_{\text{latent}}}$, is divided along the spectral dimension into G groups, resulting in grouped sub-features, $x_g \in \mathbb{R}^{B\times (C_{\text{latent}}/G)\times H_{\text{latent}}\times W_{\text{latent}}}$. The module then extracts multi-scale contextual information through two primary branches. The first branch, referred to as the global directional context branch, captures global contextual information along the horizontal and vertical axes. For a specific channel, c, and batch index, b, within each grouped sub-feature, x_g, horizontal adaptive average pooling is first performed to generate the compressed feature $x_h(b,c,h)$. The computation is defined as follows:

$$M_c^H(h) = \frac{1}{W_{latent}} \sum_{i=1}^{W_{latent}} x_c(h,i). \tag{2}$$

Here, $x_c(h,i)$ denotes the value of the grouped sub-feature, x_g, at batch b, channel c, row h, and column i. Meanwhile, vertical adaptive average pooling is performed to obtain $x_w(b,c,w)$, which is computed as follows:

$$M_c^W(w) = \frac{1}{H_{latent}} \sum_{j=1}^{H_{latent}} x_c(j,w). \tag{3}$$

Here, $x_c(j,w)$ denotes the value of the grouped sub-feature, x_g, at batch b, channel c, row j, and column w. Subsequently, x_h and x_w are concatenated along a new dimension to form $X_{h,w} \in \mathbb{R}^{B\times C_{\text{latent}}\times G\times(H_{\text{latent}}+W_{\text{latent}})\times 1}$. After cross-dimensional information fusion via a 1×1 convolution, the features are split back into $x\prime_h$ and $x\prime_w$. These features are then passed through a Sigmoid function to generate the initial spatial weights, which are multiplied by the original grouped sub-feature, x_g, to enhance spatially salient regions. The weighted features are normalized and subjected to Schmidt orthogonal transformation, ultimately yielding the feature $\hat{x}_1 \in \mathbb{R}^{B\times(C_{\text{latent}}/G)\times H_{\text{latent}}\times W_{\text{latent}}}$. The entire process is illustrated in Formulae (4) and (5).

$$M_c = \frac{1}{H\times W}\sum_{j}^{H}\sum_{i}^{W} x_c(i,j). \tag{4}$$

The second branch focuses on extracting fine-grained local spatial context. It directly applies a 3×3 convolution to each grouped sub-feature, x_g, without undergoing Schmidt orthogonalization, resulting in the feature $x_2 \in \mathbb{R}^{B\times(C_{\text{latent}}/G)\times H_{\text{latent}}\times W_{\text{latent}}}$.

Notably, Schmidt orthogonalization is only applied in the branch that extracts global horizontal and vertical features. Specifically, the input feature, $x_1 \in \mathbb{R}^{B\times(C_{\text{latent}}/G)\times H_{\text{latent}}\times W_{\text{latent}}}$, is first reshaped along the spatial dimension into a two-dimensional matrix, after which Group Normalization is performed to normalize the feature distribution within each group. Then, Schmidt orthogonalization is applied in a group-wise manner. Formally, for each batch b and group g, the reshaped feature matrix is denoted as $X_1^{(b,g)} \in \mathbb{R}^{(C_{latent}/G)\times H_{latent}\times W_{latent}}$. The Gram–Schmidt process sequentially orthogonalizes each channel vector, x_i, against the previously obtained orthogonal basis, $\{\hat{x}_1, \ldots, \hat{x}_{i-1}\}$:

$$u_i = x_i - \sum_{j=1}^{i-1} \frac{\langle x_i, \hat{x}_j\rangle}{\langle \hat{x}_j, \hat{x}_j\rangle}\hat{x}_j, \hat{x}_i = \frac{u_i}{\| u_i \|_2}. \tag{5}$$

This ensures that the output channels satisfy the orthogonality condition $\langle \hat{x}_i, \hat{x}_j\rangle = 0 (i \neq j)$. By enforcing this decorrelation, channel redundancy is suppressed, and more independent spectral features are obtained.

Although orthogonalization introduces additional computation, the group-wise design significantly reduces the cost since it is only performed within smaller channel subsets and in one branch. This trade-off is computationally efficient while improving channel independence and enhancing the discriminative capacity of the fused features.

Once the outputs from the two branches are obtained, $\hat{x}_1$ and x_2, a cross-dimensional attention interaction mechanism is introduced to adaptively aggregate multi-scale context. The complementary features, $\hat{x}_1$ and x_2, are fused to capture interdependent spatial–spectral information.

This mechanism enhances the network's ability to focus on informative spatial regions, effectively suppressing noise and redundant information. It is especially beneficial in handling mixed or noisy pixels, allowing the network to better capture fine-grained mineralogical variations, which improves discrimination in complex mineralogical contexts.

2.1.3. Decoder

The decoder constitutes the feature reconstruction path of MSFA-Net and is symmetrical to the encoder. Its primary function is to integrate high-level semantic features processed by the multi-scale attention module with low-level detail features introduced via skip connections, progressively performing dimensional upsampling operations and ultimately reconstructing the latent features into an output, $X_{recon} \in \mathbb{R}^{B \times C_{in} \times H \times W}$, that matches the spatial dimensions of the original micro-XRF input image. The decoder in the proposed model also consists of three modules, each comprising a fully connected layer, a normalization layer, and a ReLU activation function. The core of the decoder lies in its skip connections, which directly transfer multi-level feature information from the encoder to the corresponding levels in the decoder and concatenate them along the channel dimension before the fully connected operation. This mechanism compensates for the spatial detail loss during encoder downsampling, ensuring that the mineral information recovered from the latent features is spatially accurate and of high fidelity. The computation is defined as follows:

$$\overline{M} = \text{De}(Out_{ema}, Out_{\text{De}}), \tag{6}$$

Here, $\overline{M}$ denotes the reconstructed feature representation obtained by the decoder, while $De(\cdot)$ represents the decoder function that performs progressive upsampling and feature fusion. Out_{ema} indicates the feature maps refined by the EMA module, serving as the primary input to the decoder, and Out_{De} refers to the target output dimensionality of the decoder, corresponding to the spatial dimensions of the reconstructed micro-XRF image.

Through this stacked module design and the incorporation of crucial skip connection mechanisms, the decoder can effectively upsample the learned latent representations and fuse multi-level information, ultimately producing high-quality reconstructed images that can evaluate the model's feature-learning capabilities.

2.2. *Feature Aggregation Module*

In the field of micro-XRF image analysis, the accurate identification of mineral phases has long been a significant challenge. In previous studies, such identification often relied on manual assessment and refinement of spectral attributes [27]. However, this traditional approach not only incurs substantial labor and time costs but also exhibits a limited ability to detect subtle anomalies and achieve full automation in the presence of complex and heterogeneous image backgrounds [28]. In recent years, machine learning-based geological mapping methods have demonstrated the potential for automated extraction of complex geological background information, providing new ideas for improving the efficiency and accuracy of mineral recognition [29]. To address this issue, a feature aggregation module (FAM) was designed to enable automated identification of background regions and potential anomalous regions, which are typically characterized by similar features. Given that micro-XRF images are inherently constrained by spatial resolution, mineral mixtures are inevitably present—meaning that the background often contains multiple mineral components, each usually occupying a relatively large proportion of pixels. By contrast, anomalous categories generally consist of far fewer pixels. This discrepancy in pixel count can be effectively exploited to distinguish between background and anomaly classes.

Unlike most unsupervised clustering algorithms, which require manual specification of the number of categories, this work employs the Density-Based Spatial Clustering of Applications with Noise (DBSCAN) model [30]. The module first feeds the latent features obtained from the multi-scale attention mechanism into the DBSCAN model to produce an initial clustering label map. The resulting category labels are then ranked in descending order by their corresponding pixel counts, and a reasonable pixel threshold, denoted as ξ, is applied to effectively separate background classes from anomaly classes. This separation process can be mathematically expressed as

$$D(\xi, K_i) = \begin{matrix} K_i \in K_{normal}, & S_a(K_i) > \xi, \\ K_i \in K_{anomaly}, & S_a(K_i) \leq \xi. \end{matrix} \quad (7)$$

In Formula (7), K_i denotes the pixel set of the i-th category, and the total number of categories is n. K_{normal} represents the background classes, with a total of m categories, while $K_{anomaly}$ is defined as the anomaly classes, with a total of n-m categories. Here, the $S_a(\cdot)$ function calculates the pixel count, and ξ is a key threshold parameter used to determine whether the current class belongs to the anomaly class or the background class.

2.3. Loss Function

To further enhance the model's sensitivity to anomalous regions, this study incorporates SAM into the reconstruction loss function [31]. SAM measures the angle between the original and reconstructed pixels in the high-dimensional spectral space, thereby reflecting the similarity of their spectral shapes. For the i-th pixel, with original spectral vector x_i and reconstructed spectral vector $\hat{x}_i$, the spectral angle Θ_i is calculated as follows:

$$\theta_i = \arccos\left(\frac{x_i \cdot \hat{x}_i}{\| x_i \| \cdot \| \hat{x}_i \|}\right). \quad (8)$$

Based on the DBSCAN clustering labels and SAM values, this study proposes a weighted reconstruction loss function to enhance the model's differentiated attention toward various types of regions during training. Specifically, the loss function is defined as follows:

$$\mathcal{L}_{\text{weighted}} = \frac{1}{N}\sum_{i=1}^{N} \omega_i \cdot \| x_i - \hat{x}_i \|^2, \quad (9)$$

where N denotes the total number of pixels in the image; x_i and $\hat{x}_i$ represent the original and reconstructed spectral vectors of the i-th pixel, respectively; and ω_i is the weighting coefficient corresponding to that pixel. The weight, ω_i, is defined as

$$\omega_i = D_i + \lambda \cdot \theta_i, \quad (10)$$

where $D_i \in \{0, 1\}$ is determined by the DBSCAN clustering result: $D_i = 1$ if the i-th pixel is clustered as background; otherwise, $D_i = 0$ if it is classified as a potential anomaly. Θ_i is the normalized spectral angle value used to quantify the spectral deviation during reconstruction, and λ is a hyperparameter that regulates the contribution of the spectral angle to the total loss.

2.4. Anomaly Detection

After the above operations, the trained multi-scale feature aggregation network is first used to test each pixel, generating the corresponding latent feature representation. The residual between the original micro-XRF image and the image reconstructed by the decoder is then calculated. Finally, the Mahalanobis distance is introduced to measure the degree of abnormality in the residual features of each pixel, which is expressed as

$$R(y_l) = \sqrt{(y_l - \mu)^T \Gamma^{-1} (y_l - \mu)}, \quad (11)$$

where $R(\cdot)$ denotes the Mahalanobis distance function, y_l is the residual feature of the test pixel, and μ and Γ represent the mean vector and covariance matrix of the residual features for the entire micro-XRF image, respectively.

2.5. Evaluation Indices

This study evaluates the anomaly detection performance of the algorithms using quantitative metrics widely adopted in anomaly detection research, including the Receiver Operating Characteristic (ROC) curve [32], the Area Under the ROC Curve (AUC) score [33], and a background anomaly separation map [34]. The detection probability (P_d) and false alarm rate (P_f) are critical parameters for generating ROC curves; under the same detection probability, a lower false alarm rate indicates superior algorithm performance. Additionally, the AUC score quantitatively reflects the overall quality of the ROC curve. The background–anomaly separation boxplots provide an intuitive qualitative perspective based on data distribution characteristics.

Microscopic comparative analysis of mineral thin sections is a qualitative evaluation method. By visually comparing the anomaly score maps generated by the algorithms with expert-identified and annotated mineral thin sections under a microscope, the correspondence between detected anomalous regions and actual anomalous mineral distributions can be intuitively assessed [35].

3. Materials

3.1. Sample Description

The samples analyzed in this study were collected from the Liqingdi lead–zinc–silver deposit, located in Chayouqianqi, Inner Mongolia. The deposit lies in the eastern segment of the Inner Mongolia Terrane, along the northern margin of the North China Platform. The investigated specimens comprise thin sections of porphyritic granite and carbonate rocks. The principal ore minerals are sphalerite, galena, and pyrite. Galena occurs mainly as disseminated grains within calcite- and quartz-rich matrices, exhibiting dissolution, interpenetrative, and inclusion textures with pyrite and sphalerite. It also displays metasomatic replacement along the margins and fractures of brecciated sphalerite, forming distinctive rim structures. Post-fragmentation galena commonly appears as breccia fragments or cementing material. Sphalerite predominantly shows anhedral granular to euhedral crystal forms, with most grains displaying cataclastic features. Fractured sphalerite is frequently replaced by galena along cleavage planes and grain boundaries. Pyrite is characterized by penetrative, dissolved, and encapsulated textures within galena and is often enclosed within recrystallized quartz overgrowths [36]. The principal gangue minerals are quartz and carbonate-group minerals.

3.2. Data Preparation

The data were acquired using the M6 JETSTREAM high-resolution, large-area micro-XRF imaging spectrometer developed by Bruker, Germany. The instrument's core component features a high-performance rhodium-target micro-focus X-ray source. The scanning parameters used for data acquisition were as follows: spot size of 50 μm, step size of 50 μm, excitation voltage of 50 kV, excitation current of 600 μA, and pixel dwell time of 15 ms.

To comprehensively evaluate the detection performance of the proposed anomaly detection algorithm and comparative methods, a dataset consisting of real mineral micro-XRF images collected from different locations was selected for experimentation. Concurrently, a corresponding anomalous mineral annotation dataset was constructed on a website. As shown in Figure 3, the experimental micro-XRF images contain 256 spectral bands.

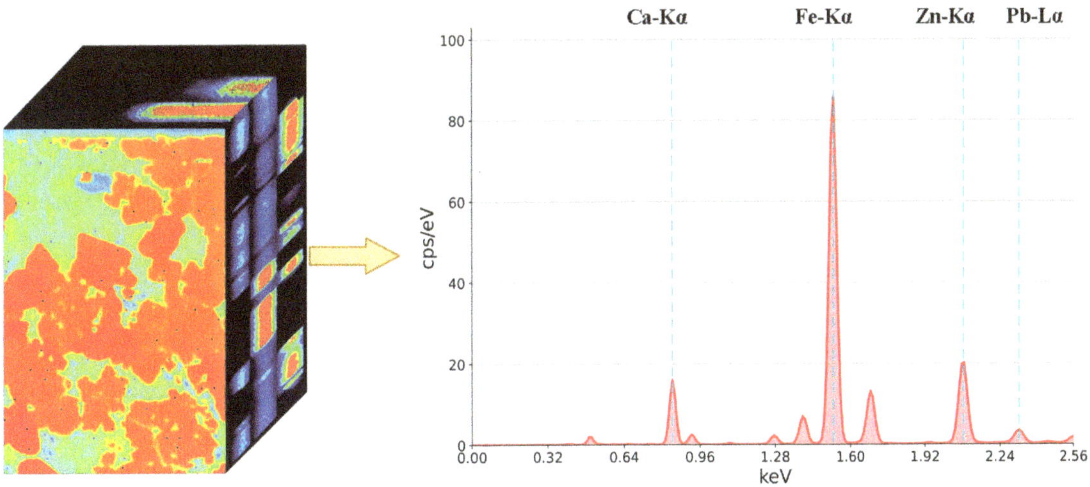

Figure 3. Data structure of a micro-XRF image and its spectral curve.

4. Results and Discussion

4.1. Detection Performance

To comprehensively evaluate the performance advantages of the proposed MSFA-Net model for micro-XRF anomaly detection, we implemented RX [37] and AE [38] for comparison with MSFA-Net. The RX detector is a classical statistical anomaly detection algorithm that characterizes the global background distribution and measures the spectral deviation of each pixel using the Mahalanobis distance. By contrast, AE is an unsupervised deep learning approach that reconstructs the input through a low-dimensional latent representation, where anomalies are identified based on reconstruction errors reflecting spectral discrepancies. By incorporating these two widely adopted baselines, the performance gains achieved by MSFA-Net can be ascribed to its architectural innovations rather than differences in preprocessing or evaluation settings. The hyperparameters of MSFA-Net used in our experiments are detailed in Table A1 in the Appendix A.

4.1.1. Comparisons of Detection Maps

Figure 4 presents a visual comparison of the detection results obtained by the three algorithms on micro-XRF images. It can be observed that both AE and MSFA-Net demonstrate superior detection performance, accurately identifying potential anomalous regions within the images. By contrast, the traditional RX, which relies on the Gaussian distribution assumption, fails to effectively detect meaningful anomalies in the complex mineral backgrounds, resulting in poor detection outcomes. Further inspection reveals that AE is prone to false positives during anomaly detection, as exemplified by the misclassification of non-anomalous regions as anomalies in the LQD_1 and LQD_2 images. Although MSFA-Net also generates a small number of false positives, its overall detection results are more stable and exhibit greater robustness. Notably, MSFA-Net not only effectively highlights anomalous regions but also preserves the integrity of their edges and spatial structures, which is beneficial for subsequent tasks such as region segmentation and mineral recognition.

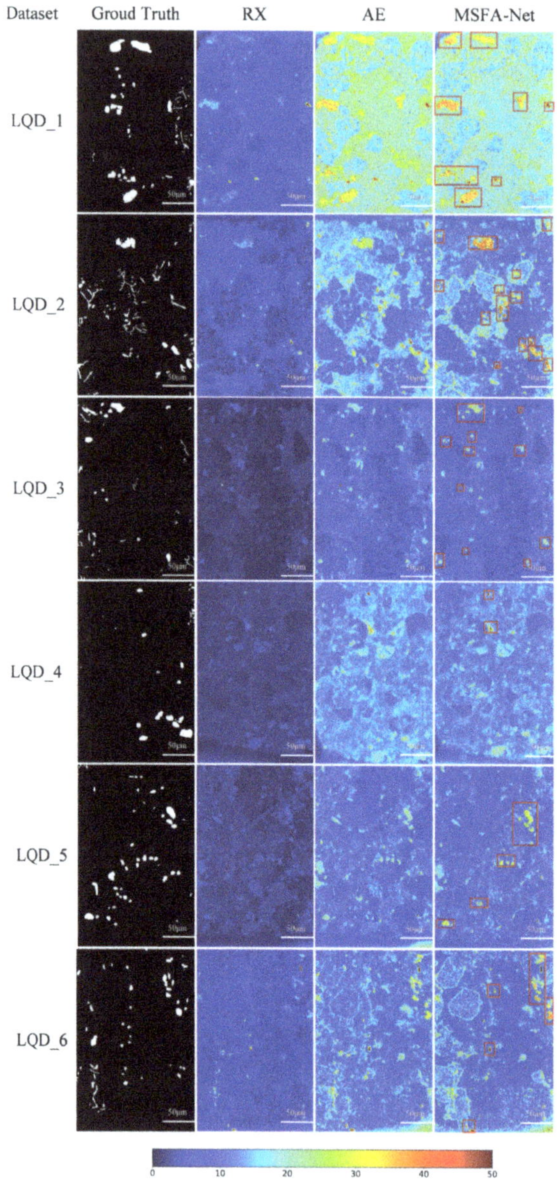

Figure 4. Anomaly detection results of different algorithms on the datasets. The red boxes indicate regions identified as anomalies by the models.

4.1.2. Comparisons of AUC and ROC

As shown in Figure 5, the experimental results demonstrate that the MSFA-Net model significantly outperforms comparative algorithms in anomaly detection on micro-XRF images. Specifically, as shown in Figure 6, the AUC values of MSFA-Net on six datasets (LQD_1 to LQD_6) are 0.852, 0.821, 0.639, 0.812, 0.790, and 0.773, respectively, all of which surpass the corresponding performance metrics of AE and RX. For instance, on the LQD_1 dataset, MSFA-Net achieves an AUC of 0.852, showing a notable improvement over AE's 0.836 and RX's 0.806. In the LQD_2 dataset, this advantage is further expanded, with MSFA-

Net attaining an AUC of 0.821, significantly exceeding AE's 0.720 and RX's 0.710. Despite the increased complexity of anomalous regions in other datasets, MSFA-Net consistently maintains stable detection capability, achieving relatively optimal AUC performance across multiple sub-images.

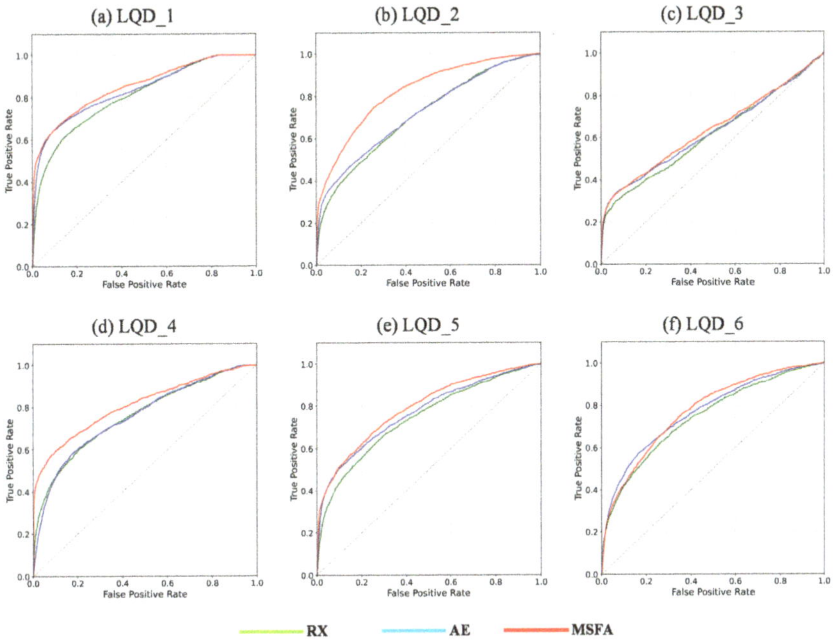

Figure 5. ROC curves of RX, AE, and MSFA-Net on the datasets.

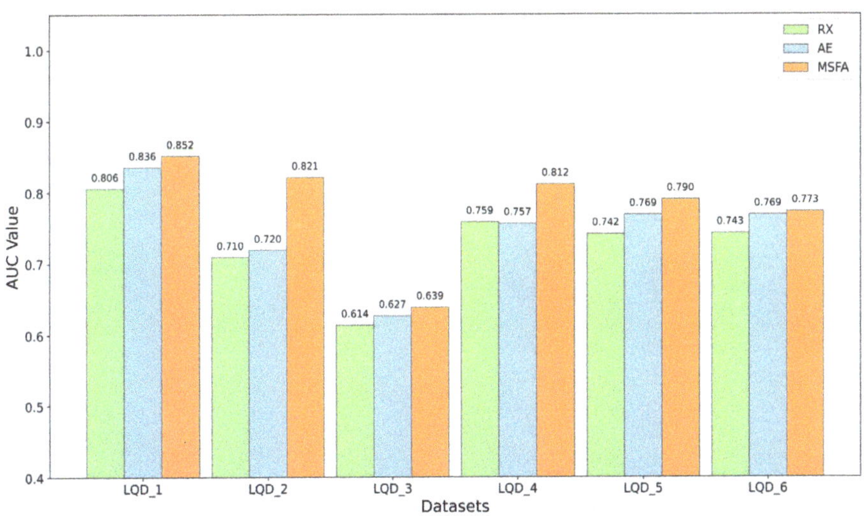

Figure 6. AUC of the datasets.

These results strongly validate the structural advantages of MSFA-Net in anomaly detection. The model not only reconstructs normal backgrounds with higher accuracy but also generates stronger reconstruction error responses in anomalous regions, thereby markedly enhancing anomaly discriminability. By contrast, although AE possesses certain reconstruction abilities, its optimization objective focuses on minimizing errors over all pixels, making it difficult to effectively concentrate on anomalous areas, which leads to performance bottlenecks across multiple datasets (e.g., an AUC of only 0.720 on LQD_2). Meanwhile, RX, relying on linear statistical properties and covariance structures, lacks the capacity to model deep spectral features, resulting in generally mediocre AUC performance.

4.1.3. Comparisons of the Separability Map

As illustrated in Figure 7, all three algorithms maintain low background pixel value distributions within the micro-XRF datasets, indicating their ability to suppress background pixels to some extent. However, notable differences exist among the methods in terms of separation between anomalies and background, as well as anomaly score distributions. Specifically, RX exhibits the most concentrated background score distribution across all datasets, demonstrating strong background suppression. Nevertheless, its anomaly scores tend to be relatively low, resulting in weaker separation between anomalies and backgrounds and limiting its ability to accurately detect subtle anomalies. AE demonstrates an enhanced response to anomalous regions in certain datasets but suffers from more dispersed background score distributions, which increases the risk of false positives. By contrast, MSFA-Net achieves low overall background pixel scores while producing higher anomaly pixel scores than the other methods, establishing a clear boundary between anomalies and background. This superior discriminative ability indicates that MSFA-Net effectively suppresses background interference and enhances anomalous pixel regions, thereby validating that the feature aggregation module improves the model's ability to distinguish anomalies from background.

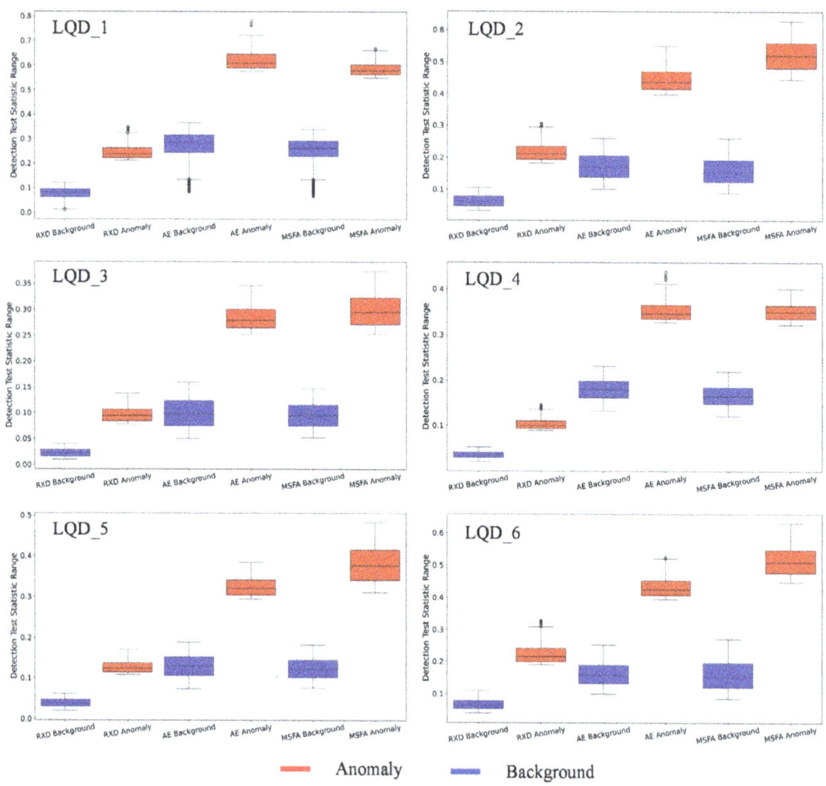

Figure 7. Background anomaly separation map of the datasets.

4.2. Discussion

To further explore the material composition of the anomalous regions identified by the algorithm, a comparative analysis was conducted between the mineral classification maps generated by SAM, the mineral polished sections, and the detection maps produced by MSFA-Net, as shown in Figure 8. Two representative samples, LQD_2 and LQD_4, were selected for this analysis. The polished mineral sections provide real optical images of the minerals under the microscope, with red boxes indicating manually annotated anomalous regions used as a reference for evaluation.

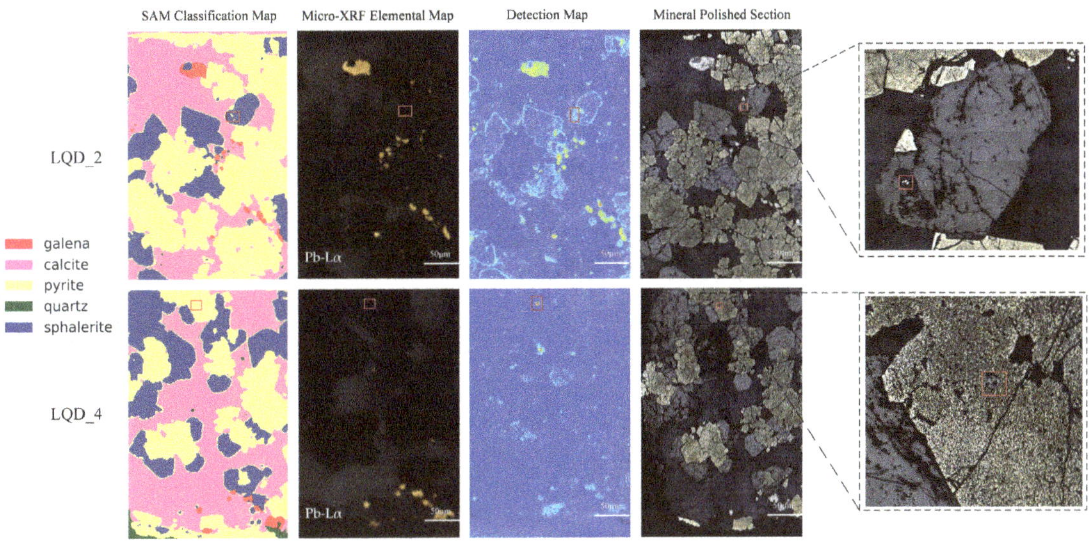

Figure 8. Microscopic comparison and validation.

The comparison reveals a strong spatial correlation between the high-response regions in the detection maps and most of the fine-grained galena observed in the polished mineral sections under 100× magnification. These samples contain multiple mineral species, with galena exhibiting complex intergrowth and interaction patterns with other minerals such as calcite and pyrite. We hypothesize that these intricate mineral associations induce significant local variations in elemental composition, causing the micro-XRF spectral signatures in these regions to deviate from those of the surrounding matrix or single-phase mineral areas.

Further analysis of the Pb-Lα distribution in the micro-XRF elemental maps shows that, due to the limited spatial resolution of the instrument, the Pb signal in the red-boxed regions is not clearly discernible. However, a comparison between the detection maps and the mineral polished sections reveals the presence of small galena grains within these regions. This finding is significant, indicating that MSFA-Net can, to some extent, overcome the spatial resolution limitations of micro-XRF instruments. By learning global spectral patterns, the model is capable of precisely localizing anomalous regions even when individual elemental maps lack clear signals.

Notably, the SAM classification maps further highlight the limitations of traditional supervised methods. SAM relies on a pre-constructed spectral library, and its performance is inherently constrained by the completeness and representativeness of that library. For instance, in sample LQD_2, the SAM map identifies only pyrite within the labeled region and fails to detect the fine-grained galena anomaly. In LQD_4, SAM also fails to identify the small galena grains. By contrast, the anomaly detection model requires no prior knowledge and can capture subtle, unexpected spectral variations through reconstruction error and latent spatial distributions, enabling the discovery of anomalies that conventional classification approaches often miss. In conclusion, MSFA-Net offers an effective solution for micro-XRF image analysis, enabling the proactive discovery and precise localization of small or rare mineral anomalies that are often overlooked in conventional mineral classification workflows, which may carry important geological implications.

4.3. Ablation Study

To assess the effectiveness of the proposed framework, we performed an ablation study to investigate the contribution of the MOA module and its key components. Four experimental settings were designed: No_Attention (no attention mechanism), No_Groups (group size is 1), No_Orthogonal (no orthogonal transformation), and the complete model, MSFA-Net.

As shown in Table 1, MSFA-Net consistently outperformed the other variants across all six datasets (LQD_1 to LQD_6), achieving the highest or near-highest AUC scores of 0.8517, 0.8214, 0.6389, 0.8116, 0.7905, and 0.7735, respectively. By contrast, the No_Attention model showed a notable performance drop on LQD_3 and LQD_4, with AUCs of 0.6329 and 0.7520, respectively, indicating that the MOA module is essential for capturing discriminative spectral–spatial features that distinguish anomalous mineral regions from complex background signals. Under the No_Groups setting, the AUC on LQD_3 decreased to 0.6310, suggesting that channel grouping enables more stable and expressive feature extraction by aggregating related spectral channels and reducing noise influence. The No_Orthogonal model yielded the lowest performance on LQD_3, with an AUC of only 0.6184, demonstrating the importance of orthogonal transformation in reducing spectral redundancy, enhancing channel independence, and improving feature separability for subtle anomalies.

Table 1. AUC results of the ablation study.

Module	LQD_1	LQD_2	LQD_3	LQD_4	LQD_5	LQD_6
No_Attention	0.8508	0.7851	0.6329	0.7520	0.7902	0.7645
No_Groups	0.8495	0.7878	0.6310	0.7516	0.7902	0.7727
No_Orthogonal	0.8498	0.8165	0.6184	0.7526	0.7896	0.7676
MSFA-Net	0.8517	0.8214	0.6389	0.8116	0.7905	0.7735

As shown in Figure 9, the anomaly detection results for MSFA-Net exhibit a clear advantage within the highlighted red box regions. Although the No_Attention, No_Orthogonal, and No_Groups configurations can partially distinguish background and anomalous areas, they are inferior to the complete MOA module in terms of detection accuracy and comprehensive anomaly characterization. For instance, in the LQD_3 dataset, the anomalous regions detected by No_Attention and No_Orthogonal show relatively low intensity, while the No_Groups configuration produces substantial background noise. In the LQD_4 dataset, only MSFA-Net yields high-intensity and well-defined anomaly regions within the red box, demonstrating more precise and reliable detection performance. These observations indicate that the combination of multi-scale context capture, channel grouping, and orthogonal feature compression in the MOA module directly contributes to the model's ability to effectively highlight and isolate subtle or mixed mineral anomalies.

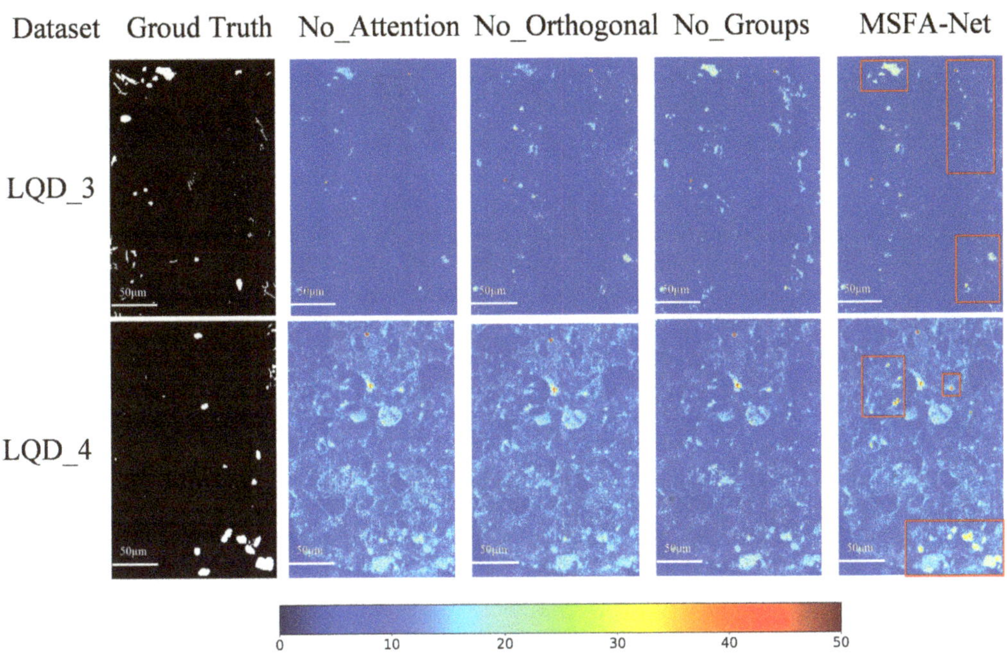

Figure 9. Detection results for MSFA-Net under different experimental settings. The red boxes indicate regions where MSFA-Net demonstrates clear advantages in anomaly detection.

In conclusion, the ablation study confirms the effectiveness and necessity of the MOA module and its internal mechanisms in enhancing unsupervised anomaly detection performance on micro-XRF images.

5. Conclusions

This study presents MSFA-Net, a deep learning-based multi-scale feature aggregation network designed to address key challenges in micro-XRF imaging, including limited spatial resolution, large data volumes, and high noise levels. Built upon an autoencoder framework, MSFA-Net incorporates an orthogonal spectral attention module and a density-based clustering strategy to achieve end-to-end unsupervised anomaly detection without requiring manual labels or external spectral libraries.

Extensive experiments on real micro-XRF datasets demonstrate that MSFA-Net consistently outperforms traditional methods such as RX and AE, achieving higher AUC scores and better localization of subtle anomalies. Comparative analysis with microscope-annotated mineral slices shows that MSFA-Net can effectively identify fine-grained anomalies that are indistinguishable in conventional elemental maps, partially overcoming the resolution limits of micro-XRF instruments. The main results are as follows:

1 Improved detection of subtle anomalies: The proposed multi-scale orthogonal attention module effectively captures global and local spatial–spectral contexts, significantly enhancing the network's ability to detect weak mineral anomalies in mixed pixels.

2 Breaking through spatial resolution limitations: By deep modeling reconstruction errors, MSFA-Net can accurately localize fine-scale anomalies that are not visible in elemental distribution maps, enabling sub-resolution anomaly detection.

3 Reduced reliance on manual labels: A weighted reconstruction loss is designed based on DBSCAN clustering and spectral angle divergence, allowing the model to assign greater attention to potential anomalous regions during training, thus improving detection performance with minimal dependence on manual labeling.

In summary, MSFA-Net provides an effective and scalable solution for unsupervised mineral anomaly detection in micro-XRF images. By integrating multi-scale feature aggregation with orthogonal attention, it effectively captures both global and local spatial–spectral information, enhancing sensitivity to fine-grained anomalies. Future work will explore its application in more complex scenarios, such as hyperspectral core scanning and 3D geochemical imaging, as well as the development of adaptive clustering strategies to further improve robustness and generalization, contributing to intelligent deep resource identification and geological process modeling.

Supplementary Materials: The following supporting information can be downloaded at: https://www.mdpi.com/article/10.3390/min15090970/s1, Folder S1: Data folder containing LQD_1 to LQD_6.npy (micro-XRF image datasets). Folder S2: Ground Truth folder containing gt_1 to gt_6.npy (manually annotated ground truth maps for micro-XRF anomaly detection).

Author Contributions: Conceptualization, Y.L., W.J. and M.Z.; Methodology, Y.L., W.J. and M.Z.; Software, W.J.; Validation, Y.L. and Y.Z.; Formal analysis, Y.L.; Investigation, Y.L., W.J. and M.Z.; Resources, M.Z. and J.Y.; Data curation, J.Y.; Writing—original draft, Y.L.; Writing—review and editing, W.J., M.Z., Y.Z., J.Y., K.Q. and Q.C.; Visualization, Y.L.; Supervision, M.Z., Y.Z. and K.Q.; Project administration, M.Z. and Q.C.; Funding acquisition, M.Z., J.Y. and Q.C. All authors have read and agreed to the published version of the manuscript.

Funding: This work was supported by the Deep Earth probe and Mineral Resources Exploration–National Science and Technology Major Project (Grant No. 2024ZD1001205-06), the National Natural Science Foundation of China (Grant No. JBA013001), and the Fundamental Research Funds for the Central Universities (Grant No. 2652022061).

Data Availability Statement: The original contributions presented in this study are included in the article/Supplementary Material. Further inquiries can be directed to the corresponding author.

Conflicts of Interest: The authors declare no conflicts of interest.

Appendix A. Hyperparameter Settings of MSFA-Net

To ensure reproducibility, the detailed hyperparameter settings of the proposed MSFA-Net are summarized in Table A1. These parameters were selected based on preliminary experiments and previous works on hyperspectral and micro-XRF anomaly detection. Unless otherwise stated, all experiments in this study were trained under the same configuration.

Table A1. Hyperparameters of MSFA-Net. Notably, *bandwidth*, *eps*, and *min_samples* correspond to the DBSCAN clustering module.

Hyperparameter	Value
Epochs	150
Batch size	36,000
Learning rate	0.0015
N_clusters	4
Latent_layer_dim	64
Anomal_prop	0.015
Bandwidth	0.3
Eps	0.1
Min_samples	25

References

1. Gao, Z.; Zhu, X.; Sun, J.; Luo, Z.; Bao, C.; Tang, C.; Ma, J. Spatial Evolution of Zn-Fe-Pb Isotopes of Sphalerite within a Single Ore Body: A Case Study from the Dongshengmiao Ore Deposit, Inner Mongolia, China. *Min. Depos.* **2018**, *53*, 55–65. [CrossRef]
2. Gregory, D.D.; Large, R.R.; Halpin, J.A.; Steadman, J.A.; Hickman, A.H.; Ireland, T.R.; Holden, P. The Chemical Conditions of the Late Archean Hamersley Basin Inferred from Whole Rock and Pyrite Geochemistry with Δ33S and δ34S Isotope Analyses. *Geochim. Cosmochim. Acta* **2015**, *149*, 223–250. [CrossRef]
3. Hurai, V.; Huraiová, M.; Konečný, P. REE Minerals as Geochemical Proxies of Late-Tertiary Alkalic Silicate ± Carbonatite Intrusions Beneath Carpathian Back-Arc Basin. *Minerals* **2021**, *11*, 369. [CrossRef]
4. Wang, B.; Ding, Z.; Bao, Z.; Song, M.; Zhou, J.; Lv, J.; Wang, S.; Zhang, Q.; Liu, C. Mesozoic Magmatic and Geodynamic Evolution in the Jiaodong Peninsula, China: Implications for the Gold and Polymetallic Mineralization. *Minerals* **2022**, *12*, 1073. [CrossRef]
5. Boltshauser, B.E.; Zaffarana, C.B.; Gallastegui, G.; Orts, D.L.; Molina, J.F.; Poma, S.M.N.; González, V.R. Petrogenetic Evolution and Thermobarometry of the Late Jurassic La Hoya Pluton, Early Stages of the North Patagonian Batholith, Southwestern Argentina. *Int. J. Earth Sci.* **2023**, *112*, 1687–1716. [CrossRef]
6. Balaram, V. Advances in Analytical Techniques and Applications in Exploration, Mining, Extraction, and Metallurgical Studies of Rare Earth Elements. *Minerals* **2023**, *13*, 1031. [CrossRef]
7. Yang, J.; Zhang, Z.; Cheng, Q. Resolution Enhancement in Micro-XRF Using Image Restoration Techniques. *J. Anal. At. Spectrom.* **2022**, *37*, 750–758. [CrossRef]
8. Wu, L.; Bak, S.; Shin, Y.; Chu, Y.S.; Yoo, S.; Robinson, I.K.; Huang, X. Resolution-Enhanced X-Ray Fluorescence Microscopy via Deep Residual Networks. *Npj Comput. Mater.* **2023**, *9*, 43. [CrossRef]
9. Nikonow, W.; Rammlmair, D. Automated Mineralogy Based on Micro-Energy-Dispersive X-Ray Fluorescence Microscopy (μ-EDXRF) Applied to Plutonic Rock Thin Sections in Comparison to a Mineral Liberation Analyzer. *Geosci. Instrum. Method. Data Syst.* **2017**, *6*, 429–437. [CrossRef]
10. Liang, J.; Sun, Y.; Lebedev, M.; Gurevich, B.; Nzikou, M.; Vialle, S.; Glubokovskikh, S. Multi-Mineral Segmentation of Micro-Tomographic Images Using a Convolutional Neural Network. *Comput. Geosci.* **2022**, *168*, 105217. [CrossRef]
11. Barker, R.D.; Barker, S.L.L.; Wilson, S.; Stock, E.D. Quantitative Mineral Mapping of Drill Core Surfaces I: A Method for *μ* XRF Mineral Calculation and Mapping of Hydrothermally Altered, Fine-Grained Sedimentary Rocks from a Carlin-Type Gold Deposit. *Econ. Geol.* **2021**, *116*, 803–819. [CrossRef]
12. Tang, J.; Wang, W.; Yuan, C. A New Anisotropic Singularity Algorithm to Characterize Geo-Chemical Anomalies in the Duolong Mineral District, Tibet, China. *Minerals* **2023**, *13*, 988. [CrossRef]
13. Wang, W.; Yuan, C.; Tang, J.; Ren, X.; Zhao, J. Enhancing Deep Orebody Prediction and Localization through the Revelation of Geochemical Primary Halo Patterns in Drill Holes. *Appl. Geochem.* **2024**, *171*, 106100. [CrossRef]
14. Liu, C.; Wang, W.; Tang, J.; Wang, Q.; Zheng, K.; Sun, Y.; Zhang, J.; Gan, F.; Cao, B. A Deep-Learning-Based Mineral Prospectivity Modeling Framework and Workflow in Prediction of Porphyry–Epithermal Mineralization in the Duolong Ore District, Tibet. *Ore Geol. Rev.* **2023**, *157*, 105419. [CrossRef]
15. Lou, W.; Zhang, D.; Bayless, R.C. Review of Mineral Recognition and Its Future. *Appl. Geochem.* **2020**, *122*, 104727. [CrossRef]
16. Kim, J.J.; Ling, F.T.; Plattenberger, D.A.; Clarens, A.F.; Lanzirotti, A.; Newville, M.; Peters, C.A. SMART Mineral Mapping: Synchrotron-Based Machine Learning Approach for 2D Characterization with Coupled Micro XRF-XRD. *Comput. Geosci.* **2021**, *156*, 104898. [CrossRef]
17. Xu, Y.; Zhang, L.; Du, B.; Zhang, L. Hyperspectral Anomaly Detection Based on Machine Learning: An Overview. *IEEE J. Sel. Top. Appl. Earth Obs. Remote Sens.* **2022**, *15*, 3351–3364. [CrossRef]
18. Lv, S.; Zhao, S.; Li, D.; Pang, B.; Lian, X.; Liu, Y. Spatial–Spectral Joint Hyperspectral Anomaly Detection Based on a Two-Branch 3D Convolutional Autoencoder and Spatial Filtering. *Remote Sens.* **2023**, *15*, 2542. [CrossRef]
19. Chen, J.; Xu, F.; Zeng, T.; Li, X.; Chen, S.; Yu, J. MSFA: Multi-stage Feature Aggregation Network for Multi-label Image Recognition. *IET Image Process.* **2024**, *18*, 1862–1877. [CrossRef]
20. Sun, T.; Wang, W.; Ma, W. MSFA-Net: A Multi-Scale Feature Aggregation Network with Mixed Attention for Industrial Defect Detection. *SSRN Preprint* **2025**. [CrossRef]
21. Cheng, X.; Huo, Y.; Lin, S.; Dong, Y.; Zhao, S.; Zhang, M.; Wang, H. Deep Feature Aggregation Network for Hyperspectral Anoma ly Detection. *IEEE Trans. Instrum. Meas.* **2024**, *73*, 5033016. [CrossRef]
22. Wang, Q.; Li, F.; Jiang, X.; Wu, S.; Xu, M. On-Stream Mineral Identification of Tailing Slurries of Tungsten via NIR and XRF Data Fusion Measurement Techniques. *Anal. Methods* **2020**, *12*, 3296–3307. [CrossRef] [PubMed]
23. Wang, W.; Liu, Z.; Tang, J.; Yuan, C. An Enhanced Strategy for Geo-Exploratory Data Analysis to Facilitate the Discovery of New Mineral Deposits. *J. Geochem. Explor.* **2024**, *258*, 107411. [CrossRef]
24. Wang, W.; Pei, Y.; Cheng, Q.; Wang, W. Local Singularity Spectrum: An Innovative Graphical Approach for Analyzing Detrital Zircon Geochronology Data in Provenance Analysis. *Fractal Fract.* **2024**, *8*, 64. [CrossRef]

25. Ouyang, D.; He, S.; Zhang, G.; Luo, M.; Guo, H.; Zhan, J.; Huang, Z. Efficient Multi-Scale Attention Module with Cross-Spatial Learning. In Proceedings of the ICASSP 2023—2023 IEEE International Conference on Acoustics, Speech and Signal Processing (ICASSP), Rhodes Island, Greece, 4–10 June 2023; IEEE: Rhodes Island, Greece, 2023; pp. 1–5.
26. Salman, H.; Parks, C.; Swan, M.; Gauch, J. *OrthoNets: Orthogonal Channel Attention Networks 2023*; IEEE: Los Angeles, CA, USA, 2023.
27. Wang, Q.; Zhang, X.; Tang, B.; Ma, Y.; Xing, J.; Liu, L. Lithology Identification Technology Using BP Neural Network Based on XRF. *Acta Geophys.* **2021**, *69*, 2231–2240. [CrossRef]
28. Long, T.; Zhou, Z.; Hancke, G.; Bai, Y.; Gao, Q. A Review of Artificial Intelligence Technologies in Mineral Identification: Classification and Visualization. *J. Sens. Actuator Netw.* **2022**, *11*, 50. [CrossRef]
29. Wang, W.; Xue, C.; Zhao, J.; Yuan, C.; Tang, J. Machine Learning-Based Field Geological Mapping: A New Exploration of Geological Survey Data Acquisition Strategy. *Ore Geol. Rev.* **2024**, *166*, 105959. [CrossRef]
30. Duan, L.; Xu, L.; Guo, F.; Lee, J.; Yan, B. A Local-Density Based Spatial Clustering Algorithm with Noise. *Inf. Syst.* **2007**, *32*, 978–986. [CrossRef]
31. Salles, R.D.R.; De Souza Filho, C.R.; Cudahy, T.; Vicente, L.E.; Monteiro, L.V.S. Hyperspectral Remote Sensing Applied to Uranium Exploration: A Case Study at the Mary Kathleen Metamorphic-Hydrothermal U-REE Deposit, NW, Queensland, Australia. *J. Geochem. Explor.* **2017**, *179*, 36–50. [CrossRef]
32. Kerekes, J. Receiver Operating Characteristic Curve Confidence Intervals and Regions. *IEEE Geosci. Remote Sens. Lett.* **2008**, *5*, 251–255. [CrossRef]
33. Tan, K.; Hou, Z.; Wu, F.; Du, Q.; Chen, Y. Anomaly Detection for Hyperspectral Imagery Based on the Regularized Subspace Method and Collaborative Representation. *Remote Sens.* **2019**, *11*, 1318. [CrossRef]
34. Chang, C.-I.; Chiang, S.-S. Anomaly Detection and Classification for Hyperspectral Imagery. *IEEE Trans. Geosci. Remote Sens.* **2002**, *40*, 1314–1325. [CrossRef]
35. Tusa, L.; Andreani, L.; Khodadadzadeh, M.; Contreras, C.; Ivascanu, P.; Gloaguen, R.; Gutzmer, J. Mineral Mapping and Vein Detection in Hyperspectral Drill-Core Scans: Application to Porphyry-Type Mineralization. *Minerals* **2019**, *9*, 122. [CrossRef]
36. Li, B.; Yang, Y.; Wang, J.; Song, Z. Mesozoic Metallogenic Characteristics and Ore-Controlling Factors in the Southeastern Inner Mongolia, China. *Arab. J. Geosci.* **2022**, *15*, 1368. [CrossRef]
37. Reed, I.S.; Yu, X. Adaptive Multiple-Band CFAR Detection of an Optical Pattern with Unknown Spectral Distribution. *IEEE Trans. Acoust. Speech Signal Process.* **1990**, *38*, 1760–1770. [CrossRef]
38. Wang, S.; Wang, X.; Zhang, L.; Zhong, Y. Auto-AD: Autonomous Hyperspectral Anomaly Detection Network Based on Fully Convolutional Autoencoder. *IEEE Trans. Geosci. Remote Sens.* **2022**, *60*, 5503314. [CrossRef]

minerals

MDPI

Article

Source and Precipitation Process of Gold in the Linglong Gold Deposit, Jiaodong Peninsula: Constraints from Trace Elements of Pyrite and S-Pb Isotopes

Fei Ren [1], Zheng-Jiang Ding [1,*], Zhong-Yi Bao [1], Jun-Wei Wang [1], Shun-Xi Ma [1], Tao Niu [1], Kai-Qiang Geng [1], Bin Wang [1], Chao Li [2,3], Gui-Jie Li [4] and Shan-Shan Li [1,5]

1 No. 6 Geological Team of Shandong Provincial Bureau of Geology and Mineral Resources, Ministry of Natural Resources Technology Innovation Center for Deep Gold Resources Exploration and Mining, Shandong Engineering Research Center of Application and Development of Big Data for Deep Gold Exploration, Weihai 264209, China; renfei0627@163.com (F.R.); gybzy2005@126.com (Z.-Y.B.); lywhwjw@163.com (J.-W.W.); ma527030@126.com (S.-X.M.); ntjls@126.com (T.N.); gengkq@126.com (K.-Q.G.); wangbinjlu@163.com (B.W.); shanshanli@cugb.edu.cn (S.-S.L.)
2 National Research Center for Geoanalysis, Beijing 100037, China; re-os@163.com
3 Key Laboratory of Re-Os Isotope Geochemistry, China Geological Survey, Beijing 100037, China
4 Zhaoyuan Engineering Construction Service Center, Zhaoyuan 265400, China; zyjslgj@163.com
5 School of Science, China University of Geosciences (Beijing), Beijing 100083, China
* Correspondence: ytdzhj@126.com

Abstract

Jiaodong Gold Province is a globally rare giant gold cluster, with ongoing debates regarding its metallogenic material sources and mineralization mechanisms. This study focuses on the Linglong quartz-vein-type gold deposit within the Zhaoping Fault Zone, conducting in situ trace element and S-Pb isotope analyses of pyrite from different mineralization stages. The trace element characteristics were investigated to explore the sources of metallogenic materials, the evolution of ore-forming fluids, and the mechanisms of gold precipitation. The main findings are as follows: (1) In the Linglong gold deposit, gold primarily enters the pyrite lattice as a solid solution (Au^+) through Au-As coupling. From the Py1 to Py3 stages, Co and Ni contents significantly decrease, while Cu, As, Au, and polymetallic element contents continuously increase. Additionally, Cu mainly replaces Fe^{2+} in the form of Cu^{2+}, whereas Pb predominantly exists as micro inclusions of galena. (2) The S isotope (Py1: $\delta^{34}S$ = +7.60‰–+8.25‰, Py2: $\delta^{34}S$ = +6.15‰–+8.15‰, Py3: $\delta^{34}S$ = +6.90‰–+9.10‰) and Pb isotope ($^{206}Pb/^{204}Pb$ = 16.95–17.715, $^{207}Pb/^{204}Pb$ = 15.472–15.557, $^{208}Pb/^{204}Pb$ = 37.858–38.394) systems collectively constrain the ore-forming materials such that they are dominated by metasomatized enriched lithospheric mantle, with simultaneous mixing of crustal materials. (3) The ore-forming fluid underwent a continuous evolution process characterized by persistently decreasing temperatures and a transition from mantle-dominated to crust–mantle mixed sources. The Py1 stage was predominantly composed of mantle-derived magmatic fluids uncontaminated by crustal materials, representing a high-temperature, closed environment. In the Py2 stage, the fluid system transitioned to an open system with the incorporation of crustal materials. Through coupled substitution of "$As^{3+} + Au^+ \rightarrow Fe^{2+}$" and dissolution–reprecipitation processes, gold was initially activated and enriched. During the Py3 stage, pyrite underwent dissolution–reprecipitation under tectonic stress and fluid activity, promoting extraordinary element enrichment and serving as the primary mechanism for gold precipitation. Concurrently, bismuth–tellurium melt interactions further facilitated the precipitation of gold minerals.

 https://doi.org/10.3390/min15111220

Keywords: pyrite; trace elements; S-Pb isotope; source of ore-forming materials; Linglong gold deposit; Jiaodong

1. Introduction

The Jiaodong gold ore concentration area is located on the southeastern margin of the North China Craton, with proven gold resource reserves approaching 6000 tons, making it a rare giant gold province in China and even the world [1–3]. The Jiaodong region hosts dozens of large and super large gold deposits, along with over a hundred medium and small gold occurrences [4–6]. The gold deposits in this region are genetically distinct from typical orogenic or magmatic hydrothermal deposits, hence termed as "Jiaodong type" gold deposits [7,8].

Regarding the sources of the ore-forming fluids and materials, there are currently four main controversial models: (1) orogenic gold deposits, proposing that the ore-forming fluids were derived from metamorphic dehydration [9,10]; (2) magmatic hydrothermal gold deposits: emphasizing the contribution of Mesozoic granites [11]; (3) subducted slab metasomatized mantle type gold deposits (Jiaodong type gold deposits), suggesting that the ore-forming materials originated from the Paleo-Pacific slab and the overlying mantle wedge [12,13]; (4) craton destruction model gold deposits, attributing the mineralization to hydrous basaltic magmatism [14]. Although these models are based on different geological evidence and theoretical reasoning, no consensus has been reached yet.

As the primary gold-bearing mineral permeating the entire mineralization process in the Jiaodong gold deposits, pyrite often exhibits polygenerational characteristics. Its complex structure and internal variations in trace elements can effectively reflect the properties and evolutionary processes of ore-forming fluids, as well as the mechanisms of gold enrichment and precipitation [15–17]. Although previous studies have been conducted, particularly in situ elemental and isotopic research on pyrite in the Jiaodong gold deposits within the region, exploring the sources of ore-forming materials, mineralization processes, hydrothermal evolution, genesis of deposits, and influencing factors of gold enrichment and mineralization [18–23], there remains a lack of unified understanding regarding the regularities of the mineralization system.

The Linglong gold deposit, a typical quartz-vein type, is located in the Zhaoping fault zone of the northwestern Jiaodong ore concentration area. Its mineralization process involves four consecutive stages: Py1 (pyrite–sericite–quartz), Py2 (quartz–pyrite), Py3 (quartz–polymetallic sulfide), and quartz–carbonate. This study focuses on Py1-Py3 pyrites, conducting high-resolution in situ microanalysis to measure trace element contents. Combined with sulfur–lead isotopes, it investigates trace element occurrence/enrichment mechanisms, constrains ore-forming material sources and fluid evolution, and provides geochemical insights into deposit genesis.

2. Regional Geology

The Jiaodong region is located on the eastern side of the Tanlu Fault and the western side of the Pacific subduction zone. It is mainly composed of two tectonic units: the Sulu terrane in the east and the Jiaobei terrane in the west, with the Wulian–Qingdao–Yantai Fault serving as the suture zone (Figure 1A). The Sulu terrane is an ultra-high-pressure metamorphic belt, while the Jiaobei terrane is divided into the Jiaobei Uplift in the north and the Jiaolai Basin in the south [24]. Mesozoic tectonomagmatic activity was intense, with Late Jurassic and Early Cretaceous magmatism widespread, including granite bodies and intermediate basic dykes [25,26].

Jiaodong has widespread Precambrian basement metamorphic rocks, including the Neoarchean Jiaodong Group (TTG gneiss dominated) in the Jiaobei Block and Proterozoic Jingshan, Fenzishan, and Penglai Groups (metamorphic sedimentary rock dominated) (Figure 1B) [27,28]. Mesozoic large-scale gold mineralization made Jiaodong rich in gold resources [29–31].

The tectonic characteristics of Jiaodong are mainly manifested as regional fault structures, dominated by NE-NNE and NW-NNW trending structures. Among them, the NE-NNE trending faults consist of extensional faults, which have derived a series of secondary faults, and gold mineralization is mainly controlled by the NE-NNE trending faults [32]. Eastward from the Yishu Fault, the NE-NNE-trending Sanshandao, Jiaojia, and Zhaoping Faults occur sequentially. Widely developed NE structures, formed by Early Mesozoic North China–Yangtze plate collision, consist of metamorphic basement folds and secondary faults, with EW faults later reworked. As secondary faults of the Yishu Fault, NE-NNE faults (including west to east Sanshandao, Jiaojia, Zhaoyuan–Pingdu, and Muru Faults) control nearly all proven gold mineralization and have experienced multiple compression extension episodes.

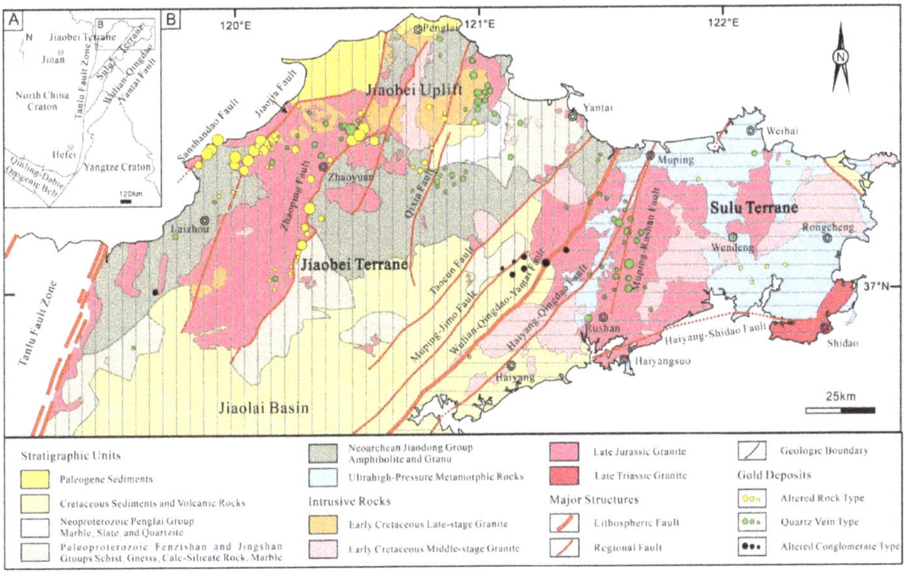

Figure 1. (**A**) Simplified Tectonic Map of Jiaodong; (**B**) Simplified geological map of the Jiaodong area (after [33]).

Intrusive rocks are extensively developed in the area, primarily comprising three types of granite: the Linglong type, Guojialing type, and Aishan type [34]. Linglong-type granite is mainly composed of the Linglong biotite granite and the Luanjiahe monzogranite. It is developed between the Zhaoping Fault and the Jiaojia Fault and is widely exposed. Zircon U-Pb dating by LA-ICP-MS indicates that the Linglong-type granite formed mainly between 166 and 149 Ma [35]. Guojialing-type granite shows an intrusive contact relationship with the Neoarchean Jiaodong Group and the Linglong-type granite. It is generally distributed in the Shangzhuang and Guojialing areas of Zhaoyuan, with an overall nearly E-W trend [36]. Zircon U-Pb ages show its emplacement occurred between 133 and 126 Ma, formed by the partial melting of the Precambrian metamorphic basement of the lower crust [37]. Aishan-type granite is mainly exposed in eastern Jiaodong and consists of undeformed alkaline granites. Zircon U-Pb ages indicate that the Aishan-type granite was emplaced mainly between 120 and 108 Ma and has a crust–mantle mixed origin [38,39]. Mafic dikes are widely distributed but typically small in volume. Zircon U-Pb dating shows that these basaltic dikes were emplaced between 130 and 110 Ma, derived primarily from low-degree partial melting of the lithospheric mantle [40].

3. Deposit Geology

The Linglong gold deposit is located in the northwestern part of the Jiaodong Peninsula, along the northern segment of the Zhaoyuan–Pingdu Fault. The mining area spans from Dongfeng in the east to Oujiakuang in the west, and from Taishang in the south to Dazhuangzi in the north. With a prospective gold reserve exceeding 1000 tons, it ranks as a world-class super-large gold deposit. The main exposed strata in the mining area consist of the Neoarchean Jiaodong Group and the Cenozoic Quaternary System. Jiaodong Group is predominantly composed of amphibolites, biotite gneisses, and schists (Figure 2), which occur as relatively small-scale layered or lenticular bodies. From the Paleoproterozoic to the Early Paleozoic, the metamorphic and deformational processes underwent a complex evolution, transitioning from ductile shear zones to brittle fracture zones, and further to mylonitization.

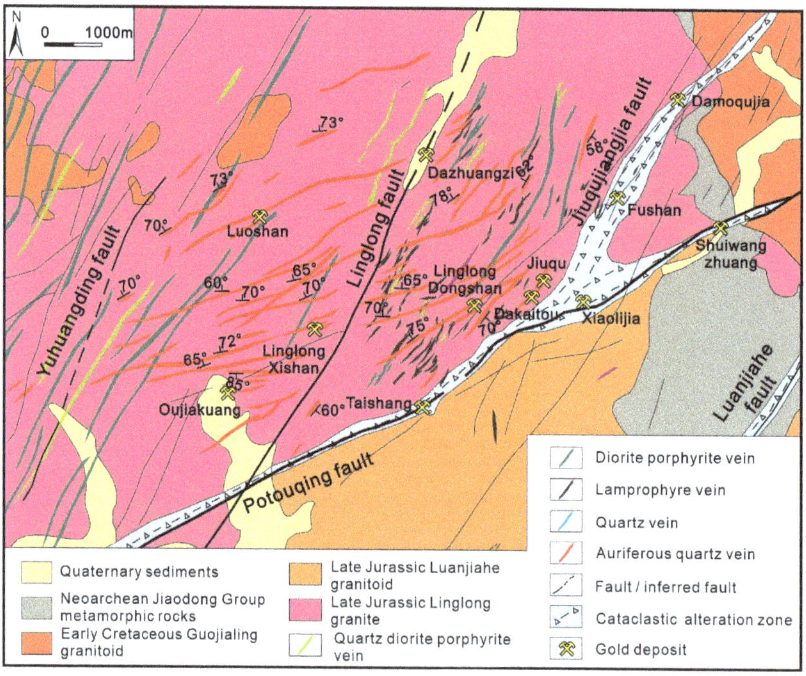

Figure 2. Simplified geological map of the Linglong gold deposit (after [41]).

The structural framework of the Linglong gold deposit is primarily controlled by regional faults. These faults have been superimposed and compounded during multiple tectonic movements, forming a series of secondary faults with varying scales and dense distribution. Gold mineralization is closely associated with these ore-controlling faults, and ore enrichment zones are typically located within fault zones or at their intersections. The emplacement of orebodies is mainly controlled by first-order faults, with second-order faults developed in their footwalls, including the ore-forming Potouqing Fault and the post-ore-forming Linglong Fault. The most developed secondary faults in the mining area are the northeast–north–northeast trending ones, which are relatively small in scale, mainly distributed on both sides of the first-order faults, and belong to ore-forming structures. Compared with other ore-controlling structures, this set of faults exhibits significant control and modification effects on gold mineralization, showing a zonal distribution in the central and eastern parts of the mining area.

Mesozoic magmatism is highly developed within the mining area. The Linglong-type, Guojialing-type, and Luanjiahe-type granites are the main ore-hosting rocks in this region, with the Linglong-type granite being dominant. According to previous studies, starting from the northwest of the Potouqing Fault, the lithology gradually transitions from intermediate to basic as the frequency of dark dikes increases [42]. In the Linglong gold deposit, dikes and ore bodies often exhibit similar temporal and spatial characteristics, suggesting they may be products of cogenetic differentiation.

The ore types in the Linglong gold deposit mainly include altered rock type and quartz-vein-type ores. Altered rock type ores are dominated by pyrite-sericite rock and pyrite-sericitized granite (Figure 3A), while quartz-vein-type ores are represented by quartz-pyrite veins and quartz–polymetallic sulfide veins (Figure 3B,C). The two ore types exhibit distinct structural characteristics: the former primarily shows disseminated or stockwork-disseminated structures, while the latter commonly develops veinlet and stockwork structures. In terms of mineral assemblages, they share similarities, with metallic minerals dominated by pyrite, accompanied by sulfides such as chalcopyrite, galena, and sphalerite (Figure 3D,E); gangue minerals mainly consist of quartz, sericite, and calcite. Ore textures are diverse, including euhedral–subhedral granular texture, cataclastic texture, and interstitial texture; ore structures mainly include massive, stockwork, and banded structures. Based on their occurrence, gold-bearing minerals are classified into three types: intergranular gold distributed between pyrite and other mineral grains, inclusion gold enclosed within sulfide crystals, and fracture gold filling in sulfide fractures (Figure 3G–I).

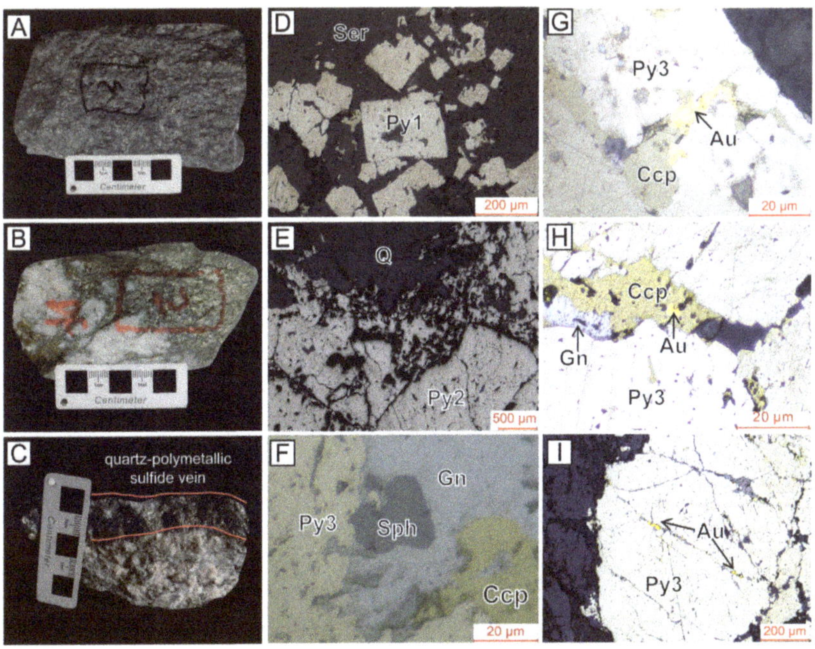

Figure 3. Characteristics of ore samples and metallic mineral assemblages in the Linglong gold deposit. (**A**) Pyrite–quartz–sericite stage ore; (**B**) Quartz–pyrite stage ore; (**C**) Quartz–polymetallic

sulfide stage ore; (**D**) Euhedral to subhedral pyrite from the Py1 stage; (**E**) Subhedral to anhedral pyrite from the Py2 stage, with dissolution textures formed at the edges of pyrite; (**F**) Metallic sulfide mineral assemblage from the Py3 stage; (**G–I**) Different occurrences of native gold ((**G**) Intercrystalline gold; (**H**) Encapsulated gold; (**I**) Fracture gold). Py: Pyrite, Gn: Galena, Sph: Sphalerite, Apy: Arsenopyrite, Au: Native gold.

The Linglong gold deposit exhibits typical wall-rock alteration zoning, whose spatial distribution generally shows symmetry centered around ore veins. From the ore vein to the surrounding rocks on both sides, the alteration zones are sequentially: silicification zone, pyrite-sericite zone, pyrite-sericitized granite zone, potassic-altered granite zone, and finally transitioning to unaltered biotite granite. Among them, K-feldspathization is distributed in the outer distal wall rocks, formed earlier than the sericitization stage, and the intensity of potassic alteration gradually weakens with increasing distance from the main fault. In contrast, sericitization alteration is extremely developed in the deposit, mainly distributed in the area adjacent to the main fault, and is a product of early ore-forming hydrothermal processes.

Based on the orebody occurrence, ore mineral assemblage characteristics, wall-rock alteration features, and vein cross-cutting relationships, the metallogenic process of the Linglong gold deposit is divided into 4 stages (Figure 4).

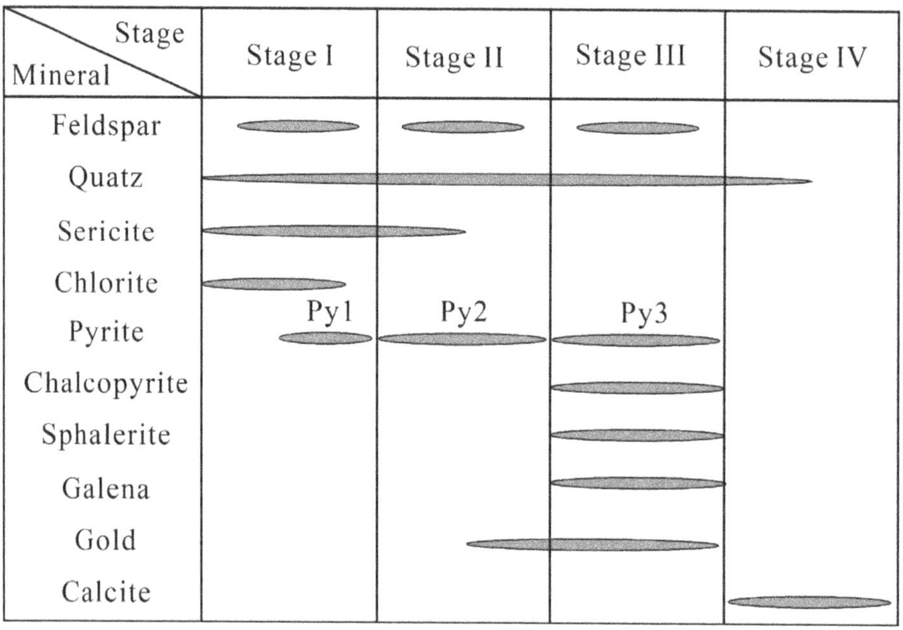

Figure 4. Mineral paragenesis sequence of the Linglong gold deposit.

Stage I (Pyrite–Sericite–Quartz Stage): The sericite–quartz rock or pyrite–sericite–quartz rock characterized by the mineral assemblage of pyrite + quartz + sericite forms the early stage of mineralization. Pyrite is mostly distributed in euhedral–subhedral granular texture with well-developed crystals. Quartz shows dynamic recrystallization, and some pyrite is stretched, indicating that this stage experienced ductile deformation.

Stage II (Quartz–Pyrite Stage): As the middle mineralization stage, it is characterized by smoky gray quartz and widely exposed quartz–pyrite veins. The mineral assemblage is dominated by quartz and pyrite, with a small amount of chalcopyrite. Pyrite is mostly distributed as medium-coarse-grained irregular massive aggregates, partially as subhedral crystals, and occasionally undergoes late dissolution and replacement. Quartz is mostly smoky gray, medium-fine-grained subhedral crystals.

Stage III (Quartz–Polymetallic Sulfide Stage): This is the middle-late mineralization stage and also the late stage of gold mineralization, characterized by the development of polymetallic sulfide veins and smoky gray quartz veins. The mineral assemblage includes grayish-white quartz, pyrite, chalcopyrite, galena, sphalerite, etc. Metallic sulfides such as pyrite occur as medium-fine-grained aggregates, often cutting through the earlier quartz veins to form massive or banded ores. Quartz is light grayish-white, medium-coarse-grained euhedral–anhedral crystals. Native gold is mainly developed in the interstices of pyrite crystals, coexisting with polymetallic sulfides, occurring as intercrystalline gold.

Stage IV (Quartz–Carbonate Stage): As the late mineralization stage, it is characterized by a large number of calcite veins and low-gold-bearing quartz, with low sulfide content and a small amount of pyrite. Almost no gold mineralization occurs in this stage.

4. Samples and Analytical Methods

4.1. LA-ICP-MS Analysis of Trace Elements in Pyrite

Trace element concentrations in pyrite were determined using laser ablation inductively coupled plasma mass spectrometry (LA-ICP-MS) at the National Research Center of Geoanalysis of China.

The analyses were performed on an ICP-MS (Agilent Technologies, Inc., Santa Clara, CA, USA) equipped with a 193 nm ArF excimer laser ablation system (Fremont, CA, USA). Argon gas was used as the supplementary gas, which was mixed with the carrier gas via a T-junction before entering the ICP. High cleaning efficiency was achieved under high flow rates of He (0.9 L/min) and Ar (0.9 L/min). For each analysis, a homogeneous laser spot with a diameter of 30 μm was employed, operating at 7 Hz with an energy density of 2 J/cm^2, with a measurement duration of 20 s and a total run time of 40 s. Reference materials NIST 610 and NIST 612 were used as external standards to construct calibration curves (National Institute of Standards and Technology, Gaithersburg, MD, USA). Based on MASS-1 (USGS), the analytical error for trace elements in sulfide minerals was less than 30%. A set of standard samples was analyzed after every 10 sample spot analyses.

4.2. LA-MC-ICP-MS Analysis of Sulfur Isotopes in Pyrite

In situ sulfur isotope analysis of pyrite was conducted at the National Research Center for Geoanalysis of China. The pyrite samples selected for testing were derived from the polymetallic sulfide stage of the Linglong gold deposit. Through processing of the test data, sulfur isotope composition data and related parameters of sulfides in the minerals were obtained. The laser ablation system employed was an NWR 193 nanosecond laser (Fremont, CA, USA), and the analysis was performed using a multi-collector inductively coupled plasma mass spectrometer (Thermo Scientific NEPTUNE Plus MC-ICP-MS) manufactured by Thermo Fisher Scientific (Karlsruhe, Germany). Helium was used as the carrier gas during laser ablation with a flow rate of 600 mL/min. The analysis uses a single-point mode with laser conditions of a large beam spot (44 μm) and low frequency (2 Hz), and the laser energy density is fixed at 5 J/cm^2. The system is integrated with a signal smoothing device to ensure signal stability under low frequency [43]. The Neptune Plus mass spectrometer is equipped with 9 Faraday cups and adopts a medium resolution mode (approximately 7000) to eliminate the influence of interference peaks. It uses three Faraday cups (L3, C, and H3) to statically receive ^{32}S, ^{33}S, and ^{34}S signals simultaneously. A single data acquisition is set with an integration time of 0.131 s, and 200 sets of data are continuously acquired to form a complete signal sequence of approximately 27 s. In this study, laboratory standard samples were used, and the testing sequence of "Standard Substance–Sample–Standard Substance" (SSB) was adopted for cross-validation and data correction.

4.3. LA-MC-ICP-MS Analysis of Galena Pb Isotopes

In this study, in situ Pb isotope experiments were conducted using LA-MC-ICP-MS at the National Research Center for Geoanalysis of China. Samples were selected from gold-bearing quartz-vein ores of the Linglong gold deposit and subjected to pretreatment procedures including numbering and wiping. For in situ Pb isotope analysis, a femtosecond laser ablation system (ASIJ-200–343 nm, Applied Spectra Inc., West Sacramento, CA, USA) was employed with a spot diameter of 2 μm, laser ablation energy of 80%, and a 2 μm line scan mode, with an ablation duration of approximately 15 s. Analyses were performed using a multi-collector inductively coupled plasma mass spectrometer (Thermo Scientific NEPTUNE Plus MC-ICP-MS). To optimize analytical quality, signal smoothing technology and a mercury removal device were introduced in the experiment, which helps improve signal stability and testing accuracy, while reducing gas background noise and mercury interference in samples. The mass spectrometer is equipped with 9 Faraday cups, capable of simultaneously receiving lead isotope signals including ^{208}Pb, ^{207}Pb, ^{206}Pb, and ^{204}Pb, as well as signals of ^{205}Tl, ^{203}Tl, and ^{202}Hg. A single-standard Tl solution introduced via Aridus II membrane desolvation technology is mixed with ablation-generated aerosols before entering the ICP, and the ^{205}Tl/^{203}Tl ratio is used for real-time mass fractionation correction of lead isotopes.

5. Results

5.1. Trace Element Characteristics of Pyrite

Fifty-seven pyrite grains were analyzed using LA-ICP-MS in this study, including 12 points for Py1, 20 for Py2, and 25 for Py3. The test data are presented in Supplementary Table S1. Based on the pyrite analysis results, trace element boxplots for the three types of pyrite were generated (Figure 5). Analytical results show that concentrations of elements such as Co, Ni, Cu, Zn, As, Te, Ag, Sb, Au, Bi, and Pb are generally above the detection limit, with only some data points falling below it.

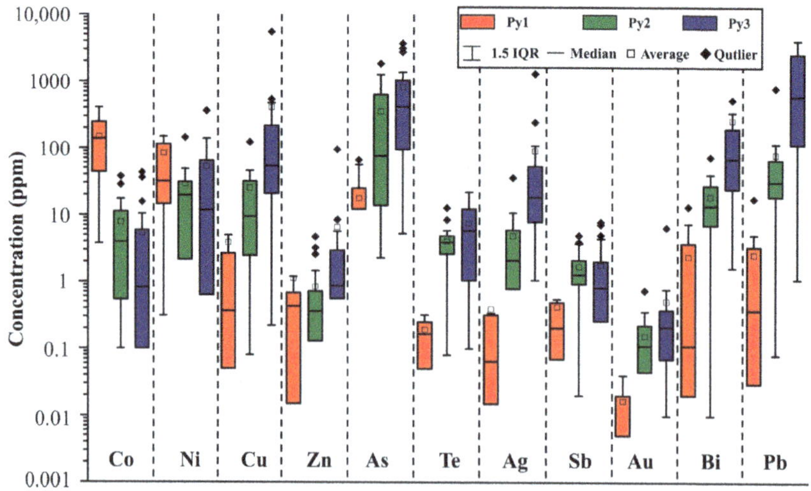

Figure 5. Box plots of trace element concentrations for the three pyrite generations in the Linglong gold deposit.

The composition of Py1 shows relatively high concentrations of Co (median 137.87 ppm) and Ni (median 31.53 ppm), with Co/Ni ratios ranging from 0.57 to 17.19. Au concentrations range from 0.01 to 0.09 ppm, while As concentrations range from 0.77 to 66.28 ppm. The concentrations of Cu, Zn, Ag, Bi, and Pb are mostly less than 1 ppm. Compared with Py1, Py2 has significantly lower concentrations of Co (median 3.94 ppm) and Ni (median 19.51 ppm). Au concentrations in Py2 are less than 0.80 ppm, and the concentrations of As, Cu, Zn, Te, Ag, Bi, and Pb are all higher than those in Py1. Compared with Py1 and Py2, Py3 has relatively lower contents of Ni and Co. The gold concentrations in Py3 range from 0.01 to 6.52 ppm, and the concentrations of As, Cu, Zn, Te, Ag, Bi, and Pb are all higher than those in both Py1 and Py2.

5.2. Sulfur Isotope Characteristics

LA-MC–ICP–MS sulfur isotope analysis was conducted on pyrite from three mineralization stages of the Linglong gold deposit (Table 1). The results show a range of $\delta^{34}S$ values from +6.15‰ to +9.10‰. The $\delta^{34}S$ values of Py1 stage pyrite range from +7.60‰ to +8.25‰, with an average of +7.9‰ (6 measurement points). The $\delta^{34}S$ values of Py2 stage pyrite range from +6.15‰ to +8.15‰, with an average of +7.3‰ (8 measurement points). The $\delta^{34}S$ values of Py3 stage pyrite range from +6.90‰ to +9.10‰, with an average of +7.9‰ (10 measurement points).

Table 1. ^{34}S values of three stages of pyrite in the Linglong gold deposit.

Sample	Analyzed Minerals (Stage)	$\delta^{34}S$ (‰, VCDT)	2σ
19LL17-1	Pyrite (Py1)	+7.80	0.10
19LL17-2	Pyrite (Py1)	+7.75	0.09
19LL17-3	Pyrite (Py1)	+8.10	0.14
19LL17-4	Pyrite (Py1)	+8.25	0.08
19LL37-1	Pyrite (Py1)	+7.70	0.11
19LL37-2	Pyrite (Py1)	+7.60	0.10
D007B1-1	Pyrite (Py2)	+8.15	0.13
D007B1-2	Pyrite (Py2)	+7.90	0.09
D007B1-3	Pyrite (Py2)	+8.05	0.08
D007B4-1	Pyrite (Py2)	+7.25	0.14
D007B4-2	Pyrite (Py2)	+7.55	0.10
D007B4-3	Pyrite (Py2)	+6.40	0.07
D012B1-1	Pyrite (Py2)	+6.15	0.11
D012B1-2	Pyrite (Py2)	+6.70	0.09
670-B1-3-1	Pyrite (Py3)	+8.21	0.10
670-B1-3-2	Pyrite (Py3)	+7.64	0.09
670-B1-3-3	Pyrite (Py3)	+7.37	0.14
670-B3-1	Pyrite (Py3)	+6.90	0.08
670-B3-2	Pyrite (Py3)	+8.40	0.11
670-B3-3	Pyrite (Py3)	+7.88	0.09
670-B3-4	Pyrite (Py3)	+8.10	0.10
670-B3-5	Pyrite (Py3)	+7.90	0.27
670-B4-1	Pyrite (Py3)	+7.30	0.08
670-B4-2	Pyrite (Py3)	+9.10	0.09

5.3. Lead Isotope Characteristics

Pb isotope test results and previous research data are shown in Supplementary Table S3. The $^{206}Pb/^{204}Pb$ ratios range from 16.95 to 17.715, with an average of 17.268 and a range of 0.765; $^{207}Pb/^{204}Pb$ ratios range from 15.472 to 15.557, with an average of 15.500 and a range of 0.085; $^{208}Pb/^{204}Pb$ ratios range from 37.858 to 38.394, with an average of 38.067 and a range of 0.536. The μ values ($^{238}U/^{204}Pb$) range from 9.38 to 9.54, which are higher than the μ values of normal lead (8.686–9.238); the ω values ($^{232}Th/^{204}Pb$) range from 39.73 to 45.13, exceeding the ω values of normal lead (35.55 ± 0.59); Th/U ratios range from 4.06 to 4.58, indicating relative enrichment of thorium-lead in this mining area. From the early mineralization stage to the main mineralization stage, the values of $^{206}Pb/^{204}Pb$, $^{207}Pb/^{204}Pb$, and $^{208}Pb/^{204}Pb$ all reach their maximum values during the main mineralization stage, suggesting possible contamination by highly radiogenic lead in the main mineralization stage.

6. Discussion

6.1. Occurrence State of Trace Elements in Pyrite

Trace elements in sulfides primarily occur in three forms: (1) solid solutions entering the crystal lattice through isomorphism; (2) nanoscale mineral inclusions; (3) microscale mineral inclusions [44]. Among these, isomorphic substitution is a universal mechanism for the occurrence of trace elements and serves as a carrier for numerous trace elements [45]. Correlation analysis of trace elements can effectively distinguish their occurrence forms [46].

The chemical properties of gold indicate that under hydrothermal conditions, it mainly exists in pyrite as solid solution gold (Au^+) and nanoparticulate gold (Au^0). Au^+ most likely enters pyrite lattices through Au-As coupling, involving the chemical process:

$$Fe(S,As)_2 + 2Au(HS)^0 \rightarrow Fe(S,As)_2 + Au_2S + H_2S$$

During this process, As substitutes for S, distorting the pyrite lattice and allowing Au^+ to enter [47]. Reich defined the solubility curve of solid solution Au in As-bearing pyrite [48]. In the Au-As binary diagram (Figure 6a), points above the gold solubility curve represent Au present as nanoscale gold minerals, whereas points below the curve indicate Au in solid solution form. All Au-As data points of pyrite from the Linglong gold deposit fall below the gold solubility curve, demonstrating that Au primarily exists in solid solution form. Au and As show no clear correlation in Py1, but exhibit a significant positive correlation in Py2 and Py3, indicating distinct coupling behaviors between Au and As during different mineralization stages.

In the Au-Ag binary diagram (Figure 6b), the data points exhibit a scattered distribution, suggesting possible variations in the occurrence forms of Ag across different mineralization stages. During the early mineralization stage (Py1), Ag content is relatively low and shows no clear correlation with Au, whereas in the later mineralization stages (Py2 and Py3), the correlation between Ag and Au becomes more pronounced.

Au shows no significant correlation with Sb in the three stages of pyrite (Figure 6c), indicating that the occurrence mechanism of Sb in pyrite differs from that of Au. Its enrichment may be influenced by changes in the redox conditions or pH of the fluids. Furthermore, Sb is more likely to exist as independent mineral phases rather than entering the pyrite lattice through isomorphic substitution.

In hydrothermal gold deposits, bismuth-chalcogen compounds (primarily S and Te) are often closely associated with gold [49,50]. In the Au-Bi and Au-Te binary diagrams (Figure 6d,e), the positive correlations in the Py2 and Py3 stages indicate a closer association of gold with bismuth and tellurium.

The occurrence mechanism of Cu in pyrite is jointly controlled by the redox state of ore-forming fluids and the Cu/Au ratio. In a strongly oxidizing environment, when the Cu/Au ratio is close to 1, Cu typically replaces Fe^{2+} in the crystal lattice in the form of [Au^{3+} + Cu^+] complexes; when the system is under reducing conditions and the Cu/Au ratio is significantly higher than 1 (usually ranging from 1 to 100), Cu elements directly replace Fe^{2+} in the pyrite crystal lattice in the form of Cu^{2+} single ions [45]. Determination of fluid inclusions has shown that the ore-forming fluid of the Linglong gold deposit is in a reducing environment [51]. Combined with the characteristic that the Cu/Au ratios of pyrite in all three stages are greater than 1 (Figure 6f), it can be inferred that the copper in pyrite in the study area is mainly derived from the direct lattice replacement of Fe^{2+} by Cu^{2+}.

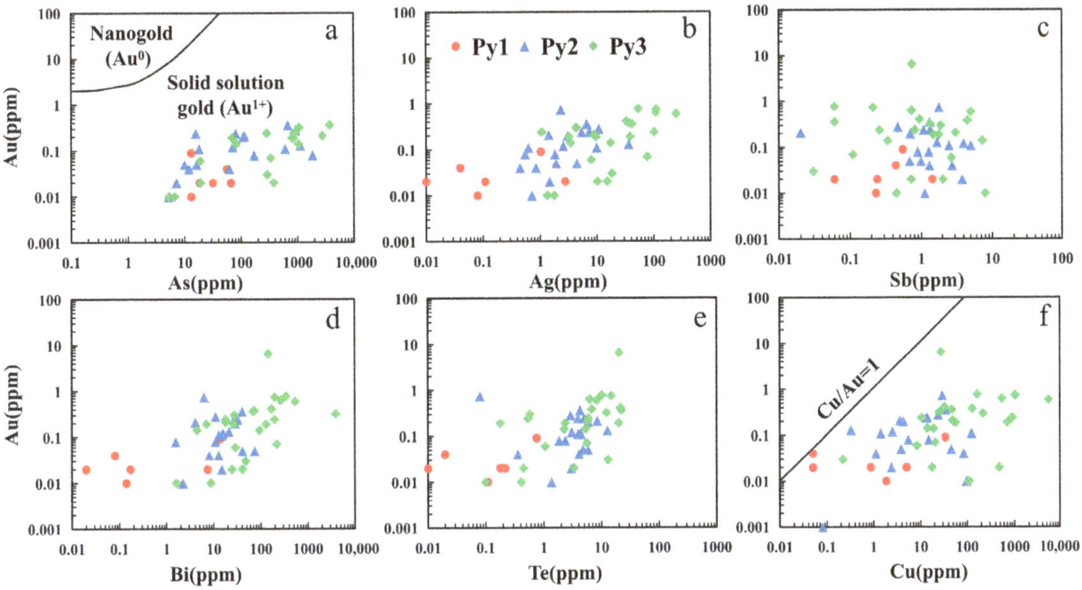

Figure 6. Binary diagrams of Au versus other trace elements in pyrite from the Linglong gold deposit. (**a**) Plot of Au vs. As; (**b**) Plot of Au vs. Ag; (**c**) Plot of Au vs. Sb; (**d**) Plot of Au vs. Bi; (**e**) Plot of Au vs. Te; (**f**) Plot of Au vs. Cu.

The ionic radius of Pb^{2+} is 1.19Å, and that of Fe^{2+} is 0.74Å [52]. Therefore, without a specific crystal chemical environment, Pb is difficult to enter the pyrite lattice and often occurs in pyrite as lead-bearing mineral inclusions [53]. Pb has a high positive correlation with Ag and Bi (Figure 7a,b), indicating that Pb may occur in pyrite as galena inclusions.

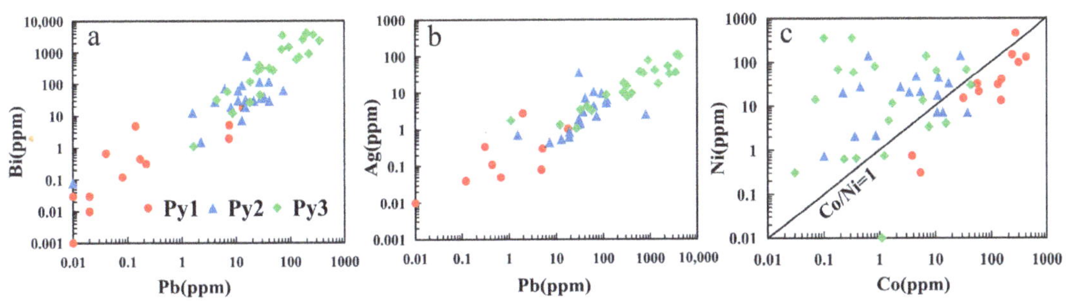

Figure 7. Trace element binary diagrams of pyrite from the Linglong gold deposit. (**a**) Pb-Bi diagram; (**b**) Pb-Ag diagram; (**c**) Co-Ni diagram.

6.2. Sources of Ore-Forming Materials

Previous studies have proposed various genetic models for the sources of gold ore-forming materials in the Jiaodong Gold Province, including Archean metamorphic basement rocks, Mesozoic granites, the subducted Paleo-Pacific Plate and its overlying metasomatized mantle, mixed subducted continental crustal materials, or basic magmas derived from partial melting of metasomatized lithospheric mantle [54–61]. Based on S-Pb isotope geochemical data and combined with regional geological settings, this paper evaluates the above models and further investigates the most likely material source of gold deposits in the Jiaodong area.

The migration and precipitation of gold are closely related to hydrogen sulfide, while sulfur is generally preserved in deposits as metal sulfides. The $\delta^{34}S$ values are typically influenced by temperature, fO_2, and pH [62–64]. Thermodynamic simulations of equilibrium sulfur isotope fractionation show that no significant sulfur isotope fractionation occurs between sulfides and fluids when the temperature drops to 300–250 °C, oxygen fugacity is under pyrite-pyrrhotite buffer conditions, and pH ranges from 3 to 4 [65]. For the Linglong gold deposit, the temperature during the main metallogenic stage ranges from 250 to 300 °C, with metallic sulfides mainly including pyrite, pyrrhotite, sphalerite, chalcopyrite, and galena, and pH ranging from 3 to 5 [66]. Thus, the sulfur isotope characteristics of sulfides can reflect the sulfur isotope composition of ore-forming fluids [66,67].

The $\delta^{34}S$ values of various mineralization types in the different gold metallogenic belts of the Jiaodong gold deposits exhibit generally consistent, high values (Figure 8, Supplementary Table S2, Zhaoping belt: $\delta^{34}S$ = +3.50‰–+9.10‰; Qixia belt: $\delta^{34}S$ = +4.40‰–+8.80‰; Muru belt: $\delta^{34}S$ = +3.50‰–+13.00‰; Jiaojia belt: $\delta^{34}S$ = +5.30‰–+12.72‰; Sanshandao belt: $\delta^{34}S$ = +4.67‰–+12.60‰). These values overlap with the $\delta^{34}S$ ranges of the main host rocks (Mesozoic granites and Precambrian metamorphic rocks) and intermediate-mafic dikes in the Jiaodong gold deposits (Figure 8, Supplementary Table S2, Mesozoic granites: $\delta^{34}S$ = +3.4‰–+9.9‰; intermediate-basic dikes: $\delta^{34}S$ = +3.4‰–+9.4‰). The Jiaodong Group metamorphic rocks cannot be the source layer for the mineralization. The metamorphism of the Jiaodong Group metamorphic rocks occurred nearly 2 billion years earlier than the mineralization event, and the water and related ore-forming elements within them had already been released through intense metamorphic dehydration and devolatilization. Furthermore, more compelling evidence comes from multiple sulfur isotope studies, which reveal that the ore-related pyrites in Jiaodong exhibit relatively constant $\Delta^{33}S$ values (approximately 0‰), with no signals of mass-independent fractionation of sulfur detected [55]. Therefore, the sulfur could not have been derived from the Archean metamorphic basement, also ruling out Mesozoic granites formed by the remelting of the Archean metamorphic basement as a source.

The $\delta^{34}S$ values of pyrite from the Linglong gold deposit overlap with those of intermediate-mafic dikes, and some orebodies of the Linglong gold deposit occur within or are distributed on both sides of lamprophyre dikes, suggesting a potential deep-seated connection between them. However, compared to the volume of gold orebodies, mafic dikes are relatively limited and are more likely to act as carriers rather than sources of the ore-forming materials. The $\delta^{34}S$ values of sulfur derived from the upper mantle are typically concentrated around 0 ± 3‰, which is significantly lower than the $\delta^{34}S$ values of the various metallogenic belts [68]. Therefore, the upper mantle cannot be the source of the ore-forming materials. The high $\delta^{34}S$ values of pyrite from the mineralization stage of the Linglong gold deposit may result from the mixing of seawater or magma oxidation. However, there is no evidence of marine sedimentation from the Paleozoic to Mesozoic on the Jiaodong Peninsula [69]. Even with appropriate consideration of sulfur isotope fractionation related to transport, the formation of the Linglong gold deposit still requires a sulfur source with relatively high $\delta^{34}S$ values. A comparison of sulfur isotope data among the Jiaodong gold deposits, seawater sulfate, and global sedimentary-orogenic gold deposits shows that the $\delta^{34}S$ values of pyrite from the Linglong gold deposit are much higher than those of global sedimentary-orogenic gold deposits from the Jurassic to Early Cretaceous (Figure 8). Thus, it is not the reduced sulfide from seawater sulfate in the sediments of the Mesozoic Paleo-Pacific Plate subduction that provided the high $\delta^{34}S$ fluids. This effectively negates the direct derivation of sulfur and associated gold from the Paleo-Pacific subduction system.

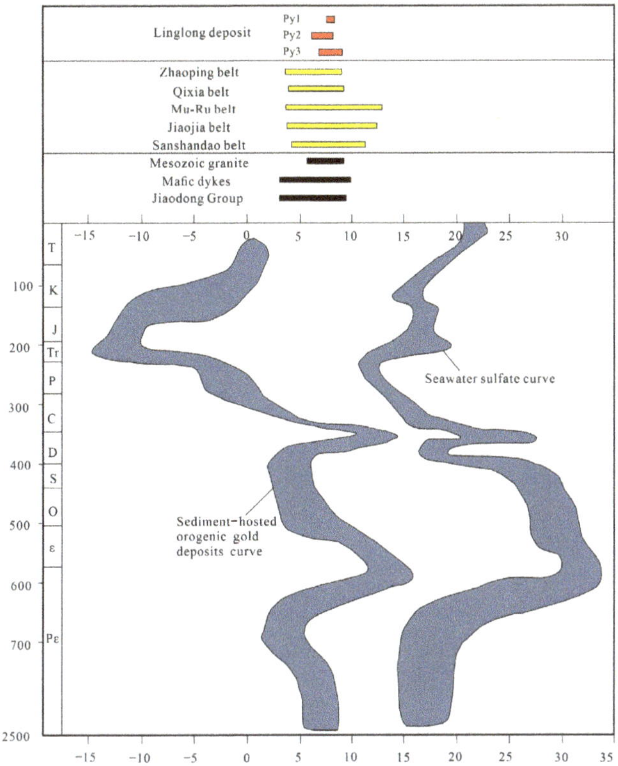

Figure 8. $\delta^{34}S$ values of sulfur isotopes in Jiaodong granites, metamorphic rocks, mafic dykes, and gold ores, as well as sulfur isotope compositions of sulfides in global sedimentary-orogenic gold deposits, in relation to geological ages based on temporal variations in marine sulfate curves (The sulfur isotope data are cited from the references in Liang et al. (2023) [66], Mao et al. (2005) [70], Zhu et al. (2019) [71], Yuan et al. (2019) [72], Xue et al. (2018) [73], Chen et al. (2010) [74], Yang et al. (2010) [75], Mills et al. (2015) [69], Li et al. (2015) [76], Liu et al. (2019) [77], Guo et al. (2009) [78], Feng et al. (2018, 2019) [79,80], Wang et al. (2013) [81], Zhang et al. (2014) [82], Huang (1994) [83] and Deng et al. (2020) [56]. For detailed data, please refer to Supplementary Table S2).

The high $\delta^{34}S$ values of auriferous pyrite require anomalously heavy sulfur source regions. Qiu et al. proposed that fluids, gold, and most sulfur were derived from the devolatilization of the Mesozoic westward-subducting Paleo-Pacific Plate and overlying sediments [55]. Subduction-related metamorphism induces the decomposition of pyrite into pyrrhotite, releasing H_2S and Au into the generated fluids. Plate-derived fluids metasomatize the overlying mantle, enriching volatiles including S and Au while preserving crustal $\delta^{34}S$ signatures. Deng et al. found that the $\delta^{34}S$ values of the Jiaodong gold deposits overlap with those of global Proterozoic sedimentary-orogenic gold deposits [56]. In the Late Neoproterozoic sedimentary rocks of the Yangtze Craton, diagenetic pyrite—dominantly hosted in black shales and carbonates—exhibits anomalously high $\delta^{34}S$ values. Globally, Au-rich diagenetic pyrite in Neoproterozoic sedimentary rocks constitutes an important source of gold and sulfur for orogenic gold deposits. Neoproterozoic rocks from the northern Yangtze Craton were subducted beneath the North China Craton; devolatilization of these rocks likely caused metasomatic enrichment of the mantle lithosphere, and the mixing of sedimentary heavy sulfur with mantle sulfur explains the $\delta^{34}S$ values of pyrite in the Jiaodong gold deposits. Thus, the metasomatically enriched lithospheric mantle represents a relatively reliable source of S, while the incorporation of crust-derived S information adds complexity to the interpretation of ore-forming material sources.

In addition, as shown in Figure 8, the overall distribution range of $\delta^{34}S$ values of pyrite from the three metallogenic stages is relatively narrow. The Py2 and Py3 stages exhibit the minimum and maximum values, respectively, while the Py1 stage lies between the two. The narrow distribution of $\delta^{34}S$ values suggests that the sulfur source of the ore-forming fluids maintained high consistency throughout the metallogenic process, which is mainly derived from the enriched metasomatized lithospheric mantle. The extreme values in the Py2 and Py3 stages may reflect minor material exchange between fluids and crustal materials during the middle to late metallogenic stages. Such minor fluctuations have not altered the dominance of the overall sulfur source, but rather indicate that the incorporation of crustal sulfur might be a gradual process, slightly intensified in the late metallogenic stage.

Lead isotopes are well-studied tools for tracing ore sources, mantle–crust interactions, and deposit genesis [84]. Their composition remains stable during physical/chemical processes (except radioactive decay or mixing), making them reliable for tracking material sources. However, low-salinity ore-forming fluids carry limited lead, which can be overwhelmed by other sources during migration, potentially complicating interpretations [56].

Linglong gold deposit lead isotopes show concentrated ratios, with rare outliers likely from radiogenic Pb contamination. Low μ (7.98–8.64) and ω (31.87–37.47) values indicate non-radiogenic Pb (Table 2), suggesting a mantle origin or minimal radiogenic Pb input. Single-stage model ages (652–940 Ma, avg. 804 Ma) predate host rock emplacement (166–149 Ma) and mineralization (120 Ma), but postdate Jiaodong Group protolith (3.4–2.6 Ga) and metamorphism (~1.8 Ga) [2,35,85]. This implies an ancient non-radiogenic Pb reservoir, not directly derived from host rocks or syn-mineral magmatism. Lead isotopes from Linglong and other Zhaoping Fault Zone deposits exhibit high-slope, mixed-source trends. Most data plot below the S-K 1975 curve and above the metasomatized lithospheric mantle (Figure 9), indicating a mixed source of crust and/or metasomatized lithospheric mantle.

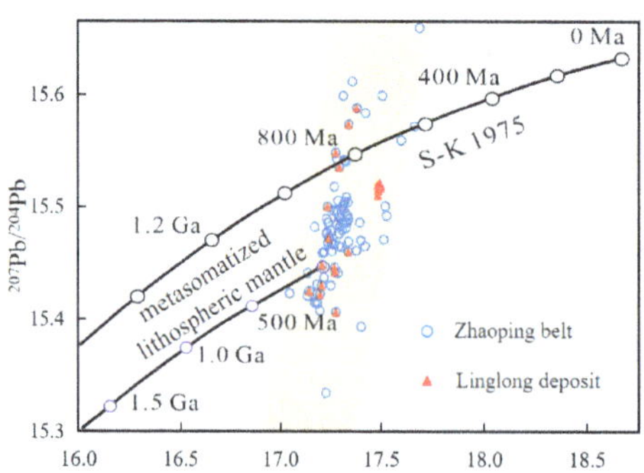

Figure 9. Pb isotope evolution line, two-stage crustal growth curve (S-K 1975) was proposed by Stacey and Kramers [86]. The ore Pb isotope data are cited from the references in Zhang et al. (1993) [87], Zhang et al. (2016) [59], Yuan et al. (2019) [72], Li et al. (1990) [88] and Deng et al. (2020) [56]. For detailed data, please refer to Supplementary Table S3.

Table 2. Pb isotope data of ore galena from the Linglong gold deposit.

Sample	Analyzed Minerals	$^{206}Pb/^{204}Pb$	$^{207}Pb/^{204}Pb$	$^{208}Pb/^{204}Pb$	T_{CDT}/Ma	μ	ω
670-B1-3-1	Galena	17.491	15.518	38.075	729	8.440	34.710
670-B1-3-2	Galena	17.494	15.518	38.078	727	8.450	34.720
670-B1-3-3	Galena	17.494	15.517	38.083	726	8.450	34.740
670-B1-3-4	Galena	17.492	15.516	38.080	726	8.440	34.730
670-B1-3-5	Galena	17.493	15.521	38.085	731	8.440	34.750
670-B1-3-6	Galena	17.492	15.518	38.083	728	8.440	34.740
670-B2-1-1	Galena	17.487	15.516	38.062	729	8.440	34.660
670-B2-1-2	Galena	17.493	15.520	38.083	730	8.440	34.740
670-B2-1-3	Galena	17.492	15.520	38.082	730	8.440	34.740
670-B2-1-4	Galena	17.490	15.517	38.080	728	8.440	34.730
670-B2-1-5	Galena	17.493	15.521	38.084	731	8.440	34.750
670-B2-1-6	Galena	17.495	15.519	38.084	727	8.450	34.750
670-B2-1-7	Galena	17.487	15.514	38.063	727	8.440	34.660
670-B2-1-8	Galena	17.495	15.519	38.077	727	8.450	34.720
670-B2-1-9	Galena	17.494	15.518	38.058	727	8.450	34.640
670-B2-1-10	Galena	17.494	15.518	38.087	727	8.450	34.760
670-B4-1-1	Galena	17.484	15.510	38.042	725	8.440	34.580
670-B4-1-2	Galena	17.494	15.516	38.085	724	8.450	34.750
670-B4-1-3	Galena	17.486	15.516	38.064	730	8.440	34.660
670-B4-1-4	Galena	17.495	15.518	38.083	726	8.450	34.740
670-B4-1-5	Galena	17.494	15.520	38.080	729	8.450	34.730
670-B4-1-6	Galena	17.495	15.519	38.085	727	8.450	34.750
670-B4-1-7	Galena	17.494	15.517	38.082	726	8.450	34.740
670-B4-1-8	Galena	17.238	15.472	37.883	857	8.180	33.930
670-B4-2-1	Galena	17.481	15.519	38.040	737	8.430	34.570
670-B4-2-2	Galena	17.206	15.448	37.829	853	8.150	33.720
670-B4-2-3	Galena	17.340	15.573	38.151	893	8.290	35.020
670-B4-2-4	Galena	17.270	15.442	37.808	801	8.210	33.630
670-B4-2-5	Galena	17.275	15.406	37.731	758	8.220	33.320
670-B4-2-6	Galena	17.203	15.430	37.751	836	8.150	33.400
670-B4-2-7	Galena	17.195	15.422	37.731	833	8.140	33.320
670-B5-1-1	Galena	17.379	15.588	38.271	882	8.330	35.500
670-B5-1-2	Galena	17.337	15.460	37.866	774	8.280	33.870
670-B5-1-3	Galena	17.267	15.446	37.785	808	8.210	33.540
670-B5-1-4	Galena	17.292	15.535	38.061	886	8.240	34.650
670-B5-1-5	Galena	17.233	15.500	38.008	890	8.180	34.440
670-B5-1-6	Galena	17.142	15.425	37.720	874	8.080	33.280
670-B5-1-7	Galena	17.274	15.548	38.154	912	8.220	35.030

In the $^{207}Pb/^{204}Pb$ vs. $^{206}Pb/^{204}Pb$ and $^{208}Pb/^{204}Pb$ vs. $^{206}Pb/^{204}Pb$ tectonic discrimination diagrams (Figure 10), the Pb isotopic composition of the Linglong deposit overlaps with most Jiaodong gold deposits, Precambrian metamorphic rocks, and the Early Cretaceous Guojialing granite (with crust–mantle mixed sources). Significant overlap also occurs with the Late Jurassic Linglong granite (derived from ancient lower crust remelting) and Early Cretaceous mafic dikes (from enriched lithospheric mantle melting). This indicates a potential genetic link between Linglong, other Zhaoping Fault Zone deposits, and these geological units. The Pb isotopic data plot between the lower crust and orogenic belt fields, closer to the lower crust. However, the extensive overlap makes pinpointing the ultimate Pb source ambiguous.

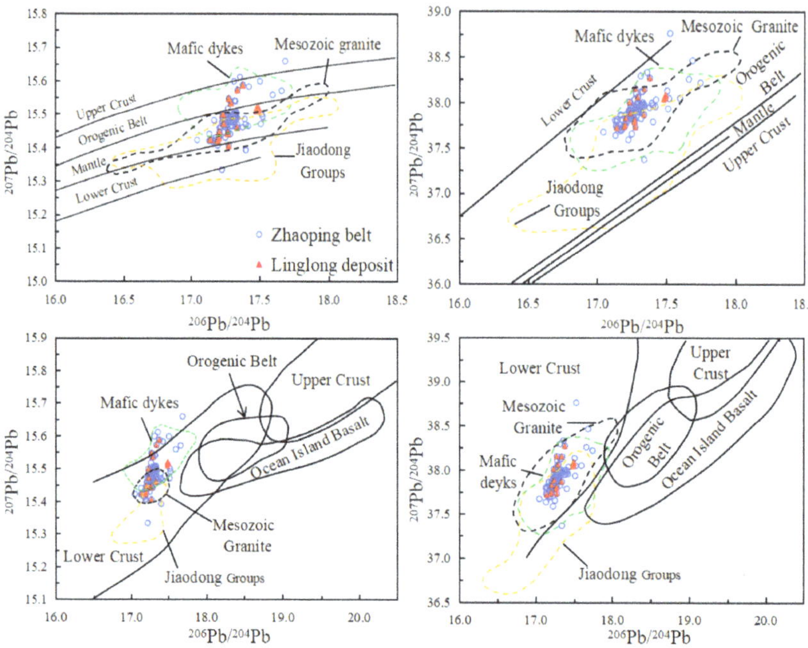

Figure 10. Tectonic diagrams of $^{207}Pb/^{204}Pb$ vs. $^{206}Pb/^{204}Pb$ and $^{208}Pb/^{204}Pb$ vs. $^{206}Pb/^{206}Pb$ models. The Pb isotope data of the host rocks are cited from the references in Yang et al. (1986) [89], Chen et al. (1994) [90], Hou et al. (2006) [91], Tan et al. (2012) [92], Ma et al. (2014) [34] and Deng et al. (2020) [56]. For detailed data, please refer to Supplementary Table S3.

6.3. Evolution of Ore-Forming Fluids and Mechanism of Gold Precipitation

As the main gold-bearing mineral throughout the entire mineralization process in the Jiaodong gold deposits, the multi-generational structure and trace element composition of pyrite directly record the evolution of ore-forming fluid properties, mineralization processes, and gold precipitation mechanisms [18,93–96]. Both Co and Ni are siderophile elements that can replace Fe through isomorphism, so the Co/Ni ratio can indicate changes in ore-forming conditions [97]. Generally, a smaller Co/Ni ratio represents a lower formation temperature of pyrite [98,99]. Pyrites of different genetic types show significant differences in Co/Ni ratios: those of sedimentary origin are usually less than 1 [100], while those of magmatic-hydrothermal origin are generally greater than 1 [98,101,102]. The Co/Ni ratios of the Py1 stage are basically greater than 1, whereas 78% of the Co/Ni ratios in the Py2 and Py3 stages are less than 1 (Figure 7c; Supplementary Table S1), indicating that the ore-forming fluid underwent a temperature-decreasing process from the Py1 to Py3 stages.

Pyrite in the Py1 stage occurs as euhedral–subhedral granular crystals (Figure 3D) with smooth crystal surfaces, showing stable crystallization characteristics of a magmatic-hydrothermal system. Its high Co and Ni contents and Co/Ni ratio (>1) are consistent with the characteristics of pyrite in mantle-derived magmatic-hydrothermal deposits. Combined with the relatively low contents of Cu, Zn, and As, it indicates that the primary fluid in the early mineralization stage may have been dominated by mantle-derived magmatic fluid, uncontaminated by crustal materials [103]. The low Au (<0.09 ppm) and As (<66.28 ppm) contents and the absence of Au-As correlation (Figure 6a) suggest that Au mainly entered the crystal lattice as a dispersed solid solution (Au^{+}) without significant precipitation and enrichment [48]. At this time, the fluid had a high temperature and stable chemical composition, belonging to a stable and closed ore-forming environment.

Pyrite in the Py2 stage develops obvious dissolution textures at the edges (Figure 3E). Intermittent fluctuations in fluid composition caused by multiple hydrothermal events generally lead to stage-wise precipitation of pyrite, accompanied by irregular boundaries and corrosion structures between alternating zones, indicating the reworking characteristics of dissolution-reprecipitation of early crystals by fluids [104]. In this stage, Co and Ni contents are significantly lower than those in Py1, with most Co/Ni ratios less than 1, while the contents of Cu, Zn, As, Te, Ag, Au, Bi, and Pb are significantly higher than those in Py1. This reflects a decreased proportion of mantle-derived fluid, a transition of the fluid system from closed to open, and the leaching and mixing of crustal materials, which diluted the mantle-derived signal [103]. The increased As content may be attributed to As in sedimentary rocks being released through fluid leaching and enriched in pyrite. Due to the larger ionic radius of As compared to S and Fe, the crystal lattice of As-rich pyrite exhibits obvious distortion. As^{3+} enters the pyrite lattice through the coupling mechanism of "$As^{3+} + Au^{+} \rightarrow Fe^{2+}$", forming an "arsenic-bearing gold-pyrite" solid solution. While promoting elements such as Au to enter the pyrite lattice, it also facilitates dissolution-reprecipitation by late-stage fluids [105], thereby leading to the activation, migration, enrichment, and precipitation of Au [106,107].

Pyrite in the Py3 stage is characterized by fracture-filling and porous textures, with native gold and metal sulfides commonly filling the fractures, showing typical features of dissolution-reprecipitation [108]. At this stage, the contents of Cu, Zn, Au, and Ag in pyrite reach their peak values. Under tectonic stress, dislocation slip and creep of pyrite create a relatively open space at the interface between dislocated and undislocated areas in pyrite crystals, which is the most favorable site for Au activation and re-enrichment [105]. Invisible gold and other trace elements fixed inside pyrite can be reactivated during late metamorphic-deformation processes and/or water-rock reactions and reprecipitated at the grain edges or microfractures of pyrite [93]. The brittle fracturing may be caused by the accumulation of dislocation sliding, indicating characteristics of low-temperature deformation [109], which is consistent with the trend of gradual temperature decrease in the ore-forming fluid. The porous and fracture-filling textures formed by pyrite dissolution-reprecipitation are direct products of the transformation of ore-forming fluid from the "migration stage" to the "precipitation stage" and are also the controlling factors for the peak contents of elements such as Au, Cu, Zn, and Ag. By expanding fluid–mineral interaction interfaces, providing precipitation space, and responding to abrupt changes in fluid physicochemical conditions, they synergistically promote the super-enrichment of elements. Notably, the relatively significant enrichment of Te and Bi suggests the possible existence of local melt–fluid interaction. The presence of bismuth–tellurium melt not only provides an important transport carrier for gold but also promotes gold precipitation from hydrothermal fluids by forming stable gold–tellurium compounds (such as tellurobismuthite) [49,50,110].

7. Conclusions

(1) Trace elements in pyrite show systematic variations across mineralization stages (Py1 $\rightarrow$ Py3), marked by decreasing Co and Ni and increasing Au, As, Cu, Pb, Zn, Bi, and Te. These trends reflect the continuous evolution of the ore-forming fluids. The early stage (Py1) involved high-temperature, reduced fluids dominated by mantle sources, representing a phase of gold pre-enrichment. During the main ore-forming stages (Py2–Py3), fluid temperature decreased, incorporation of crustal material increased, and water–rock reactions intensified. Gold precipitation was governed by two principal mechanisms: in Py2, Au^{+} coupled with As^{3+} and entered the pyrite lattice as a solid solution; in Py3, fluid immiscibility broke down gold complexes,

while Bi–Te melts or complexes promoted further gold enrichment and the formation of visible gold.

(2) The S isotope characteristics indicate that sulfur is mainly derived from fluids released by dehydration of the Paleo-Pacific Plate during subduction, and these fluids have metasomatized the overlying lithospheric mantle. The Pb isotope characteristics reflect the ore-forming fluid mixed with crust-derived materials during its ascent. The S-Pb isotope system jointly constrains that the ore-forming materials are mainly derived from the metasomatized lithospheric mantle, with the involvement of crustal materials.

(3) The ore-forming fluid underwent a continuous evolutionary process from early high-temperature, mantle-derived dominance to late low-temperature, crustal-derived mixing. The Py1 stage was dominated by mantle-derived magmatic fluids that were not contaminated by crustal materials. The Py2 stage was accompanied by the incorporation of crustal materials and an increase in As content, triggering the coupled substitution of "$As^{3+} + Au^{+} \rightarrow Fe^{2+}$" and dissolution–reprecipitation, which promoted the initial activation and enrichment of gold. In the Py3 stage, dissolution–reprecipitation occurred in a low-temperature, open environment, forming fractures and porous structures that facilitated the supernormal enrichment of elements, representing the main mechanism for gold precipitation. Local bismuth–tellurium melt also contributed to gold precipitation.

Supplementary Materials: The following supporting information can be downloaded at: https://www.mdpi.com/article/10.3390/min15111220/s1. Supplementary Table S1: In situ trace element data of pyrite. Supplementary Table S2: S isotope data. Supplementary Table S3: Pb isotope data. References [70–83,87–92] are cited in the Supplementary Materials.

Author Contributions: Conceived the ideas, F.R., Z.-J.D. and Z.-Y.B.; writing—original draft, F.R.; visualization, J.-W.W., S.-X.M., T.N. and K.-Q.G.; investigation, S.-X.M., T.N. and B.W.; methodology, J.-W.W.; software, C.L.; resources, G.-J.L.; writing—review and editing, S.-S.L. All authors have read and agreed to the published version of the manuscript.

Funding: This study was financially supported by the National Key R&D Program of China (2023YFC2906900), the Open Project of Technology Innovation Center for Deep Gold Resources Exploration and Mining, Ministry of Natural Resources (LDKF-2023BZX-17), the Shandong Engineering Research Center of Application and Development of Big Data for Deep Gold Exploration (SDK202214), Taishan Scholar Program of Shandong (grant tstp20240847).

Data Availability Statement: The original contributions presented in this study are included in the article. Further inquiries can be directed to the corresponding author.

Acknowledgments: We thank Shandong Gold Mining (Linglong) Co., Ltd. for their help in providing samples. We also deeply thank the anonymous reviewers and editors for their helpful comments and suggestions.

Conflicts of Interest: The authors declare no conflicts of interest.

References

1. Goldfarb, R.J.; Santosh, M. The dilemma of the Jiaodong gold deposits: Are they unique? *Geosci. Front.* **2014**, *5*, 139–153. [CrossRef]
2. Deng, J.; Qiu, K.F.; Wang, Q.F.; Goldfarb, R.; Yang, L.Q.; Zi, J.W.; Geng, J.Z.; Ma, Y. In situ dating of hydrothermal monazite and implications for the geodynamic controls on ore formation in the Jiaodong gold province, eastern China. *Econ. Geol.* **2020**, *115*, 671–685. [CrossRef]
3. Qiu, K.F.; Deng, J.; Sai, S.X.; Yu, H.C.; Tamer, M.T.; Ding, Z.J.; Yu, X.F.; Goldfarb, R. Low-temperature thermochronology for defining the tectonic controls on heterogeneous gold endowment across the Jiaodong peninsula, eastern China. *Tectonics* **2023**, *42*, e2022TC007669. [CrossRef]

4. Deng, J.; Wang, C.; Bagas, L.; Carranza, E.J.M.; Lu, Y. Cretaceous–Cenozoic tectonic history of the Jiaojia Fault and gold mineralization in the Jiaodong Peninsula, China: Constraints from zircon U–Pb, illite K–Ar, and apatite fission track thermochronometry. *Miner. Depos.* **2015**, *50*, 987–1006. [CrossRef]
5. Zhu, R.; Sun, W. The big mantle wedge and decratonic gold deposits. *Sci. China Earth Sci.* **2021**, *64*, 1451–1462. [CrossRef]
6. Wang, C.Y.; Wei, B.; Tan, W.; Wang, Z.; Zeng, Q. The distribution, characteristics and fluid sources of lode gold deposits: An overview. *Sci. China Earth Sci.* **2021**, *64*, 1463–1480. [CrossRef]
7. Zhai, M.G.; Santosh, M. Metallogeny of the North China Craton: Link with secular changes in the evolving Earth. *Gondwana Res.* **2013**, *24*, 275–297. [CrossRef]
8. Deng, J.; Wang, Q.F.; Zhang, L.; Xue, S.C.; Liu, X.F.; Yang, L.; Yang, L.Q.; Qiu, K.F.; Liang, Y.Y. Metallogenetic model of Jiaodong-type gold deposits, eastern China. *Sci. China Earth Sci.* **2023**, *66*, 2287–2310. [CrossRef]
9. Yang, L.Q.; Deng, J.; Guo, L.N.; Wang, Z.L.; Li, X.Z.; Li, J.L. Origin and evolution of ore fluid, and gold-deposition processes at the giant Taishang gold deposit, Jiaodong Peninsula, eastern China. *Ore Geol. Rev.* **2016**, *72*, 585–602. [CrossRef]
10. Goldfarb, R.J.; Baker, T.; Dubé, B.; Groves, D.I.; Gosselin, P. *Distribution, Character, and Genesis of Gold Deposits in Metamorphic Terranes*; One Hundredth Anniversary Volume; Society of Economic Geologists: Littleton, CO, USA, 2005.
11. Nie, F.J.; Jiang, S.H.; Liu, Y. Intrusion-related gold deposits of North China Craton, People's Republic of China. *Resour. Geol.* **2004**, *54*, 299–324. [CrossRef]
12. Deng, J.; Liu, X.F.; Wang, Q.F.; Pan, R.G. Origin of the Jiaodong-type Xinli gold deposit, Jiaodong Peninsula, China: Constraints from fluid inclusion and C–D–O–S–Sr isotope compositions. *Ore Geol. Rev.* **2015**, *65*, 674–686. [CrossRef]
13. Qiu, K.F.; Romer, R.L.; Long, Z.Y.; Williams-Jones, A.E.; Yu, H.C.; Turner, S.; Wang, Q.F.; Li, S.S.; Zhang, J.Y.; Duan, H.R. The role of an oxidized lithospheric mantle in gold mobilization. *Sci. Adv.* **2024**, *10*, eado6262. [CrossRef] [PubMed]
14. Lin, L.; Chao, L.; Qing, L.; Wen, Y.M.; Quan, Z.J.; Rong, L.S.; Santosh, M.; Feng, S.J.; Feng, Z.H. Indicators of decratonic gold mineralization in the North China Craton. *Earth-Sci. Rev.* **2022**, *228*, 103995. [CrossRef]
15. Chao, W.W.; Ye, H.S.; Tang, X.W.; Zhang, Z.Q.; Wang, H.S. Textural, trace element, and sulfur isotope analyses of pyrite from the Yindongpo deposit, East Qinling Orogen: Implications for gold mineralization. *Ore Geol. Rev.* **2023**, *163*, 105796. [CrossRef]
16. Liu, Y.; Zhao, J.H.; Jiao, Y.Q.; Liu, Z.R.; Zhou, M.F. Constraints of in-situ S-isotopic compositions of pyrite on the genesis of the Bayinqinggeli sandstone-hosted uranium deposit, Ordos Basin, Northern China. *Ore Geol. Rev.* **2023**, *161*, 105608. [CrossRef]
17. Wei, Y.J.; Yang, L.Q.; Qiu, K.F.; Wang, S.R.; Ren, F.; Dai, Z.H.; Li, D.P.; Shan, W.; Li, Z.S.; Wang, J.H.; et al. Geology, mineralogy and pyrite trace elements constraints on gold mineralization mechanism at the giant Dayingezhuang gold deposit, Jiaodong Peninsula, China. *Ore Geol. Rev.* **2022**, *148*, 104992. [CrossRef]
18. Li, X.H.; Fan, H.R.; Yang, K.F.; Hollings, P.; Cai, Y.C. Pyrite textures and compositions from the Zhuangzi Au deposit, southeastern North China Craton: Implication for ore-forming processes. *Contrib. Mineral. Petrol.* **2018**, *173*, 73. [CrossRef]
19. Tang, L.; Zhao, Y.; Zhang, S.T.; Sun, L.; Hu, X.K.; Sheng, Y.M.; Zeng, T. Origin and evolution of a porphyry-breccia system: Evidence from zircon U-Pb, molybdenite Re-Os geochronology, in situ sulfur isotope and trace elements of the Qiyugou deposit, China. *Gondwana Res.* **2021**, *89*, 88–104. [CrossRef]
20. Li, S.S.; Zang, C.J.; Li, L.; Li, S.R.; Tao, W.; Cheng, X. Geochemistry, isotopes, and morphology of coarse-grained pyrite from the Jinqingding gold deposit, Jiaodong Peninsula: Implications for episodic ore-forming fluid evolution. *Ore Geol. Rev.* **2025**, *184*, 106742. [CrossRef]
21. Zhang, M.; Tan, J.; Zhao, Y.Y.; Yan, Z.Q.; Zhao, S.S.; Lu, C.Y.; Liu, X.Y.; Zhang, Y. Metallogenic mechanism of Ankou gold deposit in the Qixia-Penglai Gold Belt, Jiaodong Peninsula, China: Constraints from sericite Ar-Ar geochronology, H-O isotope, and in-situ trace element of pyrite. *Ore Geol. Rev.* **2025**, *178*, 106471. [CrossRef]
22. Xue, S.M.; Liang, Y.Y.; Shu, L.; Xue, W.H.; Zhang, C.X.; Shen, C.H.; He, B.; Wang, H.Y.; Fang, Y.Y. Genesis of the Jiuqu gold deposit, Jiaodong gold province, China: Constraints on the texture, mineral geochemistry, and sulfur isotope of pyrite. *Acta Geochim.* **2024**, *44*, 631–647. [CrossRef]
23. Fang, Y.Y.; Liang, Y.Y.; Xia, R.; Shu, L.; He, B.; Xue, W.H.; Zhang, C.X.; Wang, H.Y.; Xue, S.M. Geochemistry of pyrite from the Jiaojia gold deposit, Jiaodong Peninsula, north China craton: Implications for source of ore-forming fluids and gold precipitation. *Appl. Sci.* **2025**, *15*, 1548. [CrossRef]
24. Mao, J.W.; Wang, Y.T.; Li, H.M.; Pirajno, F.; Zhang, C.Q.; Wang, R.T. The relationship of mantle-derived fluids to gold metallogenesis in the Jiaodong Peninsula: Evidence from D–O–C–S isotope systematics. *Ore Geol. Rev.* **2008**, *33*, 361–381. [CrossRef]
25. Bor-Ming, J.; Yi, L.D.; Sheng, W.Y.; Biao, S.; Shan, W.J. Archean crustal evolution of the Jiaodong Peninsula, China, as revealed by zircon SHRIMP geochronology, elemental and Nd-isotope geochemistry. *Am. J. Sci.* **2008**, *308*, 232–269. [CrossRef]
26. Qiu, K.F.; Romer, R.L.; Long, Z.Y.; Yu, H.C.; Turner, S.; Wan, R.Q.; Li, X.Q.; Gao, Z.Y.; Deng, J. Potassium isotopes as a tracer of hydrothermal alteration in ore systems. *Geochim. Cosmochim. Acta* **2024**, *368*, 185–196. [CrossRef]
27. Tang, J.; Zheng, Y.F.; Wu, Y.B.; Gong, B.; Zha, X.P.; Liu, X.M. Zircon U–Pb age and geochemical constraints on the tectonic affinity of the Jiaodong terrane in the Sulu orogen, China. *Precambrian Res.* **2007**, *161*, 389–418. [CrossRef]

28. Feng, Y.C.; Deng, J.; Yu, H.C.; Qiang, F.; Cui, T.; Wei, S.; Sheng, L.Z.; Li, S.S. Gold occurrence and its indicative significanceto mineralization process in Linglong gold district, Jiaodong gold province. *Acta Petrol. Sin.* **2023**, *39*, 377–392.
29. Goldfarb, R.J.; Groves, D.I. Orogenic gold: Common or evolving fluid and metal sources through time. *LITHOS* **2015**, *233*, 2–26. [CrossRef]
30. Wang, Q.F.; Yang, L.; Zhao, H.S.; Groves, D.I.; Weng, W.J.; Xue, S.C.; Li, H.J.; Dong, C.Y.; Yang, L.Q.; Li, D.P.; et al. Towards a universal model for orogenic gold systems: A perspective based on Chinese examples with geodynamic, temporal, and deposit-scale structural and geochemical diversity. *Earth-Sci. Rev.* **2021**, *224*, 103861. [CrossRef]
31. Wang, Q.W.; Zhao, H.S.; Yang, L.; Groves, D.I.; Han, J.L.; Qiu, K.F.; Li, D.P.; Liu, Z.; Zhao, R.; Deng, J. Formation of the giant Cretaceous Jiaodong-type orogenic gold province of the North China Craton: A consequence of lithospheric multi-layer reworking. *Geosci. Front.* **2025**, *167*, 102047. [CrossRef]
32. Yang, L.Q.; Deng, J.; Wang, Z.L.; Zhang, L.; Guo, L.N.; Song, M.C.; Zheng, X.L. Mesozoic gold metallogenic system of the Jiaodong gold province, eastern China. *Acta Geol. Sin.* **2014**, *30*, 2447–2467.
33. Zhang, L.; Weinberg, R.F.; Yang, L.Q.; Groves, D.I.; Sai, S.X.; Matchan, E.; Phillips, D.; Kohn, B.P.; Miggins, D.P.; Liu, Y. Mesozoic Orogenic Gold Mineralization in the Jiaodong Peninsula, China: A Focused Event at 120 +/− 2 Ma During Cooling of Pregold Granite Intrusions. *Am. J. Gastroenterol.* **2020**, *115*, 415–441. [CrossRef]
34. Ma, L.; Jiang, S.Y.; Hofmann, A.W.; Dai, B.Z.; Hou, M.L.; Zhao, K.D.; Chen, L.H.; Li, J.W.; Jiang, Y.H. Lithospheric and asthenospheric sources of lamprophyres in the Jiaodong Peninsula: A consequence of rapid lithospheric thinning beneath the North China Craton? *Geochim. Cosmochim. Acta* **2014**, *124*, 250–271. [CrossRef]
35. Zhao, R.; Wang, Q.F.; Deng, J.; Santosh, M.; Liu, X.F.; Liang, Y.Y.; Cheng, H.Y. Characterizing episodic orogenesis and magmatism in eastern China based on detrital zircon from the Jiaolai Basin. *Am. J. Sci.* **2019**, *319*, 500–525. [CrossRef]
36. Li, H.K.; Li, Y.F.; Geng, K.; Zhuo, C.Y.; Zhang, Y.B.; Liang, T.T. Study on the Orogenic Type Gold Deposits in Eastern Shandong Province. *Geotecton. Metallog.* **2011**, *35*, 533–542. [CrossRef]
37. Hou, M.L.; Jiang, Y.H.; Jiang, S.Y.; Ling, H.F.; Zhao, K.D. Contrasting origins of late Mesozoic adakitic granitoids from the northwestern Jiaodong Peninsula, east China: Implications for crustal thickening to delamination. *Geol. Mag.* **2007**, *144*, 619–631. [CrossRef]
38. Yun, T.H.; Zheng, J.P.; Yu, C.M.; Ping, X.Q.; Ren, H.W. Multistage crust–mantle interactions during the destruction of the North China Craton: Age and composition of the Early Cretaceous intrusions in the Jiaodong Peninsula. *LITHOS* **2014**, *190–191*, 52–70.
39. Charles, N.; Augier, R.; Gumiaux, C.; Monié, P.; Chen, Y.; Faure, M.; Zhu, R. Timing, duration and role of magmatism in wide rift systems: Insights from the Jiaodong Peninsula (China, East Asia). *Gondwana Res.* **2013**, *24*, 412–428. [CrossRef]
40. Deng, J.; Liu, X.F.; Wang, Q.F.; Dilek, Y.; Liang, Y.Y. Isotopic characterization and petrogenetic modeling of Early Cretaceous mafic diking—Lithospheric extension in the North China craton, eastern Asia. *Geol. Soc. Am. Bull.* **2017**, *129*, 1379–1407. [CrossRef]
41. Yang, L.Q.; Deng, J.; Goldfarb, R.J.; Zhang, J.; Gao, B.F.; Wang, Z.L. 40Ar/39Ar geochronological constraints on the formation of the Dayingezhuang gold deposit: New implications for timing and duration of hydrothermal activity in the Jiaodong gold province, China. *Gondwana Res.* **2014**, *25*, 1469–1483. [CrossRef]
42. Steefel, C.; Lichtner, P. Multicomponent reactive transport in discrete fractures: II: Infiltration of hyperalkaline groundwater at Maqarin, Jordan, a natural analogue site. *J. Hydrol.* **1998**, *209*, 200–224. [CrossRef]
43. Fu, J.L.; Hu, Z.C.; Zhang, W.; Yang, L.; Liu, Y.S.; Li, M.; Zong, K.Q.; Gao, S.; Hu, S.H. In situ sulfur isotopes ($\delta^{34}S$ and $\delta^{33}S$) analyses in sulfides and elemental sulfur using high sensitivity cones combined with the addition of nitrogen by laser ablation MC-ICP-MS. *Anal. Chim. Acta* **2016**, *911*, 14–26. [CrossRef] [PubMed]
44. Belousov, I.; Large, R.R.; Meffre, S.; Danyushevsky, L.V.; Steadman, J.; Beardsmore, T. Pyrite compositions from VHMS and orogenic Au deposits in the Yilgarn Craton, Western Australia: Implications for gold and copper exploration. *Ore Geol. Rev.* **2016**, *79*, 474–499. [CrossRef]
45. Voute, F.; Hagemann, S.; Evans, N.; Villanes, C. Sulfur isotopes, trace element, and textural analyses of pyrite, arsenopyrite and base metal sulfides associated with gold mineralization in the Pataz-Parcoy district, Peru: Implication for paragenesis, fluid source, and gold deposition mechanisms. *Miner. Depos.* **2019**, *54*, 1077–1100. [CrossRef]
46. Large, R.R.; Bull, S.W.; Maslennikov, V.V. A Carbonaceous Sedimentary Source-Rock Model for Carlin-Type and Orogenic Gold Deposits. *Econ. Geol. Bull. Soc. Econ. Geol.* **2011**, *106*, 331–358. [CrossRef]
47. Deditius, A.P.; Reich, M.; Kesler, S.E.; Utsunomiya, S.; Chryssoulis, S.L.; Walshe, J.; Ewing, R.C. The coupled geochemistry of Au and As in pyrite from hydrothermal ore deposits. *Geochim. Cosmochim. Acta* **2014**, *140*, 644–670. [CrossRef]
48. Reich, M.; Kesler, S.E.; Utsunomiya, S.; Palenik, C.S.; Chryssoulis, S.L.; Ewing, R.C. Solubility of gold in arsenian pyrite. *Geochim. Cosmochim. Acta* **2005**, *69*, 2781–2796. [CrossRef]
49. Voudouris, P.C.; Spry, P.G.; Mavrogonatos, C.; Sakellaris, G.-A.; Bristol, S.K.; Melfos, V.; Fornadel, A.P. Bismuthinite derivatives, lillianite homologues, and bismuth sulfotellurides as indicators of gold mineralization in the Stanos shear-zone related deposit, Chalkidiki, northern Greece. *Can. Mineral.* **2013**, *51*, 119–142. [CrossRef]

50. Tooth, B.; Ciobanu, C.L.; Green, L.; O'Neill, B.; Brugger, J. Bi-melt formation and gold scavenging from hydrothermal fluids: An experimental study. *Geochim. Cosmochim. Acta* **2011**, *75*, 5423–5443. [CrossRef]
51. Zhang, Z.; Yu, C.; Wu, Z.D.; Wang, H.Y.; Xu, Z.T.; Liang, Y.Y. Geological characteristics and genesis of Linglong Gold Deposit in Jiaodong. *GOLD* **2024**, *45*, 74–79.
52. Shannon, R.D. Revised effective ionic radii and systematic studies of interatomic distances in halides and chalcogenides. *Found. Crystallogr.* **1976**, *32*, 751–767. [CrossRef]
53. Mathieu, L. Detecting magmatic-derived fluids using pyrite chemistry: Example of the Chibougamau area, Abitibi Subprovince, Québec. *Ore Geol. Rev.* **2019**, *114*, 103127. [CrossRef]
54. Yang, L.Q.; Deng, J.; Wang, Z.L.; Guo, L.N.; Li, R.H.; Groves, D.I.; Danyushevsky, L.V.; Zhang, C.; Zheng, X.L.; Zhao, H. Relationships Between Gold and Pyrite at the Xincheng Gold Deposit, Jiaodong Peninsula, China: Implications for Gold Source and Deposition in a Brittle Epizonal Environment. *Econ. Geol. Bull. Soc. Econ. Geol.* **2016**, *111*, 105–126. [CrossRef]
55. Qiu, K.F.; Deng, J.; Laflamme, C.; Long, Z.Y.; Wan, R.Q.; Moynier, F.; Yu, H.C.; Zhang, J.Y.; Ding, Z.J.; Goldfarb, R. Giant Mesozoic gold ores derived from subducted oceanic slab and overlying sediments. *Geochim. Cosmochim. Acta* **2023**, *343*, 133–141. [CrossRef]
56. Deng, J.; Wang, Q.; Santosh, M.; Liu, X.; Liang, Y.; Yang, L.; Zhao, R.; Yang, L. Remobilization of metasomatized mantle lithosphere: A new model for the Jiaodong gold province, eastern China. *Miner. Depos.* **2020**, *55*, 257–274. [CrossRef]
57. Groves, D.I.; Santosh, M. The giant Jiaodong gold province: The key to a unified model for orogenic gold deposits? *Geosci. Front.* **2016**, *7*, 409–417. [CrossRef]
58. Deng, J.; Yang, L.-Q.; Groves, D.I.; Zhang, L.; Qiu, K.-F.; Wang, Q.-F. An integrated mineral system model for the gold deposits of the giant Jiaodong province, eastern China. *Earth-Sci. Rev.* **2020**, *208*, 103274. [CrossRef]
59. Zhang, C.; Liu, Y.; Liu, X.D.; Feng, J.Q.; Huang, T.; Zhang, Q.; Wang, X.D. Hydrothermal alteration of the Xincheng gold deposit, northwestern Jiaodong, China. *Acta Geol. Sin.* **2016**, *32*, 2433–2450.
60. He, D.Y.; Qiu, K.F.; Simon, A.C.; Pokrovski, G.S.; Yu, H.C.; Connolly, J.A.; Li, S.S.; Turner, S.; Wang, Q.F.; Yang, M.F. Mantle oxidation by sulfur drives the formation of giant gold deposits in subduction zones. *Proc. Natl. Acad. Sci. USA* **2024**, *121*, e2404731121. [CrossRef]
61. Zhang, J.Y.; Qiu, K.F.; Yin, R.; Long, Z.Y.; Feng, Y.C.; Yu, H.C.; Gao, Z.Y.; Deng, J. Lithospheric mantle as a metal storage reservoir for orogenic gold deposits in active continental margins: Evidence from Hg isotopes. *Geology* **2024**, *52*, 423–428. [CrossRef]
62. Ohmoto, H. Systematics of Sulfur and Carbon Isotopes in Hydrothermal Ore Deposits. *Econ. Geol.* **1972**, *67*, 551–578. [CrossRef]
63. Hutchison, W.; Finch, A.A.; Boyce, A.J. The sulfur isotope evolution of magmatic-hydrothermal fluids: Insights into ore-forming processes. *Geochim. Cosmochim. Acta* **2020**, *288*, 176–198. [CrossRef]
64. Hastie, E.; Kontak, D.; Lafrance, B.; Petrus, J.; Sharpe, R.; Fayek, M. Evaluating geochemical discriminants in Archean gold deposits: A Superior province perspective with an emphasis on the Abitibi greenstone belt. *Econ. Geol.* **2023**, *118*, 123–155. [CrossRef]
65. Lan, T.G.; Wang, H.; Fan, H.R.; Ulrich, T.; Hu, H.L.; Chen, Y.W.; Shu, L. Gold endowment and unloading along pathway for giant gold mineralization: Insights from spatiotemporal variations of in-situ pyrite geochemistry and gold fineness from the Jiaodong gold deposits, north China Craton. *Gondwana Res.* **2023**, *118*, 58–75. [CrossRef]
66. Liang, Y.Y.; Shu, L.; Ma, P.Y.; Zhang, C.X.; Ma, Y.; Khan, M.; Shen, C.H. Gold source and ore-forming process of the Linglong gold deposit, Jiaodong gold province, China: Evidence from textures, mineral chemical compositions and sulfur isotopes of pyrite. *Ore Geol. Rev.* **2023**, *159*, 105523. [CrossRef]
67. Zhang, J.J.; Ding, Z.J.; Bo, J.W.; Ji, P.; Li, T.; Xin, W. In Situ Trace Element and S-Pb Isotope Study of Pyrite from the Denggezhuang Gold Deposit in the Jiaodong Peninsula—Insights into the Occurrence of Gold and the Source of Ore-Forming Materials. *Minerals* **2024**, *14*, 158. [CrossRef]
68. Labidi, J.; Cartigny, P.; Moreira, M. Non-chondritic sulphur isotope composition of the terrestrial mantle. *Nature* **2013**, *501*, 208–211. [CrossRef]
69. Mills, S.E.; Tomkins, A.G.; Weinberg, R.F.; Fan, H.-R. Implications of pyrite geochemistry for gold mineralisation and remobilisation in the Jiaodong gold district, northeast China. *Ore Geol. Rev.* **2015**, *71*, 150–168. [CrossRef]
70. Mao, J.W.; Li, H.M.; Wang, Y.T.; Zhang, C.Q.; Wang, R.T. The relationship between mantle-derived fluid and gold ore-formation in the Eastern Shandong Peninsula: Evidences from D–O–C–S isotopes. *Acta Geol. Sin.* **2005**, *79*, 839–857.
71. Zhu, Z.X.; Zhao, X.F.; Lin, Z.W.; Zhao, S.R. In Situ Trace Elements and Sulfur Isotope Analysis of Pyrite from Jinchiling Gold Deposit in the Jiaodong Region: Implications for Ore Genesis. *Earth Sci.* **2019**, *45*, 945–959.
72. Yuan, Z.Z.; Li, Z.K.; Zhao, X.F.; Sun, H.S.; Qiu, H.N.; Li, J.W. New constraints on the genesis of the giant Dayingezhuang gold (silver) deposit in the Jiaodong district, North China Craton. *Ore Geol. Rev.* **2019**, *112*, 103038. [CrossRef]
73. Xue, J.L.; Li, S.R.; Pang, Z.S.; Tao, W.; Sun, W.Y.; Chen, H.; Zhang, Y.Q. Ore-forming fluids, sources of materials in the Denggezhuang gold deposit, Jiaodong Penisula and implications for ore genesis. *Acta Petrol. Sin.* **2018**, *34*, 1453–1468.
74. Chen, H.Y.; Li, S.R.; Zhang, X.B.; Zhou, Q.F.; Zhang, Y.Q.; Liu, Z.H.; Zhang, H.F.; Wang, N. Thermoelectric character of pyrite from Jinqingding glod deposit in eastern Shandong Province and its significance. *Miner. Depos.* **2010**, *29*, 1125–1137.

75. Yang, C.F. Mineralization Characteristics of Gold Deposit in the Guocheng Region, Haiyang, Shandong Province. *Geol. Explor.* **2010**, *46*, 462–469.
76. Li, Y.J.; Li, S.R.; Santosh, M.; Liu, S.A.; Zhang, L.; Li, W.T.; Song, Y.X.; Wang, B.X. Zircon geochronology, geochemistry and stable isotopes of the Wang'ershan gold deposit, Jiaodong Peninsula, China. *J. Asian Earth Sci.* **2015**, *113*, 695–710. [CrossRef]
77. Liu, Y.Z. Characteristics of the Ore-Forming Fluid and Thermodynamic Simulation of Water-Rock Interaction at the Sanshandao Gold Deposit, Shandong Province. Ph.D. Thesis, China University of Geosciences, Beijing, China, 2019.
78. Guo, C.Y. Tectonic Setting, Magmatie Sequence and Fluid of GoldMetallogenic System of the Sanshandao-Cangshang Fault in Jiaodong, China. Ph.D. Thesis, China University of Geosciences, Beijing, China, 2009.
79. Feng, K.; Fan, H.R.; Hu, F.F.; Yang, K.F.; Liu, X.; Shangguan, Y.N.; Cai, Y.C.; Jiang, P. Involvement of anomalously As-Au-rich fluids in the mineralization of the Heilan'gou gold deposit, Jiaodong, China: Evidence from trace element mapping and in-situ sulfur isotope composition. *J. Asian Earth Sci.* **2018**, *160*, 304–321. [CrossRef]
80. Feng, K.; Fan, H.R.; Hu, F.F. Invisible Gold in Arsenian Pyrite from the High-Grade Daliuhang Gold Deposit, Jiaodong, China: Insights from LA-ICPMS Mappings. In Proceedings of the Twenty-ninth Annual Goldschmidt Conference, Barcelona, Spain, 18–23 August 2019; p. 971.
81. Wang, J.L.; Sun, F.Y.; Wang, L.; Ding, Z.J.; Huang, W.P.; Hu, W.H. Geological characteristics and genesis of Majiayao gold deposit, Qixia City, Shandong Province. *Gold* **2013**, *34*, 14–20.
82. Zhang, C.; Liu, Y.; Liu, X.D.; Feng, J.Q.; Huang, T.; Zhang, Q.; Wang, X.D. Characteristics of sulfur isotope geochemistryof the Xincheng gold deposit, Northwest Jiaodong, China. *Acta Petrol. Sin.* **2014**, *30*, 2495–2506.
83. Huang, D.Y. Sulfur Isotope Studies on the Metallogenic Series of Gold Deposits in the Jiaodong (Eastern Shandong) Area. *Miner. Depos.* **1994**, *13*, 75–87.
84. Zartman, R.; Doe, B. Plumbotectonics—The model. *Tectonophysics* **1981**, *75*, 135–162. [CrossRef]
85. Wang, L.; Qiu, Y.; McNaughton, N.; Groves, D.; Luo, Z.; Huang, J.; Miao, L.; Liu, Y. Constraints on crustal evolution and gold metallogeny in the Northwestern Jiaodong Peninsula, China, from SHRIMP U–Pb zircon studies of granitoids. *Ore Geol. Rev.* **1998**, *13*, 275–291. [CrossRef]
86. Stacey, J.S.; Kramers, J. Approximation of terrestrial lead isotope evolution by a two-stage model. *Earth Planet. Sci. Lett.* **1975**, *26*, 207–221. [CrossRef]
87. Zhang, Z.F. The typomorphic characteristics of main metallic minerals in the Luoshan gold deposit, Shandong Province. *J. Guilin Inst. Metall. Geol.* **1993**, 366–375.
88. Li, Z.L.; Zhang, L.Y.; Xiao, X.M.; Fan, B.H. Geological and Geochemical Characteristics and Genesis of the Majiayao Gold Deposit, Shandong Province. *Contrib. Geol. Miner. Resour. Res.* **1990**, *6*, 36–48.
89. Yang, S.W. A discussion on the Jiaodong Group strata, the source-bed of gold and the strata-bound features of gold ore deposits in northwest part of Jiaodong peninsula. *Contrib. Geol. Miner. Resour. Res.* **1986**, *1*, 1–12.
90. Chen, Z.S.; Zhang, L.G.; Liu, J.X.; Wang, B.C.; Xu, J.F.; Zheng, W.S. A study on lead isotope geochemical backgrounds of geological bodies in Jiaodong region. *Contrib. Geol. Miner. Resour. Res.* **1994**, *10*, 65–78.
91. Hou, M.L.; Jiang, S.Y.; Jiang, Y.H.; Ling, H.F. S-Pb isotope geochemistry and Rb-Sr geochronology of the Penglai gold field in the eastern Shangdong province. *Acta Petrol. Sin.* **2006**, *22*, 2525–2533.
92. Tan, J.; Wei, J.H.; Audétat, A.; Pettke, T. Source of metals in the Guocheng gold deposit, Jiaodong Peninsula, North China Craton: Link to early Cretaceous mafic magmatism originating from Paleoproterozoic metasomatized lithospheric mantle. *Ore Geol. Rev.* **2012**, *48*, 70–87. [CrossRef]
93. Li, Z.K.; Li, J.W.; Cooke, D.R.; Danyushevsky, L.; Zhang, L.J.; O'Brien, H.; Lahaye, Y.; Zhang, W.; Xu, H.J. Textures, trace elements, and Pb isotopes of sulfides from the Haopinggou vein deposit, southern North China Craton: Implications for discrete Au and Ag–Pb–Zn mineralization. *Contrib. Mineral. Petrol.* **2016**, *171*, 99. [CrossRef]
94. Wu, Y.F.; Li, J.W.; Evans, K.; Koenig, A.E.; Li, Z.K.; O'Brien, H.; Lahaye, Y.; Rempel, K.; Hu, S.Y.; Zhang, Z.P. Ore-forming processes of the Daqiao epizonal orogenic gold deposit, West Qinling orogen, China: Constraints from textures, trace elements, and sulfur isotopes of pyrite and marcasite, and Raman spectroscopy of carbonaceous material. *Econ. Geol.* **2018**, *113*, 1093–1132. [CrossRef]
95. Fabricio-Silva, W.; Schutesky, M.E.; Frimmel, H.E.; Fougerouse, D.; Rosière, C.A.; Caxito, F.A.; Bosco-Santos, A. Is there a specific "timing of mineralization" in gold deposits? *Ore Geol. Rev.* **2025**, *182*, 106663. [CrossRef]
96. Wu, Y.F.; Fougerouse, D.; Evans, K.; Reddy, S.M.; Saxey, D.W.; Guagliardo, P.; Li, J.W. Gold, arsenic, and copper zoning in pyrite: A record of fluid chemistry and growth kinetics. *Geology* **2019**, *47*, 641–644. [CrossRef]
97. Koglin, N.; Frimmel, H.E.; Minter, W.E.L.; Brätz, H. Reply to Reimer and Mossman. Comment on "Trace-element characteristics of different pyrite types in Mesoarchaean to Palaeoproterozoic placer deposits" by Koglin et al. (Mineralium Deposita 42: 259–280, 2010). *Miner. Depos.* **2011**, *46*, 839–840. [CrossRef]
98. Hong, T.; Xu, X.-W.; Gao, J.; Peters, S.G.; Li, J.; Cao, M.; Xiang, P.; Wu, C.; You, J. Element migration of pyrites during ductile deformation of the Yuleken porphyry Cu deposit (NW-China). *Ore Geol. Rev.* **2018**, *100*, 205–219. [CrossRef]

99. Keith, M.; Haase, K.M.; Chivas, A.R.; Klemd, R. Phase separation and fluid mixing revealed by trace element signatures in pyrite from porphyry systems. *Geochim. Cosmochim. Acta* **2022**, *329*, 185–205. [CrossRef]
100. Gregory, D.D.; Large, R.R.; Halpin, J.A.; Baturina, E.L.; Lyons, T.W.; Wu, S.; Danyushevsky, L.; Sack, P.J.; Chappaz, A.; Maslennikov, V.V.; et al. Trace Element Content of Sedimentary Pyrite in Black Shales. *Econ. Geol.* **2015**, *110*, 1389–1410. [CrossRef]
101. He, J.T.; Li, J.J.; Shi, G.Y.; Wang, C.G.; Man, R.H. Genetic Mineralogy and Geological Significance of Pyritesfrom the Muping-Rushan gold belt, Jiaodong Peninsula. *Geoscience* **2025**, *39*, 667–680.
102. Chi, N.J.; Han, Z.Z.; Shan, W.; Sun, Y.Q.; Xiong, Y.X.; Shu, L.; Liu, C.E.; Shen, Y.; Li, Z.S. Typomorphic Characteristics of Gold-bearing Pyrite in Jiaojia Fault DeepZone of Northwest Jiaodong Peninsula and Its Geological Significance. *Acta Geosci. Sin.* **2020**, *41*, 949–962.
103. McDonough, W.F.; Sun, S.-S. The composition of the Earth. *Chem. Geol.* **1995**, *120*, 223–253. [CrossRef]
104. Barker, S.L.; Hickey, K.A.; Cline, J.S.; Dipple, G.M.; Kilburn, M.R.; Vaughan, J.R.; Longo, A.A. Uncloaking invisible gold: Use of nanoSIMS to evaluate gold, trace elements, and sulfur isotopes in pyrite from Carlin-type gold deposits. *Econ. Geol.* **2009**, *104*, 897–904. [CrossRef]
105. Blanchard, M.; Alfredsson, M.; Brodholt, J.; Wright, K.; Catlow, C.R.A. Arsenic incorporation into FeS_2 pyrite and its influence on dissolution: A DFT study. *Geochim. Cosmochim. Acta* **2007**, *71*, 624–630. [CrossRef]
106. Hastie, E.C.; Kontak, D.J.; Lafrance, B. Gold remobilization: Insights from gold deposits in the Archean Swayze greenstone belt, Abitibi subprovince, Canada. *Econ. Geol.* **2020**, *115*, 241–277. [CrossRef]
107. Fougerouse, D.; Micklethwaite, S.; Ulrich, S.; Miller, J.; Godel, B.; Adams, D.T.; McCuaig, T.C. Evidence for two stages of mineralization in West Africa's largest gold deposit: Obuasi, Ghana. *Econ. Geol.* **2017**, *112*, 3–22. [CrossRef]
108. Putnis, A. Mineral replacement reactions. *Rev. Mineral. Geochem.* **2009**, *70*, 87–124. [CrossRef]
109. Yang, M.; Li, J.; Zhao, X.; Xu, H.; Zhou, Y. Electron back-scattered diffraction and LA-ICP-MS analysis of pyrite from the Dahu lodegold deposit, southern North China craton: Insights into geochemistry and distribution of trace element connection to microstructure of pyrite. *Ore Geol. Rev.* **2019**, *115*, 103164. [CrossRef]
110. Feng, Y.C.; Qiu, K.F.; Wang, D.Z.; Sha, W.J.; Li, S. Forming conditions of tellurides and their constraints on gold enrichment in Linglong gold district, Jiaodong gold province. *Acta Petrol. Sin.* **2022**, *38*, 63–77.

minerals

MDPI

Article

Genesis of High-Grade Gold Mineralization at the Guocheng Deposit, Jiaodong Peninsula: Constraints from Magnetite Geochemistry

Ning-Han Sun [1], Tao Cui [1,2,*], Rong-Zhi Zhou [3], Yu-Ying Li [1], Qiang Fu [4], Lian-Yuan Qin [5], Qian-Jie Deng [1] and Xu-Feng Wei [6,*]

1 State Key Laboratory of Geological Processes and Mineral Resources, Frontiers Science Center for Deep-Time Digital Earth, School of Earth Sciences and Resources, China University of Geosciences, Beijing 100083, China; ninghansun@foxmail.com (N.-H.S.); 1001220402@email.cugb.edu.cn (Y.-Y.L.); 2119250024@email.cugb.edu.cn (Q.-J.D.)

2 Key Laboratory of Gold Mineralization Processes and Resource Utilization Subordinated to the Ministry of Natural Resources, Shandong Key Laboratory of Mineralization Processes and Resources Utilization of Strategic Metal Minerals (Preparatory), Shandong Institute of Geological Sciences, Jinan 250013, China

3 Shandong Gold Geological Mineral Exploration Co., Ltd., Yantai 265147, China; zhourzh@sd-gold.com

4 State Key Laboratory of Process Automation in Mining & Metallurgy, Beijing Key Laboratory of Process Automation in Mining & Metallurgy, BGRIMM Technology Group, Beijing 100160, China; fuqiang@bgrimm.com

5 Shandong Yantai Xintai Gold Mining Industry Co., Ltd., Yantai 265147, China; lianyuanqin@126.com

6 Shandong Provincial No. 3 Exploration Institute of Geology and Mineral Resources, Yantai 264011, China

* Correspondence: cuitao@email.cugb.edu.cn (T.C.); wxf8701@126.com (X.-F.W.); Tel.: +86-15894604183 (T.C.); +86-18660018388 (X.-F.W.)

Abstract

The processes responsible for high-grade disseminated gold mineralization remain poorly constrained, hindering effective exploration. This study integrates petrography, BPMA, and LA-ICP-MS analysis of magnetite from marble- and granite-hosted ores with contrasting gold grades, to constrain wall-rock-induced changes in the thermodynamic environment. BPMA results show distinct mineral assemblages: granite-hosted ores are characterized by quartz (52.31%)-K-feldspar (19.65%)-sericite (9.56%)-pyrite (8.36%), whereas marble-hosted ores feature pyrrhotite (33.90%)-chlorite (27.50%)-pyrite (15.22%)-magnetite (1.94%). The closed intergrowths of magnetite with gold and sulfides, along with the magnetite Ga-V (Grant-Vaughan) discrimination diagram, indicate a hydrothermal origin for magnetite formed during the mineralization stage. Geochemical data show that marble-hosted magnetite has lower V and chalcophile element (Co, Ni, Sn, Zn) concentrations than granite-hosted magnetite. Considering the partitioning behavior of these elements in magnetite, these differences indicate magnetite crystallization under increasing oxygen fugacity (fO_2) and decreasing sulfur fugacity (fS_2). Thermodynamic modeling results demonstrate that these changes in fO_2 and fS_2 destabilized gold-sulfur complexes in the ore-forming fluid, significantly enhancing gold precipitation efficiency and ultimately leading to the formation of high-grade ores in marble.

Keywords: mineral quantitative analysis; magnetite geochemistry; gold precipitation processes; Guocheng gold deposit; Jiaodong gold province

1. Introduction

High-grade disseminated gold ores constitute an economically crucial component of many gold deposits, as their formation directly determines the overall ore grade and mining potential of a deposit [1–4]. The formation of disseminated gold ores is widely recognized to be controlled by fluid–rock interactions, through which ore-forming fluids exchange components with wall rocks, leading to thermodynamic variations such as changes in redox state, pH, and sulfur activity [5–8]. These variations can significantly influence the solubility and transportation of gold, thereby controlling the efficiency of gold precipitation [9–11]. However, because different deposits mostly form under distinct temperature, pressure, and fluid-composition conditions, it remains challenging to systematically compare gold-precipitation efficiency among disseminated gold systems [5,12,13]. Consequently, the influence of different fluid–rock interaction mechanisms on gold-precipitation efficiency and ore grade is still unclear [14–16]. This uncertainty limits our understanding of gold precipitation processes in disseminated system and impedes the advancement of exploration efforts.

The Jiaodong gold province, located on the southeastern margin of the North China Craton, is the largest gold-producing region in China, with total gold resources exceeding 5000 tonnes [17,18]. The majority of gold deposits in this region display highly consistent mineral assemblages, alteration zonation, mineralization stages, isotopic signatures and ages [19–23]. Their ore-forming fluids are widely regarded as having similar source and properties, characterized by medium-to-low temperatures (~200–400 °C), low salinities (2–10 wt% NaCl eqv.) and near-neutral to weakly acidic pH [19,24]. The Guocheng gold deposit is situated within this province, shares these common features. Previous studies on the Guocheng gold deposit indicate that mineralization occurred at 119 ± 2 Ma [25], and the ore-forming fluids were of medium–low temperature, low salinity, and weakly acidic to near-neutral properties [26–28], consistent with the mineralization peak (120 ± 2 Ma) and fluid characteristics of the Jiaodong gold deposits [29,30]. These similarities suggest that the Guocheng deposit shares the same ore-forming fluid system with other deposits in the province, and that it was formed by the fluid which ascended along the Guocheng fault. However, field investigations at Guocheng reveal a marked difference in gold grade between ores hosted in its two main wall rock (marble and granite): marble-hosted ores often display high-grade gold mineralization (gold grade typically 5–10 g/t Au), whereas granite-hosted ores generally show lower gold grades (~1.4 g/t Au) [31,32]. The significant gold-grade disparity between the two types of disseminated ores within the same metallogenic system at Guocheng gold deposit, provides ideal conditions to investigate how different fluid–rock interactions control gold-precipitation efficiency and the formation of high-grade disseminated gold ores. Nevertheless, the processes controlling the distinct contrast in gold grades between the two types of disseminated ores within the same system have not yet been convincingly explained.

In recent years, the trace element composition of magnetite has been demonstrated as an effective tracer of the thermodynamic environment during its formation [33–35]. For example, by comparing two types of magnetite in the Weizigou gold deposit, previous researchers found that the V content of early-generation magnetite (Mag-II-B) was significantly higher than that of late-generation magnetite (Mag-II-A), revealing that the ore-forming fluid evolved from relatively high oxygen fugacity in the early stage to low oxygen fugacity in the late stage [15]. At the Guocheng gold deposit, magnetite is widely developed in the gold ores and can serve as a reliable indicator for reflecting the physicochemical conditions of mineralization. To better understand the mineralogical characteristics associated with gold mineralization, we employed the BGRIMM Process Mineralogy Analyzer (BPMA)—the first automated mineralogy system independently developed in China by the Beijing General Research Institute of Mining & Metallurgy (BGRIMM Technology Group). This SEM–EDS–based technology enables rapid and quantitative characterization of mineral composition [36–38], providing an effective tool for investigating ore-forming processes [39]. In this study, we conducted an integrated investigation combining microscopic petrography, automated mineralogy (BPMA), and LA-ICP-MS trace-element analysis of magnetite from representative disseminated marble-hosted and granite-hosted ores at Guocheng gold deposit. By systematically comparing mineral assemblages and magnetite geochemical characteristics between the two ore types, we constrained the variations in thermodynamic conditions during gold precipitation, and revealed how different fluid–rock interactions controlled gold precipitation efficiency and the formation of high-grade disseminated ores at Guocheng gold deposit. The study can advance the understanding of the formation mechanisms of high-grade disseminated gold ores and provide a theoretical basis for their exploration and targeting.

2. Regional Geologic Setting

The Jiaodong Peninsula lies along the southeastern margin of the North China Craton, bounded by the Tan–Lu strike-slip fault zone to the west, the Yellow Sea to the east, and surrounded on the remaining three sides by the Bohai Sea (Figure 1A). The region is tectonically divided by the N–NE-trending Wulian–Yantai Fault Zone into the Sulu Terrane to the southeast and the Jiaobei Terrane to the northwest, the latter of which is further subdivided into the Jiaobei Uplift and the Jiaolai Basin (Figure 1B) [19]. The Sulu Terrane is mainly composed of metamorphic basement rocks overlain by Jurassic–Cretaceous volcanic–sedimentary sequences and intruded by Cretaceous magmatic bodies [17]. In contrast, the Jiaobei Terrane, occupying the northwestern to central-northern part of the peninsula, exposes extensive Neoarchean–Paleoproterozoic basement rocks of the North China Craton [20], consisting of Precambrian metamorphic sequences, Mesozoic granites, and volcanic–sedimentary assemblages. The metamorphic basement includes the Neoarchean Jiaodong Group dominated by tonalite–trondhjemite–granodiorite (TTG) gneisses [29], the Paleoproterozoic Jingshan and Fenzishan Groups characterized by gneiss, marble, calc-silicate, graphitic schist assemblages [40], and the Neoproterozoic Penglai Group composed mainly of marble, slate, and marl [41]. Spatially, the Jingshan Group is predominantly exposed east of the Wulian–Qingdao–Yantai Fault Zone. The Sulu Terrane, on the other hand, exhibits a distinct dual structural framework, with an ultrahigh-pressure metamorphic belt distributed linearly along the Wulian–Qingdao–Yantai Fault Zone and forming continuous outcrops in its eastern part [42].

Mesozoic magmatism in Jiaodong was driven by the subduction of the Paleo-Pacific Plate. It comprises two main phases: (1) emplacement of the Linglong biotite granite (160–145 Ma) and (2) emplacement of the Guojialing porphyritic granodiorite (130–122 Ma) [40,43]. During the Triassic to Jurassic, flat-slab subduction induced a compressional collisional stress regime in the North China Craton lithospheric mantle [43]. Magmatism was minimal in Jiaodong during this period, represented mainly by the Linglong granites [44]. A fundamental tectonic shift occurred in the Cretaceous with the onset of slab rollback of the Paleo-Pacific Plate. This triggered intense NW-SE lithospheric extension and large-scale thinning beneath the North China Craton [44,45]. Concurrently, widespread magmatic activity peaked around 125 Ma, producing abundant intermediate-acid intrusive rocks (notably the Guojialing suite) and volcanic assemblages [46,47]. Critically, the major gold mineralization events in Jiaodong are both temporally and genetically associated with this Early Cretaceous (ca. 130–120 Ma) magmatic–hydrothermal activity, particularly with the Guojialing intrusions and related fluid systems, wherein gold deposits are primarily hosted within the Linglong and Guojialing granites [29,43].

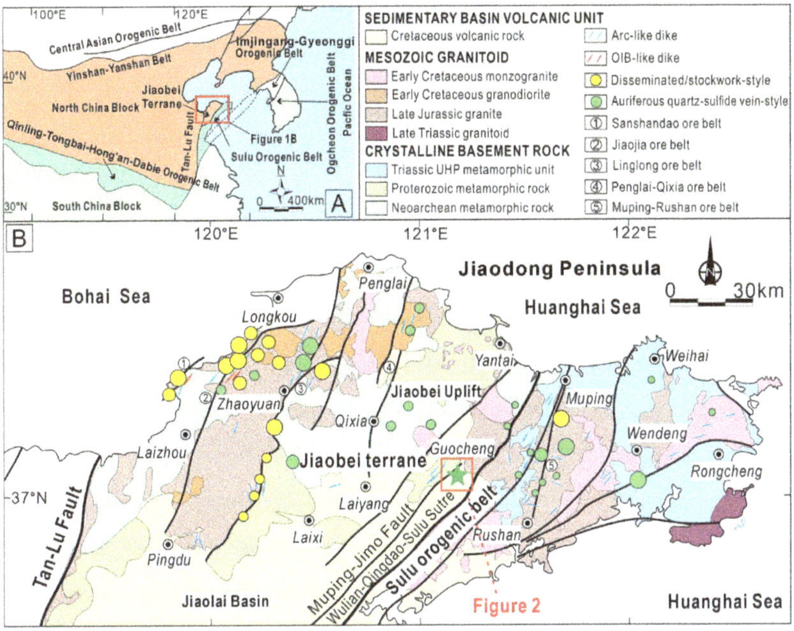

Figure 1. (**A**) Simplified geological map showing the tectonic setting of the Jiaodong gold province, China. (**B**) Geological map of the Jiaodong gold province showing the distribution of gold deposits in terms of size, modified after [29]. UHP—ultra high-pressure; OIB—ocean island basalt.

The Jiaodong region is predominantly governed by E-W and NNE-NE trending tectonic systems (Figure 1B). E-W trending structures mainly manifest as ancient basement folds, whereas the spatial distribution of gold deposits is strictly controlled by NNE-NE trending regional fault zones and their subsidiary structures (Figure 1B) [48]. First-order lithospheric-scale faults include the Wulian-Yantai Fault Zone and the Tan-Lu Fault Zone, with NNE-NE trending second- and third-order fault networks constituting their branch systems [49]. Within the region, five major fault zones develop from west to east: Sanshandao-Cangshang, Jiaojia, Zhaoyuan-Pingdu, Penglai-Qixia, and Muping-Rusha [49]. These fault systems fundamentally control gold deposit localization (Figure 1B). Among these, three major ore belts develop along the Penglai-Qixia Fault Zone in western Jiaodong [19,48], while the Muping-Rushan metallogenic belt dominates the east. The NE-NNE trending fault systems control over 90% of the region's gold resources, concentrated particularly within the three major ore belts in the west [50].

3. Deposit Geology

The Guocheng gold deposit is situated in the northeastern part of the Jiaolai Basin, occupying a key tectonic position at the junction of the Sulu ultrahigh-pressure metamorphic belt, the Jiaolai Basin, and the Jiaobei Terrane [19,51]. The exposed strata are mainly composed of Archean Jingshan Group metamorphic rocks and Mesozoic Laiyang Group sedimentary rocks (Figure 2). The Jingshan Group is dominated by marbles, which are metamorphosed carbonate rocks formed during the Paleoproterozoic regional metamorphic event associated with the Jingshan orogeny [52]. Intrusive rocks are widely distributed in the southern and eastern parts of the area, covering about half of the total surface. The dominant intrusive unit is the Late Proterozoic Muniushan adamellite pluton, which intrudes the Archean Jingshan Group with a tongue-like morphology. Mesozoic mafic to felsic dikes, including dolerite, monzonite porphyry and lamprophyre, are also common.

These bodies are mainly of Late Yanshanian age and cut through both the Jingshan Group and the Muniushan pluton.

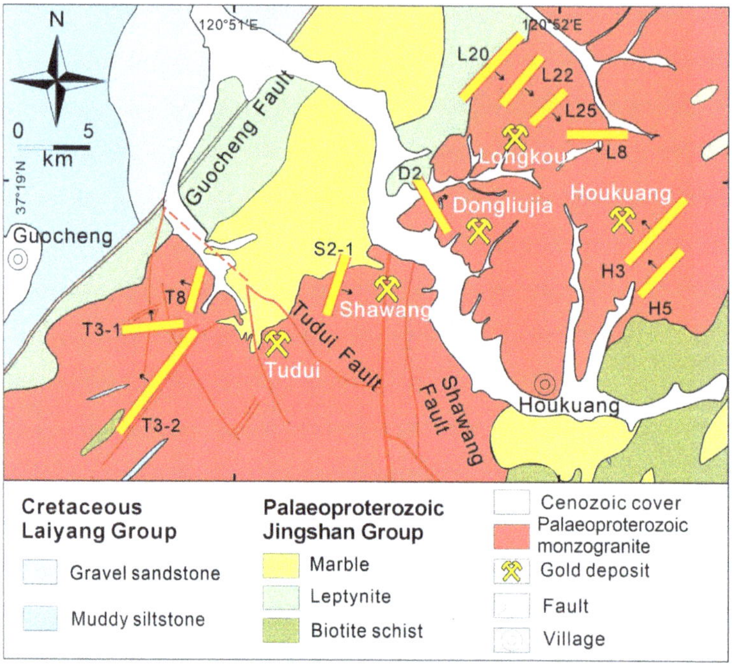

Figure 2. Simplified geological map of Guocheng gold deposit in Jiaodong, modified after [27].

The Guocheng gold deposit contains five main mines: Tuidui, Shawan, Dongliujia, Houkuang, and Longkou, which are distributed along subsidiary NE–SW-trending faults related to the Guocheng Fault (Figure 2). The gold orebodies are predominantly fault-controlled, occurring as NE–SW–striking veins and lenses within the marble of the Jingshan Group and the granite of the Muniushan pluton (Figure 2) [53]. The ore exhibits disseminated mineralization (Figure 3), characterized by a pronounced contrast in gold grade between the two main wall-rock types, with altered granites hosting relatively low-grade ores (~1.4 g/t Au) and altered marbles containing significantly higher-grade ores, typically ranging from 5 to 10 g/t Au [27]. The BGRIMM Process Mineralogy Analyzing System (BPMA) indicates that the dominant ore minerals in the Guocheng deposit include pyrite, pyrrhotite, and magnetite, accompanied by minor chalcopyrite, sphalerite, and galena (Figures 4 and 5). The gangue minerals are mainly composed of quartz, dolomite, calcite, chlorite, and sericite (Figure 5). Distinct contrasts in mineral assemblages are observed both among different ore types and among their corresponding wall rocks. The marble-hosted ores comprise pyrrhotite (33.90%), chlorite (27.5%), pyrite (15.22%), actinolite (5.82%), pyroxene (4.64%), sericite (4.47%), calcite (1.99%), magnetite (1.94%), dolomite (1.35%), and accessory minerals (1.33%), with magnetite commonly intergrown with pyrrhotite and pyrite (Figures 4A–G, 5A–C and 6C,D). In contrast, the barren marble is dominated by dolomite (72.46%) and calcite (18.85%), with minor actinolite (3.28%), chlorite (1.24%), pyrrhotite (0.33%), pyrite (0.21%), and magnetite (0.63%) (Figure 6A,B). The granite-hosted ores contain quartz (52.31%), K-feldspar (19.65%), sericite (9.56%), and pyrite (8.36%), with trace pyrrhotite (0.14%) and magnetite (<0.01%) (Figures 4H,I, 5D–F and 6G,H). The barren granite consists mainly of quartz (47.84%) and plagioclase (41.81%), with muscovite (4.24%) after feldspar and biotite alteration (Figure 6E,F).

Figure 3. Displays photographs of representative marble-hosted ores samples (**A**), granite-hosted ores (**B**), as well as the wall rocks of barren marble (**C**) and barren granite (**D**) in the Guocheng gold deposit.

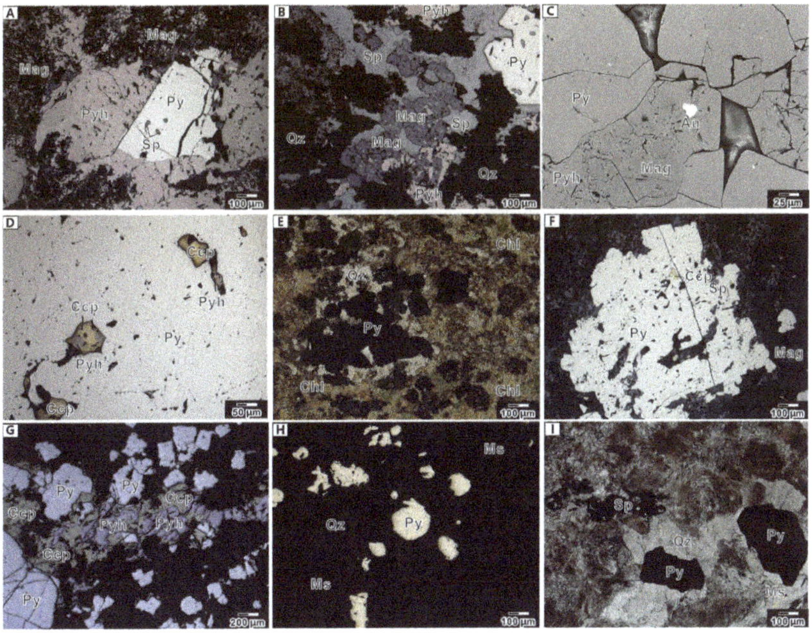

Figure 4. Micrographs of metal sulfide of marble-hosted ores samples, sulfides are dominated by Pyh and Py, accompanied by minor Sp, Mag, and Ccp; meanwhile, sulfides show close paragenetic relationships with Mag and Chl, particularly characterized by Pyh-Mag assemblages (**A–G**). Euhedral Py is closely intergrown with Ser and Qz in granite-hosted gold ores samples (**H**,**I**). Mineral abbreviations: Py = pyrite, Pyh = pyrrhotite, Mag = magnetite, Au = gold, Ccp = chalcopyrite, Sp = sphalerite, Chl = chlorite, Ms = sericite, Qz = quartz.

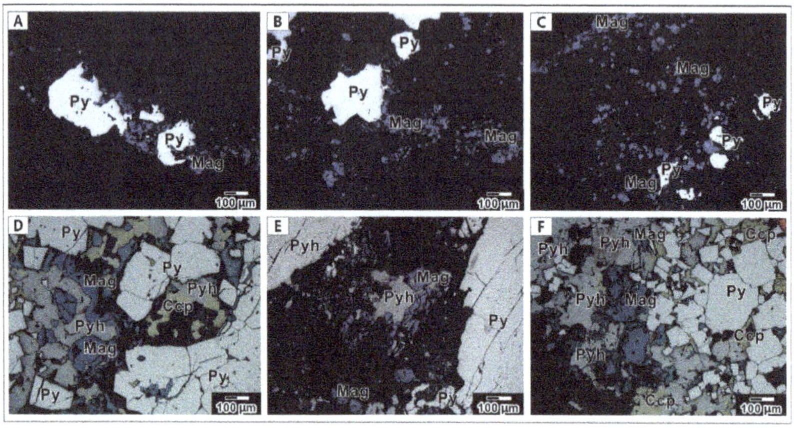

Figure 5. In marble-hosted ores, magnetite content is higher (2~5 wt%), characterized by coarser-grained and more euhedral crystals, showing close relationships with Pyh and Ccp (**A–C**). In altered granite-hosted ores, porous magnetite widely developed and filled by calcite, with low magnetite content (<0.5 wt%), exhibiting small crystal sizes and anhedral forms, commonly associated with Py (**D–F**).

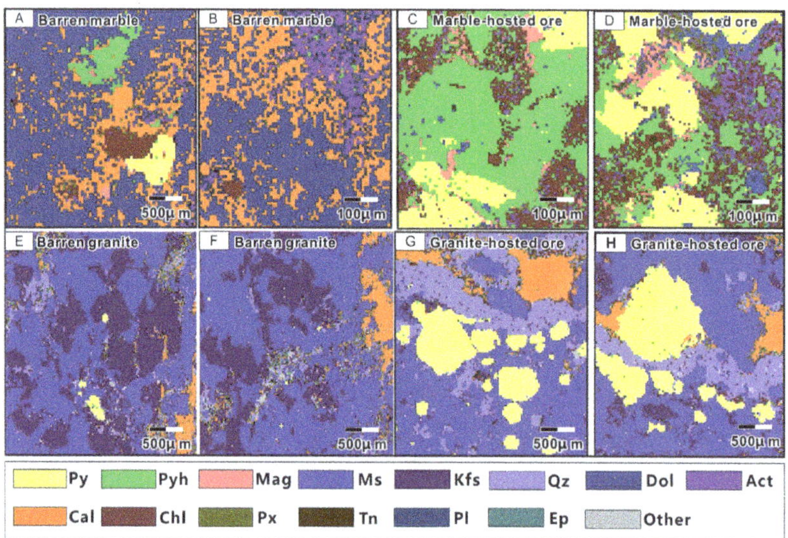

Figure 6. Advanced Mineral Identification and Characterization System-derived maps showing variations in mineral assemblages in representative samples from the study area in the Guocheng gold deposit, China. (**A**,**B**) Barren marble is dominated by dolomite (72.46%) and calcite (18.85%), with minor actinolite (3.28%) and chlorite (1.24%). (**C**,**D**) Marble-hosted ores consist mainly of pyrrhotite (33.90%), chlorite (27.50%), and pyrite (15.22%), accompanied by actinolite (5.82%), pyroxene (4.64%), sericite (4.47%), calcite (1.99%), magnetite (1.94%), and dolomite (1.35%). (**E**,**F**) Barren granite is composed predominantly of quartz (47.84%) and plagioclase (41.81%), with minor muscovite (4.24%). (**G**,**H**) Granite-hosted ores are characterized by a quartz–K-feldspar–sericite–pyrite assemblage, containing quartz (52.31%), K-feldspar (19.65%), sericite (9.56%), and pyrite (8.36%). Mineral abbreviations: Py = pyrite, Pyh = pyrrhotite, Mag = magnetite, Ms = sericite, Kfs = K-feldspar, Qz = quartz, Dol = dolomite, Act = actinolite, Cal = calcite, Chl = chlorite, Px = pyroxene, Tn = titanite, Pl = plagioclase, Ep = epidote, Other = other minerals.

4. Samples and Analysis Methods

4.1. Samples

Systematic sampling was conducted at the Guocheng gold deposit to collect both marble-hosted and granite-hosted ores, along with their respective barren wall rocks (Figure 3). This sampling strategy aimed to compare and evaluate mineralization differences between ore units hosted in distinct barren lithologies.

4.2. BGRIMM Process Mineralogy Analyzing System (BPMA)

To mitigate analyst-induced bias, representative samples of marble-hosted ore, granite-hosted ore, and their wall rocks—selected based on detailed fieldwork—were subjected to automated mineral identification testing. Analyses were performed at the State Key Laboratory of Mineral Processing (BGRIMM Technology Group, Beijing, China), using a TESCAN VEGA scanning electron microscope coupled with a Bruker energy dispersive spectrometer and the BPMA software (version 2.0). Operational parameters included high-vacuum mode, a 15 mm working distance, 20 kV accelerating voltage, and a 2 nA beam current, with backscattered electron detection. A whole-particle measurement mode was employed to automatically identify metallic sulfides. Quantitative parameters (e.g., sulfide content, particle size distribution, and dissemination complexity) were acquired under high magnification. Detailed analytical procedures follow references [38].

4.3. LA-ICP-MS Analysis of Magnetite

In situ trace element analyses of magnetite were conducted via laser ablation-inductively coupled plasma mass spectrometry (LA-ICP-MS) at the National Research Center for Geoanalysis (CAGS, Beijing, China). An Agilent 8900 ICP-MS/MS Triple Quad coupled with an NWR 193™ laser (Elemental Scientific Lasers, Bozeman, MT, USA) was utilized. Helium carrier gas transported ablated aerosols from the chamber, where they were mixed with argon and nitrogen (as a diatomic gas, N_2) to enhance sensitivity. Laser ablation employed a 6 Hz repetition rate at 5 J/cm^2 energy density with 30 μm spot diameters. A baffle-type smoothing device installed upstream of the ICP-MS minimized signal fluctuations induced by laser pulses, improving data quality. Each analysis comprised 10 s of gas blank (background) acquisition followed by 30 s of sample data acquisition. Calibration standards NIST SRM 610 and MASS-1 were analyzed after every ten samples to correct for time-dependent instrumental drift and mass bias. Data reduction utilized ICPMSDataCal 11.8 software employing specific protocols and calibration methods following [54]. A total of 59 magnetite analyses were performed, of which 56 were retained after applying strict quality control criteria, including rejection of analyses with low signal intensity, unstable baselines, or >15% relative standard deviation for key elements (e.g., Ti, V, Fe, Mn, Mg). The detection: limits for Ti, V, Ga, Mn, Co and Ni are <0.1 ppm. For Sn and Zn, the detection limits are <1 ppm; for Mg and Al, it is approximately 3 ppm. The precision and accuracy are better than 10% for most analyzed elements.

5. Results

5.1. Mineral Size Distribution and Mineral Intergrowth Assemblages

Mineral intergrowth assemblages can reveal spatial relationships and crystallization sequences between minerals, providing key insights into the mineralization process [55]. In the marble-hosted ores, pyrite most frequently intergrows with pyrrhotite (47.84%), chlorite (12.96%), and magnetite (6.42%), while pyrrhotite mainly associates with pyrite (15.97%), iron-rich chlorite (42.95%), and magnetite (4.59%) (Figures 4A–G, 5A,B and 6C,D; Table 1). In the granite-hosted ores, pyrite primarily intergrows with quartz (39.14%) and sericite

(15.01%), along with minor associations with plagioclase (6.72%) and pyrrhotite (8.06%) (Figures 4H,I, 5D–F and 6G,H; Table 1).

Table 1. Statistical table of the association and combination of metal sulfides.

Mineral	Marble-Hosted Ores		Granite-Hosted Ores
	Pyrrhotite	Pyrite	Pyrite
pyrrhotite	\	47.84	8.06
pyrite	15.97	\	\
magnetite	4.59	6.42	\
sericite	7.31	5.74	15.01
iron chlorite	42.95	12.96	2.80
feldspar	\	\	6.72
quartz	\	\	39.14
other	29.18	27.04	22.27

Sulfides such as pyrrhotite and pyrite can be classified by grain size into ultrafine (<20 μm), fine (20–100 μm), medium (104–295 μm), coarse (295–417 μm), and ultra-coarse (>417 μm) categories [56]. Grain size differences reflect variations in crystallization rate and the stability of the hydrothermal environment during mineralization [57]. In the marble-hosted ores, pyrite is predominantly medium to coarse grained, with medium grains accounting for 39.64%, coarse grains for 30.80%, ultra-coarse grains for 12.49%, while fine and ultrafine grains make up 9.49% and 7.58%, respectively (Figure 7A). Pyrrhotite shows a more scattered distribution, with medium grains comprising 29.62%, coarse grains 26.19%, ultra-coarse grains 10.62%, fine grains 26.83%, and ultrafine grains 6.74% (Figure 7B). In the granite-hosted ores, pyrite is mainly coarse to ultra-coarse grained, with medium grains accounting for 32.62%, coarse grains for 16.39%, and ultra-coarse grains for 37.28% (Figure 7C), while fine and ultrafine fractions are almost absent.

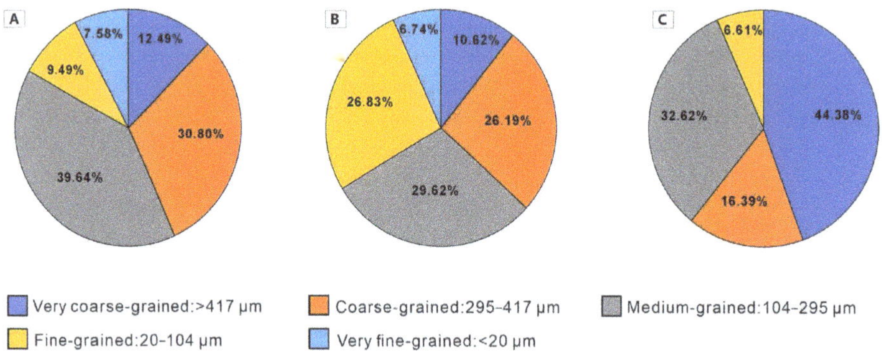

Figure 7. Statistical Table of Particle Size Distribution of Metal Sulfides in Gold Ores. (**A**) Pyrite in marble-hosted ores, predominantly medium to coarse grained. (**B**) Pyrrhotite in marble-hosted ores, showing a scattered grain-size distribution. (**C**) Pyrite in granite-hosted ores, mainly coarse to ultra-coarse grained.

5.2. LA-ICP-MS Trace Element Analysis of Magnetite

The LA-ICP-MS results of magnetite from the Guocheng gold deposit are summarized in Table 2. A total of 56 analytical spots were conducted, including 38 spots on magnetite in marble-hosted ores and 18 spots on magnetite in granite-hosted ores. Mag in marble-hosted ores exhibits larger and more euhedral crystal forms (Figure 7A–C), whereas Mag in granite-hosted ores commonly contains vugs filled with calcite cementation (Figure 7A–C).

For magnetite in marble-hosted ores, the lithophile elements show highly heterogeneous distributions. Ti concentrations are consistently below 20 ppm (average = 9.41 ppm), whereas V contents are irregular but mostly <25 ppm (average = 36.23 ppm). Al and Ga have mean concentrations of 393 ppm and 3.96 ppm, respectively. Mg is notably enriched (average = 12,030 ppm), while Mn averages 1405 ppm. The chalcophile elements in magnetite in marble-hosted ores are generally low: Co is frequently below the detection limit, with an average of 0.12 ppm for detected samples; Ni and Sn are mostly below 3 ppm, averaging 1.74 ppm and 2.78 ppm, respectively; and Zn averages 63.87 ppm.

Table 2. Trace element characteristics of magnetite analyzed by LA-ICP-MS in marble- and granite-hosted gold ores from the Guocheng deposit.

Type	Samples	Ti	V	Al	Ga	Mg	Mn	Co	Ni	Sn	Zn
Magnetite in marble-hosted ores	25GC-M-01	1.55	1.53	92.64	5.14	18,889.51	1178.56	0	0.30	0.80	420.36
	25GC-M-02	2.38	13.18	74.98	4.31	7893.42	1366.66	0	0.99	0.92	47.92
	25GC-M-03	7.48	128.52	160.68	4.73	7095.37	536.03	0.03	0.99	0.80	33.91
	25GC-M-04	9.65	44.02	308.51	4.85	10,008.75	1136.88	0.07	0.01	0.98	70.14
	25GC-M-05	2.68	21.37	125.60	6.83	3969.90	605.63	0.30	0.65	0.69	31.93
	25GC-M-06	14.70	34.86	512.36	5.41	10,867.31	981.48	0.09	0.19	0.94	77.45
	25GC-M-07	20.97	47.04	321.09	2.96	15,027.13	1039.42	0.03	6.06	1.16	48.09
	25GC-M-08	2.96	3.74	293.61	1.59	34,613.43	1334.99	0	0.34	1.29	103.06
	25GC-M-09	4.77	4.69	217.89	1.26	7811.52	1378.32	0.03	0.24	1.05	55.59
	25GC-M-10	0.60	0.53	49.97	1.94	25,943.93	1371.73	0	0.83	5.78	71.40
	25GC-M-11	0.85	1.07	29.27	1.18	12,772.48	1907.82	0	1.25	3.23	70.33
	25GC-M-12	0.70	1.02	27.67	1.44	2195.17	1146.40	0	1.26	0.37	20.75
	25GC-M-13	6.12	16.14	261.17	3.55	3874.90	1198.41	0	0.71	0.91	50.25
	25GC-M-14	2.16	2.42	167.76	2.04	1002.34	626.55	0.10	1.15	1.15	132.13
	25GC-M-15	1.95	2.33	130.12	2.05	1742.14	902.58	0	0.75	1.42	29.70
	25GC-M-16	1.23	1.83	124.46	1.09	465.84	693.22	0	0.83	1.30	25.81
	25GC-M-17	4.62	9.39	482.87	1.78	1822.56	898.60	0	3.20	2.96	22.28
	25GC-M-18	8.01	4.46	454.02	5.15	32,495.20	1669.53	0.02	0.50	2.89	113.97
	25GC-M-19	46.81	254.25	1953.28	19.38	2374.26	203.66	0.13	3.43	2.15	9.89
	25GC-M-20	5.84	144.75	1095.38	11.81	2817.30	174.57	0.04	8.66	1.97	8.68
	25GC-M-21	52.20	13.00	452.32	4.59	18,043.83	832.00	0.42	0.64	0.54	52.54
	25GC-M-22	60.29	12.82	527.37	4.44	17,130.37	2245.09	0.17	1.11	0.65	97.04
	25GC-M-23	6.43	7.60	314.82	2.68	15,909.26	910.37	0.08	0.76	0.69	49.77
	25GC-M-24	5.56	14.83	177.34	0.96	12,459.35	722.42	0	0.23	42.93	46.82
	25GC-M-25	3.05	7.72	242.24	1.42	25,015.49	1235.95	0.07	0.14	0.93	72.68
	25GC-M-26	5.94	5.60	384.56	1.18	14,140.47	430.01	0.02	4.80	0.46	44.40
	25GC-M-27	25.61	126.38	1305.63	3.58	17,423.18	1006.48	0	0.79	16.46	45.08
	25GC-M-28	9.64	7.49	492.82	3.84	20,988.47	848.98	0	0.54	0.54	77.17
	25GC-M-29	5.51	4.32	91.20	3.82	11,728.98	963.50	00	0.82	0.85	0.00
	25GC-M-30	2.73	21.21	564.11	5.96	13,807.69	3027.67	0.02	1.10	1.17	65.55
	25GC-M-31	0.70	20.12	140.46	3.95	7452.43	2585.04	0.04	0.86	0.82	36.35
	25GC-M-32	1.68	4.62	371.26	2.50	8322.82	2331.17	0.01	2.34	1.18	36.33
	25GC-M-33	7.21	13.07	339.63	2.74	18,549.06	2541.47	0	1.02	1.14	63.45
	25GC-M-34	3.04	13.62	406.69	2.93	18,903.83	3907.35	0	1.77	1.22	77.24
	25GC-M-35	5.50	29.81	432.78	5.17	7239.37	471.84	0.25	7.77	0.72	45.16
	25GC-M-36	1.07	24.56	60.22	3.61	8445.28	3839.19	0.06	2.75	0.90	25.51
	25GC-M-37	0.21	9.50	23.63	4.09	8595.96	3865.44	0.00	2.68	1.01	81.98
	25GC-M-38	15.31	303.28	1705.34	4.35	9295.10	1274.92	0.98	3.74	0.85	66.29

Table 2. *Cont.*

Type	Samples	Ti	V	Al	Ga	Mg	Mn	Co	Ni	Sn	Zn
Magnetite in granite-hosted ores	25GC-G-01	8.41	40.95	3388.67	8.24	4266.75	338.51	0.20	4.99	0.44	32.59
	25GC-G-02	3.30	66.98	0.00	0.00	0.00	0.00	0.32	1.65	0.72	0.00
	25GC-G-03	7.64	12.52	527.87	8.41	65.03	125.09	0.09	2.98	8.61	47.18
	25GC-G-04	4.02	63.44	288.65	29.14	8178.99	3209.20	2.83	3.16	1.92	261.97
	25GC-G-05	6.73	54.43	983.83	7.68	2024.78	316.57	0.75	5.08	102.85	15.84
	25GC-G-06	8.92	45.70	519.42	7.23	957.40	380.63	0.73	4.54	25.94	11.10
	25GC-G-07	7.57	46.13	13,684.61	24.44	13,563.58	338.38	0.30	8.39	0.63	59.69
	25GC-G-08	18.81	66.40	1571.77	9.23	2748.99	510.74	0.32	6.41	15.18	16.16
	25GC-G-09	2.74	34.76	263.77	5.70	2257.63	911.11	0.53	5.61	0.58	8.10
	25GC-G-10	13.31	51.00	654.10	11.92	2665.75	1952.74	1.24	3.45	112.74	13.35
	25GC-G-11	18.11	52.19	1019.86	15.62	17,227.56	2550.01	0.85	5.62	0.95	32.31
	25GC-G-12	13.99	61.56	154.75	11.49	1064.08	347.78	0.48	5.08	276.48	16.78
	25GC-G-13	6.22	48.94	1082.95	13.49	23,869.25	3654.22	2.26	4.36	0.98	161.14
	25GC-G-14	20.74	41.07	14,453.95	20.46	17,632.71	798.36	0.85	6.53	4.02	163.69
	25GC-G-15	3.69	31.34	359.86	5.49	2052.51	621.53	0.20	2.30	0.48	12.66
	25GC-G-16	4.83	39.22	2511.24	7.83	2435.40	232.85	0.39	3.22	77.15	443.82
	25GC-G-17	8.80	52.20	674.62	6.65	829.39	79.89	0.23	3.75	9.20	22.53
	25GC-G-18	6.71	30.37	212.22	8.09	222.02	289.74	0.11	3.31	36.81	19.59

In magnetite from granite-hosted ores, the lithophile elements remain low but display distinct differences compared with marble-hosted ores. Ti concentrations are similar (<20 ppm), whereas V contents are slightly higher (40–70 ppm, average = 46.21 ppm). Al and Ga are significantly enriched (average = 888 ppm and 11.17 ppm, respectively), while Mg and Mn are depleted (average = 5670 ppm and 925 ppm, respectively). Although the chalcophile elements in Mag in granite-hosted ores also exhibit low concentrations, they are notably higher than those in magnetite in marble-hosted ores: Co concentrations, though low, are detectable, averaging 0.71 ppm; Ni concentrations mostly range between 2 and 8 ppm, averaging 4.47 ppm; Sn concentrations vary widely, averaging 37.54 ppm; and Zn averages 74.36 ppm.

6. Discussion

6.1. Classification and Origins of Magnetite in Marble-Hosted Ores and Granite-Hosted Ores

Magnetite develops in diverse geological environments, and its composition is influenced by physicochemical conditions as well as by the chemistry of both fluids and wall rocks, resulting in a wide compositional range [58–61]. Consequently, comparisons of magnetite compositions between different regions or deposits are often difficult to establish. However, the different wall rocks but coeval mineralization styles at the Guocheng gold deposit provide an ideal opportunity to compare hydrothermal magnetite compositional variations controlled by the same mineralizing event under comparable physicochemical conditions. Based on its formation process, magnetite can be classified into hydrothermal and magmatic types [33,62]. In this study, petrographic observations reveal that magnetite is closely associated with pyrite, pyrrhotite and gold (Figures 4C and 5A–F), combined with the Ga–V discrimination diagrams (Figure 8A) that classify this magnetite from both ore types into the low-temperature hydrothermal category [63]. This suggests that the magnetite crystallized during the main mineralization stage and can be used to constrain the thermodynamic conditions of gold precipitation (Figure 4C).

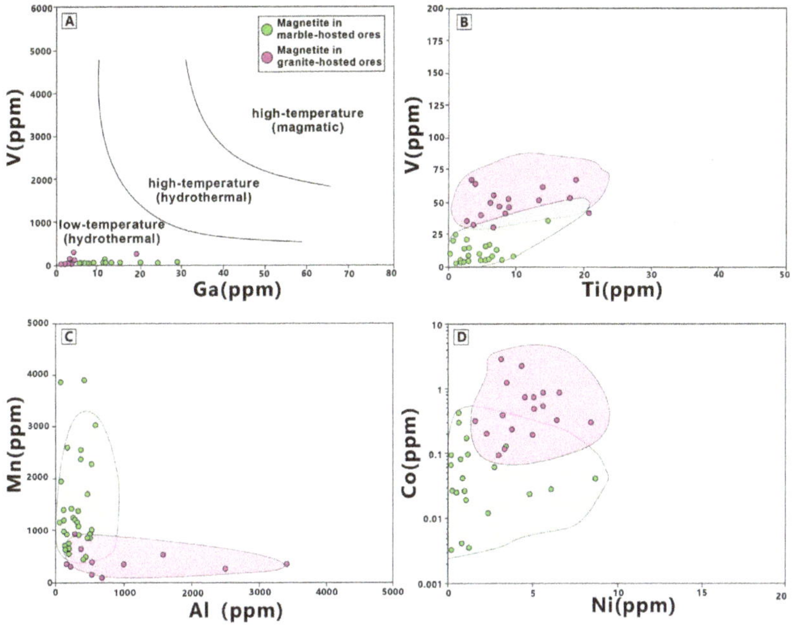

Figure 8. V vs. Ga discrimination diagram for high-temperature magmatic, high–temperature hydrothermal and low-temperature hydrothermal magnetite (**A**) modified after [63]). V vs. Ti content (**B**) and Al vs. Mn content (**C**) in magnetite for comparing formation fO_2. Co vs. Ni content (**D**) for comparing formation fS_2.

In hydrothermal systems, magnetite can form either through in-situ dissolution-recrystallization or through dissolution-migration-recrystallization [59,63]. The former preserves pseudomorphic structures inherited from precursor minerals, while the latter involves direct precipitation of minerals from hydrothermal fluids without the participation of specific precursors [64,65]. Petrographic observations reveal that magnetite exhibits two distinct textural types in both granite-hosted and marble-hosted ores. Most of these magnetite grains exhibit euhedral to subhedral crystal forms and show no evidence of replacing any precursor minerals (Figure 5), indicating that they formed via dissolution–migration–recrystallization [59]. The formation mechanism of this type of magnetite indicates that elements such as Fe were released into the fluid as chloride complexes (for example, Fe-Cl) from iron-magnesium-rich minerals (ankerite, biotite, amphibole) in wall rocks (Reactions (1)–(3)). These components were then transported along permeable pathways and eventually co-precipitated with sulfides (mainly pyrite, pyrrhotite) in favorable locations (Reaction (4)). Additionally, some magnetite replaces pyrite (Figure 5), indicating formation through replacement of precursor pyrite. This reaction process is shown in Equation (5). Notably, the primary textures of both types of magnetite are well preserved and show no visible signs of secondary alteration, such as corrosion, dissolution, or replacement by hematite (Figure 5). This suggests that their chemical compositions have likely retained primary hydrothermal signatures without significant post-mineralization alteration. Although these two types of magnetite formed through different mechanisms, both of them occur in close association with gold and sulfides (Figures 4C and 5A–F) and crystallized from the same ore-forming hydrothermal system. Therefore, they can be regarded as reliable tracers for constraining the physicochemical conditions of gold precipitation of Guocheng gold deposit.

$$Ca(Fe, Mg)(CO_3)_{2(Dol)} + 4H^+ = Ca^{2+} + Fe^{2+} + Mg^{2+} + 2CO_2\uparrow + 2H_2O_{(aq)} \quad (1)$$

$$2KFe_3AlSi_3O_{10}(OH)_{2(Bt)} + 14H^+ + 6H_2O = 6Fe^{2+} + 2K^+ + 2Al(OH)_{3(aq)} + 6H_4SiO_{4(aq)} \quad (2)$$

$$Ca_2Fe_5Si_8O_{22}(OH)_{2(Hbl)} + 14H^+ = 2Ca^{2+} + 5Fe^{2+} + 8H_4SiO_{4(aq)} \quad (3)$$

$$3FeCl^+ + 4H_2O_{(aq)} = Fe_3O_{4(Mag)} + H_{2(aq)} + 6H^+ + 3Cl^- \quad (4)$$

$$3FeS_{2(Py)} + 4H_2O = Fe_3O_{4(Mag)} + 6H_2S_{(aq)} \quad (5)$$

6.2. Thermodynamic Conditions of High-Grade Disseminated Gold Ore Formation Constrained by Magnetite Geochemistry

Disseminated mineralization is widely attributed to fluid–rock interactions [5–8]. Both the marble-hosted and granite-hosted ores at Guocheng exhibit typical disseminated mineralization (Figure 3A,C), with gold grades strictly controlled by the lithology of the surrounding rock (marble vs. granite). These observations suggest that the significant difference in ore grade is closely related to distinct fluid–rock interactions with the different wall rocks. In this section, we compare the ore-forming environments of marble-hosted and granite-hosted ores by contrasting magnetite geochemistry and mineral assemblages, and discuss variations in (1) oxygen fugacity and (2) sulfur fugacity, as well as their roles in controlling gold precipitation and ore grade. Furthermore, magnetite from both ore types exhibits similarly low Ti concentrations (mostly <20 ppm, Table 2). Given that Ti incorporation into magnetite is highly sensitive and positive correlated with temperature [63,66], this consistent low-Ti signature confirms that both two types of magnetite crystallized under similar low-temperature conditions.

(1) Oxygen fugacity (fO_2)

Magnetite from marble-hosted ores contains significantly lower vanadium (V) concentrations (mostly <25 ppm) than that from granite-hosted ores, where V contents are typically concentrated between 40 and 70 ppm (Figure 8B). Because V is highly sensitive to the redox state of hydrothermal fluids, this compositional contrast directly reflects differences in the oxygen fugacity during magnetite formation [67–69]. In fluids, V mainly exists as V^{3+} and V^{4+}, among which V^{3+} can replace Fe^{3+} in magnetite through isomorphic substitution via exchange reactions [70,71]. As oxygen fugacity increases, V^{3+} in the fluid is oxidized to V^{4+}, resulting in less V entering the magnetite lattice [67]. Therefore, the lower V concentrations in marble-hosted magnetite indicate crystallization under more oxidizing conditions than granite-hosted magnetite. Thermodynamic models have demonstrated that fluid oxidation can occur due to different degrees of fluid-rock interaction, primarily through carbonation reactions, without requiring externally derived oxidizing fluids [72]. The elevated fO_2 in the marble-hosted system likely resulted from strong fluid–rock interactions with the Fe-rich marble rocks of the Jingshan Group, which formed during the Paleoproterozoic Lomagundi-Jatuli oxidation event [52]. Consequently, the elevated fO_2 in the marble-hosted system likely resulted from strong fluid–rock interactions with the Fe-rich marble rocks of the Jingshan Group. During carbonatization, Fe^{2+} released from these wall rocks was oxidized and precipitated as ankerite and magnetite, thereby increasing the oxygen fugacity of the fluid [73,74]. This interpretation is supported by the widespread ankerite and magnetite assemblages and pervasive carbonatization observed in marble-hosted ores (Figure 6C,D). The ore-forming fluids of the Guocheng deposit belong to the medium- to low-temperature reduced H_2O-CO_2-NaCl $\pm$ CH_4 system [26,27], consistent with other Jiaodong gold deposits [19]. In these deposits, gold is mainly transported as hydrosulfide complexes (AuHS(aq) and $Au(HS)_2^-$) [75,76]. Thermodynamic modeling results demonstrate that elevated oxygen fugacity in the ore-forming fluid would lead to the destabilization of Au–S complexes, reduce gold solubility, and consequently enhance gold precipitation efficiency [24,77]. Consequently, the elevated oxygen fugacity induced by carbonate–fluid interaction is a primary control on the formation of high-grade disseminated gold ores in marble-hosted zones.

(2) Sulfur fugacity (fS_2)

The concentrations of chalcophile elements such as Co, Ni, Sn, and Zn in magnetite are mainly governed by their partitioning behavior between sulfide and oxide phases [33]. In sulfur-bearing hydrothermal systems, these elements are highly compatible with sulfide minerals but incompatible with magnetite [78–80]. As sulfides begin to crystallize from the ore-forming fluid, they preferentially incorporate these elements, thereby depleting the coexisting magnetite in chalcophile components [81]. Consequently, the lower concentrations of chalcophile elements in magnetite from marble-hosted ores, compared with those from granite-hosted ores (Table 2, Figure 8C), indicate more extensive sulfide precipitation. Considering that both ore types were derived from the same ore-forming fluid, this trend suggests a decrease in sulfur fugacity (fS_2) during fluid–rock interaction with the marble wall rocks.

This process of a decrease in sulfur fugacity is further supported by the widespread pyrrhotite–magnetite mineral assemblage observed in marble-hosted ores (Figures 4A–C, 5D–F and 6C,D). Magnetite is an oxidizing mineral, whereas pyrrhotite is a reducing mineral, their coexistence commonly indicates that the ore-forming fluids experienced complex physicochemical evolution during gold mineralization [28,77,82]. Thermodynamic modeling shows that a decrease in total sulfur concentration and an increase in fluid pH can effectively expand the upper stability limit of pyrrhotite with respect to $logfO_2$, while the decline in sulfur fugacity can broaden the stability field of magnetite, allowing pyrrhotite and magnetite to co-precipitate [24,77]. The abundant occurrence of pyrrhotite in marble-hosted ores, in contrast to its scarcity in granite-hosted ores, suggests that during fluid passage through the Jingshan Group metamorphic wall rocks, the sulfur and H^+ in fluids were continuously consumed (Reactions (2) and (7)), causing sulfide precipitation to shift from pyrite (Reaction (8)) to pyrrhotite (Reaction (9)). This decrease in sulfur fugacity can significantly reduce gold solubility, thereby destabilizing Au–S complexes and promoting rapid precipitation (Figure 4C) [83,84]. This decrease in sulfur fugacity thus represents a key factor controlling the formation of high-grade marble-hosted ores.

$$CaCO_{3(Cal)} + 2H^+ = Ca^{2+} + CO_2 \uparrow + H_2O_{(aq)} \tag{6}$$

$$FeCl^+ + 2H_2S_{(aq)} = FeS_{2(Py)} + Cl^- + H_{2(aq)} + 2H^+ \tag{7}$$

$$FeCl^+ + H_2S_{(aq)} + 4H_2O_{(aq)} = FeS_{(Pyh)} + 2H^+ + Cl^- \tag{8}$$

In summary, comparative data on magnetite geochemistry and mineral assemblages indicates that marble-hosted ores formed under increasing oxygen fugacity and decreasing sulfur fugacity. This dual evolution drove large-scale co-precipitation of pyrrhotite and magnetite, while a decrease in dissolved sulfur and an increase in fluid oxidation state destabilized Au-S complexes (e.g., AuHS(aq) and $Au(HS)_2^-$), reducing gold solubility and promoting efficient gold precipitation [85], which ultimately formed the high-grade disseminated ores in the marble-hosted zones (Figure 9).

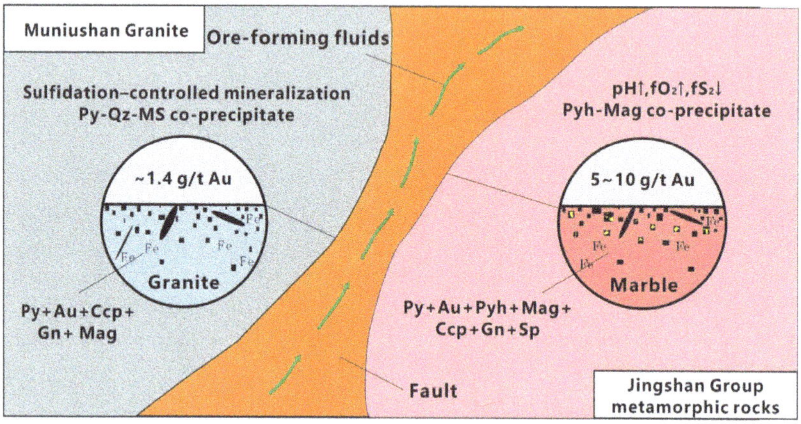

Figure 9. Ore-forming models of different host rocks in Guocheng gold deposit.

7. Conclusions

(1) Magnetite in the Guocheng gold deposit is of hydrothermal origin and was formed during the main mineralization stage. Petrographic observations indicate that most magnetite crystallized via dissolution–migration–recrystallization, while a minor proportion formed through in situ dissolution–recrystallization. The close association of all magnetite with gold and sulfides, together with their classification into the low-temperature hydrothermal category via Ga–V diagrams, indicates that they reliably record the ore-forming environment of gold precipitation.

(2) A systematic comparison of magnetite geochemistry between the two ore types reveals distinct mineralization environments. Specifically, the lower V and chalcophile element (Co, Ni, Sn, Zn) concentrations in marble-hosted magnetite, demonstrate that it crystallized under increasing oxygen fugacity (fO_2) and decreasing sulfur fugacity (fS_2) compared to its granite-hosted counterpart.

(3) The pronounced contrast in gold grades between the two ore types is closely related to the variations in fluid thermodynamic conditions induced by different fluid–rock reactions. In the marble-hosted ores, coupled carbonate dissolution and sulfide precipitation increased fO_2 and decreased fS_2, driving a mineralogical transition from pyrite to pyrrhotite–magnetite co-precipitation. This transition destabilized gold–sulfur complexes, enhanced gold precipitation efficiency, ultimately leading to the formation of high-grade disseminated ores.

Author Contributions: Conceptualization, T.C. and X.-F.W.; methodology, Q.F.; software, Y.-Y.L.; validation, L.-Y.Q. and R.-Z.Z.; formal analysis, N.-H.S. and Q.-J.D.; investigation, N.-H.S., T.C., L.-Y.Q., R.-Z.Z. and X.-F.W.; resources, T.C. and X.-F.W.; data curation, N.-H.S., Y.-Y.L. and R.-Z.Z.; writing—original draft preparation, N.-H.S.; writing—review and editing, T.C. and X.-F.W.; visualization, N.-H.S., Q.-J.D. and R.-Z.Z.; supervision, T.C. and X.-F.W.; project administration, T.C.; funding acquisition, T.C. and X.-F.W. All authors have read and agreed to the published version of the manuscript.

Funding: This research was funded by the National Key R&D Program of China (grant 2023YFC2906900); the Open Project of Technology Innovation Center for Deep Gold Resources Exploration and Mining, Ministry of Natural Resources (No. LDKF-2023BZX-04); the Joint Project of Ministry of Natural Resources and Shandong Province (grant 2024ZRBSHZ130); the Key Laboratory of Gold Mineralization Processes and Resource Utilization Subordinated to the Ministry of Natural Resources and Shandong Key Laboratory of Metallogenic Geological Process and Resources Utilization (grant KFKT202414);

the Technology Research Project of No. 6 Geological Team of Shandong Provincial Bureau of Geology and Mineral Resources (grant LDKJ20240101); the Frontiers Science Center for Deep-Time Digital Earth (grant 2652023001); the Taishan Scholar Program of Shandong (tstp20240847).

Data Availability Statement: The data presented in this study are available within the article.

Acknowledgments: We are grateful to Hao-Cheng Yu and Shan-Shan Li for their field guidance and constructive discussions. We are grateful to Peng-Yue Yu for the assistance with the sample preparation and analysis. We also deeply thank the anonymous reviewers and editors for their helpful comments and suggestions.

Conflicts of Interest: Authors Qiang Fu was employed by the BGRIMM Technology Group, Rong-Zhi Zhou was employed by the Shandong Gold Geological Mineral Exploration Co., Ltd., and Lian-Yuan Qin was employed by the Shandong Yantai Xintai Gold Mining Industry Co., Ltd. The remaining authors declare that the research was conducted in the absence of any commercial or financial relationships that could be construed as a potential conflict of interest.

References

1. Lunt, D.; Weeks, T. Process flowsheet selection. *Dev. Miner. Process.* **2005**, *15*, 73–96.
2. Gao, P.; Zhang, Y.-J.; Liu, Z.-P.; Wang, P.-F.; Li, J.-Y.; Li, S.-B. Polymetallic metallogenic law and mineral resources prediction in Shanghuangqi Banner area in North Hebei province. *Contrib. Geol. Miner. Resour. Res.* **2024**, *39*, 35–44.
3. Huang, H.-G.; Li, Z.; Liu, X.-L.; Zhang, C.-G.; Liu, L.-Q. Geological characteristics and genesis of Chafang gold deposit in Fanshi County, Shanxi Province. *Contrib. Geol. Miner. Resour. Res.* **2024**, *39*, 169–176.
4. Qiu, K.-F.; Yu, H.-C.; Deng, J.; McIntire, D.; Gou, Z.-Y.; Geng, J.-Z.; Chang, Z.-S.; Zhu, R.; Li, K.-N.; Goldfarb, R. The giant Zaozigou Au-Sb deposit in West Qinling, China: Magmatic-or metamorphic-hydrothermal origin? *Miner. Depos.* **2020**, *55*, 345–362. [CrossRef]
5. Petrella, L.; Thébaud, N.; Evans, K.; LaFlamme, C.; Occhipinti, S. The role of competitive fluid-rock interaction processes in the formation of high-grade gold deposits. *Geochim. Cosmochim. Acta* **2021**, *313*, 38–54. [CrossRef]
6. Herzog, M.; LaFlamme, C.; Beaudoin, G.; Barré, G.; Martin, L.; Savard, D. Fluid-rock sulfidation reactions control Au-Ag-Te-Bi precipitation in the Val-d'Or orogenic gold vein field (Abitibi subprovince, Canada). *Miner. Depos.* **2024**, *59*, 1039–1064. [CrossRef]
7. Farsang, S.; Zajacz, Z. Sulfur species and gold transport in arc magmatic fluids. *Nat. Geosci.* **2024**, *18*, 98–104. [CrossRef]
8. Lang, X.-C.; Han, X.-B.; Zhi, Y.-D.; Li, P.-L.; Wu, Q.-H. Geological characteristics and ore-forming mechanism of Au deposits in Kanggur area, East Tianshan mountainous district, Xinjiang Autonomous Region. *Contrib. Geol. Miner. Resour. Res.* **2024**, *39*, 177–185.
9. Zhu, Y.; An, F.; Tan, J. Geochemistry of hydrothermal gold deposits: A review. *Geosci. Front.* **2011**, *2*, 367–374. [CrossRef]
10. Liu, W.; Etschmann, B.; Testemale, D.; Hazemann, J.-L.; Rempel, K.; Müller, H.; Brugger, J. Gold transport in hydrothermal fluids: Competition among the Cl^-, Br^-, HS^- and NH_3 (aq) ligands. *Chem. Geol.* **2014**, *376*, 11–19. [CrossRef]
11. Li, H.; Wang, Q.; Weng, W.; Dong, C.; Yang, L.; Wang, X.; Deng, J. Co-precipitation of gold and base metal sulfides during fluid boiling triggered by fault-valve processes in orogenic gold deposits. *Ore Geol. Rev.* **2022**, *149*, 105090. [CrossRef]
12. Wang, J.-Q.; Han, J.-H. Study on ore material of Au-minerals in Au deposit in surrounding of Bieluwu Cu deposit in Sunituyue Banner, Inner Mongolia. *Contrib. Geol. Miner. Resour. Res.* **2025**, *40*, 27–32.
13. Qiu, K.-F.; Deng, J.; Yu, H.-C.; Rasbury, T.; Tang, Y.-W.; Zhu, R.; Zhang, P.-C.; Goldfarb, R. The Zaozigou orogenic gold-antimony deposit, West Qinling Orogen, China: Structural controls on multiple mineralization events. *Geol. Soc. Am. Bull.* **2024**, *136*, 4218–4232. [CrossRef]
14. Zhou, Y.; Wen, Z.; Liu, Y.; Wu, J.; Huang, B.; He, H.; Luo, Y.; Fan, P.; Wang, X.; Liu, X.; et al. The Contribution of Carbonaceous Material to Gold Mineralization in the Huangjindong Deposit, Central Jiangnan Orogen, China. *Minerals* **2024**, *14*, 1042. [CrossRef]
15. Gao, H.; Chen, J.; Leng, C.; Hu, Y.; Xie, H.; Li, Z. Ore-forming mechanism of Huxu Au-dominated polymetallic deposit in the Dongxiang Basin, South China: Constraints from in-situ trace elements and S–Pb isotopes of pyrite. *Acta Geochim.* **2024**, *43*, 1223–1240. [CrossRef]
16. Wang, J.-Y.; Zhang, L.; Yang, L.-Q.; Sun, S.-C.; Li, R.-H.; Gao, X.; Olin, P.; Shu, L.; Zhang, Q.-B.; Chen, X.-G.; et al. Ore-forming fluid evolution and gold precipitation mechanism at Huangjindong gold deposit, southern China: Insights from fluid inclusions and trace elements in quartz. *Ore Geol. Rev.* **2025**, *178*, 106444. [CrossRef]
17. Deng, J.; Yang, L.Q.; Li, R.H.; Groves, D.I.; Santosh, M.; Wang, Z.L.; Sai, S.X.; Wang, S.R.; Li, S. Regional structural control on the distribution of world-class gold deposits: An overview from the Giant Jiaodong Gold Province, China. *Geol. J.* **2018**, *54*, 378–391. [CrossRef]

18. He, D.-Y.; Qiu, K.-F.; Simon, A.C.; Pokrovski, G.S.; Yu, H.-C.; Connolly, J.A.; Li, S.-S.; Turner, S.; Wang, Q.-F.; Yang, M.-F.; et al. Mantle oxidation by sulfur drives the formation of giant gold deposits in subduction zones. *Proc. Natl. Acad. Sci. USA* **2024**, *121*, e2404731121. [CrossRef] [PubMed]
19. Deng, J.; Yang, L.-Q.; Groves, D.I.; Zhang, L.; Qiu, K.-F.; Wang, Q.-F. An integrated mineral system model for the gold deposits of the giant Jiaodong province, eastern China. *Earth Sci. Rev.* **2020**, *208*, 103274. [CrossRef]
20. Qiu, K.-F.; Goldfarb, R.J.; Deng, J.; Yu, H.-C.; Gou, Z.-Y.; Ding, Z.-J.; Wang, Z.-K.; Li, D.-P. Chapter 35: Gold Deposits of the Jiaodong Peninsula, Eastern China. In *Geology of the World's Major Gold Deposits and Provinces;* Society of Economic Geologists: Littleton, CO, USA, 2020; pp. 753–774.
21. Qiu, K.F.; Deng, J.; Sai, S.X.; Yu, H.C.; Tamer, M.T.; Ding, Z.J.; Yu, X.F.; Goldfarb, R. Low-temperature thermochronology for defining the tectonic controls on heterogeneous gold endowment across the Jiaodong peninsula, eastern China. *Tectonics* **2023**, *42*, e2022TC007669. [CrossRef]
22. Qiu, K.-F.; Romer, R.L.; Long, Z.-Y.; Yu, H.-C.; Turner, S.; Wan, R.-Q.; Li, X.-Q.; Gao, Z.-Y.; Deng, J. Potassium isotopes as a tracer of hydrothermal alteration in ore systems. *Geochim. Cosmochim. Acta* **2024**, *368*, 185–196. [CrossRef]
23. Zhang, J.-Y.; Qiu, K.-F.; Yin, R.; Long, Z.-Y.; Feng, Y.-C.; Yu, H.-C.; Gao, Z.-Y.; Deng, J. Lithospheric mantle as a metal storage reservoir for orogenic gold deposits in active continental margins: Evidence from Hg isotopes. *Geology* **2024**, *52*, 423–428. [CrossRef]
24. Fan, H.; Lan, T.; Li, X.; Santosh, M.; Yang, K.; Hu, F.; Feng, K.; Hu, H.; Peng, H.; Zhang, Y. Conditions and processes leading to large-scale gold deposition in the Jiaodong province, eastern China. *Sci. China Earth Sci.* **2021**, *64*, 1504–1523. [CrossRef]
25. Li, J.; Zhang, L.-P.; Li, C.-Y.; Jiang, M.-Y. Rb-Sr isochron age of the Guocheng gold deposit in the Jiaodong Peninsula, Shandong. *Geol. China* **2020**, *47*, 894–895.
26. Wu, J.; Chen, Y.; Zheng, C.; Li, H.; Yonezu, K.; Tang, Y.; Zong, Q. Genesis of the Longkou–Tudui gold deposit, Jiaodong Peninsula, eastern China: Constraints from zircon U-Pb dating, fluid inclusion studies and C–H–O–S stable isotopes. *Ore Geol. Rev.* **2021**, *139*, 104449. [CrossRef]
27. Cui, T.; Yu, H.-C.; Jowitt, S.M.; Li, Z.-S.; Wei, X.-L.; Zou, S.-H.; Bo, J.-W.; Pei, L.; Li, S.-S.; Nassif, M.T.; et al. Chlorite composition records the formation of increased oxygen fugacity-induced disseminated gold mineralization within the Guocheng gold deposit, China. *Geol. Soc. Am. Bull.* **2025**, *137*, 4786–4800. [CrossRef]
28. Yang, L.-Q.; Deng, J.; Wang, Z.-L.; Guo, L.-N.; Li, R.-H.; Groves, D.I.; Danyushevsky, L.V.; Zhang, C.; Zheng, X.-L.; Zhao, H. Relationships between gold and pyrite at the Xincheng gold deposit, Jiaodong Peninsula, China: Implications for gold source and deposition in a brittle epizonal environment. *Econ. Geol.* **2016**, *111*, 105–126. [CrossRef]
29. Deng, J.; Qiu, K.-F.; Wang, Q.-F.; Goldfarb, R.; Yang, L.-Q.; Zi, J.-W.; Geng, J.-Z.; Ma, Y. In situ dating of hydrothermal monazite and implications for the geodynamic controls on ore formation in the Jiaodong gold province, eastern China. *Econ. Geol.* **2020**, *115*, 671–685. [CrossRef]
30. Zhang, L.; Weinberg, R.F.; Yang, L.-Q.; Groves, D.I.; Sai, S.-X.; Matchan, E.; Phillips, D.; Kohn, B.P.; Miggins, D.P.; Liu, Y. Mesozoic orogenic gold mineralization in the Jiaodong Peninsula, China: A focused event at 120 ± 2 Ma during cooling of pregold granite intrusions. *Econ. Geol.* **2020**, *115*, 415–441. [CrossRef]
31. Tan, J.; Wei, J.; Li, Y.; Fu, L.; Li, H.; Shi, W.; Tian, N. Origin and geodynamic significance of fault-hosted massive sulfide gold deposits from the Guocheng–Liaoshang metallogenic belt, eastern Jiaodong Peninsula: Rb–Sr dating, and H–O–S–Pb isotopic constraints. *Ore Geol. Rev.* **2015**, *65*, 687–700. [CrossRef]
32. Bo, J.-W.; Ding, Z.-J.; Deng, J.; Sun, F.-Y.; Qiu, K.-F.; Song, M.-C. Metallogenic characteristics and mineralization of supergiant gold deposits (223 t) in the northeastern margin of the Jiaolai Basin, Shandong Province: A review. *China Geol.* **2025**, *8*, 806–842.
33. Dare, S.A.S.; Barnes, S.-J.; Beaudoin, G.; Méric, J.; Boutroy, E.; Potvin-Doucet, C. Trace elements in magnetite as petrogenetic indicators. *Miner. Depos.* **2014**, *49*, 785–796. [CrossRef]
34. Huang, X.-W.; Sappin, A.-A.; Boutroy, É.; Beaudoin, G.; Makvandi, S. Trace Element Composition of Igneous and Hydrothermal Magnetite from Porphyry Deposits: Relationship to Deposit Subtypes and Magmatic Affinity. *Econ. Geol.* **2019**, *114*, 917–952. [CrossRef]
35. Zeng, L.-P.; Zhao, X.-F.; Spandler, C.; Hu, H.; Hu, B.; Li, J.-W.; Hu, Y. Origin of high-Ti magnetite in magmatic-hydrothermal systems: Evidence from iron oxide-apatite (IOA) deposits of eastern China. *Econ. Geol.* **2022**, *117*, 923–942.
36. Yang, J.; Qiu, Y. Comparative study on mineral dissemination characteristics of phosphate ores by X-ray micro computed tomography and BGRIMM Process Mineralogy Analysis. *Sci. Rep.* **2022**, *12*, 21122. [CrossRef]
37. Li, Y.; Kou, J.; Sun, C.; Wang, P.; Wang, X. Enhancing Copper Leaching from Refractory Copper Oxide Ore Using Organic Cationic Surfactant. *Separations* **2025**, *12*, 212. [CrossRef]
38. Wen, L.-G.; Jia, M.-X.; Wang, Q.; Fu, Q.; Zhao, J.-J. A new SEM-based automated mineralogy system: BPMA and its application prospects in mining industry. *Nonferrous Met.* **2021**, *2*, 12–23.
39. Feng, Y.-C.; Deng, J.; Yu, H.-C.; Fu, Q.; Cui, T.; Shan, W.; Li, Z.-S.; Li, S.-S. Gold occurrence and its indicative significance to mineralization process in Linglong gold district, Jiaodong gold province. *Acta Petrol. Sin.* **2023**, *39*, 377–392. [CrossRef]

40. Wang, L.; Qiu, Y.; McNaughton, N.; Groves, D.; Luo, Z.; Huang, J.; Miao, L.; Liu, Y. Constraints on crustal evolution and gold metallogeny in the Northwestern Jiaodong Peninsula, China, from SHRIMP U–Pb zircon studies of granitoids. *Ore Geol. Rev.* **1998**, *13*, 275–291. [CrossRef]
41. Tan, J.; Wei, J.; He, H.; Su, F.; Li, Y.; Fu, L.; Zhao, S.; Xiao, G.; Zhang, F.; Xu, J.; et al. Noble gases in pyrites from the Guocheng-Liaoshang gold belt in the Jiaodong province: Evidence for a mantle source of gold. *Chem. Geol.* **2018**, *480*, 105–115. [CrossRef]
42. Qiu, K.-F.; Romer, R.L.; Long, Z.-Y.; Williams-Jones, A.E.; Yu, H.-C.; Turner, S.; Wang, Q.-F.; Li, S.-S.; Zhang, J.-Y.; Duan, H.-R.; et al. The role of an oxidized lithospheric mantle in gold mobilization. *Sci. Adv.* **2024**, *10*, eado6262. [CrossRef] [PubMed]
43. Goldfarb, R.J.; Santosh, M. The dilemma of the Jiaodong gold deposits: Are they unique? *Geosci. Front.* **2014**, *5*, 139–153. [CrossRef]
44. Zheng, Y.; Xu, Z.; Zhao, Z.; Dai, L. Mesozoic mafic magmatism in North China: Implications for thinning and destruction of cratonic lithosphere. *Sci. China Earth Sci.* **2018**, *61*, 353–385. [CrossRef]
45. Zhu, R.; Xu, Y. The subduction of the west Pacific plate and the destruction of the North China Craton. *Sci. China Earth Sci.* **2019**, *62*, 1340–1350. [CrossRef]
46. Wu, F.-Y.; Yang, J.-H.; Xu, Y.-G.; Wilde, S.A.; Walker, R.J. Destruction of the North China Craton in the Mesozoic. *Annu. Rev. Earth Planet. Sci.* **2019**, *47*, 173–195. [CrossRef]
47. Zhang, J.-H.; Gao, S.; Ge, W.-C.; Wu, F.-Y.; Yang, J.-H.; Wilde, S.A.; Li, M. Geochronology of the Mesozoic volcanic rocks in the Great Xing'an Range, northeastern China: Implications for subduction-induced delamination. *Chem. Geol.* **2010**, *276*, 144–165. [CrossRef]
48. Hacker, B.R.; Wallis, S.R.; McWilliams, M.O.; Gans, P.B. 40Ar/39Ar Constraints on the tectonic history and architecture of the ultrahigh-pressure Sulu orogen. *J. Metamorph. Geol.* **2009**, *27*, 827–844. [CrossRef]
49. Goldfarb, R.J.; Groves, D.I.; Gardoll, S. Orogenic gold and geologic time: A global synthesis. *Ore Geol. Rev.* **2001**, *18*, 1–75. [CrossRef]
50. Qiu, Y.; Groves, D.I.; McNaughton, N.J.; Wang, L.-G.; Zhou, T. Nature, age, and tectonic setting of granitoid-hosted, orogenic gold deposits of the Jiaodong Peninsula, eastern North China craton, China. *Miner. Depos.* **2002**, *37*, 283–305. [CrossRef]
51. Tan, J.; Wei, J.; Audétat, A.; Pettke, T. Source of metals in the Guocheng gold deposit, Jiaodong Peninsula, North China Craton: Link to early Cretaceous mafic magmatism originating from Paleoproterozoic metasomatized lithospheric mantle. *Ore Geol. Rev.* **2012**, *48*, 70–87. [CrossRef]
52. Chen, Y.; Li, H.; Gu, S.; Lash, G.G.; Zheng, C.; Chen, S.; Li, D.; Elatikpo, S.M.; Duan, L. Marble trace element and C-O isotope geochemistry of the Paleoproterozoic Jingshan Group, North China: Insights into BIF formation during the Lomagundi-Jatuli Event. *Precambrian Res.* **2023**, *395*, 107152. [CrossRef]
53. Tan, J.; Wei, J.; Guo, L.; Zhang, K.; Yao, C.; Lu, J.; Li, H. LA-ICP-MS zircon U-Pb dating and phenocryst EPMA of dikes, Guocheng, Jiaodong Peninsula: Implications for North China Craton lithosphere evolution. *Sci. China Ser. D Earth Sci.* **2008**, *51*, 1483–1500. [CrossRef]
54. Nadoll, P.; Koenig, A.E. LA-ICP-MS of magnetite: Methods and reference materials. *J. Anal. At. Spectrom.* **2011**, *26*, 1872–1877. [CrossRef]
55. Zhang, L.; Qiu, K.; Hou, Z.; Pirajno, F.; Shivute, E.; Cai, Y. Fluid-rock reactions of the Triassic Taiyangshan porphyry Cu-Mo deposit (West Qinling, China) constrained by QEMSCAN and iron isotope. *Ore Geol. Rev.* **2021**, *132*, 104068. [CrossRef]
56. Blott, S.J.; Pye, K. Particle size scales and classification of sediment types based on particle size distributions: Review and recommended procedures. *Sedimentology* **2012**, *59*, 2071–2096. [CrossRef]
57. Reich, M.; Deditius, A.; Chryssoulis, S.; Li, J.-W.; Ma, C.-Q.; Parada, M.A.; Barra, F.; Mittermayr, F. Pyrite as a record of hydrothermal fluid evolution in a porphyry copper system: A SIMS/EMPA trace element study. *Geochim. Cosmochim. Acta* **2013**, *104*, 42–62. [CrossRef]
58. Palma, G.; Reich, M.; Barra, F.; Ovalle, J.T.; Del Real, I.; Simon, A.C. Thermal evolution of Andean iron oxide–apatite (IOA) deposits as revealed by magnetite thermometry. *Sci. Rep.* **2021**, *11*, 18424. [CrossRef]
59. Wen, G.; Li, J.-W.; Hofstra, A.H.; Koenig, A.E.; Lowers, H.A.; Adams, D. Hydrothermal reequilibration of igneous magnetite in altered granitic plutons and its implications for magnetite classification schemes: Insights from the Handan-Xingtai iron district, North China Craton. *Geochim. Cosmochim. Acta* **2017**, *213*, 255–270. [CrossRef]
60. Knipping, J.L.; Bilenker, L.D.; Simon, A.C.; Reich, M.; Barra, F.; Deditius, A.P.; Wälle, M.; Heinrich, C.A.; Holtz, F.; Munizaga, R. Trace elements in magnetite from massive iron oxide-apatite deposits indicate a combined formation by igneous and magmatic-hydrothermal processes. *Geochim. Cosmochim. Acta* **2015**, *171*, 15–38. [CrossRef]
61. Hassanpour, S.; Rajabpour, S. Magmatic–hydrothermal evolution of the Anjerd Cu skarn deposit, NW Iran: Perspectives on mineral chemistry, fluid inclusions and stable isotopes. *Ore Geol. Rev.* **2020**, *117*, 103269. [CrossRef]

62. Anderson, J.L.; Barth, A.P.; Wooden, J.L.; Mazdab, F. Thermometers and Thermobarometers in Granitic Systems. *Rev. Mineral. Geochem.* **2008**, *69*, 121–142. [CrossRef]
63. Nadoll, P.; Angerer, T.; Mauk, J.L.; French, D.; Walshe, J. The chemistry of hydrothermal magnetite: A review. *Ore Geol. Rev.* **2014**, *61*, 1–32. [CrossRef]
64. Schirra, M.; Zajacz, Z.; Müller, D. Magnetite geochemistry of giant alkalic-type epithermal gold deposits: Insights into the magmatic evolution of mineralized alkaline systems. *Geol. Soc. Lond. Spec. Publ.* **2024**, *551*, 385–408. [CrossRef]
65. Hu, H.; Li, J.-W.; Lentz, D.; Ren, Z.; Zhao, X.-F.; Deng, X.-D.; Hall, D. Dissolution–reprecipitation process of magnetite from the Chengchao iron deposit: Insights into ore genesis and implication for in-situ chemical analysis of magnetite. *Ore Geol. Rev.* **2014**, *57*, 393–405. [CrossRef]
66. Toplis, M.; Carroll, M. An experimental study of the influence of oxygen fugacity on Fe-Ti oxide stability, phase relations, and mineral—Melt equilibria in ferro-basaltic systems. *J. Petrol.* **1995**, *36*, 1137–1170. [CrossRef]
67. Dupuis, C.; Beaudoin, G. Discriminant diagrams for iron oxide trace element fingerprinting of mineral deposit types. *Miner. Depos.* **2011**, *46*, 319–335. [CrossRef]
68. Toplis, M.J.; Corgne, A. An experimental study of element partitioning between magnetite, clinopyroxene and iron-bearing silicate liquids with particular emphasis on vanadium. *Contrib. Mineral. Petrol.* **2002**, *144*, 22–37. [CrossRef]
69. Bordage, A.; Balan, E.; de Villiers, J.P.R.; Cromarty, R.; Juhin, A.; Carvallo, C.; Calas, G.; Sunder Raju, P.V.; Glatzel, P. V oxidation state in Fe–Ti oxides by high-energy resolution fluorescence-detected X-ray absorption spectroscopy. *Phys. Chem. Miner.* **2011**, *38*, 449–458. [CrossRef]
70. Reguir, E.P.; Chakhmouradian, A.R.; Halden, N.M.; Yang, P.; Zaitsev, A.N. Early Magmatic and Reaction-Induced Trends in Magnetite from the Carbonatites of Kerimasi, Tanzania. *Can. Mineral.* **2008**, *46*, 879–900. [CrossRef]
71. Righter, K.; Leeman, W.P.; Hervig, R.L. Partitioning of Ni, Co and V between spinel-structured oxides and silicate melts: Importance of spinel composition. *Chem. Geol.* **2006**, *227*, 1–25. [CrossRef]
72. Evans, K.A. A test of the viability of fluid–wall rock interaction mechanisms for changes in opaque phase assemblage in metasedimentary rocks in the Kambalda-St. Ives goldfield, Western Australia. *Miner. Depos.* **2009**, *45*, 207–213. [CrossRef]
73. Hodkiewicz, P.F.; Groves, D.I.; Davidson, G.J.; Weinberg, R.F.; Hagemann, S.G. Influence of structural setting on sulphur isotopes in Archean orogenic gold deposits, Eastern Goldfields Province, Yilgarn, Western Australia. *Miner. Depos.* **2008**, *44*, 129–150. [CrossRef]
74. Feng, K.; Fan, H.-R.; Ulrich, T.; Yang, K.-F.; Hu, F.-F.; Liu, X. Contribution of Precambrian basements to the Mesozoic ore-fluid system: An illustration using the Majiayao gold deposit, Jiaodong, China. *Ore Geol. Rev.* **2021**, *139*, 104447. [CrossRef]
75. Pokrovski, G.S.; Akinfiev, N.N.; Borisova, A.Y.; Zotov, A.V.; Kouzmanov, K. Gold speciation and transport in geological fluids: Insights from experiments and physical-chemical modelling. *Geol. Soc. Lond. Spec. Publ.* **2014**, *402*, 9–70. [CrossRef]
76. Williams-Jones, A.E.; Bowell, R.J.; Migdisov, A.A. Gold in Solution. *Elements* **2009**, *5*, 281–287. [CrossRef]
77. Hu, H.-L.; Fan, H.-R.; Liu, X.; Cai, Y.-C.; Yang, K.-F.; Ma, W.-D. Two-stage gold deposition in response to H_2S loss from a single fluid in the Sizhuang deposit (Jiaodong, China). *Ore Geol. Rev.* **2020**, *120*, 103450. [CrossRef]
78. Wang, Y.; Lai, J.; Cao, Y.; Brzozowski, M. Compositional variations of magnetite in different sulfide ore types in the Jinchuan Ni-Cu-PGE sulfide deposit, NW China: Insights into the mineralizing processes of conduit-type systems. *Chem. Geol.* **2023**, *637*, 121679. [CrossRef]
79. Wen, G.; Li, J.-W.; Hofstra, A.H.; Harlov, D.E.; Zhao, X.-F.; Lowers, H.A.; Koenig, A.E. Trace element fractionation in magnetite as a function of Fe depletion from ore fluids at the Baijian Fe-(Co) skarn deposit, eastern China: Implications for Co mineralization in Fe skarns. *Am. Mineral.* **2024**, *109*, 1657–1669. [CrossRef]
80. Simpson, B.; Fitzherbert, J.; Moltzen, J.; Baillie, I.; Cox, B.; Huang, H. Magnetite trace element characteristics and their use as a proximity indicator to the Avoca Tank Cu-Au prospect, Girilambone copper province, New South Wales, Australia. *Miner. Depos.* **2024**, *59*, 169–187. [CrossRef]
81. Duran, C.J.; Barnes, S.-J.; Mansur, E.T.; Dare, S.A.; Bédard, L.P.; Sluzhenikin, S.F. Magnetite chemistry by LA-ICP-MS records sulfide fractional crystallization in massive nickel-copper-platinum group element ores from the Norilsk-Talnakh mining district (Siberia, Russia): Implications for trace element partitioning into magnetite. *Econ. Geol.* **2020**, *115*, 1245–1266. [CrossRef]
82. Wen, B.-J.; Fan, H.-R.; Hu, F.-F.; Liu, X.; Yang, K.-F.; Sun, Z.-F.; Sun, Z.-F. Fluid evolution and ore genesis of the giant Sanshandao gold deposit, Jiaodong gold province, China: Constrains from geology, fluid inclusions and H–O–S–He–Ar isotopic compositions. *J. Geochem. Explor.* **2016**, *171*, 96–112. [CrossRef]
83. Hayashi, K.-I.; Ohmoto, H. Solubility of gold in NaCl-and H_2S-bearing aqueous solutions at 250–350 °C. *Geochim. Cosmochim. Acta* **1991**, *55*, 2111–2126. [CrossRef]

84. Gibert, F.; Pascal, M.-L.; Pichavant, M. Gold solubility and speciation in hydrothermal solutions: Experimental study of the stability of hydrosulphide complex of gold (AuHS) at 350 to 450 C and 500 bars. *Geochim. Cosmochim. Acta* **1998**, *62*, 2931–2947. [CrossRef]
85. Pokrovski, G.S.; Kokh, M.A.; Guillaume, D.; Borisova, A.Y.; Gisquet, P.; Hazemann, J.-L.; Lahera, E.; Del Net, W.; Proux, O.; Testemale, D.; et al. Sulfur radical species form gold deposits on Earth. *Proc. Natl. Acad. Sci. USA* **2015**, *112*, 13484–13489. [CrossRef] [PubMed]

minerals

MDPI

Article

Fluorite Composition Constraints on the Genesis of the Weishan REE Deposit, Luxi Terrane

Yi-Xue Gao [1], Shan-Shan Li [1,2,*], Chuan-Peng Liu [3,*], Ming-Qian Wu [1,2], Zhen Shang [3], Ze-Yu Yang [1], Xin-Yi Wang [1] and Kun-Feng Qiu [1]

[1] State Key Laboratory of Geological Processes and Mineral Resources, Frontiers Science Center for Deep-Time Digital Earth, School of Earth Sciences and Resources, China University of Geosciences, Beijing 100083, China; gaoyixue2020@email.cugb.edu.cn (Y.-X.G.); aria.wu@uwindsor.ca (M.-Q.W.); zeyuyang@email.cugb.edu.cn (Z.-Y.Y.); h_ai110@126.com (X.-Y.W.); kunfengqiu@cugb.edu.cn (K.-F.Q.)

[2] Shandong Key Laboratory of Mineralization Processes and Resources Utilization of Strategic Metal Minerals (Preparatory), Key Laboratory of Gold Mineralization Processes and Resource Utilization Subordinated to the Ministry of Land and Resources, Shandong Institute of Geological Sciences, Jinan 250013, China

[3] Shandong Provincial Lunan Geology and Exploration Institute (Shandong Provincial Bureau of Geology and Mineral Resources No. 2 Geological Brigade), Rare Mineral Exploration and Comprehensive Utilization Engineering Research Center of Shandong Province, Jining 272100, China; shangxiaohuai@yeah.net

* Correspondence: lishanshan199811@163.com (S.-S.L.); liuchuanpeng666@163.com (C.-P.L.); Tel.: +86-180-3218-6796 (S.-S.L.); +86-138-6991-2915 (C.-P.L.)

Abstract

Fluorite, a key accessory mineral associated with rare earth element (REE) deposits, exerts a significant influence on REE migration and precipitation through complexation, adsorption, and lattice substitution within fluorine-bearing fluid systems. It therefore provides a valuable archive for constraining REE enrichment processes. The Weishan alkaline–carbonatite-related REE deposit, the third-largest LREE deposit in China, is formed through a multistage magmatic–hydrothermal evolution of the carbonatite system. However, limited mineralogical constraints on REE enrichment and precipitation have hindered a comprehensive understanding of its metallogenic processes and exploration potential. Here, cathodoluminescence imaging and LA-ICP-MS trace element analyses were conducted on fluorite of multiple generations from the Weishan deposit to constrain the physicochemical conditions of mobility and precipitation mechanisms of this REE deposit. Four generations of fluorite are recognized, recording progressive evolution of the ore-forming fluids. Type I fluorite, which coexists with bastnäsite and calcite, is LREE-enriched and exhibits negative Eu anomalies, indicating precipitation from high-temperature, weakly acidic, and reducing fluids. Type II fluorite occurs as overgrowths on Type I, while Type III fluorite replaces Type II fluorite, with both displaying LREE depletion and MREE-Y enrichment, consistent with cooling during continued hydrothermal evolution. Type IV fluorite, which is interstitial between calcite grains and associated with mica, is formed under low-temperature, oxidizing conditions, reflecting REE exhaustion and the terminal stage of fluorite precipitation. Systematic shifts in REE patterns among the four generations track progressive cooling of the system. The decreasing trend in La/Ho and Tb/La further suggests that these fluorites record dissolution–reprecipitation events and associated element remobilization during fluid evolution.

Keywords: Weishan REE deposit; precipitation mechanism of fluorite; transportation of REE; ore-forming process; Luxi Terrane

1. Introduction

Rare earth elements (REEs) constitute a strategic class of critical metals, and elucidating the controls on their origin, spatial distribution, and ore-forming processes remains a foremost frontier in contemporary geological research [1–6]. REE deposits form in a wide variety of geological settings, including carbonatites, alkaline–peralkaline igneous complexes, hydrothermal systems, and weathering-related ion-adsorption deposits [3,6–13]. These deposits commonly contain substantial coexisting resources of fluorite, apatite, and other mineral [14–17]. Fluorite is a ubiquitous vein-forming mineral that occurs widely in magmatic–hydrothermal, sedimentary–diagenetic, and metamorphic systems, and it is also frequently developed in critical metal deposits overprinted by hydrothermal activity. As a key geochemical carrier of REEs during mineralization, fluorite exerts strong control on fluid–mineral interactions and on the enrichment and transport of rare earth elements. Globally, fluorite-bearing REE deposits are diverse. Typical examples include (i) carbonatite-type REE deposits, such as the Bayan Obo deposit in Inner Mongolia, the Mianning-Dechang REE belt in Sichuan, and the Weishan REE deposit in Shan-dong, China, as well as the Kızılcaören REE deposit in Turkey and the Okorusu REE deposit in Namibia [16,18–22], and (ii) alkaline–peralkaline granite-related REE deposits, as exemplified by the deposit of the Gallinas Mountains in the United States and the Strange Lake deposit in Canada [23–27].

REE deposits may originate from magmatic, hydrothermal, or supergene processes [13,28–32], and the fluorite commonly associated with such deposits is typically closely linked to hydrothermal fluids. These fluids may derive directly from magmatic systems or be related to sedimentary architectures [23,29,33–36]. It is known that REEs in the Earth's crust predominantly occur as trivalent ions (REE^{3+}), which are characterized by large ionic radii and lithophile behavior. Their mobility in diverse geological media is significantly influenced by coordinating ligands, temperature, pH, and redox conditions. Within magmatic and hydrothermal systems, F-rich fluids are capable of effectively complexing, transporting, and concentrating REEs to economically significant levels [37–40]. Consequently, fluorite is not only an important mineral in REE mineralization but also a valuable geochemical indicator for deciphering the evolution and metallogenic mechanisms of magmatic–hydrothermal systems, and it is widely employed as a geochemical exploration tool. In F-bearing fluid systems, fluorite influences the migration and precipitation of REEs through complexation, adsorption, and lattice substitution [33,41–44]. Its coupled behavior with REEs is reflected not only in elemental abundances and distribution patterns but also in the thermodynamic and kinetic evolution of ore-forming systems. As ore-forming fluids evolve from high-temperature magmatic regimes to low-temperature hydrothermal environments, the precipitation of fluorite frequently marks a critical stage during which REEs transition from the fluid phase to the solid phase [18,45,46]. The Ca^{2+} in the fluorite lattice can be substituted by REEs, such as Sr^{2+}, Na^{+} Y^{3+}, and other ions, rendering fluorite highly sensitive to REE uptake and release under varying geological conditions [33,47,48]. In settings characterized by multiphase hydrothermal activity, fluorite may serve as both a reservoir for REEs and as a medium for REE redistribution during recrystallization or reprecipitation. Therefore, elucidating the role of fluorite in modulating REE mineralization is of fundamental significance for reconstructing the evolutionary pathways of fluid-mediated ore-forming systems, understanding the mechanisms of elemental migration, and constraining the physicochemical conditions of deposit formation.

https://doi.org/10.3390/min16010069

The Weishan REE deposit is the third-largest light rare earth deposit in China [22], with an average grade of 3.25 wt% and total resources exceeding 1.0 Mt. Its genesis is closely associated with an alkaline–carbonatite complex [13,29]. Although previous studies have made significant progress in elucidating the petrogenesis of the alkaline rocks, the isotopic characteristics, the geochronological framework of magmatism and mineralization, and the sources of ore-forming fluids, a systematic understanding of the mechanisms governing REE enrichment and migration remains lacking. As one of the pervasive minerals in the system, fluorite represents an ideal carrier for investigating the geochemical behavior of REEs in carbonatite. Based on detailed field investigations and comprehensive laboratory analyses, this study focuses on fluorite–bastnäsite assemblages developed within the Weishan carbonatite. By integrating petrographic observations with geochemical data from fluorite, we aim to elucidate the precipitation mechanisms of fluorite, the physicochemical conditions governing REE mineralization, and the implications of ore-forming processes. The results are intended to provide new mineralogical constraints on the metallogenic mechanism of the Weishan REE deposit in the Luxi Terrane and to advance the understanding of REE mineralization.

2. Regional Geology

The North China Block (NCB) is one of the oldest and most well-preserved cratonic blocks in the world (Figure 1A). Its geological evolution involved the sequential development of an early continental core, crustal growth, amalgamation of micro-continental blocks, and extensive magmatic, metamorphic, and cratonization processes [49–56]. Subsequently, the NCB underwent multiple episodes of subduction-related metasomatism and lithospheric thinning, accompanied by multiphase, large-scale metallogenic events that produced a wide range of strategic mineral resources, including iron, copper, gold, and rare metals [13,29,57–65]. The Luxi Terrane, situated along the southeastern margin of the NCB (Figure 1B), is structurally bounded by four major fault zones: the Fengpei, Qihe–Guangrao, Liaocheng–Lankao, and Yishui fault zones. The area is characterized by multiple phases of deformation and repeated magmatic activity. Exposed strata include the Archean Yishui Group, the Neoarchean Taishan Group, Cambrian–Ordovician carbonate and clastic sequences, Carboniferous–Permian units, Triassic strata, Jurassic–Cretaceous clastic and volcanic rocks, and Quaternary deposits. Intense tectonic activity in the Luxi Terrane has produced a well-developed array of planar and linear structures, including numerous ductile shear zones. Multiple sets of faults with varying orientations occur, among which NW- and NE-trending systems are the most prominent. These regional faults exert primary control over metallogenesis and Mesozoic (Yanshanian) magmatism. The Neoarchean magmatic suite comprises gabbro, diorite, and granodiorite, whereas Mesozoic magmatism is dominated by syenite porphyry, monzonitic porphyry, granite, and carbonatite. These intrusive bodies display diverse morphological characteristics and occur as plutons, stocks, and dikes. Their spatial and temporal distribution is closely linked to the regional fault architecture, reflecting long-standing and vigorous deep-seated tectono-magmatic processes beneath the Luxi Terrane.

 https://doi.org/10.3390/min16010069

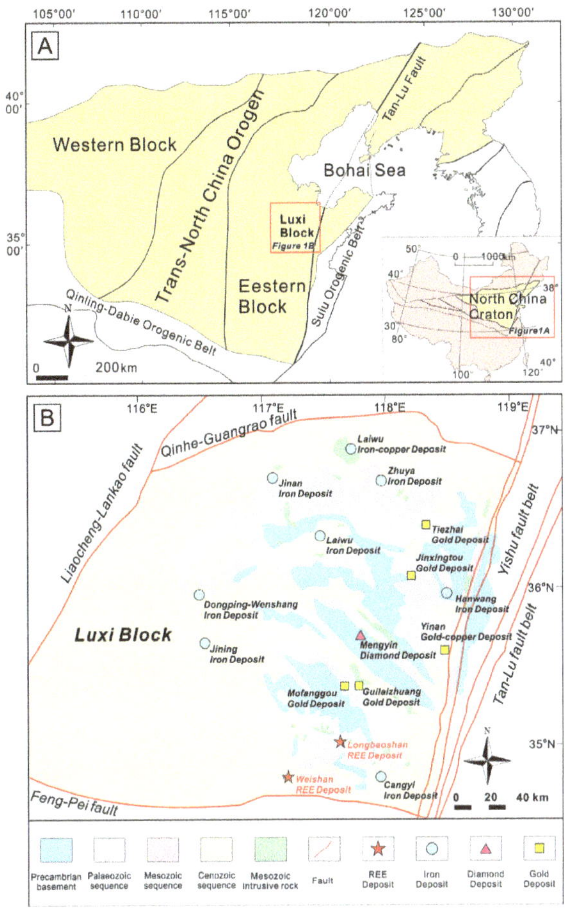

Figure 1. (**A**) Schematic map showing the position of the Luxi Block (modified after [66]). (**B**) Simplified geologic map showing the location of the Weishan area in the Luxi Terrane (modified after [21]).

3. Deposit Geology

The Weishan REE deposit is located in the southern margin of the Luxi Terrane (Figure 1B) and represents the largest REE deposit in the region, as well as the third-largest light rare earth production base in China. The exposed basement rocks within the mining area consist of biotite plagioclase gneisses of the Archean Taishan Group (Shancaoyu Formation). Apart from limited outcrops along local topographic highs, most of the basement is overlain by Quaternary sediments (Figure 2). The distributions of Mesozoic intrusive rocks and carbonatite are predominantly controlled by NW- and NE-trending faults. Magmatic activity in the area is well developed and dominated by Yanshanian syenite (quartz syenite and aegirine–augite quartz syenite) and alkaline granites, collectively forming the Weishan alkaline complex (Figure 2). The exposed area of the complex is approximately 0.5 km^2. It extends in a NE-SW orientation and intrudes the Archean gneisses along sharp contacts, with local occurrences of variable degrees of alkaline metasomatism along the contacts. The carbonatite intrudes along structures within both the Archean metamorphic basement and the Mesozoic alkaline intrusions. Their contacts with the host rocks are well defined. Individual orebodies typically range from 20 to 540 m in length and from 10 cm to 10 m in width, exhibiting relatively continuous mineralization. Based on mineral assemblages, earlier studies subdivided the

 https://doi.org/10.3390/min16010069

carbonatite of the Weishan deposit into REE-bearing quartz–barite–carbonatite veins, REE-bearing aegirine granite porphyry, aegirine-bearing veins, and monazite-bearing apatite veins [21,22]. The dominant REE minerals are bastnäsite-(Ce) and synchysite-(Ce), with minor occurrences of parisite-(Ce), strontianite phases, alkali REE-bearing phase, ancylite-(Ce), monazite, and apatite-(Ce). These REE minerals commonly coexist with quartz, barite, fluorite, calcite, and mica [67]. The REE ore is characterized by diverse and spatially heterogeneous mineral assemblages. In addition to the principal REE minerals and associated gangue minerals, metallic minerals such as pyrite, galena, and magnetite also occur. Other minerals present include chlorite, arfvedsonite, ankerite, titanite, and rutile.

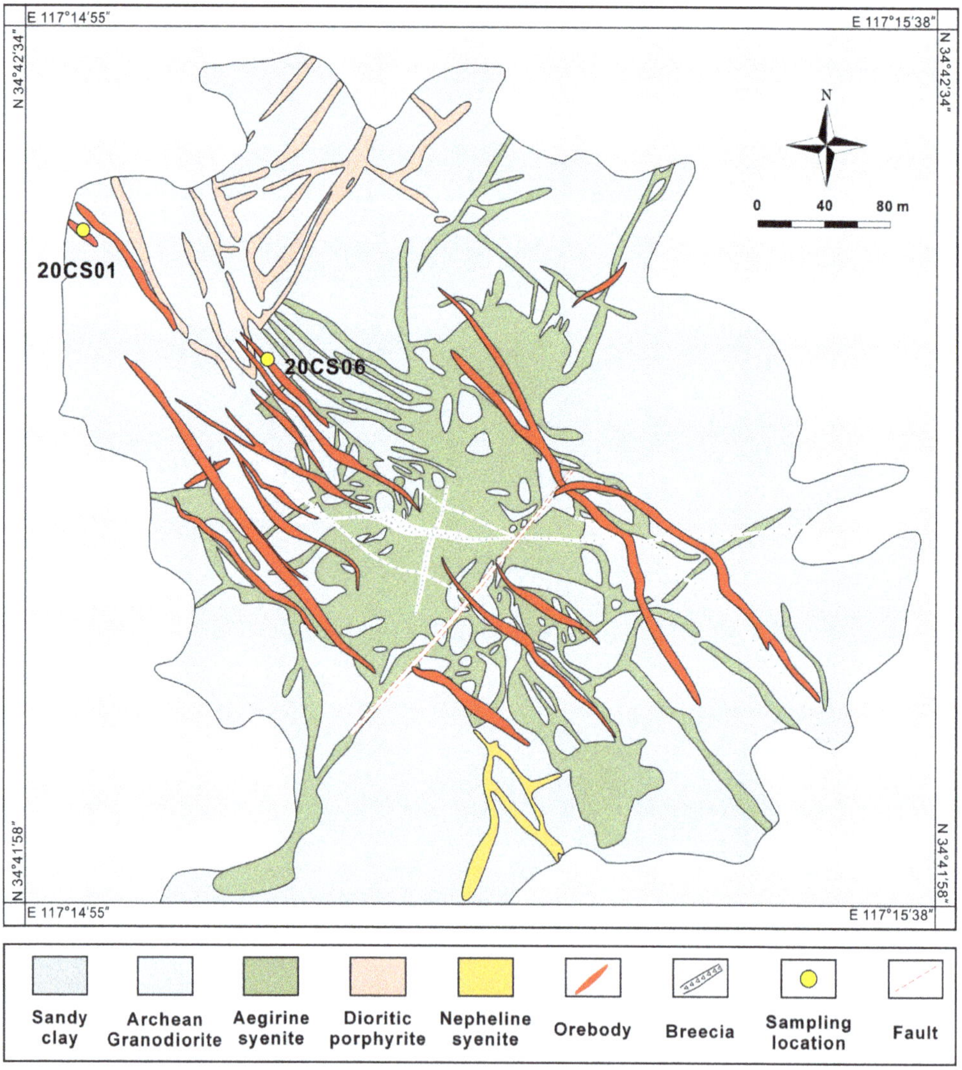

Figure 2. Simplified geological map of the Weishan REE deposit (modified after [21]).

https://doi.org/10.3390/min16010069

4. Samples and Analytical Methods

4.1. Sample Descriptions

The fluorite samples investigated in this study were collected from the REE-bearing barite–carbonatite veins of the Weishan REE deposit (Figure 3). Sample 20CS01, consisting of calcite–bastnäsite–fluorite assemblages, contains the REE minerals bastnäsite-(Ce) and synchysite-(Ce), accompanied by gangue minerals such as calcite and barite. The bastnäsite-(Ce) mineral exists as prismatic, dark brown crystals with grain sizes of 1–5 mm, whereas synchysite-(Ce) appears as dark green grains that replace bastnäsite-(Ce). The fluorite mineral is purple, displays a pegmatitic texture, and coexists with calcite, with grain sizes ranging from 1 to 3 mm. In sample 20CS06, which is composed of calcite–bastnäsite–mica assemblages, bastnäsite-(Ce) exists as fine-grained, lath-like crystals intimately intergrown with calcite and mica.

Figure 3. Representative hand specimen photograph of fluorite–REE ores from the Weishan REE deposit. (**A**,**B**) Type I and Type II fluorite from sample 20CS01 coexist with calcite, bastnäsite and barite. (**C**) Type III fluorite from sample 20CS06 coexist with quartz. (**D**) Type VI fluorite from sample 20CS06 occur in the boundary of orebody and wall-rock.

4.2. Analytical Methods

The samples were crushed to a size range of 40–80 mesh and separated by magnetic and heavy-liquid separation. Fluorite grains were selected using a binocular microscope before epoxy mounting. In order to obtain the inner structure and relative surface, the grains were polished down to half of the section prior to cleaning in the ultrasonic washing machine. Microscopic petrographic observations, cathodoluminescence (CL) imaging, and backscattered-electron (BSE) imaging of the fluorite samples were conducted at the China

 https://doi.org/10.3390/min16010069

University of Geosciences (Beijing) and the Research Institute of Petroleum Exploration and Development, PetroChina (Beijing, China). The analyses were performed using a CLF-2 cathodoluminescence system (BII, Pickering, ON, Canada) coupled with a Zeiss Axio Imager multifunctional microscope (Carl Zeiss AG, Jena, Germany). Operating conditions were 12.5 V accelerating voltage and 900 mA beam current. Grain microtextures including cracks, zonation, and inclusion were checked by the CL and BSE images, which were produced by the JXA-8800 electron microprobe (EPMA, JEOL Ltd., Tokyo, Japan) at operating conditions of 20 kV and 20 nA. Suitable sites were chosen within crystal grains for fluorite geochemistry analysis.

Major elements of fluorite were obtained with a JXA-iHP200F electron microprobe (EPMA, JEOL Ltd., Tokyo, Japan) installed at the Institute of Mineral Resources at the Chinese Academy of Geological Sciences. The working conditions of the test are as follows: The acceleration voltage is 15 kV. The acceleration current is 20 nA, and the beam spot diameter is 5 μm. All test data were processed by ZAF correction. Natural minerals or synthetic oxides are used as standards for testing elements. The trace element analyses were performed using the LA-ICP-MS (NWR193 laser-ablation microprobe from Elemental Scientific Lasers LLC, Bozeman, MT, USA; and attached to a Analytik Jena M90, Jena, Germany) apparatus at Yanduzhongshi Geological Analysis Laboratories Co., Ltd. (Beijing, China). Fluorite was sampled on 30 μm spots using the laser at 8 Hz and a density of approximately 10 J/cm^2. A flow of He carrier gas at a rate of 0.55 L/min carried particles ablated by the laser out of the chamber, allowing them to be mixed with Ar gas and then carried to the plasma torch. The unknown elemental contents in the fluorite grains were identified by the external standard (NIST610 glass standard; see Supplementary Table S2). The Ca concentrations determined via EPMA analyses were used as the internal standard for correcting the LA-ICP-MS data. The signal intensity of the ^{44}Ca measured at each spot was used to correct for matrix effects and to calculate the absolute concentrations of all trace elements. Measured trace elements include ^{39}K, ^{43}Ca, ^{88}Sr, ^{89}Y, ^{137}Ba, ^{139}La, ^{140}Ce, ^{141}Pr, ^{146}Nd, ^{147}Sm, ^{153}Eu, ^{157}Gd, ^{159}Tb, ^{163}Dy, ^{165}Ho, ^{166}Er, ^{169}Tm, ^{172}Yb, ^{175}Lu, ^{85}Rb, ^{232}Th, and ^{238}U, with a measuring frequency of 0.18 s for each element.

5. Results

5.1. Petrography of Fluorite

The REE-bearing quartz–barite–carbonatite veins of the Weishan deposit are composed primarily of calcite, bastnäsite-(Ce), and synchysite-(Ce), which are commonly associated with fluorite and barite. The ores exhibit pegmatitic textures or disseminated structures (Figure 3). Fluorite displays moderate to high vitreous luster; is generally translucent, brittle; and is characterized by uneven fracture surfaces. Fluorite crystals typically occur as cubic forms, with octahedral crystals being rare. The crystal surfaces are relatively rough, and both cleavage and fractures are well developed. In some cases, fluorite occurs as irregular granular fillings between bastnäsite-(Ce), calcite, and barite (Figure 4). The color of fluorite is heterogeneous, ranging from purple to green; among these, purple fluorite is the most widespread and appears in nearly all fluorite-bearing veins. Based on morphological features and mineral assemblages, fluorite in the Weishan deposit can be classified into four generations.

https://doi.org/10.3390/min16010069

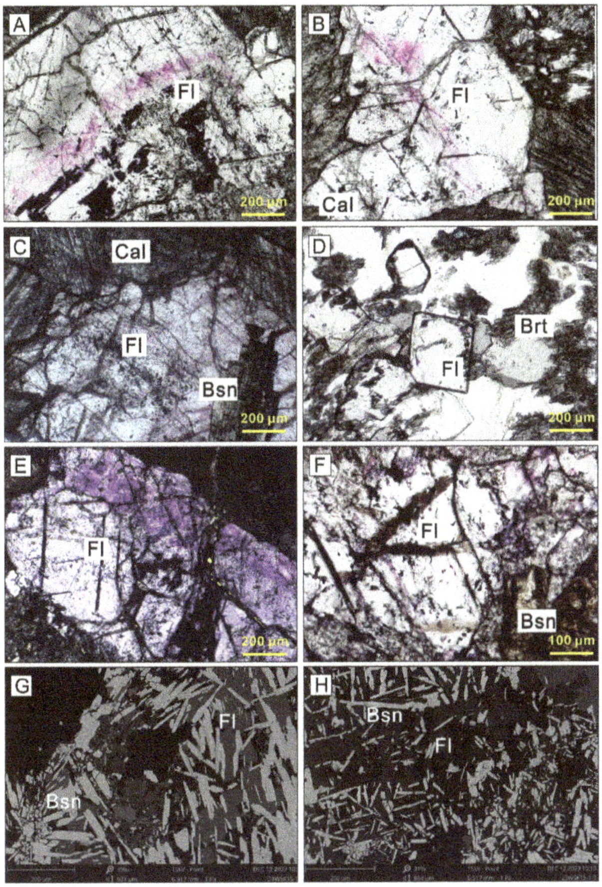

Figure 4. Petrographic characteristics of fluorite. (**A**) Subhedral to anhedral fluorite grains exhibiting deep purple coloration, coexisting with calcite and replaced by late-stage barite, and showing anhedral textures. (**B**) Subhedral fluorite crystals containing well-developed dissolution pits. (**C**) Fluorite associated with calcite and bastnäsite-(Ce) displaying distinct dissolution channels. (**D**) Euhedral fluorite grains showing peripheral dissolution pits and intergrowth with barite. (**E**,**F**) Elongated dissolution channels and corrosion pits within fluorite crystals that are locally filled by late-stage minerals. (**G**,**H**) Fluorite backscattered-electron (BSE) images show fine-grained, euhedral fluorite coexisting with acicular bastnäsite-(Ce), along with a sharp, planar contact between the two phases. Fl = fluorite; Cal = calcite; Bsn = bastnäsite; Brt = barite.

Type 1 fluorite shows anhedral morphologies with highly variable grain sizes (200–1000 μm). This generation of fluorite coexists with calcite and bastnäsite-(Ce) and is partially replaced by late-stage barite and calcite, producing distinct replacement textures (Figure 4A–C,E,G-H). Some grains exhibit pronounced dissolution features, including irregularly corroded crystal outlines, internal dissolution pits, and dissolution channels subsequently filled by later minerals (Figure 5A,B,D). Cathodoluminescence (CL) imaging reveals blue luminescence, with a few irregular internal growth zones preserved (Figure 5D). Type 2 fluorite shows subhedral forms (Figure 4F) and locally occurs as overgrowths along the margins of Type 1 fluorite. In CL images, it appears bright blue, with relatively regular oscillatory zoning. Calcite commonly fills cleavage planes within fluorite crystals. The contact boundaries between Type 1 and Type 2 fluorites are irregular (Figure 5B,C). Irregular

https://doi.org/10.3390/min16010069

triangular dissolution pits occur within the grains, accompanied by elongated etch traces along cleavage surfaces (Figure 5F). Type 3 fluorite exhibits euhedral–subhedral forms and develops along the outer rims of earlier fluorite grains. It replaces the margins of earlier fluorite through metasomatic processes. Its distinct CL characteristics (Figure 5E,F) reflect compositional differences relative to the earlier fluorite generations. Type 4 fluorite occurs near the contact zone between the carbonatite bodies and the Archean metamorphic basement, and it is associated with mica and calcite (Figure 5G,H). This generation consists of euhedral granular crystals adjacent to calcite, along with clean, transparent interiors. In CL images, Type 4 fluorite displays pale blue luminescence, exhibits relatively homogeneous internal features, and shows well-developed oscillatory zoning.

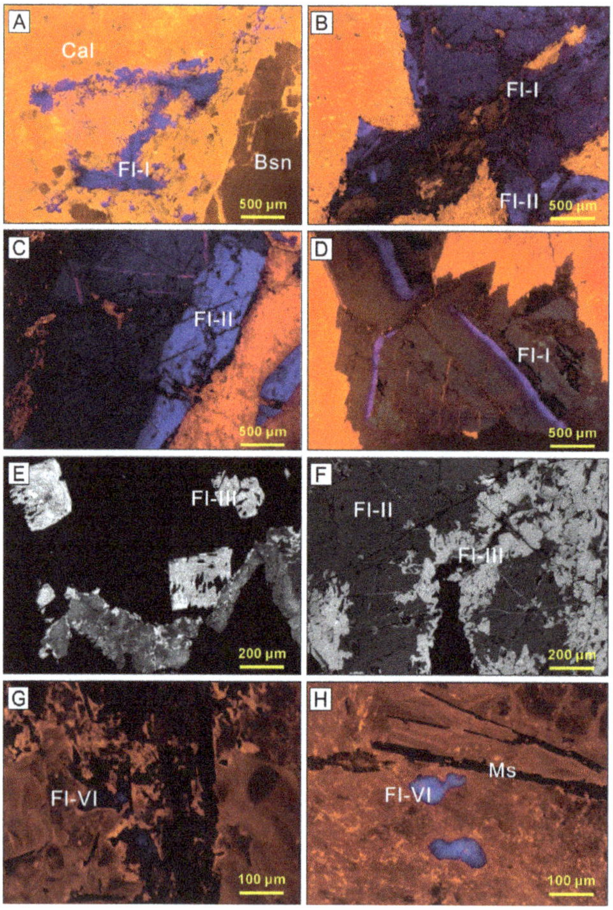

Figure 5. Cathodoluminescence characteristics of fluorite. (**A**,**B**) Type 1 anhedral fluorite coexisting with bastnäsite-(Ce) and replaced by calcite showing metasomatic textures; dissolution channels are developed within the crystals, with remnants of growth zoning preserved. (**C**,**D**) Elongated dissolution channels and corrosion pits within fluorite crystals filled by late-stage minerals; the contact between Type 1 and Type 2 fluorites is irregular. (**E**,**F**) SEM-CL images show that Type 3 fluorite developed along the margins of Type 2 fluorite, exhibiting distinct cathodoluminescence characteristics; minor dissolution pits occur along the crystal edges of Type 3 fluorite. (**G**,**H**) Type 4 fluorite coexisting with mica and filling interstitial spaces between calcite grains.

 https://doi.org/10.3390/min16010069

5.2. Geochemical Composition

The present study conducts a systematic in situ geochemical investigation of four generations of fluorites (Types I–IV) from the Weishan REE deposit, with analytical results summarized in Supplementary Table S1. Major element analyses on representative fluorite samples were conducted by systematic EPMA. The results indicate consistently high fluorite purity across all spots, with CaO contents of 68.40–73.39 wt% and F contents of 46.37–54.55 wt%. Such compositions indicate that the investigated fluorite samples are of exceptionally high purity and have experienced only minimal late-stage alteration or incorporation of impurities. Consequently, their trace element signatures can be regarded as robust proxies for the composition and physicochemical characteristics of the original precipitating fluids. Type I fluorite is characterized by a distinctive "high-Th, low-Sr" signature. Th contents are relatively elevated (~16 ppm), whereas Sr contents are comparatively low (~3959 ppm), resulting in low Sr/Th ratios. Si contents are notably high (~5150 ppm), while Na and K occur at moderate levels. Type II fluorite differs markedly from Type I fluorite. Th contents decrease significantly (average < 5 ppm), whereas Sr contents increase sharply (average > 5000 ppm), defining a "low-Th, high-Sr" pattern. Na and K concentrations display large fluctuations, which may reflect the presence of alkali-rich fluid inclusions or K-Na mineral inclusions. Type III fluorite is the most widely developed type in the Weishan deposit and shows relatively homogeneous trace element compositions at the hand specimen scale, although local internal variability is present. Sr contents are generally high (commonly 5000–8000 ppm), while Th contents remain low (mostly < 1 ppm). Na and K contents are overall lower than in Type II fluorite, but Si exhibits considerable variation. Locally elevated Ba concentrations in some spots indicate transient increases in Ba activity in the coexisting fluid. Type IV fluorite is distinguished from Type III fluorite by its behavior in high-field-strength elements (HFSEs). Elements such as Zr, Hf, Nb, and Ta, which are detectable in some Type III fluorite analyses, occur at very low levels or below detection limits in Type IV fluorite. Sr contents remain relatively high, but Y contents drop sharply to ~100 ppm.

The analytical results and characteristic parameters of rare earth elements in fluorite are presented in Supplementary Table S1. Considerable variations in total REE contents are observed among the four fluorite generations. Type 1 and Type 2 fluorites show markedly higher REE contents than Type 3 and Type 4 fluorites. There is also a clear correlation between fluorite color and REE abundance: darker purple fluorite generally contains lower total REE concentrations. The patterns of REE distributions differ significantly among the four types. Type 1 and Type 2 fluorites display pronounced LREE–HREE fractionation and strong LREE enrichment; among these, Type 1 fluorite contains higher concentrations of La, Ce, Pr, and Nd compared to Type 2 fluorite. In contrast, Type 3 fluorite is depleted in both LREE and HREE but exhibits relative enrichment in MREE. Type 4 fluorite shows the lowest total REE content and displays only slight MREE enrichment.

The REE distribution patterns of the four generations of fluorites exhibit pronounced distinctions. Type 1 fluorite displays a strongly right-sloping chondrite-normalized REE pattern, which is characterized by significant LREE enrichment (Figure 6A). In contrast, Type 2 fluorite shows a relatively flat pattern with only a slight rightward inclination, indicating weaker LREE enrichment (Figure 6B). Type 3 fluorite is marked by a left-sloping pattern with pronounced MREE enrichment (Figure 6C), whereas Type 4 fluorite exhibits an overall flat REE distribution (Figure 6D). Eu anomalies also vary among the fluorite types. Type 1 fluorite generally exhibits weak to moderate negative Eu anomalies (Figure 6A). In the Y/Ho–La/Ho diagram (Figure 7A), all analyses define a broadly consistent Y/Ho trend. Type I fluorite plots are seen at the far right due to its exceptionally high La contents. The Y/Ho ratios of all fluorite types are similar and slightly higher than the chondritic

 https://doi.org/10.3390/min16010069

value. In the Tb/Ca-Tb/La diagram, Type I fluorite falls within the pegmatitic–magmatic field. From Type II to Type IV, increasing Tb/La ratios accompanied by variations in Tb/Ca concentrations produces a band-like distribution within the hydrothermal field, outlining an evolutionary trajectory from early, REE-rich, and strongly fractionated fluids to late, REE-poor, and weakly fractionated fluids. Both positive and negative Eu anomalies occur in Type 2 and Type 3 fluorites, while Type 4 fluorite is characterized by a distinct positive Eu anomaly. Ce anomalies show similarly variable behavior: Type 1 and Type 3 fluorites display both positive and negative Ce anomalies, whereas Type 2 and Type 4 fluorites consistently show positive Ce anomalies (Figure 8).

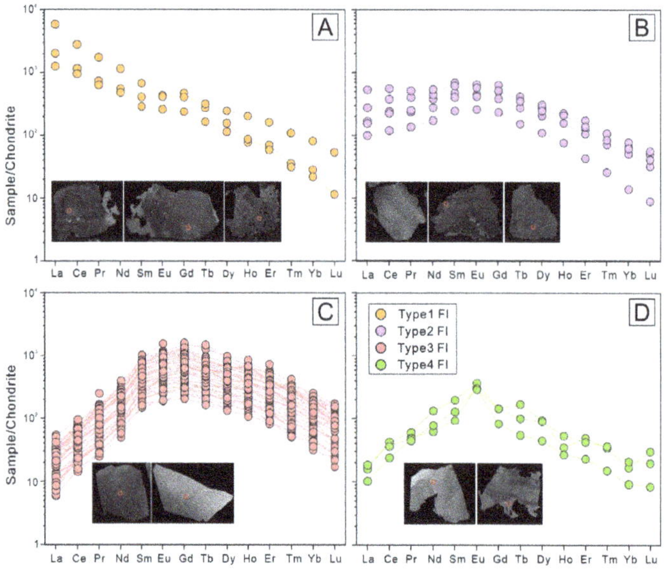

Figure 6. Chondrite-normalized REE patterns of fluorites from different generations. (**A**) Type 1 fluorite show strongly LREE enrichment REE. (**B**) Type 2 fluorite shows a relatively flat REE pattern. (**C**) Type 3 fluorite is marked by a left-sloping pattern with MREE enrichment. (**D**) Type 4 fluorite exhibits an overall flat REE distribution.

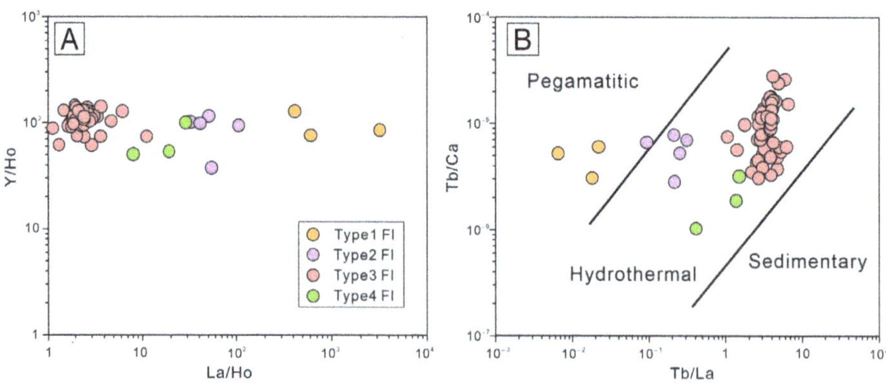

Figure 7. Tb/Ca vs. Tb/La and Y/Ho vs. La/Ho diagrams of fluorites from different generations (modified after [68]). (**A**) La/Ho versus Y/Ho diagram illustrating variations in REE fractionation among different fluorite. (**B**) Tb/La versus Tb/Ca diagram used to discriminate sedimentary, hydrothermal, and pegmatitic fluid sources.

https://doi.org/10.3390/min16010069

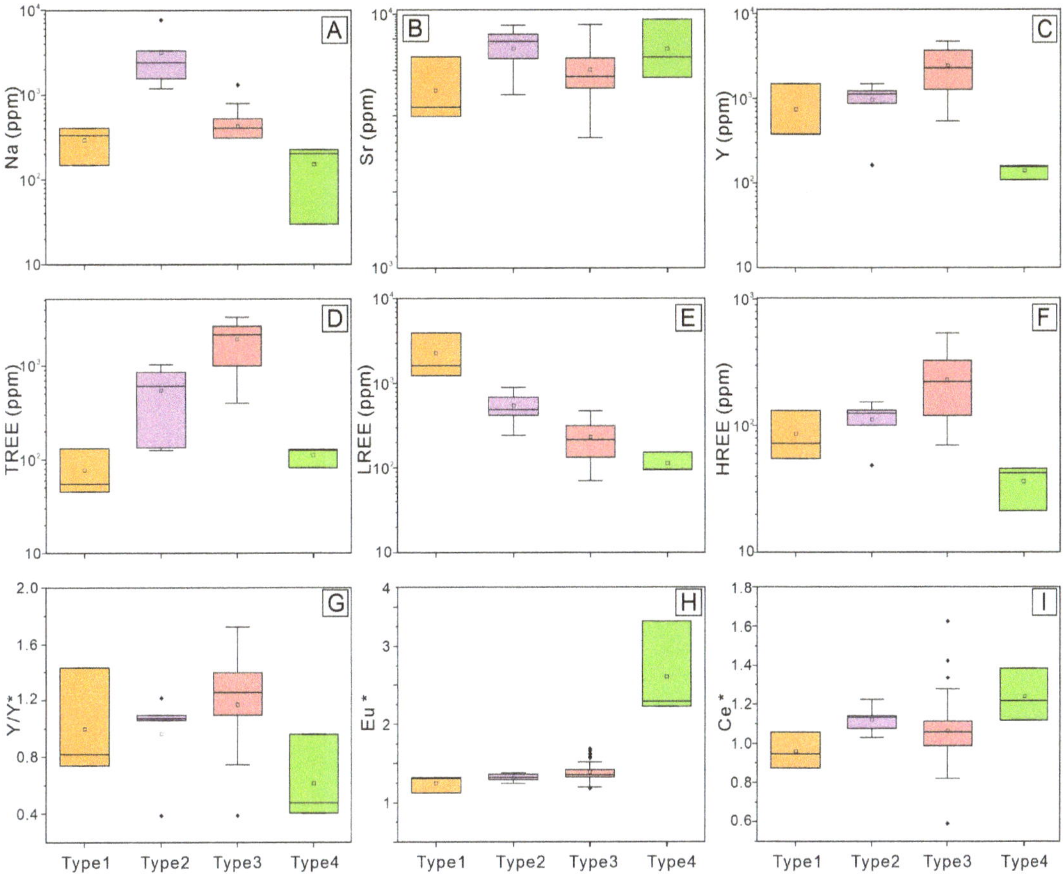

Figure 8. Box diagram of trace elements of fluorite from different generations. (**A**) Na contents are generally higher in Type II and lower in Type IV. (**B**) Sr contents are overall high, with relatively elevated median values in Type IV and Type II. (**C**) Y contents are markedly higher in Type III, whereas Type IV shows comparatively low values. (**D–F**) TREE is highest in Type III and lower in Type I and Type IV. LREE shows pronounced enrichment in Type I, while Type III and Type IV are comparatively lower. HREE is highest in Type III and lowest in Type IV. (**G**) Y/Y* varies among fluorite, with generally lower values in Type IV. (**H**) Eu* indicates a stronger positive Eu anomaly in Type IV (higher median values than other types). (**I**) Ce* values are mostly near unity, with slightly higher values in Type IV.

6. Discussion

6.1. Precipitation Mechanisms of Fluorite

In general, the co-precipitating mineral assemblages exert a significant influence on the distribution of rare earth elements (REEs) in fluorite, particularly in the presence of bastnäsite-(Ce), monazite, and calcite [9,42,69,70]. In the Weishan REE deposit, minerals co-precipitating with Type 1 fluorite include bastnäsite-(Ce) and calcite (Figure 4F), and the REE distribution patterns of fluorite exhibits pronounced LREE enrichment. This feature is consistent with the REE patterns reported for magmatic-stage calcite and bastnäsite-(Ce) [71,72], indicating that Type 1 fluorite may have formed during the early stages of magmatic–hydrothermal evolution. The low Tb/La and high Tb/Ca ratios further support a magmatic–pegmatitic origin for this fluorite generation (Figure 7). Petrographically, Type 1 fluorite locally pre-

 https://doi.org/10.3390/min16010069

serves replacement textures attributable to later hydrothermal overprinting, supporting the interpretation as an early-crystallized phase (Figure 5A). Type 2 fluorite displays generally lower total REE contents and is characterized by slight LREE depletion, modest MREE enrichment, and HREE depletion. Compared with Type 1 fluorite, Type 2 fluorite shows markedly decreased LREE concentrations and relatively elevated MREE and HREE concentrations (Figure 8). In the Tb/La-Tb/Ca diagram, Type 2 fluorite plots in a transitional field between pegmatitic and hydrothermal origins (Figure 7), indicating crystallization later than Type 1 fluorite. Type 2 fluorite crystals contain irregular dissolution pits, consistent with dissolution–reprecipitation processes. Within the study area, Type 3 and Type 4 fluorites commonly coexist with quartz, barite, and pyrite. As these associated minerals lack stoichiometric calcium, their influence on REE partitioning in fluorite is limited. Type 3 and Type 4 fluorites display broadly similar REE patterns, along with LREE and HREE depletion and relative enrichment in MREE, likely reflecting precipitation from evolved hydrothermal fluids in the later stages of the system. Petrographically, Type 3 fluorite occurs as euhedral crystals that display relatively homogeneous cathodoluminescence, whereas Type 4 fluorite occurs as fine-grained, anhedral aggregates filling interstices between calcite grains, suggesting that Type 4 fluorite postdates Type 3 fluorite. Taken together, mineralogical and geochemical evidence indicates a well-defined paragenetic sequence for fluorite in the Weishan REE deposit: Type 1 → Type 2 → Type 3 → Type 4. These successive generations document the progressive transition and evolution of the ore-forming system from high-temperature magmatic conditions to low-temperature hydrothermal regimes, reflecting systematic changes in fluid composition, temperature, and pH conditions over time.

According to the classical framework [46], fluorite precipitation can occur under three principal geological conditions: (i) fluid–rock interaction, (ii) fluid mixing accompanied by changes in chemical composition, and (iii) cooling and/or pressure reduction in Ca-F-saturated fluids. Magmatic fluorite commonly forms in highly evolved, F- and REE-enriched peraluminous granites, ongonites, pegmatites; fluorite-rich rhyolites; or selected alkaline magmatic systems [73–77]. Fluorite can crystallize in fluoride–calcium or Ca- and F-rich mafic fluorosilicate melts that result from magmatic fluoride–silicate liquid immiscibility [73,78]. Its crystallization requires fluorite saturation in the melt, a threshold controlled primarily by the melt composition and prevailing physicochemical conditions [75,76]. In this study, Type I fluorite is distinguished by anomalously high ΣREE contents (>2500 ppm) and a distinctive "high-Th, low-Sr" signature, in sharp contrast to the low-ΣREE, Sr-rich compositions typical of Types II–IV hydrothermal fluorite. This contrast raises the possibility that Type I fluorite represents a product most closely linked to a magmatic end-member fluid and may therefore preserve attributes that are transitional toward magmatic fluorite. The exceptionally high ΣREE and pronounced LREE enrichment of Type I fluorite are consistent with precipitation from a highly evolved, F-rich granitic silicate melt or from an incipient supercritical fluid exsolved from such a melt. Thorium, a strongly incompatible element, becomes concentrated in highly fractionated magmas; the elevated Th concentration in Type I fluorite may thus reflect inheritance from an evolved magmatic reservoir. Based on global REE-deposit models, mineralization in this district may involve silicate–carbonate melt immiscibility and silicate melt immiscibility for fluoride-silicate and fluoride-calcium (F-Ca or fluoritic) melts [75,76,79]. Because fluoride melts can dissolve and complex LREEs and Th efficiently, the high ΣREE and Th concentrations in Type I fluorite are best explained by precipitation from an early, high-temperature, hypersaline fluid exsolved from fluoride–calcium (F-Ca) and other fluoride-rich melts produced by such immiscibility.

During fluid–rock interaction, adsorption of REEs onto mineral surfaces commonly manifests as an ion exchange or substitution process [80,81]. However, the fluorite generations (Type 1–Type 3) developed in the Weishan REE deposit predominantly occur in pegmatitic habits, reflecting low-reactivity surface areas. Consequently, adsorption is unlikely to have exerted a major influence on REE enrichment. The widespread presence of calcite, barite, and REE fluorocarbonates in the deposit suggests that ligands such as F^-, CO_3^{2-}, and SO_4^{2-} were enriched within the fluid phase, implying that complexation played a more dominant role than adsorption in governing fluorite precipitation. The transition from Type 1 fluorite to Type 2 fluorite is marked by systematic changes in REE fractionation. Type 2 fluorite is characterized by LREE depletion coupled with relative enrichment of MREE and HREE (Figure 8D–F). LREE depletion in fluorite is commonly associated with element remobilization and recrystallization, processes controlled by differences in complex stability, evolving physicochemical conditions, and fluid compositions [47,68].

As LREEs are preferentially adsorbed on mineral surfaces during fluid migration, tectonically enhanced remobilization and recrystallization can therefore further deplete LREEs in fluorite [68,82]. Experimental petrology has demonstrated that fluorite often contains submicron REE-rich microphases that are more soluble than fluorite itself, producing heterogeneous REE distributions [47,83]. Selective dissolution of these microphases can preferentially reduce LREE concentrations while largely preserving HREE. Combined with the La/Ho-Y/Ho discriminant diagram, these observations suggest that coupled dissolution and recrystallization contributed to the formation of Type 2 fluorite. From Type 2 fluorite to Type 3 fluorite, MREE and HREE concentrations increase further, whereas LREE concentrations remain low (Figure 8D–F). Previous work has shown that, in F-rich hydrothermal systems, chloride and sulfate complexes act as effective ligands for REE transport [84]. In the Weishan deposit, however, the widespread precipitation of fluorite and bastnäsite-(Ce) suggests that REE mobility during the hydrothermal stage was dominated by fluoride complexation. At temperatures of 100–250 °C, REEs occur mainly as monofluoride complexes $[REEF]^{2+}$, whose stability increases with atomic number [80,85]. Fractionation between Y and Ho is primarily controlled by fluid geochemistry rather than the fluid source: Y forms relatively stable difluoride complexes $[YF_2]$, whereas Dy and Ho are dominated by monofluoride complexes [86], accounting for the positive Y anomalies in Type 2 and Type 3 fluorites. Because HREEs and Y form more stable fluoride complexes, they are preferentially retained and enriched in the fluid. Where complexation dominates, progressive cooling coupled with continued fluorite precipitation amplifies LREE–HREE fractionation and yields increasingly distinct REE patterns [47,87]. As shown in Figure 7, the marked decrease in La/Ho from Type 2 fluorite to Type 3 fluorite, together with near-constant Y/Ho, constitutes a characteristic signature of element reactivation and mineral reprecipitation [80,81]. The reactivation process is also reflected in Tb/La ratios: increased ligand to metal ratios during remobilization preferentially deplete La, whereas Tb remains comparatively stable [68]. Collectively, these observations suggest that Type 3 fluorite formed via partial dissolution and subsequent reprecipitation of earlier fluorite generations. Type 4 fluorite exhibits the lowest REE concentrations among all generations, implying diminished stability of LREE–F complexes and near-exhaustion of REEs in the fluid during its formation. Under these conditions, the fluid became supersaturated with respect to CaF_2, and fluorite precipitated mainly as late-stage infill between calcite grains. This generation therefore records the terminal phase of hydrothermal fluid evolution in the Weishan REE deposit.

6.2. Physicochemical Conditions of Mineralization

Variations in ore–fluid composition and physicochemical conditions (e.g., temperature and pH) during fluorite precipitation are commonly recorded by fluorite REE systematics, providing key constraints on the depositional environment [47,48,80]. The REE-Y signa-

https://doi.org/10.3390/min16010069

tures of hydrothermal fluorite are therefore widely used to trace its fluid composition and evolutionary history, with distinct evolution pathways expressed as characteristic partitioning patterns [33,47,48]. Precipitation temperature can be assessed, in part, from the presence or absence of Eu anomalies [47,80,87]. In hydrothermal fluids, the Eu^{2+}/Eu^{3+} ratio is controlled by redox equilibrium and decreases sharply below ~250 °C under oxygen fugacity typical of the shallow lithosphere [88,89]. At higher temperatures, Eu is dominated by Eu^{2+}; owing to its relatively large ionic radius, Eu^{2+} substitutes poorly for Ca^{2+} in the fluorite lattice and thus tends to remain in the fluid or be adsorbed onto mineral surfaces [88]. With cooling, Eu^{2+} is oxidized to Eu^{3+}, whose smaller ionic radius allows more efficient incorporation into fluorite [44,47,87]. Where hydrothermal fluids are sourced from mildly reducing, carbonatite-related systems and inherit positive Eu anomalies from the protolith, fluorite with positive Eu anomalies is expected to precipitate only at low temperatures (<250 °C). Previous studies further indicate that fluorite displaying positive Eu anomalies commonly forms at temperatures below ~200 °C [47,48].

Both Eu and Ce anomalies (Eu* and Ce*) are widely used as indicators of the redox state during fluorite precipitation. In the Weishan REE deposit, Eu* and Ce* values vary considerably among fluorite generations, suggesting fluctuations in the physicochemical conditions of the ore-forming fluids. Type 1 fluorite consistently exhibits negative Eu anomalies, indicating formation temperatures above 200 °C. Its range of Ce* values, from positive to negative, primarily reflects fluctuations in the redox conditions (fO_2) of the parental fluid during its evolution. Type 2 fluorite shows mixed Eu anomalies but consistently positive Ce anomalies. Type 3 fluorite shows weakly positive Eu anomalies and variable Ce anomalies, whereas Type 4 fluorite is characterized by uniformly strong positive Eu and Ce anomalies. Overall, these trends suggest sustained fluorite precipitation during progressive cooling, accompanied by a shift from reducing to increasingly oxidizing conditions. The presence of CO_2-rich or sulfate-bearing fluid inclusions in later fluorite generations, compared to earlier CH_4- or H_2S-bearing inclusions, would chemically attest to an increase in fluid oxidation state [90]. Cerium anomalies in fluorite that coexists with bastnäsite-(Ce) are best interpreted in terms of mineral–fluid fractionation rather than redox reactions alone. Preferential partitioning of Ce relative to La and Pr into fluorocarbonates depletes the residual fluid in Ce such that later fluorite progressively develops negative Ce anomalies [33,91]. This framework accounts for the lower LREE contents of Types II–III relative to Type I and for the locally negative Ce anomalies in Types II–III fluorite, consistent with main- to late-stage mineralization. By contrast, the combination of comparatively stable Na-K, low Sr, high Th, and elevated Si in Type I fluorite is most compatible with an early, less-evolved magmatic fluid signature that predates large-scale fluorocarbonate precipitation. Additional trace element systematics support this interpretation. High Sr and Ba, low Th, and strongly variable Na/K levels in Types II–III point to changing sulfate activity and episodic increases in fO_2 during the principal mineralizing interval [17,23]. The coupled co-variation in REE, Na, and Ce* levels is consistent with lattice incorporation of Ce^{3+} via charge-balanced substitution (e.g., $REE^{3+} + Na^{+} \leftrightarrow 2Ca^{2+}$, [14]), rather than submicroscopic inclusions. Type IV fluorite, which is restricted to the contact with Neoarchean plagioclase-hornblende gneiss, is characterized by pronounced positive Eu and Ce anomalies together with low total REE levels, indicating a late overprint by a compositionally distinct fluid. The positive Eu anomaly is best explained by efficient leaching of the metamorphic basement by an alkaline, reducing fluid, which mobilizes Eu preferentially as Eu^{2+} under low-fO_2 conditions [16,23]. Under the same reducing conditions, Ce is stabilized and transported predominantly as Ce^{3+} (rather than oxidized to relatively immobile Ce^{4+}), allowing Ce to be effectively remobilized and carried by the fluid. A fenitizing, alkali-, and volatile-rich fluid interacting with wall rocks, followed by cooling and mixing with

 https://doi.org/10.3390/min16010069

residual acidic fluids and/or CO_2 loss, would decrease pH, drive CaF_2 supersaturation, and precipitate Type IV fluorite. Its positive Eu-Ce signature is consistent with a wall-rock Eu contribution superimposed on a carbonatite-related Ce imprint.

Experimental petrology further demonstrates that fluid pH and salinity are key controls of fluorite dissolution–precipitation equilibria and REE solubility. Under constant-salinity conditions, REE solubility increases with decreasing pH; at a constant pH, the addition of NaCl dramatically enhances REE concentrations in the solution by several orders of magnitude [83]. In acidic to weakly acidic, high-salinity fluids, REE fluorides readily approach saturation, consistent with the relatively high solubility of fluorite under these conditions. Consequently, F^- exhibits higher chemical activity in acidic fluids relative to alkaline fluids. Under weakly acidic to alkaline conditions, shifts in REE/Ca and REE/F ratios can promote precipitation of LREE-enriched fluorides. At high temperatures and weakly acidic pH values, REE/Ca and REE/F ratios are positively correlated, and fluorite typically shows HREE depletion—consistent with the Type 1 fluorite signature. In acidic-fluid conditions, fluorite tends to exhibit LREE depletion and Y enrichment, matching the REE patterns of Type 2 and Type 3 fluorites. These observations suggest that Type 1 fluorite in the Weishan deposit precipitated from relatively high-temperature, weakly acidic, and moderately reducing fluids, whereas Type 2 and Type 3 fluorites crystallized from lower-temperature, more acidic fluids under redox conditions evolving from reducing to weakly oxidizing.

6.3. Implications for REE Mineralization

Previous studies on the Weishan REE deposit have focused on the petrogenesis of the alkaline–carbonatite complex, the source of ore-forming materials, the timing of magmatic and hydrothermal processes, and the mechanisms of REE precipitation, leading to several contrasting genetic models. Based on C-O isotopic compositions of carbonatites, some scholars demonstrated that the Weishan carbonatites are of igneous origin and were derived from CO_2-rich silicate magmas through liquid immiscibility [71]. Combined with the geochemical characteristics of apatite, these authors further proposed that the ore-forming carbonatites originated from a metasomatized lithospheric mantle and that REE minerals precipitated during the magmatic–hydrothermal transition. In contrast, the other authors, using in situ Sr-Nd isotopes of apatite, argued that the REE-rich syenite–carbonatite complex in Weishan resulted from crust–mantle interactions, with ore-forming materials sourced predominantly from the asthenosphere [92]. Zhao et al. (2023) suggested, based on the trace element composition of calcite, that the ore-forming fluid was derived from carbonate-rich melts exsolved from an alkaline magma, emphasizing that alkali-rich components played a crucial role in the liberation and transport of REEs [72]. They further proposed that REE mineralization was controlled principally by alkaline magmatism rather than carbonatite magmatism and that extensive REE mineral precipitation was triggered by fluid boiling.

Despite divergent interpretations of source characteristics and metallogenic processes at Weishan, there is broad agreement that the ore-forming fluids were F-rich. However, existing models are unable to explain the pervasive association of fluorite with REE minerals. Although previous study inferred extensive fluorite precipitation during the hydrothermal stage from mineral paragenesis, the source and genetic mechanism of fluorite remained unresolved [72]. Assuming that F was sourced from fluids exsolved from an alkaline magma, such fluids would be expected to be alkali-rich but calcium-poor. This expectation is inconsistent with fluid-inclusion evidence from quartz and barite that recorded Ca-bearing daughter minerals. Furthermore, if magmatic fluids carried both Ca and F, the very low solubility of fluorite would favor immediate precipitation upon fluid exsolution, which

 https://doi.org/10.3390/min16010069

conflicts with field relationships at Weishan, where abundant pegmatitic fluorite coexists with late-stage hydrothermal minerals such as barite and celestine. These inconsistencies suggest that the F-rich fluids responsible for widespread fluorite precipitation were more likely derived from Ca-rich carbonatite magmas, from fluid–rock interactions with Ca-rich wall rocks, or from mixing with external fluids. In this regard, some scholars argued that volatile-rich carbonatite magmas at high temperatures can sustain high activities of Ca and F and that the high-salinity fluids exsolved from such magmas can simultaneously concentrate F^- and Ca^{2+}, thereby promoting fluorite precipitation [93]. Some researchers proposed that the interaction between F-rich carbonatite fluids and carbonaceous wall rocks can generate abundant REE- and Sr-rich fluorites [18]. Therefore, the F-rich ore-forming fluids in the Weishan REE deposit were most likely derived from Ca-rich carbonatite magmas. During magmatic–hydrothermal evolution, fluorine and calcium activities were dynamically modulated by changing physicochemical conditions, producing successive generations of fluorite. This interpretation is consistent with published fluid-inclusion constraints: CO_2-bearing inclusions in quartz–fluorite veins have salinities of ~5 wt% and homogenization temperatures of 210–248 °C, indicating a medium- to low-temperature hydrothermal system of relatively low salinity [90]. Petrography further supports these trends (Figures 4 and 5). Early Type 1 fluorite, interpreted to have formed at higher temperatures under weakly acidic conditions, commonly displays etch pits and dissolution channels along cleavage planes, features conducive to element remobilization. In contrast, late-stage fluorite precipitated at lower temperatures under more acidic conditions typically preserves fine-scale dissolution textures along crystal margins, recording shifts in fluid chemistry and precipitation kinetics.

Fluorine is a key volatile in REE mineralization, exerting first-order control on mineral assemblages during REE deposition and buffering the chemistry of ore-forming fluids. Fluoride may occur as ore and gangue minerals and as daughter minerals in fluid inclusions, thereby influencing REE mobility and precipitation. Fluid inclusion studies and experimental petrology demonstrate that granitic magmas can undergo fluoride–silicate immiscibility during emplacement, promoting preferential partitioning of REEs into fluoride melts. In hydrothermal systems, F promotes the formation of REE complexes, enabling REEs to be transported as complex ions and markedly increasing their solubility [23,40,83,84]. Experimental studies on hydrothermal fluorite further show that at high temperatures, LREE-F_2^+ complexes are more stable than MREE-F_2^+ complexes [52]. Accordingly, during early high-temperature fluid evolution, REEs are transported chiefly as LREE–F_2^+ complexes, producing LREE-enriched early-stage fluorite. With cooling, the stability of LREE–F_2^+ complexes decrease, whereas MREE– and HREE–fluoride complexes become relatively more stable, leading to progressive MREE and HREE enrichment in late-stage fluorite. Thermodynamic modeling additionally suggests that, in natural F-rich systems, REEs are transported mainly as chloride and sulfate complexes in acidic to weakly acidic fluids, whereas fluoride and phosphate, given the extremely low solubility of REE fluorides and phosphates, function primarily as precipitation ligands. In near-neutral to alkaline fluids, hydroxyl and/or carbonate complexes likely play a greater role in REE precipitation [84,94,95]. In the Weishan REE deposit, the coexistence of bastnäsite-(Ce), barite, celestine, fluorite, and calcite indicates that the ore-forming system is enriched in F^-, CO_3^{2-}, and SO_4^{2-}, which likely served as the principal ligands governing REE transport in the hydrothermal fluids. REE mobility in hydrothermal solutions is controlled by both the stability of aqueous REE complexes and the solubility of REE-bearing minerals. Experimental studies on fluorite dissolution reveal that REE activation and mobility are also influenced by fluorite solubility. Between 100 and 250 °C, salinity and pH exert stronger controls on fluorite solubility than temperature [83]. Fluorite solubility, in turn, regulates REE release to the solution and the

https://doi.org/10.3390/min16010069

formation of fluoride phases. Given the extremely low solubility of fluorite, once fluid conditions cross the threshold for CaF_2 precipitation, F^- activity declines rapidly. This decrease destabilizes REE–fluoride complexes and promotes bastnäsite precipitation via the following reactions:

$$REEF_3{}^0 + CO_3{}^{2-} = \text{bastnäsite} + 2F^- \quad (1)$$

$$REEF_3{}^0 + HCO_3{}^- = \text{bastnäsite} + 2F^- + H^+ \quad (2)$$

$$F^- + H^+ = HF^0 \quad (3)$$

In Reactions (1) and (2), consumption of F^- drives the system toward bastnäsite precipitation. In Reactions (2) and (3), generation of H^+ further lowers the activities of Ca^{2+} and F^-, thereby promoting fluorite dissolution and explaining the enhanced solubility of fluorite under acidic conditions. In summary, the fluorite and REE minerals developed in the Weishan REE deposit were most likely precipitated from residual carbonatite-derived fluids rich in Ca and CO_2, along with subordinate F and H_2O and appreciable dissolved REEs. As the temperature and pressure decreased, fluorite began to precipitate, leading to a rapid decline in F activity. The ensuing increase in acidity partially dissolved early-stage fluorite, enhanced REE mobilization, and destabilized REE-F complexes, thereby triggering extensive precipitation of REE fluorocarbonates. Subsequent re-enrichment of the residual fluid in Ca^{2+} and F^- promoted renewed, large-scale fluorite precipitation during the late stage, producing the multi-episodic fluorite–REE mineral assemblages observed in the deposit. This study aims to reconstruct the complete evolution of the Weishan REE–fluorite system, from deep magmatic processes to shallow hydrothermal activity, using fluorite and its geochemical record as primary tracers. Within a carbonatite-related mineral system ultimately rooted in a mantle source, fluorite-bearing REE orebodies represent one expression of this evolution under a specific range of physicochemical conditions. By contrast, coeval REE-rich (or other element-enriched) orebodies that lack fluorite may reflect the same ore-forming fluid operating under different temperature–pressure–pH–ligand regimes. Together, these mineralization styles are interpreted to constitute a single, integrated, and spatially zoned carbonatite ore-forming system.

7. Conclusions

(1) Four generations of fluorite are distinguished in the Weishan REE deposit, which occur in a temporal sequence of Type I → Type II → Type III → Type IV, corresponding to the early magmatic–hydrothermal stage, the hydrothermal stage, and the late, low-temperature stage.

(2) Paragenetic progression reflects a decrease in temperature, an increase in fluid pH, a gradual increase in oxygen fugacity, and extensive mineral dissolution–reprecipitation processes that facilitated REE redistribution within the system.

(3) Precipitation of fluorite is a process likely governed by variations in Ca^{2+} and F^- activities within a fluorine-rich carbonatite fluid, as well as by dissolution–reprecipitation mechanisms. By regulating the activity of fluorine, buffering the chemical conditions of the hydrothermal system, and potentially controlling co-precipitation with REE fluorocarbonates, fluorite is inferred to play a key, and possibly indispensable role in REE enrichment. These features provide essential mineralogical constraints for establishing genetic models of carbonatite REE deposits and for guiding future exploration strategies.

https://doi.org/10.3390/min16010069

Supplementary Materials: The following supporting information can be downloaded at: https://www.mdpi.com/article/10.3390/min16010069/s1. The Supplementary Table contains the geochemical composition of different generations of fluorite in the Weishan REE deposit. Table S1 Geochemical composition of different generations of fluorite in Weishan REE deposit. Table S2 Geochemical composition of standard samples. Table S3 Major Element composition of fluorite from the Weishan REE deposit.

Author Contributions: Conceived the ideas, Y.-X.G., S.-S.L. and C.-P.L.; map compilation, Z.-Y.Y. and X.-Y.W.; data curation, formal analysis, and investigation, Y.-X.G., M.-Q.W., Z.S., Z.-Y.Y., X.-Y.W., K.-F.Q. and S.-S.L.; writing, Y.-X.G. and S.-S.L. All authors have read and agreed to the published version of the manuscript.

Funding: This study was financially supported by The Open Project of Weihai Key Laboratory of Energy and Mineral Resources Investigation and Evaluation (No.LDKF-2023WH-02), Open Research Projects from Shandong Provincial Lunan Geology and Exploration Institute and Shandong Engineering Research Center of Rare Elements Exploration and Comprehensive Utilization (LNY202301), Open Research Fund Program of Key Laboratory of Gold Mineralization Processes and Resource Utilization Subordinated to the Ministry of Natural Resources, Shandong Key Laboratory of Metallogenic Geological Process and Resources Utilization (KFKT202403), Young Elite Scientists Sponsorship Program of BAST (BYESS2024122), the Fundamental Research Funds for the Central Universities (2-9-2023-055).

Data Availability Statement: The dataset is presented directly in the present study.

Acknowledgments: We are grateful to De-Jian Li for providing valuable field guidance and engaging in constructive discussions. We appreciate Ya-Qi Huang for assistance with figure preparation. We are deeply grateful to the anonymous reviewers and editors for their insightful comments and helpful suggestions.

Conflicts of Interest: The authors declare no conflicts of interest.

References

1. Qiu, K.; Yu, H.; Wu, M.; Geng, J.; Ge, X.; Gou, Z.; Taylor, R.D. Discrete Zr and REE mineralization of the Baerzhe rare-metal deposit, China. *Am. Mineral.* **2019**, *104*, 1487–1502. [CrossRef]
2. Wu, M.; Samson, I.M.; Qiu, K.; Zhang, D. Concentration Mechanisms of Rare Earth Element-Nb-Zr-Be Mineralization in the Baerzhe Deposit, Northeast China: Insights from Textural and Chemical Features of Amphibole and Rare Metal Minerals. *Econ. Geol.* **2021**, *116*, 651–679. [CrossRef]
3. Xie, Y.; Xia, J.; Cui, K.; Qu, Y.; Liang, P.; Zhong, R. Rare earth elements deposits in China: Spatio-temporal distribution and ore-forming processes. *Chin. Sci. Bull.* **2020**, *65*, 3794–3808. [CrossRef]
4. Xie, Y.; Verplanck, P.L.; Hou, Z.; Zhong, R. *Mineral Deposits of China*; Society of Economic Geologists: Littleton, CO, USA, 2019; pp. 509–552.
5. Fan, H.; Niu, H.; Li, X.; Yang, K.; Yang, Z.; Wang, Q. The types, ore genesis and resource perspective of endogenic REE deposits in China. *Chin. Sci. Bull.* **2020**, *65*, 3778–3793. [CrossRef]
6. Chakhmouradian, A.R.; Zaitsev, A.N. Rare Earth Mineralization in Igneous Rocks: Sources and Processes. *Elements* **2012**, *8*, 347–353. [CrossRef]
7. Beard, C.D.; Goodenough, K.M.; Borst, A.M.; Wall, F.; Siegfried, P.R.; Deady, E.A.; Pohl, C.; Hutchison, W.; Finch, A.A.; Walter, B.F.; et al. Alkaline-Silicate REE-HFSE Systems. *Econ. Geol.* **2023**, *118*, 177–208. [CrossRef]
8. Chakhmouradian, A.R.; Wall, F. Rare Earth Elements: Minerals, Mines, Magnets (and More). *Elements* **2012**, *8*, 333–340. [CrossRef]
9. Liu, Y.; Chakhmouradian, A.R.; Hou, Z.; Song, W.; Kynický, J. Development of REE mineralization in the giant Maoniuping deposit (Sichuan, China): Insights from mineralogy, fluid inclusions, and trace-element geochemistry. *Mineral. Depos.* **2018**, *54*, 701–718. [CrossRef]
10. Smith, M.P.; Moore, K.; Kavecsánszki, D.; Finch, A.A.; Kynicky, J.; Wall, F. From mantle to critical zone: A review of large and giant sized deposits of the rare earth elements. *Geosci. Front.* **2016**, *7*, 315–334. [CrossRef]
11. Xu, C.; Kynicky, J.; Chakhmouradian, A.R.; Li, X.; Song, W. A case example of the importance of multi-analytical approach in deciphering carbonatite petrogenesis in South Qinling orogen: Miaoya rare-metal deposit, central China. *Lithos* **2015**, *227*, 107–121. [CrossRef]

12. Xu, C.; Kynický, J.; Smith, M.P.; Kopriva, A.; Brtnický, M.; Urubek, T.; Yang, Y.; Zhao, Z.; He, C.; Song, W. Origin of heavy rare earth mineralization in South China. *Nat. Commun.* **2017**, *8*, 14598. [CrossRef]
13. Qiu, K.-F.; Deng, J.; Li, S.-S.; Jowitt, S.; Hetherington, C.J.; Balen, D. Roles and perspectives of A- and I-type magmas in rare earth element and gold mineralization. *GSA Bull.* **2023**, *136*, 1238–1250. [CrossRef]
14. Broom-Fendley, S.; Styles, M.T.; Appleton, J.D.; Gunn, G.; Wall, F. Evidence for dissolution-reprecipitation of apatite and preferential LREE mobility in carbonatite-derived late-stage hydrothermal processes. *Am. Mineral.* **2016**, *101*, 596–611. [CrossRef]
15. Makin, S.A.; Simandl, G.; Marshall, D. Fluorite and its potential as an indicator mineral for carbonatite related rare earth element deposits. In *Geological Fieldwork 2013, British Columbia Ministry of Energy and Mines, British Columbia Geological Survey Paper 2014-1*; British Columbia Geological Survey: Victoria, BC, Canada, 2014; pp. 207–212.
16. Öztürk, H.; Altuncu, S.; Hanilçi, N.; Kasapçı, C.; Goodenough, K.M. Rare earth element-bearing fluorite deposits of Turkey: An overview. *Ore Geol. Rev.* **2019**, *105*, 423–444. [CrossRef]
17. Xu, C.; Taylor, R.N.; Li, W.; Kynicky, J.; Chakhmouradian, A.R.; Song, W. Comparison of fluorite geochemistry from REE deposits in the Panxi region and Bayan Obo, China. *J. Asian Earth Sci.* **2012**, *57*, 76–89. [CrossRef]
18. Bühn, B.; Rankin, A.H.; Schneider, J.; Dulski, P. The nature of orthomagmatic, carbonatitic fluids precipitating REE,Sr-rich fluorite: Fluid-inclusion evidence from the Okorusu fluorite deposit, Namibia. *Chem. Geol.* **2002**, *186*, 75–98. [CrossRef]
19. Hou, Z.; Tian, S.; Xie, Y.; Yang, Z.; Yuan, Z.; Yin, S.; Yi, L.; Fei, H.; Zou, T.; Bai, G.; et al. The Himalayan Mianning–Dechang REE belt associated with carbonatite–alkaline complexes, eastern Indo-Asian collision zone, SW China. *Ore Geol. Rev.* **2009**, *36*, 65–89. [CrossRef]
20. Smith, M.P.; Campbell, L.S.; Kynicky, J. A review of the genesis of the world class Bayan Obo Fe–REE–Nb deposits, Inner Mongolia, China: Multistage processes and outstanding questions. *Ore Geol. Rev.* **2015**, *64*, 459–476. [CrossRef]
21. Yu, X.; Tang, H.; Han, Z.; Li, C. Geological Characteristics and Origin of Rare Earth Elements Deposits Related with Alkaline Rock in the Chishan-Longbaoshan Area, Shandong Province. *Acta Geol. Sin.* **2012**, *84*, 407–417.
22. Peishan, Z.; Xueming, Y.; Kejie, T. *Mineralogy and Geology of Rare Earths in China*; Vision Sports Publishing: Surrey, UK, 1996.
23. Williams-Jones, A.E.; Samson, I.M.; Olivo, G.R. The Genesis of Hydrothermal Fluorite-REE Deposits in the Gallinas Mountains, New Mexico. *Econ. Geol.* **2000**, *95*, 327–341. [CrossRef]
24. Vasyukova, O.V.; Williams-Jones, A.E.; Blamey, N.J.F. Fluid evolution in the Strange Lake granitic pluton, Canada: Implications for HFSE mobilisation. *Chem. Geol.* **2016**, *444*, 83–100. [CrossRef]
25. Gysi, A.P.; Williams-Jones, A.E.; Collins, P. Lithogeochemical Vectors for Hydrothermal Processes in the Strange Lake Peralkaline Granitic REE-Zr-Nb Deposit. *Econ. Geol.* **2016**, *111*, 1241–1276. [CrossRef]
26. Gysi, A.P.; Williams-Jones, A.E. Hydrothermal mobilization of pegmatite-hosted REE and Zr at Strange Lake, Canada: A reaction path model. *Geochim. Cosmochim. Acta* **2013**, *122*, 324–352. [CrossRef]
27. Siegel, K.; Vasyukova, O.V.; Williams-Jones, A.E. Magmatic evolution and controls on rare metal-enrichment of the Strange Lake A-type peralkaline granitic pluton, Québec-Labrador. *Lithos* **2018**, *308–309*, 34–52. [CrossRef]
28. Qiu, K.-F.; Romer, R.L.; Long, Z.-Y.; Yu, H.-C.; Turner, S.; Wan, R.-Q.; Li, X.-Q.; Gao, Z.-Y.; Deng, J. Potassium isotopes as a tracer of hydrothermal alteration in ore systems. *Geochim. Cosmochim. Acta* **2024**, *368*, 185–196. [CrossRef]
29. Qiu, K.-F.; Long, Z.-Y.; Romer, R.L.; Halama, R.; Williams-Jones, A.E.; Yu, H.-C.; Li, S.-S.; Wu, M.-Q.; Deng, J. The mantle source of REE-rich alkaline silicate magmas can be enriched by continent-derived sediment subduction. *Nat. Commun.* **2025**, *16*, 11191. [CrossRef] [PubMed]
30. Anenburg, M.; Broom-Fendley, S.; Chen, W. Formation of Rare Earth Deposits in Carbonatites. *Elements* **2021**, *17*, 327–332. [CrossRef]
31. Andrade, F.R.D.; Möller, P.; Lüders, V.; Dulski, P.; Gilg, H.A. Hydrothermal rare earth elements mineralization in the Barra do Itapirapuã carbonatite, southern Brazil: Behaviour of selected trace elements and stable isotopes (C, O). *Chem. Geol.* **1999**, *155*, 91–113. [CrossRef]
32. Li, M.Y.H.; Zhou, M.-F. Hyper-enrichment of heavy rare earth element in highly evolved granites through multiple hydrothermal mobilization. *Am. Mineral.* **2024**, *109*, 1945–1959. [CrossRef]
33. Cherniak, D.J.; Zhang, X.Y.; Wayne, N.K.; Watson, E.B. Sr, Y, and REE diffusion in fluorite. *Chem. Geol.* **2001**, *181*, 99–111. [CrossRef]
34. Magyarosi, Z. Late-magmatic processes in the St. Lawrence Granite: Implications for fluorite mineralization. *J. Geochem. Explor.* **2022**, *239*, 107014. [CrossRef]
35. Alvin, M.P.; Dunphy, J.M.; Groves, D.I. Nature and genesis of a carbonatite-associated fluorite deposit at Speewah, East Kimberley region, Western Australia. *Mineral. Petrol.* **2004**, *80*, 127–153. [CrossRef]
36. Huang, Y.-Q.; Wu, M.-Q.; Germain, B.; Yu, H.-C.; Qiao, B.-X.; Zhao, Z.-G.; Qiu, K.-F. Geodynamic setting and ore formation of the Younusisayi thorium deposit in the Altyn orogenic belt, NW China. *Ore Geol. Rev.* **2022**, *140*, 104552. [CrossRef]
37. Agangi, A.; Kamenetsky, V.S.; McPhie, J. The role of fluorine in the concentration and transport of lithophile trace elements in felsic magmas: Insights from the Gawler Range Volcanics, South Australia. *Chem. Geol.* **2010**, *273*, 314–325. [CrossRef]

 https://doi.org/10.3390/min16010069

38. Pan, Y.; Fleet, M.E. Rare earth element mobility during prograde granulite facies metamorphism: Significance of fluorine. *Contrib. Mineral. Petrol.* **1996**, *123*, 251–262. [CrossRef]
39. Salvi, S.; Fontan, F.o.; Monchoux, P.; Williams-Jones, A.E.; Moine, B. Hydrothermal Mobilization of High Field Strength Elements in Alkaline Igneous Systems: Evidence from the Tamazeght Complex (Morocco). *Econ. Geol.* **2000**, *95*, 559–576. [CrossRef]
40. Tagirov, B.; Schott, J.; Harrichourry, J.-C.; Salvi, S. Experimental study of aluminum speciation in fluoride-rich supercritical fluids. *Geochim. Cosmochim. Acta* **2002**, *66*, 2013–2024. [CrossRef]
41. Chebotarev, D.A.; Veksler, I.V.; Wohlgemuth-Ueberwasser, C.; Doroshkevich, A.G.; Koch-Müller, M. Experimental study of trace element distribution between calcite, fluorite and carbonatitic melt in the system $CaCO_3 + CaF_2 + Na_2CO_3 \pm Ca_3(PO_4)_2$ at 100 MPa. *Contrib. Mineral. Petrol.* **2018**, *174*, 4. [CrossRef]
42. Duan, Z.-P.; Jiang, S.-Y.; Su, H.-M.; Zhu, X.-Y.; Jiang, B.-B. Textural features and in situ trace element analysis of fluorite from the Wujianfang fluorite deposit, Inner Mongolia (NE China): Insights into fluid metasomatism and ore-forming process. *Ore Geol. Rev.* **2022**, *147*, 104982. [CrossRef]
43. Onuma, N.; Higuchi, H.; Wakita, H.; Nagasawa, H. Trace element partition between two pyroxenes and the host lava. *Earth Planet. Sci. Lett.* **1968**, *5*, 47–51. [CrossRef]
44. Shannon, R. Revised Effective Ionic Radii and Systematic Studies of Interatomic Distances in Halides and Chalcogenides. *Acta Cryst.* **1976**, *32*, 751–767. [CrossRef]
45. Bühn, B.; Schneider, J.; Dulski, P.; Rankin, A.H. Fluid–rock interaction during progressive migration of carbonatitic fluids, derived from small-scale trace element and Sr, Pb isotope distribution in hydrothermal fluorite. *Geochim. Cosmochim. Acta* **2003**, *67*, 4577–4595. [CrossRef]
46. Richardson, C.K.; Holland, H.D. Fluorite deposition in hydrothermal systems. *Geochim. Cosmochim. Acta* **1979**, *43*, 1327–1335. [CrossRef]
47. Moeller, P.; Bau, M.; Dulski, P.; Lüders, V. REE and yttrium fractionation in fluorite and their bearing on fluorite formation. In Proceedings of the Ninth Quadrennial IAGOD Symposium, Beijing, China, 12–18 August 1998; pp. 575–592.
48. Schwinn, G.; Markl, G. REE systematics in hydrothermal fluorite. *Chem. Geol.* **2005**, *216*, 225–248. [CrossRef]
49. Zhai, M.G. Tectonic evolution and metallogenesis of North China Craton. *Miner. Deposits* **2010**, *29*, 24–36. [CrossRef]
50. Wu, F.-Y.; Xu, Y.-G.; Gao, S.; Zheng, J.-P. Lithospherie thinning and destruction of the North China Craton. *Acta Petrol. Sin.* **2008**, *24*, 1145–1174.
51. Zhao, J.-H.; Zhou, M.-F. Neoproterozoic adakitic plutons in the northern margin of the Yangtze Block, China: Partial melting of a thickened lower crust and implications for secular crustal evolution. *Lithos* **2008**, *104*, 231–248. [CrossRef]
52. Zhu, R.-X.; Chen, L.; Wu, F.-Y.; Liu, J.-L. Timing, scale and mechanism of the destruction of the North China Craton. *Sci. China Earth Sci.* **2011**, *41*, 583–592. [CrossRef]
53. Liu, Y.; Gao, S.; Hu, Z.; Gao, C.; Zong, K.; Wang, D. Continental and Oceanic Crust Recycling-induced Melt–Peridotite Interactions in the Trans-North China Orogen: U–Pb Dating, Hf Isotopes and Trace Elements in Zircons from Mantle Xenoliths. *J. Petrol.* **2010**, *51*, 537–571. [CrossRef]
54. Li, S.-S.; Santosh, M.; Cen, K.; Teng, X.-M.; He, X.-F. Neoarchean convergent margin tectonics associated with microblock amalgamation in the North China Craton: Evidence from the Yishui Complex. *Gondwana Res.* **2016**, *38*, 113–131. [CrossRef]
55. Li, S.S.; Qiu, K.F.; Hernández-Uribe, D.; Gao, Y.X.; Santosh, M.; Huang, H.C.; Zheng, Z.Y.; Zhang, Z.L.; Gao, S.C. Water Recycling in the Deep Earth: Insights From Integrated μ-XRF, THz-TDS Spectroscopy, TG, and DCS of High-Pressure Granulite. *J. Geophys. Res. Solid Earth* **2023**, *128*, e2022JB025915. [CrossRef]
56. Li, S.-S.; Santosh, M.; Teng, X.-M.; He, X.-F. Paleoproterozoic arc-continent collision in the North China Craton: Evidence from the Zanhuang Complex. *Precambrian Res.* **2016**, *286*, 281–305. [CrossRef]
57. Mao, J.-W.; Xie, G.-Q.; Zhang, Z.-H.; Li, X.-F.; Wang, Y.-T.; Zhang, C.-Q.; Li, Y.-F. Mesozoic large-scale metallogenic pulses in North China and corresponding geodynamie settings. *Acta Petrol. Sin.* **2005**, *21*, 171–190.
58. Qiu, K.-F.; Deng, J.; Laflamme, C.; Long, Z.-Y.; Wan, R.-Q.; Moynier, F.; Yu, H.-C.; Zhang, J.-Y.; Ding, Z.-J.; Goldfarb, R. Giant Mesozoic gold ores derived from subducted oceanic slab and overlying sediments. *Geochim. Cosmochim. Acta* **2023**, *343*, 133–141. [CrossRef]
59. Qiu, K.-F.; Romer, R.L.; Long, Z.-Y.; Williams-Jones, A.E.; Yu, H.-C.; Turner, S.; Wang, Q.-F.; Li, S.-S.; Zhang, J.-Y.; Duan, H.-R.; et al. The role of an oxidized lithospheric mantle in gold mobilization. *Sci. Adv.* **2024**, *10*, eado6262. [CrossRef] [PubMed]
60. Long, Z.-Y.; Qiu, K.-F.; Halama, R.; Yu, H.-C.; Zhang, J.-Y.; Li, S.-S.; Han, G.-L.; Gao, Z.-Y.; Ding, Z.-J.; Deng, J. Reappraisal of potassium isotopic composition of the continental lithospheric mantle: Insights from calc-alkaline lamprophyres in the eastern North China Craton. *GSA Bull.* **2025**, *137*, 4399–4412. [CrossRef]
61. He, D.; Qiu, K.-F.; Simon, A.; Pokrovski, G.; Yu, H.-C.; Connolly, J.; Li, S.-S.; Turner, S.; Wang, Q.-F.; Yang, M.-F.; et al. Mantle oxidation by sulfur drives the formation of giant gold deposits in subduction zones. *Proc. Natl. Acad. Sci. USA* **2024**, *121*, e2404731121. [CrossRef]

 https://doi.org/10.3390/min16010069

62. Goldfarb, R.; Qiu, K.; Deng, J.; Chen, Y.; Yang, L. Orogenic gold deposits of China. In *Mineral Deposits of China*; SEG Special Publications 22; Society of Economic Geologists: Littleton, CO, USA, 2019; Chapter 8; pp. 263–324. Available online: https://pubs.geoscienceworld.org/segweb/books/edited-volume/2452/chapter-abstract/137847688/Chapter-8-Orogenic-Gold-Deposits-of-China?redirectedFrom=fulltext (accessed on 7 January 2026).
63. Qiu, K.-F.; Goldfarb, R.J.; Deng, J.; Yu, H.-C.; Gou, Z.-Y.; Ding, Z.-J.; Wang, Z.-K.; Li, D.-P. Gold deposits of the Jiaodong Peninsula, eastern China. In *Geology of the World's Major Gold Deposits and Provinces*; Society of Economic Geologists: Littleton, CO, USA, 2020.
64. Deng, J.; Yang, L.-Q.; Groves, D.I.; Zhang, L.; Qiu, K.-F.; Wang, Q.-F. An integrated mineral system model for the gold deposits of the giant Jiaodong province, eastern China. *Earth Sci. Rev.* **2020**, *208*, 103274. [CrossRef]
65. Deng, J.; Qiu, K.-F.; Wang, Q.-F.; Goldfarb, R.; Yang, L.-Q.; Zi, J.-W.; Geng, J.-Z.; Ma, Y. In situ dating of hydrothermal monazite and implications for the geodynamic controls on ore formation in the Jiaodong gold province, eastern China. *Econ. Geol.* **2020**, *115*, 671–685. [CrossRef]
66. Yang, Z.; Wu, M.; Li, S.; Yu, H.; Diao, X.; Gao, Y.; Jiadong, M.; Zhi, C.; Shang, Z.; Xi, J.; et al. The origin of glimmerite and its significance to rare earth element mineralization: Insights from the Longbaoshan deposit in North China Craton. *Ore Geol. Rev.* **2025**, *184*, 106698. [CrossRef]
67. Lan, T.-G.; Fan, H.; Hu, F.-F.; Tomkins, A.; Yang, K.-F.; Liu, Y. Multiple crust–mantle interactions for the destruction of the North China Craton: Geochemical and Sr–Nd–Pb–Hf isotopic evidence from the Longbaoshan alkaline complex. *Lithos* **2011**, *122*, 87–106. [CrossRef]
68. Moeller, P.; Parekh, P.; Schneider, H.J. The application of Tb/Ca—Tb/La abundance ratios to problems of fluorite genesis. *Miner. Depos.* **1976**, *11*, 111–116. [CrossRef]
69. Palmer, D.; Williams-Jones, A. Genesis of the carbonatite-hosted fluorite deposit at Amba Dongar, India: Evidence from fluid inclusions, stable isotopes, and whole rock-mineral geochemistry. *Econ. Geol.* **1996**, *91*, 934–950. [CrossRef]
70. Schneider, H.J.; Möller, P.; Parekh, P.P. Rare earth elements distribution in fluorites and carbonate sediments of the east-alpine mid-triassic sequences in the Nördliche Kalkalpen. *Mineral. Depos.* **1975**, *10*, 330–344. [CrossRef]
71. Wang, C.; Liu, J.; Zhang, H.; Zhang, X.; Zhang, D.; Xi, Z.; Wang, Z. Geochronology and mineralogy of the Weishan carbonatite in Shandong province, eastern China. *Geosci. Front.* **2019**, *10*, 769–785. [CrossRef]
72. Zhao, L.-Y.; Lan, T.-G.; Fan, H.-R.; Lan, J.; Wang, H.; Terry Chen, W.; Hu, H.-L.; Li, D.-J. Origin and mineralization processes of REE during magmatic-hydrothermal evolution: Insights from in-situ geochemistry of calcite from the Weishan REE deposit, eastern North China Craton. *Ore Geol. Rev.* **2023**, *162*, 105676. [CrossRef]
73. Peretyazhko, I.; Savina, E. Fluoride-Calcium (F Ca) melt in rhyolitic magma: Evidence from fluorite-rich rhyolites of the Nyalga basin, Central Mongolia. *Lithos* **2020**, *354–355*, 105348. [CrossRef]
74. Peretyazhko, I.; Savina, E.; Karmanov, N.; Dmitrieva, A. Immiscibility of fluoride–calcium and silicate melts in trachyrhyolitic magma: Data on acidic volcanic rocks from the Nyalga Basin, Central Mongolia. *Petrology* **2018**, *26*, 389–413. [CrossRef]
75. Vasyukova, O.; Williams-Jones, A.E. Fluoride–silicate melt immiscibility and its role in REE ore formation: Evidence from the Strange Lake rare metal deposit, Québec-Labrador, Canada. *Geochim. Cosmochim. Acta* **2014**, *139*, 110–130. [CrossRef]
76. Vasyukova, O.; Williams-Jones, A.E. The evolution of immiscible silicate and fluoride melts: Implications for REE ore-genesis. *Geochim. Cosmochim. Acta* **2016**, *172*, 205–224. [CrossRef]
77. Zaitsev, A.; Marks, M.; Wenzel, T.; Spratt, J.; Sharygin, V.; Strekopytov, S.; Markl, G. Mineralogy, geochemistry and petrology of the phonolitic to nephelinitic Sadiman volcano, Crater Highlands, Tanzania. *Lithos* **2012**, *152*, 66–83. [CrossRef]
78. Yang, L.; Hinsberg, V. Liquid immiscibility in the CaF2-granite system and trace element partitioning between the immiscible liquids. *Chem. Geol.* **2019**, *511*, 28–41. [CrossRef]
79. Peretyazhko, I.; Zagorsky, V.; Savina, E.; Sapozhnikov, A. Immiscibility of calcium fluoride and aluminosilicate melts in ongonite from the Ary-Bulak intrusion, Eastern Transbaikal region. *Dokl. Earth Sci.* **2007**, *413*, 315–320. [CrossRef]
80. Bau, M.; Dulski, P. Comparative study of yttrium and rare-earth element behaviours in fluorine-rich hydrothermal fluids. *Contrib. Mineral. Petrol.* **1995**, *119*, 213–223. [CrossRef]
81. Bau, M. Controls on the fractionation of isovalent trace elements in magmatic and aqueous systems: Evidence from Y/Ho, Zr/Hf, and lanthanide tetrad effect. *Contrib. Mineral. Petrol.* **1996**, *123*, 323–333. [CrossRef]
82. Shock, E.; Helgeson, H.; Sverjensky, D. Calculation of the thermodynamic and transport properties of aqueous species at high pressures and temperatures: Standard partial molal properties of inorganic neutral species. *Geochim. Cosmochim. Acta* **1989**, *53*, 2157–2183. [CrossRef]
83. Payne, M.R.; Gysi, A.P.; Hurtig, N.C. Hydrothermal fluorite solubility experiments and mobility of REE in acidic to alkaline solutions from 100 to 250 °C. *Chem. Geol.* **2023**, *617*, 121256. [CrossRef]
84. Migdisov, A.A.; Williams-Jones, A.E. Hydrothermal transport and deposition of the rare earth elements by fluorine-bearing aqueous liquids. *Mineral. Depos.* **2014**, *49*, 987–997. [CrossRef]
85. Walker, J.; Choppin, G. Thermodynamic Parameters of Fluoride Complexes of the Lanthanides. *Adv. Chem.* **1967**, *71*, 127–140. [CrossRef]

 https://doi.org/10.3390/min16010069

86. Loges, A.; Migdisov, A.; Wagner, T.; Williams-Jones, A.; Markl, G. An experimental study of the aqueous solubility and speciation of Y(III) fluoride at temperatures up to 250 °C. *Geochim. Cosmochim. Acta* **2013**, *123*, 403–415. [CrossRef]
87. Moeller, P. *Hydrogeology of Crystalline Rocks*; Springer: Berlin/Heidelberg, Germany, 2000; pp. 227–246.
88. Bilal, B. Thermodynamic study of Eu3+/Eu2+ redox reaction in aqueous solutions at elevated temperatures and pressures by means of cyclic voltammetry. *Z. Naturforsch. A* **1991**, *46*, 1108–1116. [CrossRef]
89. Sverjensky, D. Europium equilibria in aqueous solution. *Earth Planet. Sci. Lett.* **1984**, *67*, 70–78. [CrossRef]
90. Li, J.-K.; Yuan, Z.-Z.; Bai, G.; Chen, Y.-C.; Wang, D.-H.; Ying, L.-J.; Zhang, J. Ore-Forming Fluid Evolvement And Its Controlling To REE(Ag)Mneralizing in The Wshan Deposit, Shandong. *J. Miner. Petrol.* **2009**, *29*, 60–68. [CrossRef]
91. Hintzen, R.; Werner, W.; Hauck, M.; Klemd, R.; Fischer, L. Multistage fluorite mineralization in the southern Black Forest, Germany: Evidence from rare earth element (REE) geochemistry. *Eur. J. Mineral.* **2023**, *35*, 403–426. [CrossRef]
92. Zeng, X.; Li, X.; Fan, H.; Lan, T.; Lan, J.; Su, J.; Zhang, P.; Yang, K.; Zhao, X. Generation of REE-rich syenite-(carbonatite) complex through lithosphere-asthenosphere interaction: An in-situ Sr–Nd–O isotopic study of the Mesozoic Weishan pluton, Northern China. *J. Asian Earth Sci.* **2022**, *230*, 105191. [CrossRef]
93. Nasraoui, M.; Toulkeridis, T.; Clauer, N.; Bilal, E. Differentiated hydrothermal and meteoric alterations of the Lueshe carbonatite complex (Democratic Republic of Congo) identified by a REE study combined with a sequential acid-leaching experiment. *Chem. Geol.* **2000**, *165*, 109–132. [CrossRef]
94. Migdisov, A.A.; Williams-Jones, A.E. An experimental study of the solubility and speciation of neodymium (III) fluoride in F-bearing aqueous solutions. *Geochim. Cosmochim. Acta* **2007**, *71*, 3056–3069. [CrossRef]
95. Migdisov, A.A.; Williams-Jones, A.E.; Wagner, T. An experimental study of the solubility and speciation of the Rare Earth Elements (III) in fluoride- and chloride-bearing aqueous solutions at temperatures up to 300 °C. *Geochim. Cosmochim. Acta* **2009**, *73*, 7087–7109. [CrossRef]

https://doi.org/10.3390/min16010069

minerals

MDPI

Article

Apatite Geochemical Signatures of REE Ore-Forming Processes in Carbonatite System: A Case Study of the Weishan REE Deposit, Luxi Terrane

Yi-Xue Gao [1], Shan-Shan Li [1,2,*], Chuan-Peng Liu [3,*], Ming-Qian Wu [1,2], Zhen Shang [3], Yi-Zhan Sun [1], Ze-Yu Yang [1] and Kun-Feng Qiu [1]

1 State Key Laboratory of Geological Processes and Mineral Resources, Frontiers Science Center for Deep-Time Digital Earth, School of Earth Sciences and Resources, China University of Geosciences, Beijing 100083, China; gaoyixue2020@email.cugb.edu.cn (Y.-X.G.); aria.wu@uwindsor.ca (M.-Q.W.); yizhansun61@163.com (Y.-Z.S.); zeyuyang@email.cugb.edu.cn (Z.-Y.Y.); kunfengqiu@cugb.edu.cn (K.-F.Q.)

2 Shandong Key Laboratory of Mineralization Processes and Resources Utilization of Strategic Metal Minerals (Preparatory), Key Laboratory of Gold Mineralization Processes and Resource Utilization Subordinated to the Ministry of Land and Resources, Shandong Institute of Geological Sciences, Jinan 250013, China

3 Shandong Provincial Lunan Geology and Exploration Institute (Shandong Provincial Bureau of Geology and Mineral Resources No. 2 Geological Brigade), Rare Mineral Exploration and Comprehensive Utilization Engineering Research Center of Shandong Province, Jining 272100, China; shangxiaohuai@yeah.net

* Correspondence: lishanshan199811@163.com (S.-S.L.); liuchuanpeng666@163.com (C.-P.L.); Tel.: +86-18032186796 (S.-S.L.); +86-13869912915 (C.-P.L.)

Abstract

The Weishan rare earth element (REE) deposit, located in western Shandong, North China Block, is a typical carbonatite REE deposit and constitutes the third largest light REE resource in China. Its mineralization is closely related to the multi-stage evolution of a carbonatite magma–hydrothermal system. However, the mechanisms governing REE enrichment, migration, and precipitation remain insufficiently constrained from a mineralogical perspective, which hampers the understanding of the ore-forming processes and the establishment of predictive exploration models. Apatite is a pervasively developed REE phase in the Weishan deposit which occurs in multiple generations, and thus represents an ideal recorder of the magmatic–hydrothermal evolution. In this study, different generations of apatite hosted in carbonatite orebodies from the Weishan deposit were investigated using cathodoluminescence (CL), electron probe microanalysis (EPMA), and in situ LA-ICP-MS trace element analysis. Three types of apatite were identified. In paragenetic sequence, Ap-1 occurs as polycrystalline aggregates coexisting with calcite, is enriched in Na, Sr, and LREEs, and shows high $(La/Yb)_N$ ratios, suggesting crystallization from an evolved carbonatite magma. Ap-2 and Ap-3 display typical replacement textures: both contain abundant dissolution pits and dissolution channels within the grains, which are filled by secondary minerals such as monazite and ancylite, and thus exhibit characteristic features of fluid-mediated dissolution–reprecipitation during the hydrothermal stage. Ap-2 is commonly associated with barite and strontianite, whereas Ap-3 is associated with pyrite and monazite and is characterized by relatively sharp grain boundaries with adjacent minerals. From Ap-1 to Ap-3, total REE contents decrease systematically, whereas Na, Sr, and P contents increase. All three apatite types lack Eu anomalies but display positive Ce anomalies. Discrimination diagrams involving LREE-Sr/Y and log(Ce)-log(Eu/Y) indicate that apatite in the Weishan REE deposit formed during the magmatic to hydrothermal evolution of a carbonatite, and that the dissolution of early

 https://doi.org/10.3390/min16010112

magmatic apatite, followed by element remobilization and mineral reprecipitation, effectively records the progressive evolution of the ore-forming fluid.

Keywords: Weishan REE deposit; apatite geochemistry; REE ore-forming process; carbonatite system

1. Introduction

Rare earth elements (REEs) are strategic critical metals that are indispensable to modern high-technology industries, low-carbon energy technologies, national defense, and advanced manufacturing [1–4]. Among these, carbonatite-associated REE deposits, characterized by being large-scale and high-grade and a predominance of light rare earth elements (LREEs), have become the primary global source of rare earth resources, especially elements like cerium (Ce), lanthanum (La), and neodymium (Nd). Representative world-class examples of such deposits include the Bayan Obo deposit in China and the Mountain Pass deposit in the United States [5–9]. Compared to granite-associated or ion-adsorption-type REE deposits, carbonatite REE deposits stand out due to the presence of ore-forming carbonatitic magmas [10–14]. Carbonatites are widely considered to derive from low-degree partial melting of an enriched mantle source, and they are inherently rich in REEs and volatile components, such as CO_2, F, Cl, and S [15–19]. However, REE-rich primary carbonatite magma alone is generally insufficient to produce economic mineralization, implying that additional magmatic–hydrothermal processes are required to concentrate REEs to ore grades [20–22]. The process by which REEs transition from a dispersed state to highly concentrated, economically viable orebodies involves a series of complex steps, including crystallization, differentiation of magma, exsolution of immiscible fluid phases, fluid migration, and fluid–rock interactions with surrounding rocks [23–25]. Therefore, accurately understanding the REE ore-forming process, from source, through migration, to enrichment and ore formation, remains a central scientific objective in research on carbonatite-hosted REE systems and is directly relevant to exploration targeting and resource evaluation [6,26–31].

Apatite is a common REE-bearing mineral in carbonatite REE deposits, and its chemistry is highly sensitive to subtle variations in formation conditions [19,32,33]. Apatite can incorporate a large quantity of REEs, and its REE distribution patterns, anomaly features (such as Eu and Ce anomalies), and covariant relationships with other trace elements (e.g., Sr, Y) record the composition of parental melts/fluids, oxygen fugacity, crystallization temperature, and fluid involvement, making it an effective tracer of ore-forming processes [34–37]. Petrographic observations, coupled with geochemical and geochronological studies, have provided key constraints on magmatic–hydrothermal ore-forming processes, including the partitioning behavior of volatile components (e.g., F and Cl) among melt, fluid, and mineral phases [38–40]. Experimental petrology of apatite has shown that, in hydrothermal systems, apatite is not always formed directly by crystallization. Fluid-mediated dissolution–reprecipitation processes are key mechanisms that control its compositional, structural, and mineralogical variations [41–44]. These processes can mobilize the internal chemical composition of apatite without altering its original morphology, thus providing an effective pathway for the activation and re-enrichment of ore-forming elements. Some scholars have also utilized machine learning techniques to establish geochemical databases for discriminating the genesis of apatite [45,46].

https://doi.org/10.3390/min16010112

Multiple generations of apatite in the Weishan REE deposit provide a valuable archive for reconstructing mineralization; however, how their chemistry records fluid evolution and contributes to extreme REE enrichment remains poorly constrained. Addressing this issue requires detailed petrography integrated with a robust paragenetic framework. Here, we investigate the trace element geochemistry of apatite from carbonatite-associated REE mineralization, with the aim of (i) identifying the mechanisms of element transport and sequestration in apatite, (ii) tracking the evolution of ore-forming fluids, and (iii) assessing the implications for REE enrichment. Our results offer new mineralogical constraints on the ore-forming processes of the Weishan REE deposit in the Luxi terrane.

2. Regional Geology

The North China Block (NCB) is one of the Earth's most ancient and well-preserved cratonic regions (Figure 1A). Its geological development encompasses the formation of an early continental core, followed by crustal growth, the amalgamation of micro-continental blocks, and extensive magmatic, metamorphic, and cratonization processes [47–50]. In later stages, the NCB underwent several episodes of metasomatism linked to subduction and lithospheric thinning, along with multiple large-scale metallogenic events that led to the formation of various critical mineral resources, including iron, copper, gold, and rare metals [30,51–56]. The Luxi Terrane, which lies along the southeastern boundary of the NCB (Figure 1B), is defined by four major fault zones: the Fengpei, Qihe-Guangrao, Liaocheng-Lankao, and Yishui faults. The Luxi Terrane, with its eastern margin defined by the NE–SW-trending Tan-Lu Fault, preserves a complex stratigraphic record formed through multiple tectonic events, in a temporal manner, the Archean gneisses, amphibolite, and trondhjemite–tonalite–granodiorite (TTG, ca.2.5 Ga) [57], the Neoproterozoic marine clastic rocks which consist of sandstone, shale, and limestone, as well as the Palaeozoic marine sedimentary rocks including dolomite, shale, and sandstone. Quaternary sediments mainly consist of clay and mudstone. Active magmatism in this region from the Jurassic to the Cretaceous formed a large amount of alkaline and carbonatitic igneous rocks that are spatially and temporally related to REEs and Au [31,55,56,58–61]. These intrusive bodies show diverse shapes and are found as plutons, stocks, and dikes and their spatial and temporal distribution is closely tied to the regional fault structure, reflecting sustained and active tectono-magmatic processes beneath the Luxi Terrane.

https://doi.org/10.3390/min16010112

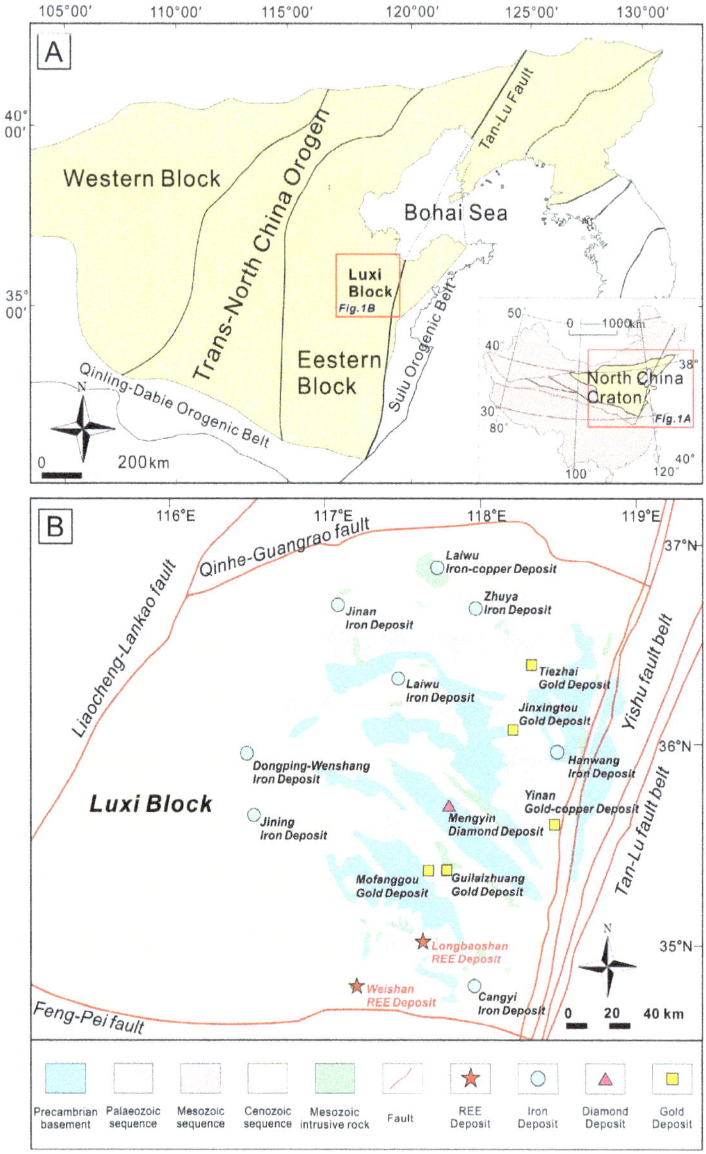

Figure 1. (**A**) Schematic map showing the position of the Luxi Block (modified after [62]). (**B**) Simplified geologic map showing the position of Weishan area in the Luxi Terrane (modified after [63]).

3. Deposit Geology

The Weishan REE deposit in the southern Luxi Terrane (Figure 1B) is a giant alkaline–carbonatite-associated deposit and represents the largest REE deposit in the region, as well as the third-largest light rare earth production base in China. Magmatic activity in this region is prominently characterized by Yanshanian syenite (including quartz syenite and aegirine-augite quartz syenite) and alkaline granites, which comprise the Weishan alkaline complex (Figure 2). The complex covers an area of approximately 0.5 km^2, extending in a NE–SW direction. It intrudes the Archean gneisses, forming distinct, sharp contacts, and displays localized alkaline metasomatic alteration along its boundaries. Based

https://doi.org/10.3390/min16010112

on mineral assemblages, earlier studies subdivided the carbonatite of the Weishan deposit into REE-bearing quartz-barite-carbonatite, REE-bearing aegirine granite porphyry, aegirine-bearing veins, and monazite-bearing apatite veins [63,64]. The dominant REE minerals are bastnaesite, with minor occurrences of parisite-(Ce), strontianite phases, silicotitanates, ancylite, monazite, and apatite-(Ce). The ore is characterized by diverse and spatially heterogeneous mineral assemblages. Geochronological data support that the broadly coeval crystallization of alkaline intrusions overextend geological time, with zircon U-Pb ages for syenites (130–122 Ma) and bastnaesite Th-Pb ages (129 Ma) for carbonatites [57,64–66]. Geochronology data also support the relatively young age of carbonatite with mica Rb-Sr of 119 Ma [67].

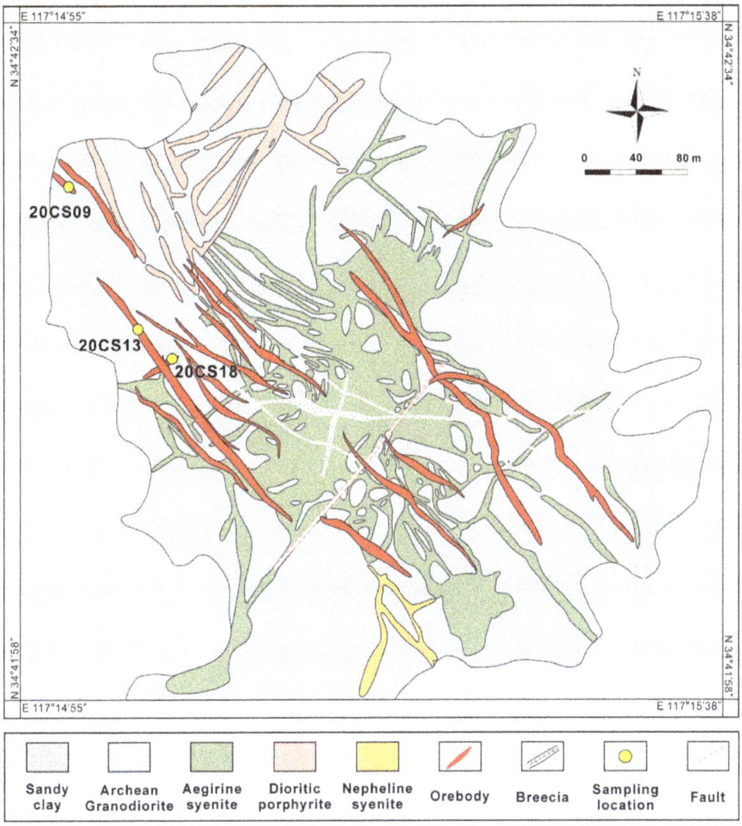

Figure 2. Simplified geological map of the Weishan REE deposit (modified after [63]).

4. Samples and Analytical Methods

4.1. Sample Descriptions

Based on previous work, this study conducted systematic sampling across different ore types within the Weishan REE deposit. Detailed petrographic observations were performed on apatite occurring in three representative ore types of contrasting mineralization (Figure 3): quartz-calcite carbonatite (20CS13), REE-barite carbonatite (20CS18), and REE-fluorite carbonatite (20CS09).

https://doi.org/10.3390/min16010112

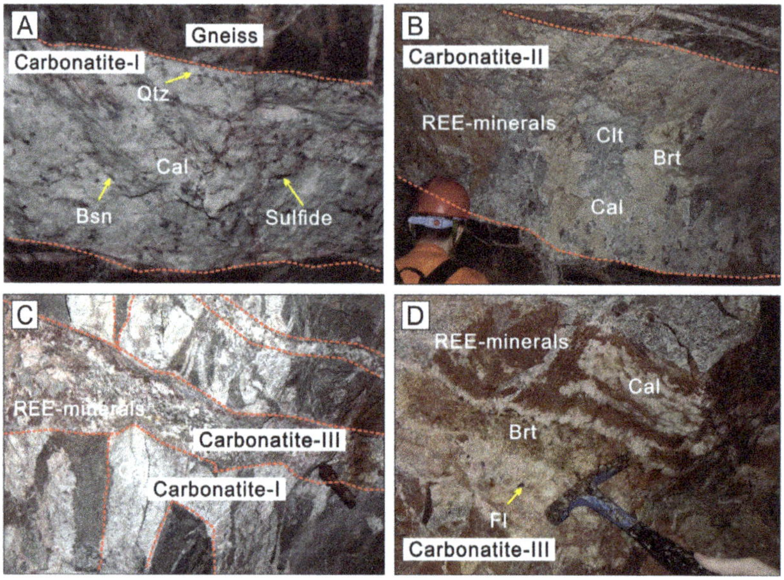

Figure 3. Carbonatites occurred in the Weishan REE deposit. (**A**) Carbonatite-I: quartz-calcite carbonatite intruded into Archean basement metamorphic rocks with a clear boundary. (**B**) Carbonatite-II: REE-barite carbonatite with quartz, REE-bearing minerals, calcite, and barite. (**C**,**D**) Carbonatite-III: REE-fluorite carbonatite. Crosscutting relationship between Carbonatite-I and Carbonatite-III shows the carbonatite sequences. Cal = calcite; Qtz = quartz; Bsn = bastnaesite; Clt = celestine; Brt = barite; Fl = fluorite.

The quartz-calcite carbonatite and its adjacent alkaline metasomatic gneiss (Figure 3A) exhibit a grayish-white color, medium-grained texture, and a veined structural fabric. Ore minerals are dominated by bastnaesite and parasite, whereas gangue minerals include calcite, quartz, fluorite, barite, celestine, and minor sulfides such as pyrite, chalcopyrite, and sphalerite. The wall rock, composed of Neoarchean gneiss, displays a flesh-red to gray-white color and a granular texture with gneissic foliation. The flesh-red portions, representing intense alkaline alteration, are characterized by a mineral assemblage of K-feldspar (50%), plagioclase (30%), quartz (15%), calcite (10%), and minor REE minerals. The gray-white portions record weaker alkaline alteration and consist of plagioclase (55%), biotite (25%), K-feldspar (10%), and quartz (5%), along with minor calcite and REE minerals. Accessory phases in both domains include rutile, zircon, and apatite (Figure 4A,C). The REE-barite carbonatite (Figure 3B) is brownish to gray-white, medium-grained, and massive in structure (20CS18). The ore minerals are dominated by synchysite-(Ce), with minor bastnaesite and calcite. Gangue minerals include calcite, celestine, barite, and quartz (Figure 4D). The REE-fluorite carbonatite (Figure 3C,D) exhibits a brownish to dark-gray coloration, medium- to coarse-grained texture, and a disseminated structure. Ore minerals consist primarily of bastnaesite, synchysite, monazite, and ancylite, whereas the gangue assemblage comprises calcite, quartz, fluorite, barite, celestine, and minor sulfides including pyrite, chalcopyrite, and sphalerite (Figure 4B).

https://doi.org/10.3390/min16010112

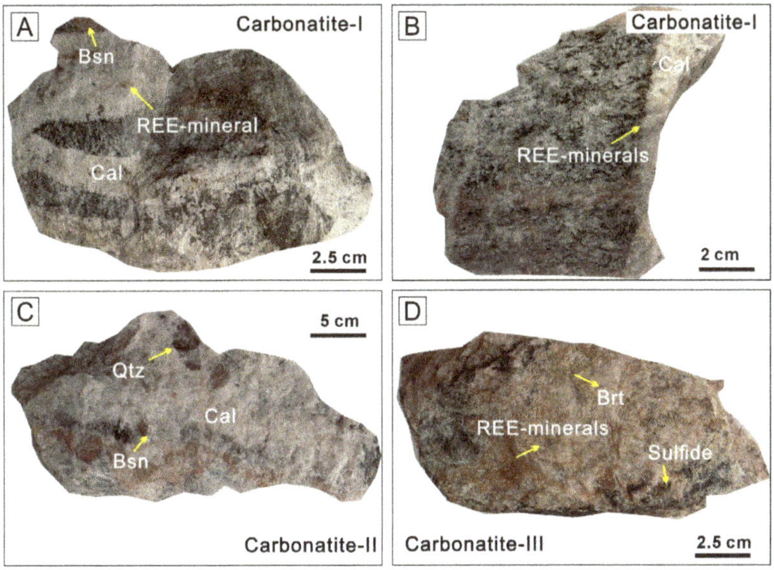

Figure 4. Representative hand specimen photograph of different carbonatites from the Weishan REE deposit. (**A**,**B**) Photograph of a hand specimen representing Carbonatite-I. The sample displays a medium-grained texture with crystals of bastnäsite, REE-bearing minerals, and calcite. (**C**) Photograph of a hand specimen representing Carbonatite-II. The sample features a coarse-grained texture with distinct crystals of quartz and calcite adjacent to a patch of bastnäsite. (**D**) Photograph of a hand specimen representing Carbonatite-III. The sample shows a fine- to medium-grained texture dominated by barite and contains patches of REE-bearing minerals in association with sulfide minerals. Cal = calcite; Bsn = bastnaesite; Brt = barite; Qtz = quartz.

4.2. Analytical Methods

Cathodoluminescence (CL): Microscopic petrographic observations, cathodoluminescence (CL) imaging, and back-scattered electron (BSE) imaging of the apatite samples were conducted at the China University of Geosciences (Beijing, China) and the Research Institute of Petroleum Exploration and Development, PetroChina (Beijing, China). The analyses were performed using a CLF-2 cathodoluminescence system (BII, Whitby, ON, Canada) coupled with a Zeiss Axio Imager multifunctional microscope (Carl Zeiss AG, Jena, Germany). Operating conditions were 12.5 V accelerating voltage and 900 mA beam current. Grain microtextures including crack, zonation, and fluid inclusion were checked by the CL and BSE images, produced by JXA-8800 electron microprobe (JEOL Ltd., Tokyo, Japan) with operating conditions of 20 kV and 20 nA. Suitable sites were chosen within crystal grains for apatite geochemistry.

EPMA: The major element concentrations in apatite crystals from the quartz-barite-carbonatite, REE-barite carbonatite, and REE-fluorite carbonatite were analyzed following thin-section carbon coating using an electron probe microanalyzer JXA-8230 (JEOL Ltd., Tokyo, Japan) at the Shandong Institute of Geological Sciences (Jinan, China) with an accelerating voltage of 20 Kv, a probe current of 20 nA, a beam size of 5 μm, and a counting time of 20 s. Different well-characterized minerals were used as standards. The reference materials selected for calibration include F (topaz), Na and Si (jadeite), Mg (forsterite), P and Ca (apatite), Ti (rutile), K (K-feldspar), Fe (hematite), Cl (halite), S (barite), Mn (manganese oxide), and REE (LaP_5O_{14}, Ce P_5O_{14}, etc.). Its OH^- content was calculated from the measured total water content of apatite.

https://doi.org/10.3390/min16010112

LA-ICP-MS: In situ laser ablation inductively coupled plasma mass spectrometry (LA-ICP-MS) was employed for apatite trace element analysis. The analyses were conducted at the National Research Center for Geoanalysis, Chinese Academy of Geological Sciences. The laser ablation system used was an NWR193HE ArF high-energy excimer laser ablation system (ESL GmbH, Bozeman, MT, USA), coupled with an Agilent 8900 ICP-MS instrument (Agilent Technologies Inc., Santa Clara, CA, USA). During laser ablation, helium served as the carrier gas and argon as the make-up gas to adjust sensitivity; both gases were homogenized using a gas homogenizer before entering the ICP. The laser spot size was 30 μm, the repetition rate was set to 6 Hz, and the laser fluence was 6 J/cm^2. Each time-resolved analysis cycle consisted of 20 s of background signal acquisition followed by 40 s of sample signal acquisition.

5. Results

5.1. Petrography and Texture of Apatite

Apatite grains from carbonatites of the Weishan REE deposit were examined using polarized light microscopy, scanning electron microscopy (SEM), and cathodoluminescence (CL) techniques. Based on the morphological features and CL color characteristics, three types of apatite were identified (Figures 5 and 6). Ap-1, collected from low-grade ores, occurs as subhedral grains (Figure 5A,B) with sharp boundaries and relatively clean interiors. The grains, 100–200 μm in size, are commonly intergrown aggregates. Under optical CL, Ap-1 displays a deep brown luminescence (Figure 6A,B), and SEM-CL shows zones of compositional heterogeneity (Figure 6E). Minor rhombohedral strontianite, monazite, and parasite occur along grain margins. The mineral assemblage includes calcite, celestine, quartz, and minor REE minerals. Ap-2, derived from the NE-trending carbonatite, occurs as anhedral to subhedral grains with diameters ranging from 300 μm to 1 mm (Figure 5C,D). Some crystals exhibit pseudo-hexagonal outlines, and their cores and rims contain abundant inclusions of ancylite and strontianite, closely intergrown with monazite. Optical CL shows a pinkish-violet emission (Figure 6C). The coexisting mineral assemblage comprises calcite, muscovite, strontianite, celestine, barite, monazite, and ancylite. Ap-3, from the NW-trending high-grade orebody, appears as anhedral grains (200–300 μm) with heavily altered rims (Figure 5E,F). The grains contain abundant REE-bearing mineral inclusions, such as bastnaesite and monazite. In transmitted-light CL images, Ap-3 exhibits pale pink luminescence (Figure 6D), and SEM-CL show patchy zones. The associated minerals include calcite, celestine, barite, quartz, fluorite, bastnaesite, and pyrite.

5.2. Trace and Rare Earth Elements Geochemistry

Major element compositions of apatite determined by in situ electron microprobe analysis are presented in Table 1. The results show distinct variations. The CaO contents of the three apatite types range from 41.66 to 44.63 wt%, 45.59 to 48.46 wt%, and 50.77 to 52.73 wt%, respectively, whereas P_2O_5 contents range from 36.66 to 37.95 wt%, 37.64 to 39.78 wt%, and 39.40 to 41.63 wt%. These systematic differences may reflect variable degrees of major element mobility during late-stage hydrothermal alteration. The Na_2O contents range from 0.86 to 1.32 wt% in Ap-1, 1.00 to 1.46 wt% in Ap-2, and 0.13 to 0.66 wt% in Ap-3. Striking contrasts are also observed in SrO contents, which vary between 4.53 and 6.35 wt% (Ap-1), 4.05 and 4.46 wt% (Ap-2), and 0.56 and 1.92 wt% (Ap-3). Notably, Na and Sr concentrations display an inverse correlation with Ca and P, suggesting coupled substitution mechanisms during hydrothermal modification. Fluorine contents in all apatite types vary between 2.85 and 4.30 wt%, whereas Cl contents are consistently low (<0.13 wt%) and show negligible inter-type differences. A weak negative correlation between F and Cl is observed (Figure 7). In the F-Cl-OH ternary diagram, all apatite samples plot close to the

F apex and secondarily toward the OH apex, indicating a predominance of fluorapatite compositions across the Weishan deposit.

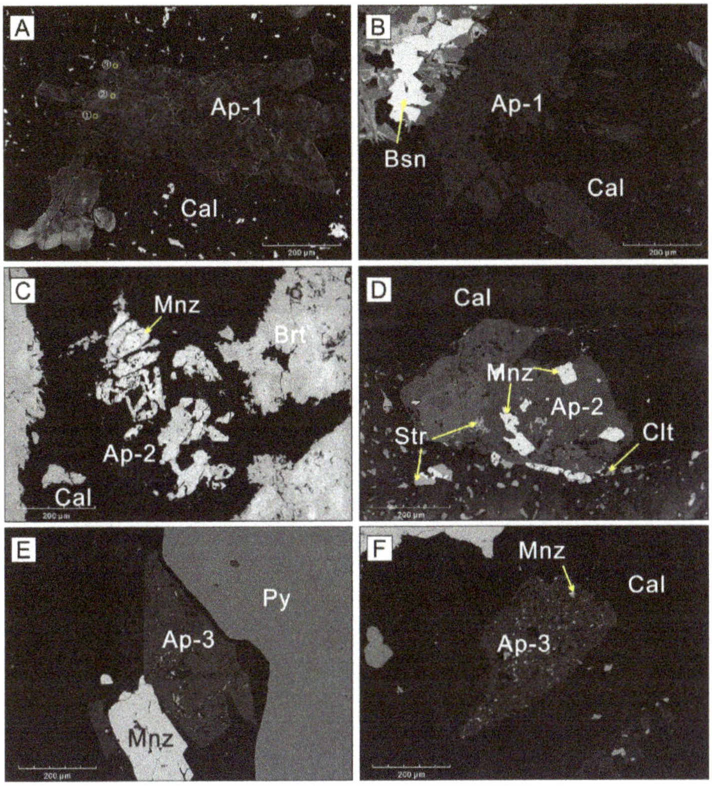

Figure 5. Representative BSE images show textures and mineral assemblages of different generations of apatite. (**A**,**B**) Type 1 apatite (Ap-1) coexists with calcite. Yellow circle and the circled numbers denote the geochemical analytical spots. (**C**,**D**) Type 2 apatite (Ap-2) occurs with celestite and barite, and develops rare earth mineral inclusions such as calcite and monazite. (**E**,**F**) Type 3 apatite (Ap-3) coexists with metasomatic textures and is infilled with monazite and pyrite. Dissolution pits and channels occur in apatite crystals. Ap = apatite; Cal = calcite; Bsn = bastnaesite; Brt = barite; Mnz = monazite; Str = Strontianite; Clt = celestine; Py = pyrite. ① = 20CS13-1-01; ② = 20CS13-1-02; ③ = 20CS13-1-03.

Trace element compositions obtained by in situ LA-ICP-MS are presented in Table 2. Thorium and uranium incorporate into the apatite lattice via $Th^{4+} = 2Ca^{2+}$ and $U^{4+} = 2Ca^{2+}$ substitution. U contents show limited variation among the three apatite types. Th contents are generally low (<75 ppm in most analyses). Ap-2 contains the lowest U, whereas Ap-1 and Ap-3 show slightly higher averages of 1.28 ppm and 4.27 ppm, respectively. Chondrite-normalized REE patterns (Figure 8) reveal marked differences in total REE abundances and distribution among the three types. Both Ap-2 and Ap-3 exhibit reduced ΣREE + Y concentrations relative to Ap-1, with averages of 67,332 ppm and 42,612 ppm, respectively. LREE contents likewise decrease in Ap-2 (39,221 ppm) and Ap-3 (40,228 ppm), slightly lower than Ap-1. In contrast, HREE contents are modestly elevated in Ap-2 (1875 ppm) and Ap-3 (1144 ppm), compared with Ap-1 (822 ppm).

 https://doi.org/10.3390/min16010112

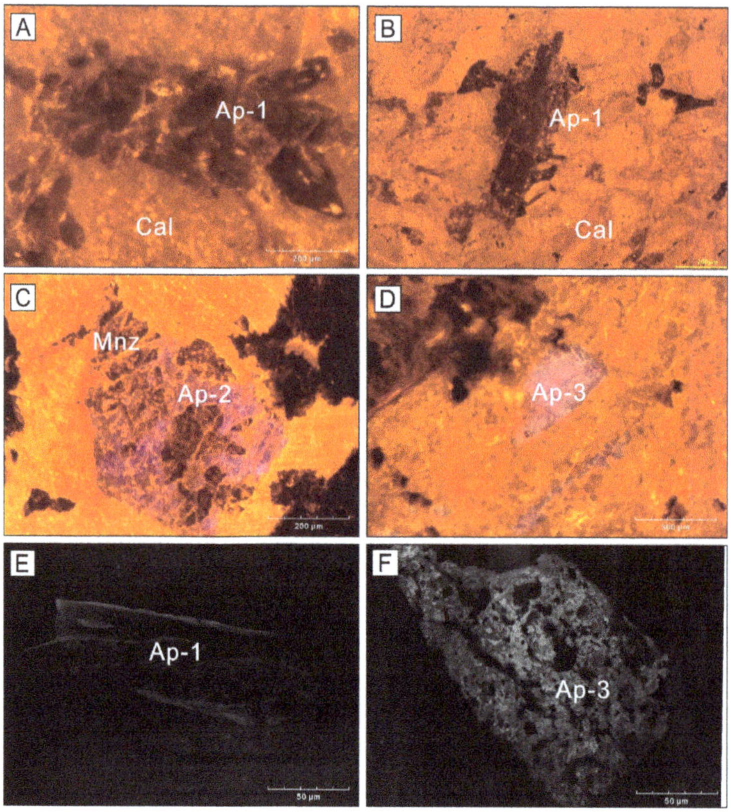

Figure 6. Cathodoluminescence characteristics of apatite. (**A**,**B**) Ap-1 exhibits a brownish luminescence, with calcite partially filling intergranular spaces. (**C**,**D**) Ap-2 and Ap-3 display a pale pink CL emission and show internally mottled textures. (**E**,**F**) SEM-CL imaging further reveals that Ap-1 contains compositionally heterogeneous domains and faint zoning, whereas Ap-3 is characterized by a strongly mottled structure. Ap = apatite; Cal = calcite; Mnz = monazite.

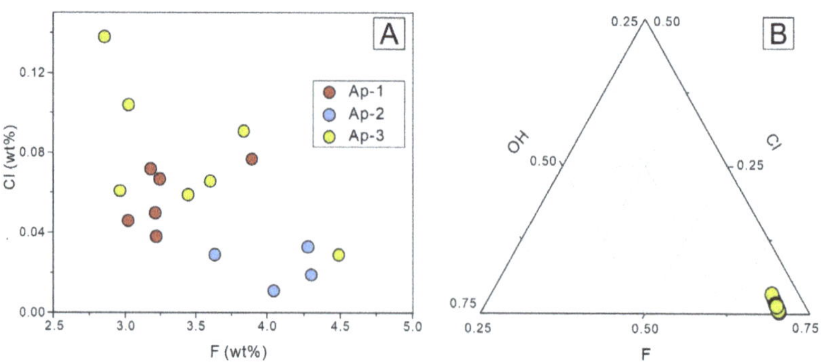

Figure 7. (**A**) Binary diagram of Cl/F content of apatite from the Weishan by EPMA. (**B**) Ternary diagram of F-Cl-OH content of apatite from the Weishan by EPMA, Ap-1 and Ap-2 are largely overlapped by those of Ap-3 in this diagram.

https://doi.org/10.3390/min16010112

Table 1. Major elements for the apatite from the Weishan carbonatite by EPMA.

Generation	Ap-1	Ap-1	Ap-1	Ap-1	Ap-1	Ap-1	Ap-1	Ap-2	Ap-2
Samples	20CS13-1-01	20CS13-1-02	20CS13-1-03	20CS13-1-04	20CS13-1-05	20CS13-1-06	20CS13-1-07	20CS18-2-01	20CS18-2-02
SrO	5.23	4.85	5.58	6.36	4.54	5.67	5.75	4.17	4.36
Y_2O_3	0.00	0.00	0.00	0.00	0.00	0.00	0.00	0.00	0.00
K_2O	0.04	0.02	0.02	0.00	0.02	0.01	0.05	0.05	0.00
CaO	42.52	42.09	41.67	44.64	43.74	43.10	42.83	46.14	47.26
MnO	0.00	0.03	0.10	0.02	0.13	0.10	0.14	0.06	0.00
FeO	0.00	0.00	0.05	0.06	0.00	0.00	0.00	0.09	0.08
F	3.21	3.18	3.89	3.22	3.02	3.24	3.00	3.63	4.28
Na_2O	1.30	1.32	1.30	0.86	1.09	0.92	1.13	1.47	1.01
MgO	0.00	0.00	0.00	0.00	0.03	0.00	0.02	0.00	0.00
Al_2O_3	0.01	0.00	0.01	0.00	0.00	0.00	0.00	0.00	0.00
BaO	0.00	0.00	0.00	0.00	0.00	0.00	0.00	0.00	0.00
Ce_2O_3	4.43	4.05	4.55	2.44	3.21	3.66	3.63	1.39	1.16
Nd_2O_3	2.16	2.01	2.12	1.19	1.51	1.69	1.95	0.00	0.00
P_2O_5	36.40	36.89	36.67	37.49	37.95	37.46	37.19	38.72	38.99
SiO_2	1.08	0.97	1.05	0.23	0.45	0.58	0.85	0.05	0.04
SO_3	0.16	0.19	0.19	0.18	0.16	0.20	0.15	0.71	0.30
Cl	0.05	0.07	0.08	0.04	0.05	0.07	0.08	0.03	0.03
OH	1.36	1.35	1.65	1.36	1.28	1.38	1.28	1.53	1.81
Total	95.22	94.32	95.59	95.35	94.63	95.30	95.47	94.97	95.70
Generation	Ap-2	Ap-2	Ap-3	Ap-3	Ap-3	Ap-3	Ap-3	Ap-3	Ap-3
Samples	20CS18-2-03	20CS18-2-04	20CS09-1-01	20CS09-1-02	20CS09-1-03	20CS09-1-04	20CS09-1-05	20CS09-1-06	20CS09-1-07
SrO	4.06	4.47	0.56	0.79	1.33	0.91	1.38	1.93	0.87
Y_2O_3	0.13	0.03	0.01	0.00	0.21	0.29	0.14	0.00	0.20
K_2O	0.00	0.00	0.09	0.13	0.10	0.07	0.00	0.02	0.05
CaO	48.47	45.59	51.28	50.77	52.03	51.96	51.45	52.74	52.67
MnO	0.00	0.02	0.00	0.00	0.00	0.00	0.13	0.13	0.00
FeO	0.01	0.00	0.06	0.10	0.08	0.00	0.13	0.09	0.09
F	4.30	4.04	3.02	2.85	3.83	2.96	3.44	4.49	3.60
Na_2O	1.04	1.38	0.36	0.66	0.50	0.53	0.44	0.52	0.13
MgO	0.02	0.00	0.02	0.01	0.03	0.00	0.01	0.01	0.01
Al_2O_3	0.00	0.00	0.01	0.02	0.00	0.00	0.03	0.00	0.02
BaO	0.07	0.00	0.01	0.00	0.06	0.01	0.07	0.00	0.02
Ce_2O_3	1.16	1.41	0.45	0.75	0.40	0.42	0.67	0.83	0.16
Nd_2O_3	0.07	0.00	0.11	0.24	0.13	0.18	0.38	0.45	0.11
P_2O_5	39.78	37.64	40.39	39.40	41.05	40.27	41.14	41.63	41.14
SiO_2	0.02	0.03	0.03	0.16	0.06	0.07	0.06	0.06	0.15
SO_3	0.27	0.87	0.19	0.21	0.25	0.46	0.28	0.34	0.07
Cl	0.02	0.01	0.10	0.14	0.09	0.06	0.06	0.03	0.07
OH	1.81	1.70	1.30	1.23	1.63	1.26	1.46	1.90	1.53
Total	97.60	93.79	95.41	95.00	98.50	96.94	98.34	101.37	97.81

https://doi.org/10.3390/min16010112

Table 2. LA-ICP-MS analyses of apatite from the Weishan REE deposit.

Apatite Type	Ap-1	Ap-1	Ap-1	Ap-1	Ap-1	Ap-2	Ap-2	Ap-2
Spot	20CS13-1-01	20CS13-1-02	20CS13-1-03	20CS13-1-04	20CS13-1-05	20CS18-2-01	20CS18-2-02	20CS18-2-03
Li	26.5	5.7	11.0	11.2	8.3	26.5	5.7	11.0
Be	0.1	0.2	0.2	0.2	0.3	0.1	0.2	0.2
B	1.7	2.5	1.9	3.0	2.1	1.7	2.5	1.9
S	640	1353	1553	1410	877	640	1353	1553
Sc	0.12	0.29	0.20	0.00	0.06	0.12	0.29	0.20
V	0.22	0.60	0.64	0.60	0.44	0.22	0.60	0.64
Cr	3.94	15.11	9.10	7.65	0.00	3.94	15.11	9.10
Co	0.03	0.54	0.41	0.56	0.22	0.03	0.54	0.41
Ni	0.00	0.00	0.00	0.94	0.00	0.00	0.00	0.00
Cu	0.09	1.01	0.88	1.14	0.64	0.09	1.01	0.88
Zn	1.24	0.91	1.24	0.13	0.52	1.24	0.91	1.24
Ga	368	232	321	254	98	368	232	321
Ge	114	88	109	96	34	114	88	109
As	77	62	74	63	47	77	62	74
Se	13	15	25	19	6	13	15	25
Rb	0.10	0.13	0.14	0.11	0.15	0.10	0.13	0.14
Sr	11,959	20,082	19,257	20,687	7413	11,959	20,082	19,257
Y	1256	1102	1288	1179	1148	1256	1102	1288
Zr	0.2	0.1	0.2	0.1	2.5	0.2	0.1	0.2
Nb	5.3	17.9	3.5	1.3	5019.8	5.3	17.9	3.5
Mo	0.3	0.9	0.3	0.7	1.1	0.3	0.9	0.3
Ag	0.0	0.0		0.0	0.1	0.0	0.0	
Ba	93	9	6	6	70	93	9	6
La	11,830	5685	8340	7159	6273	11,830	5685	8340
Ce	25,622	16,471	22,838	19,405	12,238	25,622	16,471	22,838
Pr	2623	1834	2447	2112	1347	2623	1834	2447
Nd	10,539	8257	10,760	9453	4761	10,539	8257	10,760
Sm	1245	1105	1397	1224	632	1245	1105	1397
Eu	320	287	351	318	179	320	287	351
Gd	712	654	768	717	403	712	654	768
Tb	69	62	71	65	50	69	62	71
Dy	274	241	281	259	241	274	241	281
Ho	38	35	39	36	37	38	35	39
Er	71	64	73	68	73	71	64	73
Tm	6	6	7	6	6	6	6	7
Yb	26	24	28	26	20	26	24	28
Lu	2.19	2.08	2.42	2.30	1.59	2.19	2.08	2.42
Hf	0.02	0.01	0.02	0.03	0.07	0.02	0.01	0.02
Ta	0.00	0.01	0.00	0.00	0.06	0.00	0.01	0.00
W	0.16	0.17	0.16	0.19	2.63	0.16	0.17	0.16
Au	17.7	3.2	13.3	24.0	0.6	17.7	3.2	13.3
Pb	6.3	7.9	7.8	6.8	11.6	6.3	7.9	7.8
Bi	0.0	0.3	0.2	0.2	0.1	0.0	0.3	0.2
Th	73.4	32.8	44.4	41.3	66.7	73.4	32.8	44.4
U	17.3	13.7	18.3	7.8	12.8	17.3	13.7	18.3

https://doi.org/10.3390/min16010112

Table 2. *Cont.*

Apatite Type	Ap-2	Ap-2	Ap-2	Ap-3	Ap-3	Ap-3	Ap-3	Ap-3
Spot	20CS18-2-04	20CS18-2-05	20CS18-2-06	20CS09-1-01	20CS09-1-02	20CS09-1-03	20CS09-1-04	20CS09-1-05
Li	0.0	0.0	0.0	26.5	5.7	11.0	11.2	8.3
Be	0.0	0.0	0.1	0.1	0.2	0.2	0.2	0.3
B	0.2	1.1	0.7	1.7	2.5	1.9	3.0	2.1
S	1204	1107	868	640	1353	1553	1410	877
Sc	0.00	0.17	0.02	0.12	0.29	0.20	0.00	0.06
V	0.08	0.10	0.07	0.22	0.60	0.64	0.60	0.44
Cr	0.00	0.00	0.00	3.94	15.11	9.10	7.65	0.00
Co	0.01	0.03	0.01	0.03	0.54	0.41	0.56	0.22
Ni	0.00	0.00	0.00	0.00	0.00	0.00	0.94	0.00
Cu	0.96	0.83	0.65	0.09	1.01	0.88	1.14	0.64
Zn	0.81	0.18	0.17	1.24	0.91	1.24	0.13	0.52
Ga	168	170	160	368	232	321	254	98
Ge	97	95	93	114	88	109	96	34
As	87	83	85	77	62	74	63	47
Se	50	26	23	13	15	25	19	6
Rb	0.12	0.10	0.13	0.10	0.13	0.14	0.11	0.15
Sr	45,106	47,504	45,400	11,959	20,082	19,257	20,687	7413
Y	1494	1270	1309	1256	1102	1288	1179	1148
Zr	0.0	0.1	0.1	0.2	0.1	0.2	0.1	2.5
Nb	0.0	0.0	0.0	5.3	17.9	3.5	1.3	5019.8
Mo	0.1	0.0	0.0	0.3	0.9	0.3	0.7	1.1
Ag	0.0	0.0	0.0	0.0	0.0		0.0	0.1
Ba	6	5	4	93	9	6	6	70
La	3899	4034	3958	11,830	5685	8340	7159	6273
Ce	15,903	16,319	15,681	25,622	16,471	22,838	19,405	12,238
Pr	2556	2665	2548	2623	1834	2447	2112	1347
Nd	12,462	12,493	11,960	10,539	8257	10,760	9453	4761
Sm	2151	2046	1983	1245	1105	1397	1224	632
Eu	573	531	535	320	287	351	318	179
Gd	1346	1215	1190	712	654	768	717	403
Tb	118	101	102	69	62	71	65	50
Dy	406	348	353	274	241	281	259	241
Ho	50	42	45	38	35	39	36	37
Er	81	70	73	71	64	73	68	73
Tm	7	6	6	6	6	7	6	6
Yb	26	22	24	26	24	28	26	20
Lu	2.17	1.85	1.87	2.19	2.08	2.42	2.30	1.59
Hf	0.01	0.03	0.04	0.02	0.01	0.02	0.03	0.07
Ta	0.00	0.00	0.00	0.00	0.01	0.00	0.00	0.06
W	0.02	0.03	0.01	0.16	0.17	0.16	0.19	2.63
Au	0.2	0.2	0.2	17.7	3.2	13.3	24.0	0.6
Pb	12.2	14.3	12.4	6.3	7.9	7.8	6.8	11.6
Bi	0.0	0.0	0.0	0.0	0.3	0.2	0.2	0.1
Th	9.6	5.5	7.2	73.4	32.8	44.4	41.3	66.7
U	0.2	0.2	0.1	17.3	13.7	18.3	7.8	12.8

In the Sr/Y-LREE (La-Nd) and log(Eu/Y)-log(Ce) apatite discrimination diagrams, all samples plot within the field of ultramafic–mafic igneous rocks (including carbonatites), with some Ap-3 analyses lying near the boundary between ultramafic and alkaline rocks (Figure 9). In the ΣLa-Nd-ΣSm-Ho-ΣEr-Lu ternary diagram (Figure 10), most apatite

https://doi.org/10.3390/min16010112

analyses fall within the mantle field, with several Ap-1 and Ap-2 analyses plotting along the mantle–crust boundary, suggesting that the components were dominantly derived from a mantle source. All apatite types display Eu/Eu* values of 0.99–1.08, indicating the absence of significant Eu anomalies. Ce/Ce* values between 1.03 and 1.25 show slightly positive Ce anomalies (Figure 11H,I).

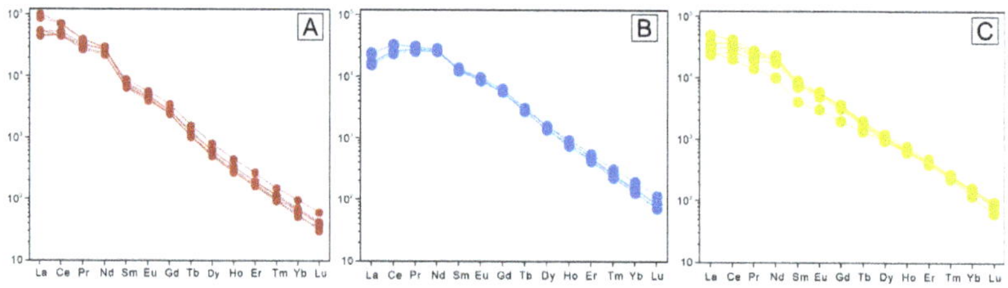

Figure 8. Chondrite-normalized REE patterns of apatite from different generations. (**A**) Type 1 apatite show higher LREE content. (**B**) Type 2 apatite show relatively flat LREE pattern. (**C**) Type 2 apatite show LREE enrichment and HREE depletion. Standardized values of chondrites are from [68] (Sun and McDonough, 1989).

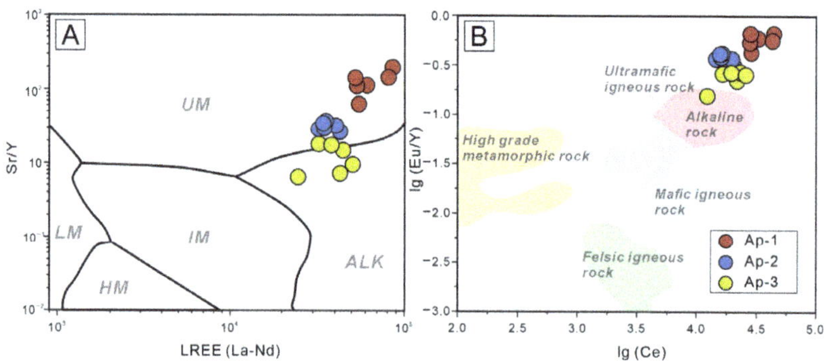

Figure 9. Discriminant diagram for genetic apatite from different geological settings (standardized values of chondrites are from [46,69]). (**A**) the discriminant diagrams of Sr/Y vs. LREE(La-Nd), (**B**) the discriminant diagrams of lg(Eu/Y) vs. lg(Ce).

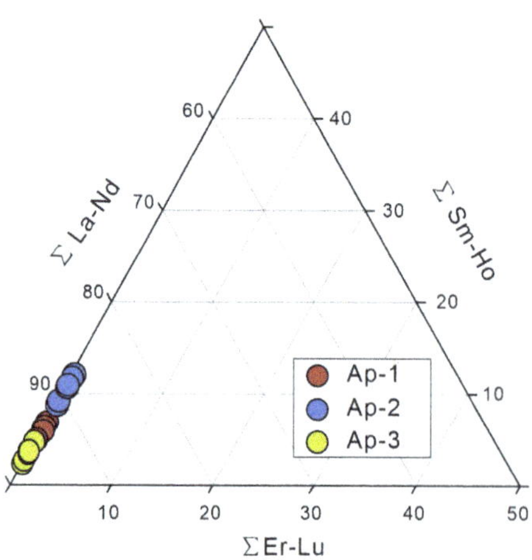

Figure 10. Ternary diagram showing ∑La-Nd-∑Sm-Ho-∑Er-Lu characteristics of apatite.

https://doi.org/10.3390/min16010112

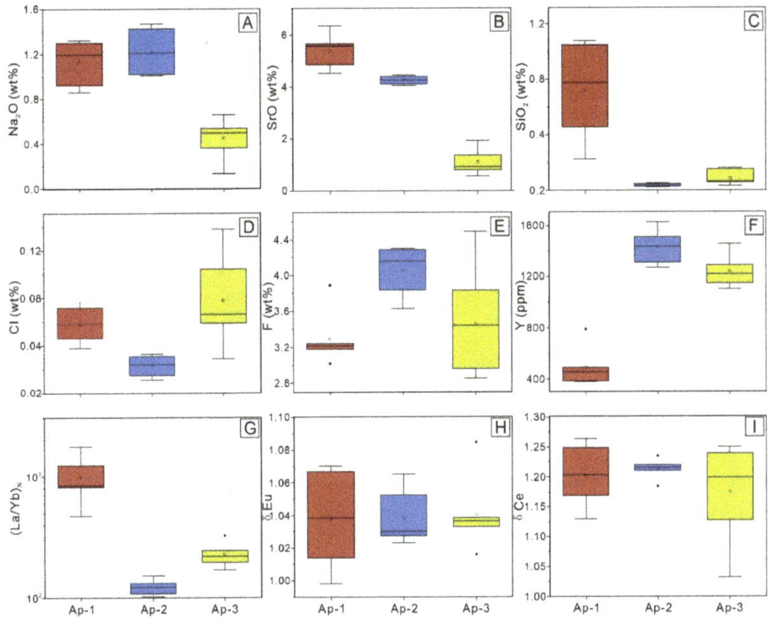

Figure 11. Box diagram of geochemical composition of apatite from different generations. (**A–F**) Major element content by EPMA. (**G–I**) Trace element content by LA-ICP-MS.

6. Discussion

6.1. Precipitation Mechanisms of Apatite

Calcium in apatite can be substituted by a variety of cations. In the system considered here, Sr^{2+} is isovalent with Ca^{2+} and therefore readily occupies Ca^{2+} sites via simple isomorphous substitution [70]. By contrast, cation substitutions involving a charge imbalance typically require coupled mechanisms, most commonly incorporating REE^{3+} or compensating anionic species such as PO_4^{3-}, CO_3^{2-}, or SO_4^{2-} to maintain electroneutrality [71–73]. In the context of this study, trivalent cations that substitute for Ca^{2+}, particularly REE^{3+} and Y^{3+}, are of primary interest. The incorporation of REE^{3+} into the apatite structure predominantly proceeds through two principal substitution mechanisms [36]:

$$REE^{3+} + Na^{+} \leftrightarrow 2Ca^{2+} \quad (1)$$

$$REE^{3+} + SiO_4^{4-} \leftrightarrow Ca^{2+} + PO4^{3-} \quad (2)$$

Given the observed correlation between REEs and Na, Na^{+} is inferred to be the main charge-balancing cation associated with REE incorporation in the apatite types examined here. Several coupled substitution mechanisms could compensate for Na^{+} incorporation, including co-substitution with trivalent cations other than REEs, SO_4^{2-}, CO_3^{2-}, or F^{-}, and in some cases the formation of vacancies [40]. Although the substitution of Ca^{2+} by trivalent cations such as Bi^{3+} and Cr^{3+} has been reported in synthetic apatite, such mechanisms are unlikely to occur in natural systems and have not been documented in apatite from other carbonatite complexes [74–76]. Substitution involving SO_4^{2-} is similarly improbable, as sulfur concentrations in the apatite examined here fall below the detection limit of EPMA.

In typical silicate alkaline–carbonatite magmatic systems, apatite is generally not among the earliest rock-forming phases to crystallize; instead, it becomes abundant only

 https://doi.org/10.3390/min16010112

after the evolving magma reaches phosphate saturation [77–80]. Experimental studies have shown that apatite solubility and its saturation temperature are jointly controlled by SiO_2 content in melt, aluminum saturation (A/CNK), Ca activity, and ambient P-T conditions. With progressive magmatic differentiation and increasing SiO_2, the solubility of P_2O_5 in the melt decreases, thereby favoring apatite crystallization at intermediate to late magmatic stages [41,81]. At Weishan, Ap-1 exhibits zones of compositional heterogeneity, whereas CL locally reveals weak banding, suggesting crystallization during magmatic fractional differentiation. Petrographically, both Ap-2 and Ap-3 show clear evidence of replacement, and contain abundant inclusions of REE minerals, mainly parasite, ancylite, and monazite. Altered apatite is characterized by lower LREE, Na, Sr, and Si contents, but slightly higher F, Cl, and HREE abundances relative to unaltered magmatic apatite (Ap-1). These systematic compositional shifts, together with the replacement textures, suggest that Ap-2 and Ap-3 formed via fluid-mediated dissolution–reprecipitation. This process, well established as a fundamental reaction mechanism, is a widely recognized mechanism which replaces a primary mineral in the presence of a reactive fluid by a compositionally modified equivalent phase and/or a new mineral phase [36,37,71]. Owing to molar volume differences and subtle solubility contrasts between the precursor and product phases, the recrystallized domains typically develop interconnected microporosity and abundant fluid inclusions [44]. During dissolution–reprecipitation, parasite inclusions precipitate within Ap-3, and LREEs are mobilized outward from the apatite, whereas the F content remains essentially unchanged. This indicates that F was supplied by late-stage fluids rather than being solely inherited from the primary apatite. Coexisting coarse fluorite further supports this inference, because in F-rich reactive fluids REEs are commonly transported as $[REEF]^{2+}$ complexes and thus can be highly mobile [82–84]. In CO_2-rich fluids, REEs also exhibit high mobility, commonly as $[REECO_3]^+$ complexes, with REE solubility increasing with fluid salinity, pH, and alkalinity [85–88]. Finally, the occurrence of calcite inclusions within barite from the Weishan orebodies indicate elevated activities of $CO_3{}^{2-}$ and Ca^{2+} in the fluid at this stage [67].

Although the Ap-3-bearing orebodies have high REE grades (up to ~20%), the total volume of REE mineral inclusions within the apatite is relatively small, suggesting that these inclusions formed largely at the expense of primary magmatic apatite (Ap-1) rather than being entirely introduced by the external fluid. The presence of fine-grained, disseminated parasite crystals in the ores further implies precipitation under relatively high temperatures followed by rapid cooling. Overall, Ap-2 and Ap-3 apatite record different intensities of fluid-mediated modification: compared with Ap-2, Ap-3 underwent more extensive element mobility and a more complete dissolution–reprecipitation overprint.

6.2. Evolution of Ore-Forming Fluid

Apatite is commonly considered a chemically robust, alteration-resistant accessory mineral in igneous and hydrothermal systems [89,90]. However, under intense late-stage hydrothermal alteration, its morphology, internal textures, chemical composition, and even fundamental physicochemical properties may be significantly modified, resulting in partial or complete transformation into altered apatite [37,80,91,92]. Studies of porphyry Cu and iron oxide–apatite (IOA) deposits show that the systematic compositional contrasts between altered and unaltered apatite can be used to constrain the source, character, and evolution of magmatic–hydrothermal fluids [93].

In the Weishan REE deposit, the three apatite types correspond to distinct stages of carbonatite magma evolution and the associated fluid activity, as evidenced by the petrographic and geochemical constraints discussed above. Although both Ap-2 and Ap-3 formed through fluid-mediated dissolution–reprecipitation during late hydrothermal

https://doi.org/10.3390/min16010112

alteration, they differ markedly in element redistribution patterns, inclusion assemblages, and paragenetic relationships. These contrasts suggest overprinting by two late-stage fluid regimes with distinct compositions and/or physicochemical conditions. Ap-1 apatite is characterized by extremely high Sr contents, elevated Sr/Y ratios, and strong LREE enrichment, consistent with crystallization from a highly evolved and strongly fractionated carbonatite melt [14,43,94–97]. During the fractional crystallization of carbonatite magma, the crystallization of apatite, calcite, and other early carbonate and phosphate phases progressively depletes the melt in compatible components, while incompatible elements such as REEs and Sr become increasingly enriched in the residual melt [20]. The high Sr/Y ratios of Ap-1 further imply advanced fractionation, because Y partitions more strongly into apatite than Sr [43,97,98]. Continued crystallization thus causes Sr/Y in the residual melt, and in apatite crystallizing from it, to increase. With further cooling and the onset of volatile saturation (H_2O, CO_2, F, Cl), an REE-rich fluid exsolves from the residual melt. At this stage, apatite begins to record transitional characteristics: Sr decreases, likely because Sr is preferentially retained in the melt and/or sequestered into early carbonate minerals during melt–fluid partitioning, while the system-wide REE budget approaches its maximum. Fluid exsolution represents a critical step in the mineralization process, as it efficiently extracts REEs from a large-volume silicate–carbonatite melt and concentrates them into a much smaller-volume fluid phase, thereby creating the fundamental geochemical precondition for the formation of economically significant REE mineralization.

Ap-2 apatite occurs in REE-barite carbonatite ores and typically forms subhedral grains in which rims and internal domains are partly infilled by strontianite, together with abundant REE-mineral precipitation dominated by ancylite and monazite. Compositionally, Ap-2 shows the same general trends in Ca, P, Sr, REE, and Na as Ap-3, but the magnitudes of these changes are significantly smaller, whereas HREE contents show only a slight increase. Together with its weaker replacement textures, these features indicate lower-intensity hydrothermal overprint and a less advanced stage of fluid-mediated dissolution–reprecipitation than in Ap-3. Although abundant strontianite fills the rims and microfractures within Ap-2, and the dominant REE mineral inclusions shift from parasite in Ap-3 to ancylite in Ap-2, the bulk Sr content of apatite decreases only modestly [72,99]. This implies that Sr in the associated assemblage was supplied largely by the fluid rather than derived solely from apatite, and that the fluid was likewise enriched in CO_3^{2}. The scarcity of fluorite in the ore and the change in REE inclusion assemblage also indicate that the F content of the fluid responsible for Ap-2 alteration was lower than that of the fluid that modified Ap-3. Monazite commonly occurs along apatite grain boundaries rather than as inclusions, consistent with precipitation during fluid–rock interaction [37,71,87,93]. This implies that the fluid was enriched in LREEs such as La, Ce, and Nd, which combined with PO_4^{3-} liberated from apatite to precipitate monazite. In contrast, Ap-3 displays much stronger evidence of chemical modification: LREE, Na, and Sr decrease markedly, whereas Ca and P increase to some extent, and F, Cl, Si, and HREE remain essentially unchanged. Combined with the distinct paragenesis of the REE-fluorite carbonatite ores relative to earlier carbonatites, these compositional features provide powerful constraints on the composition of the altering fluid. During dissolution–reprecipitation, synchysite inclusions precipitated within Ap-3, while LREEs were mobilized outward from the apatite. The near-constant F content in Ap-3 suggests relatively low free F activity in the fluid, with F preferentially complexed with REEs; consequently, REEs were transported mainly as $[REEF]^{2+}$ species.

 https://doi.org/10.3390/min16010112

6.3. Implications for REE Mineralization

The geochemical record preserved in apatite clearly demonstrates that mineralization at Weishan is not produced by a single-stage process, but rather the cumulative product of multiple episodes of fluid evolution [63,64,91]. A highly fractionated carbonatite magma provided the fundamental metal source for REE mineralization; without this stage of pre-enrichment, the REE concentrations in the original melt would have been insufficient to generate a large-scale ore deposit (Figure 2). The high REE contents of magmatic apatite provide direct evidence of this early enrichment process. The exsolution of F- and Cl-rich fluids marks the critical transition from REE "dispersion" in a large magma volume to their "concentration" in a much smaller fluid phase [78,100]. During this stage, REEs previously locked in rock-forming minerals (e.g., early apatite or silicates) or retained in the residual melt are remobilized and efficiently transferred into the fluid phase, thereby generating a REE-rich hydrothermal medium capable of ore formation. Driven by tectonic structures, these fluids migrate into favorable traps such as faults, fractures, and intrusive contacts. Within these sites, water–rock interactions, fluid boiling, and redox changes collectively trigger the rapid and voluminous precipitation of REE minerals (e.g., bastnaesite, monazite) together with hydrothermal apatite [38,39,101]. This process chain ultimately governs orebody geometry, tonnage, and grade. The widespread development of hydrothermal apatite therefore represents a major mineralization event and provides a durable mineralogical archive of the ore-forming system [73,89,102,103].

Field observations and petrographic evidence from the Weishan REE deposit indicate that the ore-forming system experienced multiple stages of metasomatic re-equilibration and elemental redistribution. During early mineralization, the system was dominated by carbonatite magma, whereas in the later stages, progressive changes in physicochemical conditions (e.g., temperature, pressure, and pH) during alkaline metasomatism with the wall rocks promoted a transition to a hydrothermal regime. As the carbonatite magma evolved to its late stage, a derivative ore-forming hydrothermal fluid was generated [104–107]. This fluid overprinted and altered the early-crystallized minerals, and through a series of complex chemical exchanges and renewed mineral crystallization, produced REE-enriched carbonatite. Geochemical evidence from apatite further suggests that, although the primitive, relatively undifferentiated carbonatite magma was already relatively enriched in REEs, subsequent magmatic–hydrothermal evolution was required to achieve ore-grade enrichment.

During this evolution, early-crystallized magmatic apatite was overprinted and partially replaced by later hydrothermal fluids, giving rise to new generations of hydrothermal apatite. Ap-1 is widely developed in quartz-calcite carbonatite at Weishan and represents a typical early carbonatite-associated REE mineralization stage; Ap-1 corresponds to apatite crystallized directly from carbonatite magma. In contrast, Ap-2 and Ap-3 occur in REE-fluorite carbonatite and REE-barite carbonatite, respectively. These latter orebodies are characterized by higher REE grades and coarser crystal sizes, and they demonstrably cut across the earlier carbonatite in the field, indicating that they are products of later fluid precipitation.

7. Conclusions

The Weishan REE deposit is a typical carbonatite-associated REE deposit in which mineralization is governed by the multi-stage evolution of a carbonatite magma–hydrothermal system. Three generations of apatite (Ap-1, Ap-2, Ap-3) from carbonatite document progressive overprinting by compositionally different fluids. Magmatic Ap-1 crystallized from carbonatite melt, whereas Ap-2 and Ap-3 formed by fluid-mediated dissolution–reprecipitation, accompanied by systematic decreases in LREE, Na, and Sr, increases in

 https://doi.org/10.3390/min16010112

Na-Sr-P (relative proportions), and incorporation of REE mineral inclusions (e.g., ancylite-, monazite, parasite), reflecting element remobilization and redistribution during hydrothermal alteration. Systematic variations in REE patterns, Sr/Y ratios, and Ce-Eu anomalies among the apatite demonstrate that the geochemistry effectively traces the transition from magmatic to hydrothermal stages, the evolution of F-CO_2-Ba-SO_4^{2-}-bearing fluids, and the associated REE enrichment and precipitation. Apatite thus provides robust mineralogical constraints on the ore-forming process of the Weishan carbonatite REE deposit and offers a useful tool for understanding and exploring similar deposits.

Author Contributions: Conceived the ideas, Y.-X.G., S.-S.L. and C.-P.L.; map compilation, Y.-Z.S.; data curation, formal analysis, and investigation, Y.-X.G., M.-Q.W., Z.S., Y.-Z.S., Z.-Y.Y., S.-S.L. and K.-F.Q.; writing, Y.-X.G. and S.-S.L. All authors have read and agreed to the published version of the manuscript.

Funding: This study was financially supported by the Open Project of Weihai Key Laboratory of Energy and Mineral Resources Investigation and Evaluation (No. LDKF-2023WH-02), the Open Research Project from Shandong Provincial Lunan Geology and Exploration Institute and Shandong Engineering Research Center of Rare Elements Exploration and Comprehensive Utilization (LNY202301), the Open Research Fund Program of Key Laboratory of Gold Mineralization Processes and Resource Utilization Subordinated to the Ministry of Natural Resources, the Shandong Key Laboratory of Metallogenic Geological Process and Resources Utilization (KFKT202403), the Fundamental Research Funds for the Central Universities (2-9-2023-055), and the Young Elite Scientists Sponsorship Program of BAST (BYESS2024122).

Data Availability Statement: The data set is presented directly in the present study.

Acknowledgments: We are grateful to Ya-Qi Huang and De-Jian Li for providing valuable field guidance and engaging in constructive discussions. We appreciate Xin-Yi Wang for assistance with figure preparation. We are deeply grateful to the anonymous reviewers and editors for their insightful comments and suggestions.

Conflicts of Interest: The authors declare no conflicts of interest.

References

1. Fan, H.-R.; Yang, K.-F.; Hu, F.-F.; Liu, S.; Wang, K.-Y. The giant Bayan Obo REE-Nb-Fe deposit, China: Controversy and ore genesis. *Geosci. Front.* **2016**, *7*, 335–344. [CrossRef]
2. Liu, Y.; Chakhmouradian, A.R.; Hou, Z.; Song, W.; Kynický, J. Development of REE mineralization in the giant Maoniuping deposit (Sichuan, China): Insights from mineralogy, fluid inclusions, and trace-element geochemistry. *Mineral. Depos.* **2018**, *54*, 701–718. [CrossRef]
3. Xie, Y.; Xia, J.; Cui, K.; Qu, Y.; Liang, P.; Zhong, R. Rare earth elements deposits in China: Spatio-temporal distribution and ore-forming processes. *Chin. Sci. Bull.* **2020**, *65*, 3794–3808. [CrossRef]
4. Chakhmouradian, A.R.; Wall, F. Rare Earth Elements: Minerals, Mines, Magnets (and More). *Elements* **2012**, *8*, 333–340. [CrossRef]
5. Smith, M.P.; Campbell, L.S.; Kynicky, J. A review of the genesis of the world class Bayan Obo Fe–REE–Nb deposits, Inner Mongolia, China: Multistage processes and outstanding questions. *Ore Geol. Rev.* **2015**, *64*, 459–476. [CrossRef]
6. Smith, M.P.; Henderson, P.; Campbell, L.S. Fractionation of the REE during hydrothermal processes: Constraints from the Bayan Obo Fe-REE-Nb deposit, Inner Mongolia, China. *Geochim. Cosmochim. Acta* **2000**, *64*, 3141–3160. [CrossRef]
7. Wei, C.; Deng, M.; Xu, C.; Chakhmouradian, A.R.; Smith, M.P.; Kynicky, J.; Song, W.; Chen, W.; Fu, B. Mineralization of the Bayan Obo Rare Earth Element Deposit by Recrystallization and Decarbonation. *Econ. Geol.* **2022**, *117*, 1327–1338. [CrossRef]
8. Yang, K.-F.; Fan, H.-R.; Santosh, M.; Hu, F.-F.; Wang, K.-Y. Mesoproterozoic carbonatitic magmatism in the Bayan Obo deposit, Inner Mongolia, North China: Constraints for the mechanism of super accumulation of rare earth elements. *Ore Geol. Rev.* **2011**, *40*, 122–131. [CrossRef]
9. Yang, K.; Fan, H.; Pirajno, F.; Li, X. The Bayan Obo (China) giant REE accumulation conundrum elucidated by intense magmatic differentiation of carbonatite. *Geology* **2019**, *47*, 1198–1202. [CrossRef]
10. Aranha, M.; Porwal, A.; Sundaralingam, M.; González-Álvarez, I.; Markan, A.; Rao, K. Rare earth elements associated with carbonatite–alkaline complexes in western Rajasthan, India: Exploration targeting at regional scale. *Solid Earth* **2022**, *13*, 497–518. [CrossRef]

 https://doi.org/10.3390/min16010112

11. Chakhmouradian, A.R. High-field-strength elements in carbonatitic rocks: Geochemistry, crystal chemistry and significance for constraining the sources of carbonatites. *Chem. Geol.* **2006**, *235*, 138–160. [CrossRef]
12. Gittins, J.; Mitchell, R.H. The genesis of calcite and dolomite carbonatite-forming magma by liquid immiscibility: A critical appraisal. *Geol. Mag.* **2023**, *160*, 1463–1480. [CrossRef]
13. Liu, Y.; Hou, Z. A synthesis of mineralization styles with an integrated genetic model of carbonatite-syenite-hosted REE deposits in the Cenozoic Mianning-Dechang REE metallogenic belt, the eastern Tibetan Plateau, southwestern China. *J. Asian Earth Sci.* **2017**, *137*, 35–79. [CrossRef]
14. Ryabchikov, I.D.; Orlova, G.P.; Senin, V.G.; Trubkin, N.V. Partitioning of rare earth elements between phosphate-rich carbonatite melts and mantle peridotites. *Mineral. Petrol.* **1993**, *49*, 1–12. [CrossRef]
15. Verplanck, P.L. The Role of Fluids in the Formation of Rare Earth Element Deposits. *Procedia Earth Planet. Sci.* **2017**, *17*, 758–761. [CrossRef]
16. Verplanck, P.L.; Hitzman, M.W. *Rare Earth and Critical Elements in Ore Deposits*; Society of Economic Geologists: Littleton, CO, USA, 2016; Volume 18.
17. Wall, F.; Rollat, A.; Pell, R. Responsible Sourcing of Critical Metals. *Elements* **2017**, *13*, 313–318. [CrossRef]
18. Xie, Y.; Verplanck, P.L.; Hou, Z.; Zhong, R. *Mineral Deposits of China*; Society of Economic Geologists: Littleton, CO, USA, 2019; pp. 509–552.
19. Zheng, X.; Liu, Y.; Smith, M.P.; Kynický, J.; Hou, Z. Carbonatitic Magma Fractionation and Contamination Generate Rare Earth Element Enrichment and Mineralization in the Maoniuping Giant REE Deposit, SW China. *J. Petrol.* **2023**, *64*, egad037. [CrossRef]
20. Mitchell, R.H. Carbonatites and Carbonatites and Carbonatites. *Can. Mineral.* **2005**, *43*, 2049–2068. [CrossRef]
21. Siegel, K.; Vasyukova, O.V.; Williams-Jones, A.E. Magmatic evolution and controls on rare metal-enrichment of the Strange Lake A-type peralkaline granitic pluton, Québec-Labrador. *Lithos* **2018**, *308–309*, 34–52. [CrossRef]
22. Tang, L.; Wagner, T.; Fusswinkel, T.; Zhang, S.-T.; Xu, B.; Jia, L.-H.; Hu, X.-K. Magmatic-hydrothermal evolution of an unusual Mo-rich carbonatite: A case study using LA-ICP-MS fluid inclusion microanalysis and He–Ar isotopes from the Huangshui'an deposit, Qinling, China. *Mineral. Depos.* **2021**, *56*, 1133–1150. [CrossRef]
23. Moore, M.; Chakhmouradian, A.R.; Mariano, A.N.; Sidhu, R. Evolution of rare-earth mineralization in the Bear Lodge carbonatite, Wyoming: Mineralogical and isotopic evidence. *Ore Geol. Rev.* **2015**, *64*, 499–521. [CrossRef]
24. O'Brien, H. *Mineral Deposits of Finland*; Elsevier: Amsterdam, The Netherlands, 2015; pp. 291–303.
25. Verplanck, P.L.; Farmer, G.L.; Kettler, R.M.; Lowers, H.A.; Johnson, C.A.; Koenig, A.E.; Blessington, M.J. Petrogenesis and rare earth element mineralization of the Elk Creek carbonatite, Nebraska, USA. *Ore Geol. Rev.* **2022**, *146*, 104953. [CrossRef]
26. Anenburg, M.; Broom-Fendley, S.; Chen, W. Formation of Rare Earth Deposits in Carbonatites. *Elements* **2021**, *17*, 327–332. [CrossRef]
27. Chakhmouradian, A.R.; Zaitsev, A.N. Rare Earth Mineralization in Igneous Rocks: Sources and Processes. *Elements* **2012**, *8*, 347–353. [CrossRef]
28. Smith, M.P.; Moore, K.; Kavecsánszki, D.; Finch, A.A.; Kynicky, J.; Wall, F. From mantle to critical zone: A review of large and giant sized deposits of the rare earth elements. *Geosci. Front.* **2016**, *7*, 315–334. [CrossRef]
29. Beard, C.D.; Goodenough, K.M.; Borst, A.M.; Wall, F.; Siegfried, P.R.; Deady, E.A.; Pohl, C.; Hutchison, W.; Finch, A.A.; Walter, B.F.; et al. Alkaline-Silicate REE-HFSE Systems. *Econ. Geol.* **2023**, *118*, 177–208. [CrossRef]
30. Long, Z.-Y.; Qiu, K.-F.; Halama, R.; Yu, H.-C.; Zhang, J.-Y.; Li, S.-S.; Han, G.-L.; Gao, Z.-Y.; Ding, Z.-J.; Deng, J. Reappraisal of potassium isotopic composition of the continental lithospheric mantle: Insights from calc-alkaline lamprophyres in the eastern North China Craton. *GSA Bull.* **2025**, *137*, 4399–4412. [CrossRef]
31. Qiu, K.-F.; Long, Z.-Y.; Romer, R.L.; Halama, R.; Williams-Jones, A.E.; Yu, H.-C.; Li, S.-S.; Wu, M.-Q.; Deng, J. The mantle source of REE-rich alkaline silicate magmas can be enriched by continent-derived sediment subduction. *Nat. Commun.* **2025**, *16*, 11191. [CrossRef]
32. Schleicher, H. In-situ Determination of Trace Element and REE Partitioning in a Natural Apatite-Carbonatite Melt System using Synchrotron XRF Microprobe Analysis. *J. Geol. Soc. India* **2019**, *93*, 305–312. [CrossRef]
33. Yu, H.-C.; Qiu, K.-F.; Hetherington, C.; Chew, D.; Huang, Y.-Q.; He, D.; Geng, J.-Z.; Xian, H.-Y. Apatite as an alternative petrochronometer to trace the evolution of magmatic systems containing metamict zircon. *Contrib. Mineral. Petrol.* **2021**, *176*, 68. [CrossRef]
34. Anenburg, M.; Mavrogenes, J.A.; Bennett, V.C. The Fluorapatite P–REE–Th Vein Deposit at Nolans Bore: Genesis by Carbonatite Metasomatism. *J. Petrol.* **2020**, *61*, egaa003. [CrossRef]
35. Apukhtina, O.B.; Kamenetsky, V.S.; Ehrig, K.; Kamenetsky, M.B.; Maas, R.; Thompson, J.; McPhie, J.; Ciobanu, C.L.; Cook, N.J. EARLY, deep magnetite-fluorapatite mineralization at the olympic dam Cu-U-Au-Ag deposit, South Australia. *Econ. Geol.* **2017**, *112*, 1531–1542. [CrossRef]
36. Broom-Fendley, S.; Styles, M.T.; Appleton, J.D.; Gunn, G.; Wall, F. Evidence for dissolution-reprecipitation of apatite and preferential LREE mobility in carbonatite-derived late-stage hydrothermal processes. *Am. Mineral.* **2016**, *101*, 596–611. [CrossRef]

37. Harlov, D.E. Apatite: A fingerprint for metasomatic processes. *Elements* **2015**, *11*, 171–176. [CrossRef]
38. Li, X.; Zhou, M.-F. Multiple stages of hydrothermal REE remobilization recorded in fluorapatite in the Paleoproterozoic Yinachang Fe–Cu–(REE) deposit, Southwest China. *Geochim. Cosmochim. Acta* **2015**, *166*, 53–73. [CrossRef]
39. Lorenz, M.; Altenberger, U.; Trumbull, R.B.; Lira, R.; López de Luchi, M.; Günter, C.; Eidner, S. Chemical and textural relations of britholite- and apatite-group minerals from hydrothermal REE mineralization at the Rodeo de los Molles deposit, Central Argentina. *Am. Mineral.* **2019**, *104*, 1840–1850. [CrossRef]
40. Pan, Y.; Fleet, M.E. Rare earth element mobility during prograde granulite facies metamorphism: Significance of fluorine. *Contrib. Mineral. Petrol.* **1996**, *123*, 251–262. [CrossRef]
41. Harlov, D.E.; Wirth, R.; Förster, H.J. An experimental study of dissolution-reprecipitation in fluorapatite: Fluid infiltration and the formation of monazite. *Contrib. Mineral. Petrol.* **2005**, *150*, 268–286. [CrossRef]
42. Harlov, D.E.; Wirth, R.; Hetherington, C.J. The relative stability of monazite and huttonite at 300–900 °C and 200–1000 MPa: Metasomatism and the propagation of metastable mineral phases. *Am. Mineral.* **2007**, *92*, 1652–1664. [CrossRef]
43. Klemme, S.; Dalpé, C. Trace-element partitioning between apatite and carbonatite melt. *Am. Mineral.* **2003**, *88*, 639–646. [CrossRef]
44. Putnis, A. Mineral Replacement Reactions. *Rev. Mineral. Geochem.* **2009**, *70*, 87–124. [CrossRef]
45. Qiu, K.-F.; Zhou, T.; Chew, D.; Hou, Z.; Müller, A.; Yu, H.-C.; Lee, R.; Chen, H.; Deng, J. Apatite trace element composition as an indicator of ore deposit types: A machine learning approach. *Am. Mineral.* **2024**, *109*, 303–314. [CrossRef]
46. Zhou, T.; Qiu, K.; Wang, Y.; Yu, H.; Hou, Z. Apatite Eu/Y-Ce discrimination diagram: A big data based approach for provenance classification. *Acta Petrol. Sin.* **2022**, *38*, 291–299.
47. Zhai, M.G. Tectonic evolution and metallogenesis of North China Craton. *Mineral. Depos.* **2010**, *29*, 24–36. [CrossRef]
48. Zhao, G.; Sun, M.; Wilde, S.A.; Sanzhong, L. Late Archean to Paleoproterozoic evolution of the North China Craton: Key issues revisited. *Precambrian Res.* **2005**, *136*, 177–202. [CrossRef]
49. Zhao, J.-H.; Zhou, M.-F. Neoproterozoic adakitic plutons in the northern margin of the Yangtze Block, China: Partial melting of a thickened lower crust and implications for secular crustal evolution. *Lithos* **2008**, *104*, 231–248. [CrossRef]
50. Li, S.-S.; Santosh, M.; Teng, X.-M.; He, X.-F. Paleoproterozoic arc-continent collision in the North China Craton: Evidence from the Zanhuang Complex. *Precambrian Res.* **2016**, *286*, 281–305. [CrossRef]
51. Mao, J.-W.; Xie, G.-Q.; Zhang, Z.-H.; Li, X.-F.; Wang, Y.-T.; Zhang, C.-Q.; Li, Y.-F. Mesozoic large-scale metallogenic pulses in North China and corresponding geodynamie settings. *Acta Petrol. Sin.* **2005**, *21*, 171–190.
52. Long, Z.-Y.; Moynier, F.; Bögels, T.; Fang, L.; Caracas, R.; Paquet, M.; Jourdan, F.; Luu, T.-H.; Rigoussen, D.; Qiu, K.-F.; et al. Impact-induced sublimation drives volatile depletion in carbonaceous meteorites. *Nat. Commun.* **2025**, *16*, 6146. [CrossRef]
53. Qiu, K.-F.; Deng, J.; Laflamme, C.; Long, Z.-Y.; Wan, R.-Q.; Moynier, F.; Yu, H.-C.; Zhang, J.-Y.; Ding, Z.-J.; Goldfarb, R. Giant Mesozoic gold ores derived from subducted oceanic slab and overlying sediments. *Geochim. Cosmochim. Acta* **2023**, *343*, 133–141. [CrossRef]
54. Qiu, K.-F.; Deng, J.; Sai, S.-X.; Yu, H.-C.; Tamer, M.T.; Ding, Z.-J.; Yu, X.-F.; Goldfarb, R. Low-Temperature Thermochronology for Defining the Tectonic Controls on Heterogeneous Gold Endowment Across the Jiaodong Peninsula, Eastern China. *Tectonics* **2023**, *42*, e2022TC007669. [CrossRef]
55. Qiu, K.-F.; Romer, R.L.; Long, Z.-Y.; Williams-Jones, A.E.; Yu, H.-C.; Turner, S.; Wang, Q.-F.; Li, S.-S.; Zhang, J.-Y.; Duan, H.-R.; et al. The role of an oxidized lithospheric mantle in gold mobilization. *Sci. Adv.* **2024**, *10*, eado6262. [CrossRef] [PubMed]
56. Qiu, K.-F.; Romer, R.L.; Long, Z.-Y.; Yu, H.-C.; Turner, S.; Wan, R.-Q.; Li, X.-Q.; Gao, Z.-Y.; Deng, J. Potassium isotopes as a tracer of hydrothermal alteration in ore systems. *Geochim. Cosmochim. Acta* **2024**, *368*, 185–196. [CrossRef]
57. Liang, Y.W.; Lai, Y.; Hu, H.; Zhang, F. Zircon U-Pb Ages and Geochemical Characteristics Study of Syenite from Weishan REE Deposit, Western Shandong. *Acta Sci. Nat. Univ. Pekin.* **2017**, *53*, 652–666. [CrossRef]
58. Qiu, K.-F.; Deng, J.; Li, S.-S.; Jowitt, S.; Hetherington, C.J.; Balen, D. Roles and perspectives of A- and I-type magmas in rare earth element and gold mineralization. *GSA Bull.* **2023**, *136*, 1238–1250. [CrossRef]
59. Long, Z.-Y.; Moynier, F.; Debret, B.; Qiu, K.-F.; Dai, W.; Sun, H.-X.; Deng, J.; Bertrand, H.; Burton, K.; Inglis, E.; et al. Heavy potassium isotopes in carbonatites reveal oceanic crust subduction as the driver of deep carbon cycling. *Sci. Adv.* **2025**, *11*, eadt1023. [CrossRef]
60. Long, Z.-Y.; Moynier, F.; Dai, W.; Zuilen, M.; Deng, J.; Qiu, K.-F. Revisiting the Eoarchean Akilia quartz-pyroxene rock with potassium isotopes: Implications for early-ocean sedimentation. *Proc. Natl. Acad. Sci. USA* **2025**, *122*, e2510228122. [CrossRef] [PubMed]
61. He, D.; Qiu, K.-F.; Simon, A.; Pokrovski, G.; Yu, H.-C.; Connolly, J.; Li, S.-S.; Turner, S.; Wang, Q.-F.; Yang, M.-F.; et al. Mantle oxidation by sulfur drives the formation of giant gold deposits in subduction zones. *Proc. Natl. Acad. Sci. USA* **2024**, *121*, e2404731121. [CrossRef]
62. Yang, Z.; Wu, M.; Li, S.; Yu, H.; Diao, X.; Gao, Y.; Jiadong, M.; Zhi, C.; Shang, Z.; Xi, J.; et al. The origin of glimmerite and its significance to rare earth element mineralization: Insights from the Longbaoshan deposit in North China Craton. *Ore Geol. Rev.* **2025**, *184*, 106698. [CrossRef]

63. Yu, X.; Tang, H.; Han, Z.; Li, C. Geological Characteristics and Origin of Rare Earth Elements Deposits Related with Alkaline Rock in the Chishan-Longbaoshan Area, Shandong Province. *Acta Geol. Sin.* **2012**, *84*, 407–417.
64. Zhao, L.-Y.; Lan, T.-G.; Fan, H.-R.; Lan, J.; Wang, H.; Terry Chen, W.; Hu, H.-L.; Li, D.-J. Origin and mineralization processes of REE during magmatic-hydrothermal evolution: Insights from in-situ geochemistry of calcite from the Weishan REE deposit, eastern North China Craton. *Ore Geol. Rev.* **2023**, *162*, 105676. [CrossRef]
65. Wang, C.; Liu, J.; Zhang, H.; Zhang, X.; Zhang, D.; Xi, Z.; Wang, Z. Geochronology and mineralogy of the Weishan carbonatite in Shandong province, eastern China. *Geosci. Front.* **2019**, *10*, 769–785. [CrossRef]
66. Zeng, X.; Li, X.; Fan, H.; Lan, T.; Lan, J.; Su, J.; Zhang, P.; Yang, K.; Zhao, X. Generation of REE-rich syenite-(carbonatite) complex through lithosphere-asthenosphere interaction: An in-situ Sr–Nd–O isotopic study of the Mesozoic Weishan pluton, Northern China. *J. Asian Earth Sci.* **2022**, *230*, 105191. [CrossRef]
67. Lan, T.G.; Fan, H.R.; Hu, F.F.; Yang, K.F.; Wang, Y. Genesis of the Weishan REE deposit, Shandong Province: Evidence from Rb-Sr isochron age, LA-MC-ICPMS Nd isotopic compositions and fluid inclusions. *Geochamica* **2012**, *40*, 428–442.
68. Sun, S.S.; McDonough, W. *Chemical and Isotopic Systematics of Oceanic Basalts: Implications for Mantle Composition and Processes*; Geological Society: London, UK, 1989; Volume 42.
69. Belousova, E.; Griffin, W.; O'Reilly, S.; Fisher, N. Apatite as an indicator mineral for mineral exploration: Trace-element compositions and their relationship to host rock type. *J. Geochem. Explor.* **2002**, *76*, 45–69. [CrossRef]
70. Schwinn, G.; Markl, G. REE systematics in hydrothermal fluorite. *Chem. Geol.* **2005**, *216*, 225–248. [CrossRef]
71. Hughes, J.; Rakovan, J. Structurally Robust, Chemically Diverse: Apatite and Apatite Supergroup Minerals. *Elements* **2015**, *11*, 165–170. [CrossRef]
72. Rakovan, J.; Hughes, J. Strontium in the apatite structure: Strontian fluorapatite and belovite-(Ce). *Can. Mineral.* **2000**, *38*, 839–845. [CrossRef]
73. Rakovan, J. Growth and Surface Properties of Apatite. *Rev. Mineral. Geochem.-REV Miner. Geochem.* **2002**, *48*, 51–86. [CrossRef]
74. Chen, W.; Simonetti, A. In-situ determination of major and trace elements in calcite and apatite, and U-Pb ages of apatite from the Oka carbonatite complex: Insights into a complex crystallization history. *Chem. Geol.* **2013**, *353*, 151–172. [CrossRef]
75. Kjarsgaard, I. Rare Earth Elements in Sovitic Carbonatites and their Mineral Phases. *J. Petrol.* **1998**, *39*, 2105–2121. [CrossRef]
76. Xu, C.; Kynicky, J.; Chakhmouradian, A.; Campbell, I.; Allen, C.M. Trace-element modeling of the magmatic evolution of rare-earth-rich carbonatite from the Miaoya deposit, Central China. *Lithos* **2010**, *118*, 145–155. [CrossRef]
77. Anenburg, M. Rare earth mineral diversity controlled by REE pattern shapes. *Mineral. Mag.* **2020**, *84*, 629–639. [CrossRef]
78. Bhattacharjee, S.; Chakrabarty, A.; Mitchell, R.H.; Patel, S.C.; Kozlov, E.N.; Fomina, E.N.; Dey, M.; Pal, S. The role of magmatic-to-carbohydrothermal processes in rare earth mineralization in Hogenakkal carbonatites, India. *Lithos* **2024**, *464–465*, 107431. [CrossRef]
79. Bodeving, S.; Williams-Jones, A.E.; Swinden, S. Carbonate–silicate melt immiscibility, REE mineralising fluids, and the evolution of the Lofdal Intrusive Suite, Namibia. *Lithos* **2017**, *268–271*, 383–398. [CrossRef]
80. Bühn, B.; Wall, F.; Le Bas, M. Rare-earth element systematics of carbonatitic fluorapatites, and their significance for carbonatite magma evolution. *Beiträage Zur Mineral. Und Petrogr.* **2001**, *141*, 572–591. [CrossRef]
81. Barker, D. Calculated silica activities in carbonatite liquids. *Contrib. Mineral. Petrol.* **2001**, *141*, 704–709. [CrossRef]
82. Migdisov, A.; Williams-Jones, A.E.; Brugger, J.; Caporuscio, F.A. Hydrothermal transport, deposition, and fractionation of the REE: Experimental data and thermodynamic calculations. *Chem. Geol.* **2016**, *439*, 13–42. [CrossRef]
83. Migdisov, A.A.; Williams-Jones, A.E. Hydrothermal transport and deposition of the rare earth elements by fluorine-bearing aqueous liquids. *Mineral. Depos.* **2014**, *49*, 987–997. [CrossRef]
84. Milani, L.; Bolhar, R.; Frei, D.; Harlov, D.E.; Samuel, V.O. Light rare earth element systematics as a tool for investigating the petrogenesis of phoscorite-carbonatite associations, as exemplified by the Phalaborwa Complex, South Africa. *Mineral. Depos.* **2017**, *52*, 1105–1125. [CrossRef]
85. Bau, M.; Dulski, P. Comparative study of yttrium and rare-earth element behaviours in fluorine-rich hydrothermal fluids. *Contrib. Mineral. Petrol.* **1995**, *119*, 213–223. [CrossRef]
86. Broom-Fendley, S.; Wall, F.; Spiro, B.; Ullmann, C.V. Deducing the source and composition of rare earth mineralising fluids in carbonatites: Insights from isotopic (C, O, 87Sr/86Sr) data from Kangankunde, Malawi. *Contrib. Mineral. Petrol.* **2017**, *172*, 96. [CrossRef]
87. Bühn, B.; Schneider, J.; Dulski, P.; Rankin, A.H. Fluid–rock interaction during progressive migration of carbonatitic fluids, derived from small-scale trace element and Sr, Pb isotope distribution in hydrothermal fluorite. *Geochim. Cosmochim. Acta* **2003**, *67*, 4577–4595. [CrossRef]
88. Harlov, D.E.; Förster, H.J. Fluid-induced nucleation of (Y+REE)-phosphate minerals within apatite: Nature and experiment. Part II. *Fluorapatite. Am. Mineral.* **2003**, *88*, 1209–1229. [CrossRef]
89. Lu, J.; Chen, W.; Ying, Y.-C.; Jiang, S.; Zhao, K. Apatite texture and trace element chemistry of carbonatite-related REE deposits in China: Implications for petrogenesis. *Lithos* **2021**, *398–399*, 106276. [CrossRef]

90. Mororó, E.A.A.; Berkesi, M.; Zajacz, Z.; Guzmics, T. Rare earth element transport and mineralization linked to fluids from carbonatite systems. *Geology* **2024**, *52*, 240–244. [CrossRef]
91. Cao, M.; Evans, N.; Hollings, P.; Cooke, D.; McInnes, B.; Qin, K.Z. Apatite Texture, Composition, and O-Sr-Nd Isotope Signatures Record Magmatic and Hydrothermal Fluid Characteristics at the Black Mountain Porphyry Deposit, Philippines. *Econ. Geol.* **2021**, *116*, 1189–1207. [CrossRef]
92. Li, H.; Hermann, J. Apatite as an indicator of fluid salinity: An experimental study of chlorine and fluorine partitioning in subducted sediments. *Geochim. Cosmochim. Acta* **2015**, *166*, 267–297. [CrossRef]
93. Hoshino, M.; Watanabe, Y.; Kon, Y. Implication of Apatite and Anhydrite for Formation of an Iron-Oxide-Apatite (IOA) Rare Earth Element Prospect, Benjamin River, Canada: Benjamin River IOA REE prospect, Canada. *Resour. Geol.* **2017**, *67*, 361–383. [CrossRef]
94. Guzmics, T.; Zajacz, Z.; Mitchell, R.H.; Szabó, C.; Wälle, M. The role of liquid–liquid immiscibility and crystal fractionation in the genesis of carbonatite magmas: Insights from Kerimasi melt inclusions. *Contrib. Mineral. Petrol.* **2015**, *169*, 17. [CrossRef]
95. Jones, A.; Genge, M.; Carmody, L. Carbonate Melts and Carbonatites. *Rev. Mineral. Geochem.* **2013**, *75*, 289–322. [CrossRef]
96. Kamenetsky, V.S.; Mitchell, R.H.; Maas, R.; Giuliani, A.; Gaboury, D.; Zhitova, L. Chlorine in mantle-derived carbonatite melts revealed by halite in the St.-Honoré intrusion (Québec, Canada). *Geology* **2015**, *43*, 687–690. [CrossRef]
97. Martin, L.H.J.; Schmidt, M.W.; Mattsson, H.B.; Guenther, D. Element Partitioning between Immiscible Carbonatite and Silicate Melts for Dry and H_2O-bearing Systems at 1–3 GPa. *J. Petrol.* **2013**, *54*, 2301–2338. [CrossRef]
98. Li, X.-C.; Fan, H.-R.; Su, J.-H.; Groves, D.I.; Yang, K.-F.; Zhao, X.-F. Giant Rare Earth Element Accumulation Related to Voluminous, Highly Evolved Carbonatite: A Microanalytical Study of Carbonate Minerals From the Bayan Obo Deposit, China. *Econ. Geol.* **2024**, *119*, 373–393. [CrossRef]
99. She, H.-D.; Fan, H.-R.; Santosh, M.; Li, X.-C.; Yang, K.-F.; Wang, Q.-W.; Wei, W.; Liu, Y.-J.; Liu, S.; Liu, S.-L. Paleozoic remelting of carbonatite in Bayan Obo (China): Further insights into the formation of a giant REE deposit. *Gondwana Res.* **2023**, *119*, 172–185. [CrossRef]
100. Chakhmouradian, A.R.; Reguir, E.P.; Zaitsev, A.N.; Couëslan, C.; Xu, C.; Kynický, J.; Mumin, A.H.; Yang, P. Apatite in carbonatitic rocks: Compositional variation, zoning, element partitioning and petrogenetic significance. *Lithos* **2017**, *274–275*, 188–213. [CrossRef]
101. Huang, L.; Li, X.; Su, J.-H.; Zeng, X.; Fan, H.; Lan, J.; Lan, T.; Yang, K.; Hu, F. Genesis of the Hujiazhuang apatite-rich carbonatite complex in North China and its implication for the REE fertility of the carbonatite system. *J. Asian Earth Sci.* **2024**, *267*, 106149. [CrossRef]
102. Pan, Y.M.; Fleet, M. Compositions of the Apatite-Group Minerals: Substitution Mechanisms and Controlling Factors. *Rev. Mineral. Geochem.-REV Miner. Geochem.* **2002**, *48*, 13–49. [CrossRef]
103. Ronsbo, J.G. Coupled substitutions involving REEs and Na and Si in apatites in alkaline rocks from the Ilimaussaq intrusion, South Greenland, and the petrological implications. *Am. Mineral.* **1989**, *74*, 896–901.
104. Chayka, I.F.; Kamenetsky, V.S.; Vladykin, N.V.; Kontonikas-Charos, A.; Prokopyev, I.R.; Stepanov, S.Y.; Krasheninnikov, S.P. Origin of alkali-rich volcanic and alkali-poor intrusive carbonatites from a common parental magma. *Sci. Rep.* **2021**, *11*, 17627. [CrossRef]
105. Duraiswami, R.; Shaikh, T. Fluid-rock interaction in the Kangankunde Carbonatite Complex, Malawi: SEM based evidence for late stage pervasive hydrothermal mineralisation. *Open Geosci.* **2014**, *6*, 476–491. [CrossRef]
106. Hou, Z.; Liu, Y.; Tian, S.; Yang, Z.; Xie, Y. Formation of carbonatite-related giant rare-earth-element deposits by the recycling of marine sediments. *Sci. Rep.* **2015**, *5*, srep10231. [CrossRef] [PubMed]
107. Walter, B.F.; Giebel, R.J.; Steele-MacInnis, M.; Marks, M.A.W.; Kolb, J.; Markl, G. Fluids associated with carbonatitic magmatism: A critical review and implications for carbonatite magma ascent. *Earth Sci. Rev.* **2021**, *215*, 103509. [CrossRef]

minerals

MDPI

Article

Genesis of the Yawan Gold Deposit, West Qinling Orogen: Insights from Calcite U-Pb Geochronology and Geochemistry of Sulfides

Chang-Qing Dong [1], Zhao-Hua Bai [2,*], Ke Yang [2], Meng-Zhen Hao [1], Jia-Yi Wang [1], Hao Zhou [1] and Jia-Nan Fu [1,*]

[1] State Key Laboratory of Geological Processes and Mineral Resources, Frontiers Science Center for Deep-Time Digital Earth, School of Earth Sciences and Resources, China University of Geosciences, Beijing 100083, China; changqingdong@163.com (C.-Q.D.); mengzhen_haocugb@163.com (M.-Z.H.); bifrost2333@gmail.com (J.-Y.W.); zh1075945441@163.com (H.Z.)

[2] Technology, Innovation Center for Gold Ore Exploration, Xi'an Mineral Resources Survey, China Geological Survey, Xi' an 710100, China; yangke051302@163.com

* Correspondence: baizhaohua206@163.com (Z.-H.B.); jiananfu.geo@gmail.com (J.-N.F.); Tel.: +86-15102909202 (Z.-H.B.); +86-18635207649 (J.-N.F.)

Abstract

The Yawan gold deposit, located in the Western Qinling Orogen, contains gold mineralisation that is predominantly controlled by approximately east-west-trending fault systems. This study integrates field geology, petrography, cathodoluminescence imaging, electron probe microanalysis of gold-bearing minerals (pyrite and arsenopyrite), and in situ laser ablation U-Pb dating of calcite to constrain the timing of mineralisation and to elucidate the mechanisms of gold enrichment. This study reveals that the deposit is significantly structurally controlled and comprises two discrete mineralisation stages: a quartz-pyrite (Py1)-arsenopyrite (Apy1)-chalcopyrite assemblage (Stage 1), and a quartz-calcite-pyrite (Py2)-arsenopyrite (Apy2)-stibnite-sphalerite-galena assemblage (Stage 2). Py1 displays distinct zonation, with rim As contents notably higher than core values, while Co and Ni contents gradually decrease from core to rim. Py2 is characterised by high As (0.00%–4.72%), low Fe/S ratios, and a porous texture, containing gold and arsenopyrite inclusions. Invisible gold occurs in lattice-bound form in both Py1 and Py2. The As-Fe-S ternary diagram of pyrite indicates that Au^+ likely entered the crystal lattice as a solid solution. Arsenopyrite geothermometry yields a mineralisation temperature of 389 ± 44 °C, and sulfur fugacity (fS_2) decreased markedly from Stage 1 to Stage 2. Combined with the S and Fe characteristics of pyrite, these features support a medium-temperature metamorphic hydrothermal environment. U-Pb dating of calcite from Stage 2 yields an age of 215.6 ± 7.1 Ma. In summary, the Yawan gold deposit belongs to the orogenic gold system, with its gold precipitation and enrichment controlled by sulfidation triggered by Late Triassic tectono-fluid activity.

Keywords: calcite U-Pb; gold enrichment mechanism; Yawan orogenic gold deposit; West Qinling orogen

1. Introduction

The West Qinling orogenic belt, situated between the North China and South China blocks, hosts > 1000 t Au and constitutes China's second largest gold metallogenic province [1–3]. The orebody is hosted in Devonian and Triassic clastic and carbonate rocks, and regional ore-forming processes are closely linked to Triassic (Indosinian) tectono-magmatic events [4,5]. Previous studies have focused on regional structural controls, distribution patterns of deposits and genetic classifications; on the basis of mineral assemblages, structural features and metal sources, the deposits have been subdivided into orogenic, Carlin-type and Carlin-like types. In contrast, gold-bearing sulfides remain insufficiently investigated and merit further detailed study [6–8].

Sulfides serve as the primary hosts for gold in orogenic systems, and their textures and paragenetic assemblages provide critical insights into gold precipitation and the evolution of ore-forming fluids [9,10]. Pyrite and arsenopyrite, in particular, exhibit diagnostic typomorphic features, contrasting formation temperatures, and distinct major and trace element compositions, making them key indicators for constraining gold occurrence, fluid chemistry, and physicochemical conditions [11–14]. Pyrite chemistry is strongly controlled by the ore-forming environment; its geochemical signature can be inverted to reconstruct element-selective partitioning and the physicochemical parameters of the fluid [15,16]. Consequently, pyrite geochemistry has been widely applied to elucidate gold mineralisation mechanisms and the redistribution of trace elements during hydrothermal–metamorphic overprinting [17–20].

The Yawan gold deposit lies in the central part of the West Qinling orogenic belt. Orebodies are hosted in Middle Devonian to Middle Carboniferous silty slate and display multiple generations of pyrite together with complex mineral parageneses [21,22]. Gold occurs predominantly as inclusions within Py2; no coeval dykes spatially associated with the orebodies have been recognised. Robust geochronological and geochemical constraints on the timing of mineralisation and on the precise state of gold occurrence are still lacking [23–29]. Here we integrate systematic field investigations with laboratory studies comprising the ore mineralogy, electron microprobe analysis, in situ laser ablation U–Pb dating of calcite and cathodoluminescence imaging of calcite and quartz to characterise gold occurrence, elucidate precipitation mechanisms and establish the timing of the main mineralisation stage. Arsenopyrite mineral thermometry is then applied to constrain the physicochemical conditions of ore formation, allowing us to evaluate the coupling between these conditions and gold mineralisation.

2. Regional Geology

The Qinling Orogen transects central China as a key segment of the Central Orogenic Belt; it links westward to the Qilian–Kunlun system and eastward to the Dabie–Sulu belt, thereby marking the suture between the South and North China blocks (Figure 1) [30–32]. Three nearly E–W sutures—Kuanping, Shangdan, and Mianlue—subdivide the Orogen into four tectonic domains: the southern margin of the North China Block, North Qinling, South Qinling, and northern margin of the South China Block [11,33–35]. These belts preserve Neoproterozoic–Late Mesozoic records of multiple subduction, accretion, and collision events [36–38]. Kuanping: a transitional zone between the North China Craton and North Qinling, dominated by greenschist facies assemblages (chlorite schist, plagioclase amphibolite, muscovite schist, marble) and ophiolitic mélange [39–41]. Shangdan: the collision boundary between North and South Qinling, characterised by Paleozoic ophiolites, subduction-related igneous rocks, and forearc sediments [42,43]. Mianlue: documents South Qinling–South China amalgamation via high-T/high-P metamorphic rocks, ophiolites, and volcanics [44]. Westward, it joins the South Kunlun suture and preserves the full Late Paleozoic–Late Triassic kinematic history of northward Mianlue plate subduction and continental collision [45,46].

https://doi.org/10.3390/min16010114

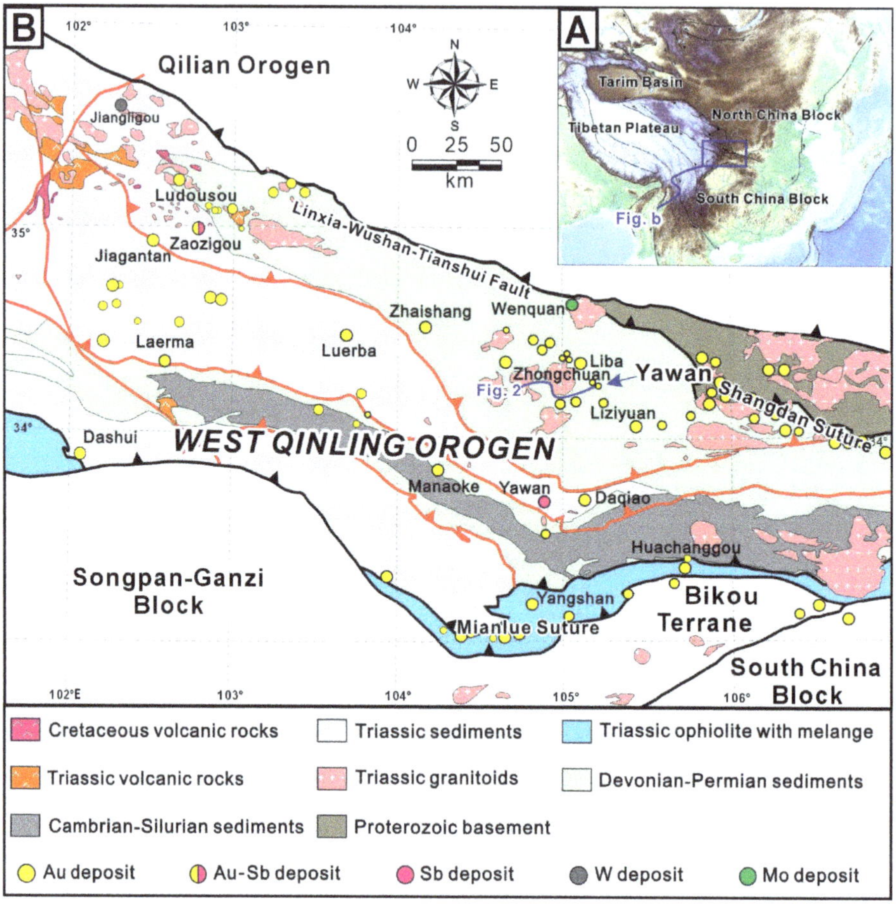

Figure 1. Geological structural position (**A**) and Simplified geological map of the West Qinling Orogen in China (**B**). (modified after [36]).

The West Qinling Orogen, the western continuation of the Qinling–Dabie–Sulu system, lies within the North–South China collision zone at the triple junction of the Circum-Pacific, Paleo Asian and Tethyan Himalayan domains [47–49]. As China's second largest Au–Sb province, its core Triassic tectono-magmatic–metallogenic events are intimately coupled to northward Paleo Tethyan slab subduction and ensuing continental collision [50,51]. Early Mianlue ocean subduction and subsequent South Qinling terrane–South China collision generated deep-seated thermal anomalies that drove large-scale orogenic hydrothermal systems [52,53]. Late Triassic orogenic and porphyry skarn mineralisation corresponds to the subduction and collision stages, respectively [54].

The Yawan deposit is situated in the central part of the western Lixian–Zhashui miogeoclinal fold belt, along the eastern exocontact of the Zhongchuan granite and within east–west-trending fault systems. The regional fault system is dominated by NW-trending structures dipping NE. Both the regional NWW-trending fault zones and their secondary fractures strictly control the spatial distribution of gold deposits [55,56].

 https://doi.org/10.3390/min16010114

3. Deposit Geology

The Yawan gold deposit is located in Luoba Township, Lixian County, Gansu Province (105°04′00″ E, 34°17′00″ N), with proven gold reserves of 5.5 tonnes and an average grade of 2.24 g/t [57,58]. The orebody is primarily hosted within the grey phyllitic silty spotted slate of the Middle Devonian Shujiaba Formation (D_2S). Seven gold orebodies have been delineated, all strictly controlled by fault structures, particularly the NWW-trending master fault, with their attitudes conforming to the fault orientation (Figure 2) [59,60]. Orebodies 1–3 have been mined out and are no longer accessible; therefore, field investigations were focused on Orebodies 4–7. The Au4 orebody is a low-grade, lens-shaped unit whose orientation aligns with the east–west trending fault structures [61]; quartz–pyrite veins are developed within north–south trending joints (Figure 3A,C). In contrast, the Au6 orebody, the principal orebody accounting for 49.65% of the total resources, is stratiform-like and extends approximately 60 m along an east–west direction. It is fault-controlled, generally dipping south at an angle of about 45° (Figure 3B), with the faults and joints filled by quartz–sulfide veins (Figure 3D).

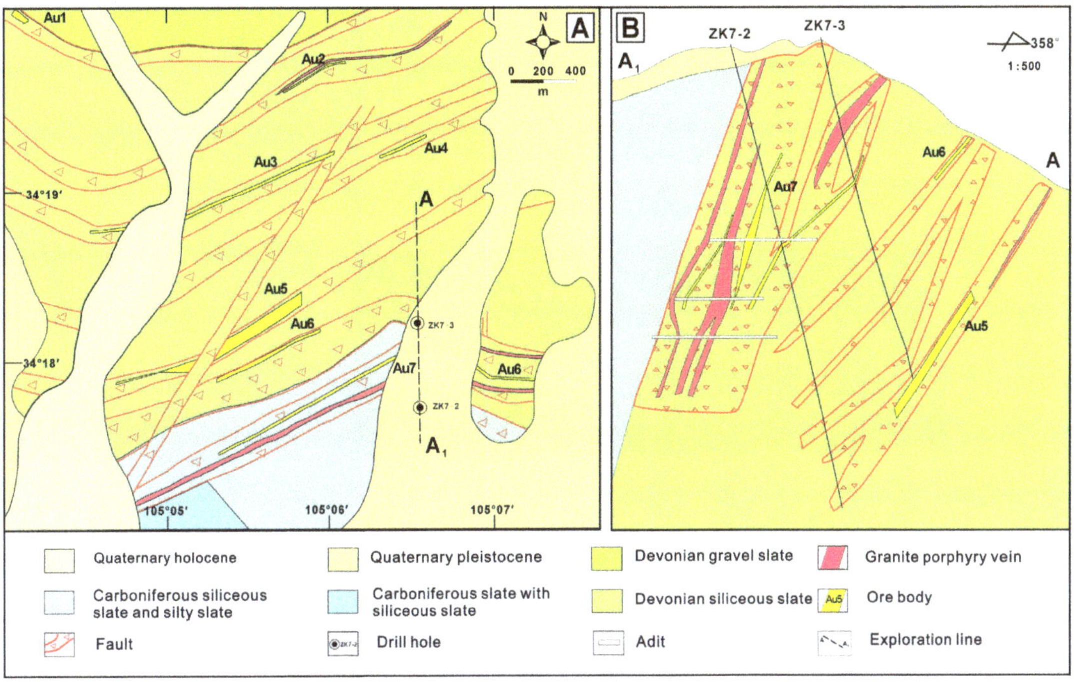

Figure 2. Simplified geological map of the Yawan gold deposit (**A**) and schematic cross-section along exploration profile A–A_1 (**B**).

The host rocks comprise grey phyllitic silty slate and carbonaceous slate [62]. Contacts between orebodies and wall rocks are generally gradational rather than sharp [63]. Alteration assemblages include limonitisation, pyritisation, chalcopyritisation, stibnitisation, galenisation, arsenopyritisation, bleaching, silicification, sericitisation, and carbonatisation. Ore types are predominantly disseminated and veinlet-band ores [11,64] (Figure 4A,B). Field investigation and optical microscopy identify the main sulfide minerals as pyrite, arsenopyrite, chalcopyrite, stibnite, galena, and sphalerite. The typical gangue mineral assemblage comprises quartz, sericite, calcite, tourmaline, monazite, zircon, and apatite [65]. Major mineralisation-related alteration types identified in the wall rock include sulfida-

 https://doi.org/10.3390/min16010114

tion, potassic alteration, sericitisation, and carbonatisation. Gold mineralisation is closely associated with quartz–calcite–sulfide veins characterised by pyrite and arsenopyrite. In cathodoluminescence (CL), calcite exhibits an orange-red glow, while quartz rims display a faint blue-violet luminescence, and cores remain dark (Figure 5A,D) [66–68].

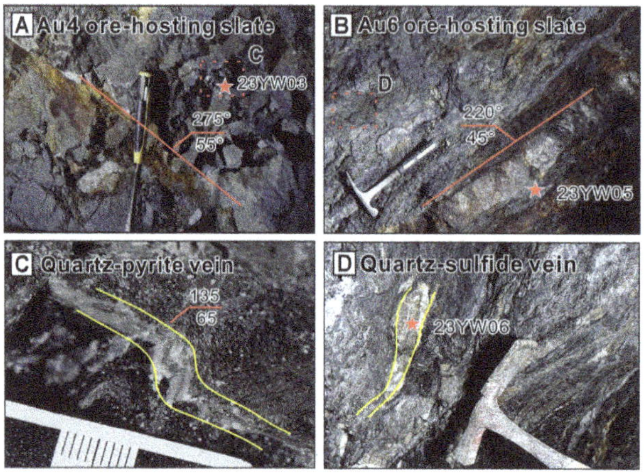

Figure 3. The low-grade Au4 orebody is dissected by a set of north–south trending joints, which are filled with quartz–pyrite veinlets (**A**,**C**). The main Au6 orebody dips shallowly to the south and exhibits faulted contacts with the country rocks. Quartz–sulfide veins, approximately 2 cm in width, are developed within the orebody (**B**,**D**).

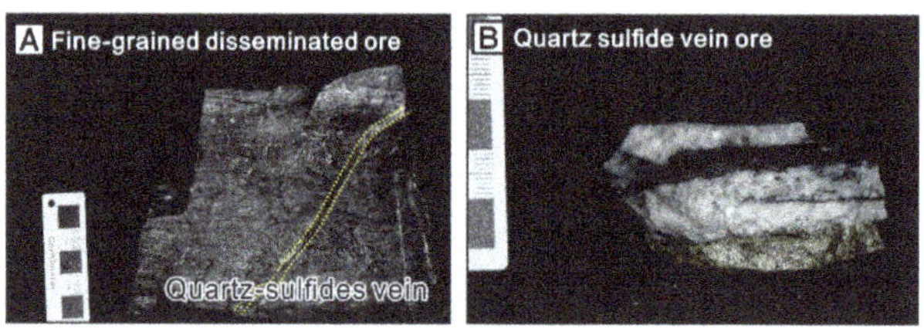

Figure 4. Disseminated ore contains quartz–sulfide veinlets (**A**), whereas vein-type ore is characterised by 2–4 cm wide quartz–pyrite veins (**B**).

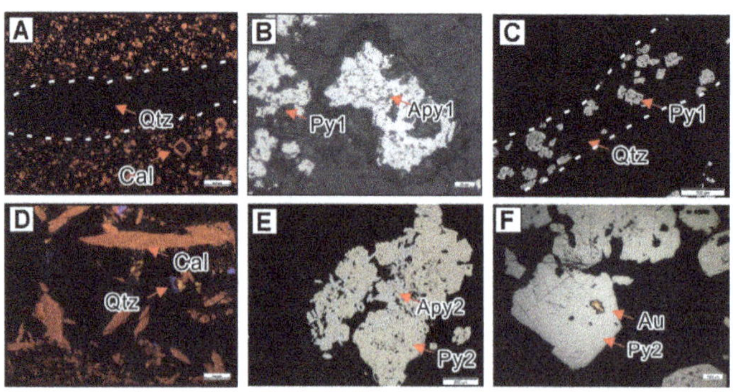

Figure 5. Microphotographs of gold-bearing sulfides in the Yawan gold deposit. The Yawan gold deposit contains two principal types of sulfide-bearing veins: quartz veins (**A**) and quartz–calcite veins (**D**). Granular Py1 occurs in quartz veins, where it is associated with short-prismatic Apy1 (**B**,**C**). Massive Py2 and acicular to clustered Apy2 are developed in quartz–calcite veins (**E**). Gold inclusions are observed within Py2 (**F**).

https://doi.org/10.3390/min16010114

The study area lies within the eastern exocontact zone of the Zhongchuan composite granite pluton (approximately 5 km from the pluton). Magmatic activity provided the primary heat source, leading to the development of contact metamorphic zones (cordierite–andalusite zone, biotite zone) and alterations like silicification and pyritisation in the exocontact zone [56,69]. The intensity of hydrothermal activity and gold mineralisation decreases with increasing distance from the pluton. Gold orebodies are spatially associated with granitic porphyry dikes, which trend sub-parallel to the orebodies. These dikes (1–3 m wide, extending 10–70 m) are emplaced along fractures and are genetically and spatiotemporally linked to the orebodies.

4. Samples and Analytical Methods

4.1. Sample Preparation

A total of 24 samples of disseminated and veinlet stockwork ores, representing different mineralisation stages, were collected from the Au4, Au5, Au6, and Au7 orebodies and drill hole ZK03 in the Yawan gold deposit for analysis. The mineral assemblages were initially identified using optical microscopy and scanning electron microscopy (SEM). Subsequently, eight selected samples from gold-bearing vein type and disseminated ores (23YW01, 23YW06, 23YW07-1, 23YW07-2, 23YW08, 23YW09-1, 23YW09-2, 23YW22) were subjected to cathodoluminescence (CL) imaging and electron probe microanalysis (EPMA) for major element composition of pyrite and arsenopyrite. Furthermore, three ore samples (23YW08, 23YW09-2, 23YW22) were selected for in situ laser ablation U-Pb dating of calcite.

4.2. In Situ U-Pb Dating of Carbonates

In situ calcite U–Pb geochronology was performed at the Research Institute of Petroleum Exploration and Development (RIPED) analytical facility, Beijing, China. Analyses were carried out on a Thermo Scientific XR multi-collector inductively coupled plasma mass spectrometer (MC-ICP-MS) coupled to a NWR 193 nm femtosecond laser ablation system (Thermo Fisher Scientific, Waltham, MA, USA). A single-spot ablation protocol was adopted, with a laser fluence of 2.35 J cm^{-2}, a pulse repetition rate of 10 Hz, and a spot diameter of 90 μm. Argon (Ar) served as the carrier gas. The carbonate reference materials XK5-5 and PS-6488, together with the WC-1 calcite standard, were used as primary and secondary U–Pb age calibration standards. The data processing was executed using Iolite V4 software.

4.3. EPMA of Pyrite and Arsenopyrite

The quantitative chemical data for pyrite and arsenopyrite were obtained with a JXA-iHP200F electron microprobe installed at the Institute of Mineral Resources at the Chinese Academy of Geological Sciences. The test working conditions are as follows: the acceleration voltage is 20 kV, the acceleration current is 20 An, and the beam spot diameter is 5 μm. All test data have been processed by ZAF correction. Standard samples were selected: jadeite (Si, Na), garnet (Al, Fe), diopside (Ca, Mg), potassium feldspar (K), rutile (Ti), rosinite (Mn), nickel (Ni), chromium trioxide (Cr), sodium aluminium silicon beryllium (Cl), and topaz (F).

5. Results

5.1. Hydrothermal Calcite U-Pb Geochronology

The U–Pb analysis was performed on a quartz–calcite vein sample (23YW09-2) containing pyrite and arsenopyrite, which represent the second stage of mineralisation. Tera–Wasserburg inverse concordia diagrams yield a lower intercept age of 215.6 ± 7.1 Ma (Stage 2 calcite, n = 60, MSWD = 2.3; Figure 6). The MSWD value indicates moderate scatter,

 https://doi.org/10.3390/min16010114

consistent with minor isotopic heterogeneity or episodic carbonate precipitation during this mineralisation stage.

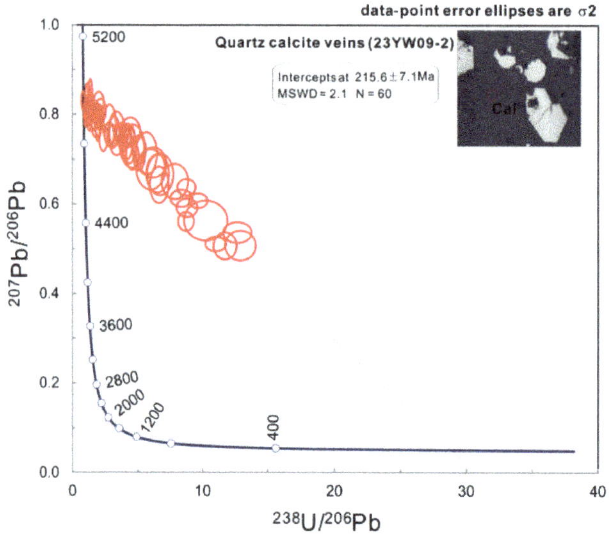

Figure 6. Tera–Wasserburg diagrams outlining the U-Pb ages of calcite.

5.2. The Texture of Gold-Bearing Minerals and the Characteristics of Major Elements

Ore microscopy identifies two principal stages of mineralisation in the Yawan gold deposit, with pyrite and arsenopyrite constituting the main gold-bearing minerals in both stages. These are categorised as Py1–Apy1 and Py2–Apy2 (Figure 5). Py1 and Apy1 occur as disseminations within altered and mineralised zones. Py1 forms subhedral to anhedral granular crystals measuring 50–600 μm in size and commonly displays growth zoning. Apy1 occurs as euhedral to subhedral short prismatic or tabular crystals, typically as individual grains in close association with Py1. In contrast, Py2 and Apy2 are mainly present in quartz–calcite veins as fine veinlets, disseminations, or massive aggregates. Py2 exhibits a subhedral, porous texture, with grain sizes ranging from 20 to 400 μm, lacks compositional zoning, and contains inclusions of stibnite, arsenopyrite, and native gold. Apy2 is euhedral, commonly forming rhombohedra, wedge-shaped crystals, or radial aggregates within fractures of Py2; no gold inclusions were observed in Apy2.

A total of 161 spot analyses were performed on the two generations of pyrite and arsenopyrite from the Yawan deposit: 32 on Py1, 92 on Py2, 15 on Apy1, and 22 on Apy2. Major element results are summarised in Appendix A Table A1.

The 32 spot analyses of Py1 show Fe contents of 45.11–46.41 wt% (avg. 45.76 wt%), S contents of 52.45–54.51 wt% (avg. 53.85 wt%), and As contents of 0.00–0.45 wt% (avg. 0.11 wt%). In comparison, the 92 analyses of Py2 yield Fe contents of 43.26–46.09 wt% (avg. 45.14 wt%), S contents of 50.14–54.71 wt% (avg. 53.02 wt%), and significantly higher As contents of 0.00–4.72 wt% (avg. 0.87 wt%). The 15 analyses of Apy1 give Fe contents of 31.47–34.50 wt% (avg. 33.57 wt%), S contents of 18.49–21.35 wt% (avg. 19.99 wt%), and As contents of 43.97–47.48 wt% (avg. 45.42 wt%). The 22 analyses of Apy2 return Fe contents of 33.60–35.46 wt% (avg. 34.42 wt%), S contents of 20.25–23.95 wt% (avg. 21.42 wt%), and As contents of 41.34–45.83 wt% (avg. 43.66 wt%). Trace elements such as Ag, Cu, Sb, Zn, and Cd are generally present in low concentrations, with most analyses below detection limits.

Gold contents also vary notably between mineral generations. Py1 contains 0.00–0.08 wt% Au (avg. 0.01 wt%), whereas Py2 shows markedly higher values of

 https://doi.org/10.3390/min16010114

0.00–1.60 wt% Au (avg. 0.40 wt%). Apy1 contains 0.00–0.14 wt% Au (avg. 0.03 wt%), and Apy2 contains 0.00–0.08 wt% Au (avg. 0.03 wt%). These results highlight a substantial difference in gold concentration between the two pyrite generations, with Py2 being significantly more enriched. In contrast, the two generations of arsenopyrite show no major variation in gold content.

The physicochemical conditions during arsenopyrite crystallisation are closely reflected by its As/S ratio, which serves as a geochemical indicator of formation temperature. Higher As/S ratios correspond to elevated crystallisation temperatures, whereas lower ratios indicate cooler conditions. At the Yawan gold deposit, the As/S ratios of Apy1 range from 2.10 to 2.56 (average 2.28), while those of Apy2 vary from 1.74 to 2.19 (average 2.04). Compared with the theoretical stoichiometric value of 2.34 for pure arsenopyrite, these measured ratios suggest a relatively lower crystallisation temperature for Apy2 than for Apy1.

Based on 37 EPMA spot analyses of major elements (Table A1), the arsenic atomic percentage (at%) in Apy1 ranges from 32.12 to 34.09 (avg. 33.10), while in Apy2 it ranges from 30.20 to 32.21 (avg. 31.20). These As values were plotted on a log fS_2–T diagram illustrating the arsenopyrite stability field (Figure 7) to constrain the formation conditions. In the disseminated ores, arsenopyrite is primarily associated with pyrite. The corresponding mineral assemblages plot within the "Apy + Py" field in the Fe–As–S phase diagram. Geothermometric calculations constrain the formation temperature of Apy1 to approximately 492 ± 32 °C, corresponding to a sulfur fugacity (fS_2) range of approximately $10^{-6.20}$ to $10^{-4.04}$. In contrast, Apy2 formed at a lower temperature of about 389 ± 44 °C, under a correspondingly lower sulfur fugacity range of roughly $10^{-8.48}$ to $10^{-6.06}$.

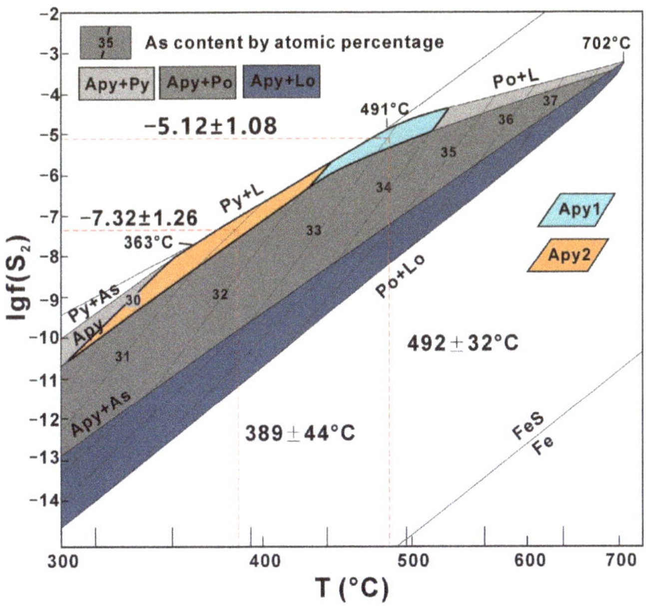

Figure 7. Lgf(S_2)-T diagram for arsenopyrite from the Yawan gold deposit (modified after [70]).

5.3. Nano-Scale Texture and Elemental Mapping of Pyrite

Based on electron probe microanalysis, individual grains of the two pyrite generations were selected for nanoscale structural and elemental mapping [71].

The results reveal that Py1 displays well developed oscillatory zoning, which provides direct evidence of periodic or pulsed changes in physicochemical conditions (e.g., tem-

 https://doi.org/10.3390/min16010114

perature, pressure, fluid composition, pH) during crystal growth. In the elemental maps, arsenic (As) shows distinct cyclical variations, with concentrations significantly higher at the grain margins than in the cores (Figure 8D). Similarly, gold (Au) is notably enriched along the rims relative to the cores (Figure 8N). Variations in the Co/Ni ratio across growth zones may further reflect the mixing of fluids from distinct sources.

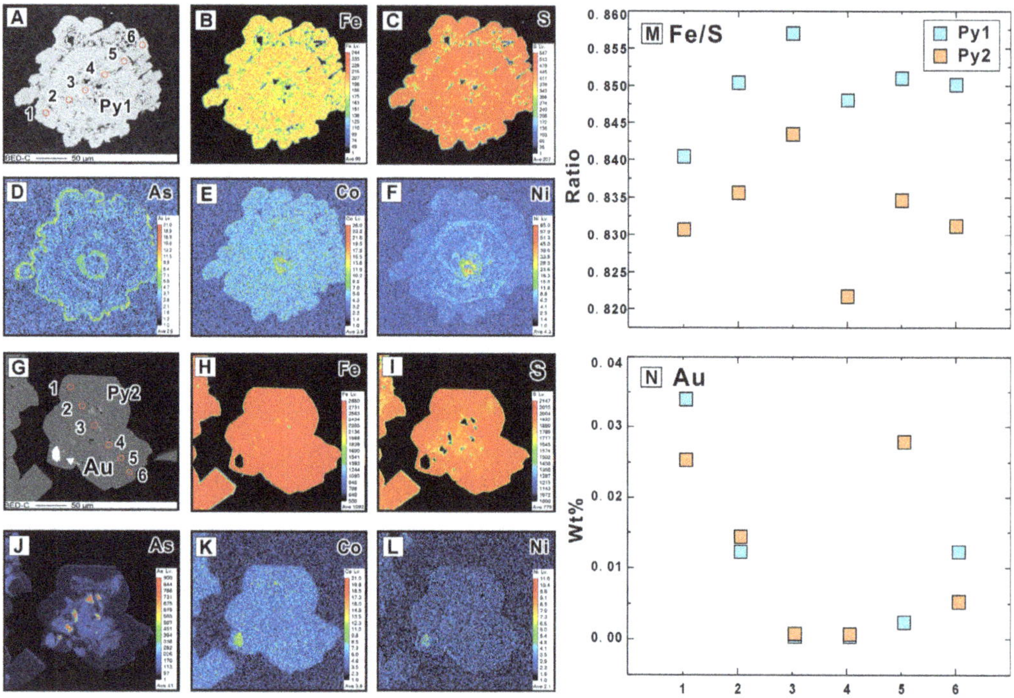

Figure 8. EPMA compositional profiles of pyrite (Py1, Py2) from the Yawan gold deposit; Py1 photomicrographs with corresponding elemental maps (**A–F**), Py2 photomicrographs with corresponding elemental maps (**G–L**), and compositional profiles (**M**,**N**).

In contrast, Py2 lacks oscillatory zoning but contains inclusions of native gold and arsenopyrite. The oscillatory zoning in Py1 indicates that the ore-forming fluids were not uniform or steady but rather comprised multiple episodic pulses, likely derived from varied sources [72].

6. Discussion

6.1. Gold Mineralisation Age of Yawan Deposit

This study employs in situ laser ablation calcite U–Pb geochronology to directly constrain the precise timing of mineralisation at the Yawan gold deposit. In contrast to conventional isotopic dating methods (e.g., zircon U–Pb), calcite precipitates directly from ore-forming fluids, and its U–Pb system can more accurately record the age of hydrothermal mineralisation events [23]. Technical refinement and the use of reference materials have enhanced the sensitivity of mass spectrometers, making the dating of hydrothermal calcite considerably less challenging. This approach addresses the dating of carbonate-associated deposits that lack conventional chronometers such as zircon, rutile, or sericite.

The obtained calcite U–Pb age of 215.6 ± 7.1 Ma tightly constrains the mineralisation of the Yawan Gold Deposit to the Late Triassic. This result is consistent with the known

https://doi.org/10.3390/min16010114

metallogenic ages of other gold deposits in the region, such as Liba (210 ± 1 Ma) and Daqiao (211 ± 1.5 Ma) [73,74]. This period corresponds to the transitional extensional stage following continental collision in the Qinling orogenic belt. Intensive tectonic activity during this time not only provided conduits for ore-forming fluids but also created favourable spaces for mineral precipitation. Although the Yawan deposit shows a close spatial relationship with the Zhongchuan pluton (220 ± 1.2 Ma), its mineralisation age is slightly younger than the emplacement age of the intrusion [75]. This suggests that gold mineralisation was not directly derived from magmatic differentiation but rather resulted from the leaching of surrounding rocks by deep metamorphic fluids or mixed fluids, driven by regional tectonism [29,61].

6.2. Parameters of the Ore-Forming Environment

Arsenopyrite geothermometry indicates that Apy1 and Apy2 represent two distinct mineralisation stages, reflecting a clear evolutionary path of the ore-forming fluid from high-temperature, high-sulfur-fugacity conditions to medium-temperature, low-sulfur-fugacity conditions [76,77]. This trend signifies a fundamental shift in the physicochemical regime of the entire mineralising system. The extensive precipitation of calcite in Stage II suggests a shift in the fluid towards more alkaline conditions; this pH increase likely served as an additional trigger for gold precipitation [78,79]. The replacement of Py2 along fractures or margins by stibnite or galena, forming residual textures, indicates an evolution in hydrothermal fluid composition (e.g., increasing Sb and Pb concentrations, decreasing temperature) [59,80]. Later-stage fluids reacted with earlier-formed Py1, dissolving it and simultaneously precipitating stibnite or galena, marking the system's evolution into a low-temperature, Sb and Pb-rich stage. Comprehensive genetic constraints, including a mineralisation temperature of 389 ± 44 °C and pyrite S/Fe ratios that are higher than those of Carlin-type deposits, lower than those of epithermal or magmatic hydrothermal systems, and comparable to metamorphic hydrothermal types, collectively indicate a medium-temperature metamorphic hydrothermal environment [80,81].

In conclusion, gold mineralisation in the Yawan deposit occurred in two distinct stages. Stage 1, under relatively high-temperature conditions, led to the formation of Py1 and Apy1. Stage 2 was characterised by water–rock interaction, which triggered a sharp decrease in temperature and sulfur fugacity. This caused the decomposition of gold–sulfur complexes, a subsequent rise in pH, and the enrichment of Sb and Pb. Under these intermediate temperature conditions, the Py2 and Apy2 assemblage co-precipitated with gold.

6.3. Indicative of Gold Precipitation Mechanisms

In orogenic gold deposits, gold occurs primarily in two modes, visible gold and invisible gold, whose spatial distribution is governed by the interplay of ore-forming stages, fluid evolution, and structural activity [5,36]. Petrographic examination of the Yawan deposit reveals that visible gold is hosted almost exclusively within pyrite, typically as encapsulated grains. Invisible gold occurs mainly in solid solution (Au^+) or as nanometre-scale metallic particles (Au^0) within polymetallic sulfides [29,82]. Single-mineral mapping and spot analyses of Py1 and Py2 show that Py1 displays distinct zonation (Figure 8A), with As content significantly higher at the rims than in the cores (Figure 8D), while Co and Ni contents gradually decrease from core to rim. Py2 exhibits no zonation but contains 20-micron gold inclusions at its margins (Figure 8G) and locally elevated As in clusters (Figure 8J). The Fe/S ratio of Py2 is generally lower than that of Py1 (Figure 8M). Gold contents vary markedly among different analytical spots within the same sulfide grain (Figure 8N), indicating an inhomogeneous distribution of gold in sulfides.

 https://doi.org/10.3390/min16010114

Gold precipitation in orogenic deposits is governed by a spectrum of mechanisms, including fluid–rock reaction, fluid mixing, phase separation, sulfidation and electrochemical adsorption. HS^- and Cl^- are the major ligands responsible for Au complexation and transport in hydrothermal systems [83,84]. Among these, sulfidation is both ubiquitous and critical in driving gold deposition. In the Yawan orogenic gold deposit, massive sulfide formation sequestered substantial sulfur from the ore fluid, producing a pronounced decrease in sulfur fugacity from the Apy1 to the Apy2 stage. This decline destabilised the Au–S complexes, triggering the precipitation of native gold. Trace-element compositions and structural diversity of sulfides can effectively trace the physicochemical conditions of mineralisation, and their chemical and structural features serve as key indicators for identifying physical–chemical processes of hydrothermal fluids [85]. In the Yawan gold deposit, Py1 and Py2 occur as fine-grained, anhedral to subhedral crystals exhibiting pyritohedral and porous textures, which confer a strong physical adsorption capacity for Au^0 [4,86]. The chemical compositions of Py1 and Py2 reveal three evolutionary trends in the ore-forming fluid (Figure 9A): first, substitution of S by As^{-1} (consistent with a solid-solution model); second, replacement of Fe^{2+} by As^{2+}; and third, replacement of Fe^{2+} by Me^{2+} (trace elements such as Cu, Co, Ni, Pb, Ag) [87]. The low Fe/S ratios also indicate isomorphous substitution during gold precipitation [56,88]. The As content in Py2 is significantly higher than that in Py1, and the S–As ratio shows a negative correlation (Figure 9B), suggesting that As substitutes for S isomorphously, inducing lattice defects and promoting the incorporation of Au^+ into the lattice as a solid solution [89]. Previous studies confirm that such invisible gold commonly occurs as sub-micron to nano-scale metallic particles [90,91]. The molecular mechanism involves specific adsorption of S-Au-S-type aqueous complexes, leading to the formation of polysulfide complexes on pyrite surfaces [92].

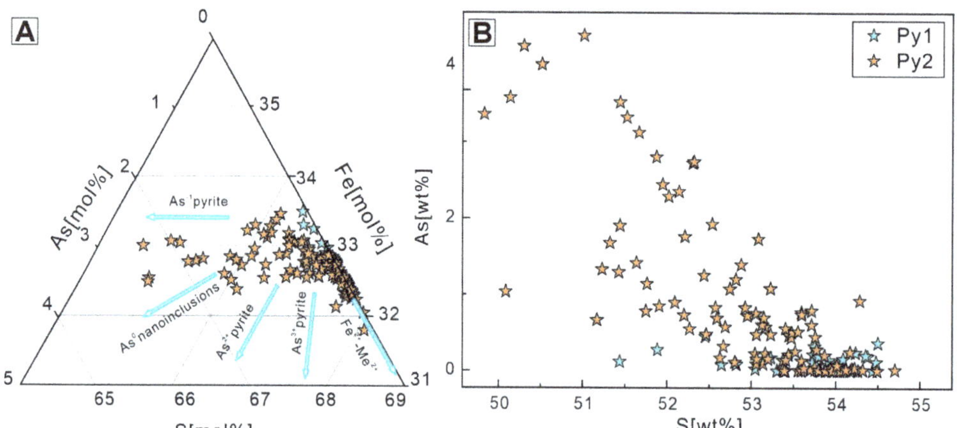

Figure 9. (**A**) Fe-As-S ternary diagram of pyrite (considering only EMPA data, n = 136). (**B**) Binary As-S plot of pyrite. (r_P = Pearson) (modified after [87]).

7. Conclusions

(1) The deposit experienced two principal mineralisation episodes: An early stage (I) precipitated a quartz–pyrite (Py1)–arsenopyrite (Apy1)–chalcopyrite assemblage, in which Py1 displays oscillatory zoning with enrichment of arsenic and gold along its rims. A main stage (II) deposited a quartz–calcite–pyrite (Py2)–arsenopyrite (Apy2)–stibnite–sphalerite–galena assemblage, characterised by unzoned Py2 that commonly contains inclusions of native gold and arsenopyrite.

https://doi.org/10.3390/min16010114

(2) Calcite U–Pb geochronology constrains the timing of mineralisation to 215.6 ± 7.1 Ma, indicating a Late Triassic age. This result is consistent within analytical uncertainty with the formation age of the regionally significant Liba gold deposit in the Western Qinling Orogen (210 ± 1 Ma), providing key chronological evidence for correlating regional mineralisation events.

(3) Integrated studies indicate that the Yawan gold deposit represents a classic orogenic gold system. Mineralisation was driven by Late Triassic tectono-hydrothermal processes, with sulfidation acting as the principal mechanism responsible for gold precipitation.

Author Contributions: Idea conception, C.-Q.D., M.-Z.H. and J.-N.F.; map compilation, Z.-H.B. and K.Y.; data curation, formal analysis, and investigation, C.-Q.D., J.-Y.W., H.Z. and J.-N.F.; writing—original draft, C.-Q.D. All authors have read and agreed to the published version of the manuscript.

Funding: This research was funded by Deep Earth Probe and Mineral Resources Exploration-National Science and Technology Major Project (2025ZD1009600), the projects of Ministry of Natural Resources, a new round of Mineral Exploration Breakthrough Strategic Action Science and Technology Support Project (grant ZKKJ202410), the Fundamental Research Funds for the Central Universities (grant 2652024008), the Frontiers Science Center for Deep-time Digital Earth (grant 2652023001), the Xi'an Center of China Geological Survey for Mineral Resource Exploration (XKD20250102).

Data Availability Statement: The data set is presented directly in the present study.

Acknowledgments: We are grateful to Hao-Cheng Yu for their field guidance and constructive discussions. We are grateful to Han-Dong Cao for the assistance with the figures. We also deeply thank the anonymous reviewers and editors for their helpful comments and suggestions.

Conflicts of Interest: The authors declare no conflicts of interest.

Appendix A

Table A1. EMPA data of pyrite and arsenopyrite of the Yawan gold deposit (wt%).

	No.	As	Sb	S	Fe	Cu	Co	Zn	Ni	Mn	Pb	Ag	Au	Total
Py1														
23YW0902-117	119	0.000	0.011	54.095	46.089	0.000	0.032	0.000	0.000	0	0.008	0	0.002	100.237
23YW0902-118	120	0.000	0.009	53.980	45.829	0.021	0.074	0.015	0.000	0	0.001	0	0.048	99.976
23YW0902-119	121	0.000	0.002	54.170	46.044	0.013	0.069	0.000	0.000	0	0.015	0	0.034	100.346
23YW0902-120	122	0.015	0.000	53.755	45.355	0.000	0.041	0.000	0.000	0	0.050	0	0.066	99.281
23YW0902-125	125	0.000	0.008	53.305	45.685	0.024	0.037	0.000	0.000	0	0.066	0	0.000	99.124
23YW0902-126	126	0.014	0.000	54.203	45.970	0.000	0.064	0.006	0.000	0	0.000	0	0.000	100.257
23YW0902-127	127	0.000	0.000	53.405	45.450	0.016	0.027	0.000	0.000	0	0.037	0	0.002	98.936
23YW0902-128	128	0.000	0.000	54.020	45.927	0.007	0.026	0.000	0.000	0	0.113	0	0.012	100.105
23YW01-1	1	0.000	0.000	54.435	46.239	0.006	0.023	0.008	0.000	0	0.009	0	0.000	100.720
23YW01-2	2	0.000	0.001	54.263	45.601	0.011	0.046	0.000	0.000	0	0.005	0	0.000	99.927
23YW01-17	17	0.082	0.000	54.052	45.859	0.000	0.050	0.004	0.046	0	0.050	0	0.000	100.142
23YW01-18	18	0.447	0.005	52.452	45.130	0.038	0.021	0.000	0.000	0	0.000	0	0.000	98.093
23YW01-19	19	0.318	0.027	53.751	45.836	0.018	0.029	0.013	0.006	0	0.006	0	0.000	100.003
23YW01-21	21	0.211	0.000	53.736	45.561	0.000	0.028	0.013	0.000	0	0.080	0	0.000	99.629
23YW1701-75	79	0.019	0.013	53.930	45.788	0.010	0.050	0.000	0.000	0	0.031	0	0.000	99.840
23YW0902-102	105	0.050	0.002	53.985	46.054	0.013	0.061	0.000	0.000	0	0.128	0	0.000	100.293
23YW0902-108	111	0.053	0.000	54.005	45.734	0.015	0.021	0.000	0.010	0	0.099	0	0.006	99.943
23YW1701-66	71	0.073	0.012	53.773	45.832	0.004	0.035	0.018	0.013	0	0.074	0	0.017	99.850
23YW0902-103	106	0.078	0.015	53.913	45.920	0.005	0.044	0.008	0.000	0	0.033	0	0.082	100.098
23YW1701-23	40	0.079	0.000	52.649	45.708	0.000	0.039	0.000	0.000	0	0.042	0	0.000	98.517
23YW1701-39	52	0.083	0.017	52.825	45.433	0.000	0.058	0.000	0.003	0	0.111	0	0.010	98.541
23YW0902-106	109	0.111	0.026	53.723	45.926	0.000	0.048	0.000	0.000	0	0.000	0	0.064	99.898
23YW1701-69	73	0.113	0.023	54.479	45.932	0.019	0.049	0.000	0.000	0	0.054	0	0.013	100.681
23YW1701-79	83	0.136	0.005	53.919	45.741	0.000	0.018	0.000	0.000	0	0.106	0	0.000	99.925
23YW0902-101	104	0.150	0.002	54.087	45.903	0.000	0.009	0.000	0.000	0	0.129	0	0.031	100.311
23YW1701-28	44	0.172	0.002	53.230	45.196	0.007	0.038	0.000	0.002	0	0.067	0	0.000	98.715
23YW0902-111	114	0.187	0.010	54.376	45.734	0.019	0.041	0.000	0.000	0	0.076	0	0.000	100.442
23YW1701-72	76	0.188	0.007	53.800	45.900	0.023	0.086	0.016	0.000	0	0.044	0	0.000	100.063
23YW1701-71	75	0.198	0.054	54.352	45.317	0.077	0.061	0.000	0.000	0	0.067	0	0.000	100.128
23YW0902-91	95	0.237	0.013	54.233	46.094	0.000	0.012	0.000	0.000	0	0.030	0	0.015	100.634
23YW0902-90	94	0.361	0.011	54.500	46.405	0.014	0.028	0.000	0.000	0	0.060	0	0.000	101.379
23YW09-2-25	67	0	0	53.822	45.111	0.022	0.108	0	0	0	0.026	0	0.026	99.116

 https://doi.org/10.3390/min16010114

Table A1. *Cont.*

	No.	As	Sb	S	Fe	Cu	Co	Zn	Ni	Mn	Pb	Ag	Au	Total
Py2														
23YW09-2-29	71	0.319	0.009	52.671	44.914	0.006	0.067	0.029	0.065	0.001	0.026	0.01	0.052	98.169
23YW09-2-30	72	0.138	0.018	53.376	45.135	0.014	0.082	0	0	0	0.079	0	0.05	98.892
23YW09-2-31	73	0.006	0.004	53.813	45.004	0	0.077	0.016	0.093	0.02	0	0	0	99.033
23YW09-2-36	78	0.112	0	52.807	43.831	0.011	0.063	0.012	0.16	0.007	0	0.066	1.604	98.673
23YW09-2-37	79	0.208	0	53.439	45.006	0.015	0.02	0.007	0.103	0	0.083	0.01	0.037	98.929
23YW09-2-38	80	0.251	0	53.520	45.144	0.016	0.039	0	0.178	0	0.009	0.012	0	99.169
23YW09-2-41	83	0.013	0	54.702	44.948	0	0.048	0.003	0.014	0.027	0	0.001	0	99.756
23YW09-2-42	84	0.043	0	53.937	44.805	0.028	0.125	0	0.079	0.008	0.044	0	0.024	99.093
23YW09-2-43	85	0.045	0.004	53.754	44.918	0.022	0.079	0.024	0	0	0.084	0.042	0.013	98.985
23YW09-2-44	86	0.029	0	53.936	45.017	0.002	0.042	0.024	0	0	0.007	0.006	0.03	99.093
23YW09-2-45	87	0	0	53.881	44.785	0	0.054	0	0.031	0	0.022	0.001	0.004	98.777
23YW09-1-5	92	0.577	0.026	53.168	44.049	0.123	0.109	0	0.314	0.012	0.019	0	0.045	98.442
23YW09-1-15	102	0	0	53.382	45.300	0	0.059	0	0	0.004	0	0	0.007	98.752
23YW09-1-16	103	0.014	0.003	53.698	45.062	0	0.066	0.011	0	0	0	0	0.004	98.858
23YW09-1-17	104	0.088	0.01	53.061	44.630	0.021	0.076	0	0.705	0.001	0	0	0	98.592
23YW06-5	113	0.222	0	53.045	44.112	0.073	0.056	0	2.025	0.008	0.056	0.015	0.015	99.628
23YW06-6	114	0.225	0.003	53.166	43.251	0.009	0.09	0.003	3.606	0.09	0.061	0.014	0.051	100.569
23YW06-8	116	0.166	0.053	52.630	43.868	0.113	0.064	0.023	2.463	0.044	0.053	0.026	0.085	99.588
23YW01-1	130	0.141	0.002	53.409	44.476	0.019	0.046	0.044	0.051	0.005	0.042	0.004	0.06	98.299
23YW01-4	133	0.005	0	53.556	44.947	0.019	0.096	0.029	0.342	0	0	0.013	0.028	99.035
23YW01-6	135	0	0	53.593	45.086	0.005	0.084	0	0.055	0	0	0	0	98.822
23YW01-8	137	0.014	0	54.194	45.243	0.01	0.1	0	0.007	0	0.032	0.006	0.065	99.671
23YW01-9	138	0.006	0	53.800	45.134	0	0.059	0	0.068	0	0	0	0.006	99.073
23YW01-12	141	0	0	53.880	44.651	0.014	0.154	0.023	1.342	0	0.058	0	0.041	100.163
23YW01-13	142	0.023	0.014	54.188	45.415	0.015	0.061	0	0.273	0	0.041	0	0.05	100.080
23YW01-14	143	0.033	0	54.223	45.691	0	0.055	0	0.027	0	0.02	0	0.076	100.125
23YW01-15	144	0	0	54.050	45.031	0.016	0.072	0	0.769	0.007	0	0	0.006	99.951
23YW01-16	145	0	0	53.956	45.167	0.01	0.08	0.004	0.051	0.009	0	0.012	0.047	99.336
23YW01-17	146	0.017	0.009	53.372	44.396	0.006	0.173	0.003	1.372	0.016	0.028	0	0.002	99.394
23YW01-18	147	0.093	0.009	53.140	44.806	0.001	0.082	0	0	0	0	0.012	0.019	98.162
23YW01-19	148	0.153	0	53.065	45.256	0.003	0.052	0	0.051	0	0.1	0.021	0.024	98.725
23YW01-21	150	0.13	0	53.595	45.510	0	0.06	0	0	0	0.009	0	0.121	99.426
23YW01-28	157	0.032	0.017	53.624	45.046	0.044	0.063	0	0	0.001	0.024	0.007	0.013	98.871
23YW01-29	158	0	0.015	53.394	44.524	0.006	0.015	0.033	0.364	0.02	0.069	0.008	0	98.448
23YW01-34	163	0.152	0.001	53.198	44.584	0	0.057	0.008	0	0.021	0	0.017	0	98.037
23YW0902-114	117	2.442	0.011	51.947	45.154	0.000	0.020	0.000	0.000	0	0.000	0	0.000	99.574
23YW0902-115	118	4.336	0.008	50.512	44.937	0.003	0.046	0.035	0.014	0	0.076	0	0.000	99.966
23YW0902-122	123	0.000	0.015	54.490	45.796	0.000	0.034	0.003	0.000	0	0.000	0	0.034	100.371
23YW0902-124	124	0.197	0.003	53.869	45.813	0.000	0.018	0.012	0.000	0	0.041	0	0.012	99.964
23YW01-3	3	0.768	0.016	53.594	45.788	0.010	0.046	0.032	0.000	0	0.044	0	0.000	100.296
23YW01-4	4	0.245	0.000	54.168	45.941	0.013	0.025	0.000	0.000	0	0.071	0	0.037	100.499
23YW01-5	5	0.010	0.000	54.319	45.914	0.006	0.046	0.000	0.000	0	0.080	0	0.053	100.429
23YW01-6	6	0.014	0.000	54.118	45.881	0.000	0.000	0.000	0.000	0	0.133	0	0.021	100.166
23YW01-7	7	0.463	0.000	53.448	45.767	0.014	0.041	0.000	0.000	0	0.027	0	0.000	99.759
23YW01-8	8	1.727	0.032	53.084	45.618	0.022	0.047	0.025	0.000	0	0.048	0	0.000	100.601
23YW01-9	9	0.276	0.004	53.774	45.640	0.028	0.046	0.000	0.000	0	0.000	0	0.069	99.837
23YW01-10	10	0.738	0.000	53.596	45.670	0.007	0.013	0.000	0.000	0	0.006	0	0.019	100.047
23YW01-11	11	2.803	0.000	51.872	44.714	0.415	0.002	0.038	0.000	0	0.000	0	0.000	99.843
23YW01-12	12	1.758	0.006	52.211	45.359	0.160	0.021	0.000	0.025	0	0.051	0	0.025	99.616
23YW01-13	13	0.751	0.008	53.054	45.902	0.006	0.041	0.013	0.000	0	0.099	0	0.031	99.905
23YW01-14	14	0.068	0.027	54.012	46.088	0.023	0.045	0.000	0.000	0	0.054	0	0.000	100.318
23YW01-15	15	1.904	0.012	51.440	44.537	0.192	0.024	0.000	0.029	0	0.046	0	0.000	98.183
23YW01-16	16	1.073	0.011	52.748	45.415	0.146	0.010	0.000	0.000	0	0.012	0	0.006	99.421
23YW01-22	22	1.078	0.019	53.226	45.248	0.089	0.052	0.017	0.008	0	0.095	0	0.000	99.833
23YW01-23	23	1.923	0.262	52.535	44.832	0.054	0.034	0.016	0.000	0	0.144	0	0.041	99.841
23YW01-24	24	2.724	0.000	52.302	44.596	0.255	0.048	0.007	0.000	0	0.058	0	0.014	100.004
23YW01-25	25	4.578	0.019	50.296	43.916	0.127	0.041	0.000	0.000	0	0.073	0	0.008	99.057
23YW1701-16	33	0.474	0.007	52.463	45.344	0.000	0.052	0.000	0.000	0	0.019	0	0.000	98.359
23YW1701-73	77	0.603	0.015	53.713	45.401	0.047	0.067	0.000	0.000	0	0.107	0	0.000	99.953
23YW1701-44	56	0.605	0.004	53.128	45.657	0.006	0.039	0.016	0.000	0	0.081	0	0.000	99.536
23YW1701-67	72	0.690	0.040	52.593	44.908	0.024	0.051	0.010	0.004	0	0.000	0	0.004	98.324
23YW0902-93	97	0.696	0.019	53.148	45.386	0.000	0.064	0.019	0.018	0	0.095	0	0.037	99.481
23YW1701-40	53	0.724	0.001	52.956	45.829	0.021	0.053	0.000	0.000	0	0.036	0	0.021	99.641
23YW1701-42	54	0.731	0.025	52.203	45.441	0.023	0.025	0.000	0.000	0	0.000	0	0.012	98.458
23YW1701-80	84	0.790	0.017	53.714	45.548	0.031	0.046	0.052	0.000	0	0.095	0	0.051	100.344
23YW1701-46	58	0.827	0.020	52.928	45.900	0.000	0.014	0.053	0.000	0	0.012	0	0.023	99.777
23YW1701-43	55	0.829	0.026	52.575	45.573	0.000	0.063	0.057	0.000	0	0.032	0	0.004	99.158
23YW1701-27	43	0.847	0.000	51.909	45.617	0.000	0.017	0.004	0.000	0	0.000	0	0.000	98.394
23YW1701-31	47	0.895	0.027	52.094	45.025	0.011	0.051	0.001	0.003	0	0.005	0	0.000	98.112
23YW0902-94	98	0.921	0.030	54.283	45.958	0.014	0.016	0.000	0.022	0	0.017	0	0.000	101.259
23YW1701-24	41	1.139	0.000	51.760	45.218	0.003	0.040	0.012	0.000	0	0.032	0	0.050	98.254
23YW1701-77	81	1.197	0.000	52.819	45.041	0.146	0.042	0.000	0.001	0	0.045	0	0.025	99.315
23YW0902-89	93	1.249	0.045	52.436	45.184	0.040	0.033	0.003	0.013	0	0.021	0	0.000	99.023
23YW1701-59	64	1.388	0.000	52.881	44.962	0.047	0.034	0.000	0.000	0	0.006	0	0.094	99.411
23YW1701-32	48	1.412	0.014	51.634	45.435	0.008	0.054	0.000	0.000	0	0.035	0	0.000	98.592
23YW1701-34	49	1.673	0.000	51.322	45.095	0.000	0.049	0.032	0.000	0	0.017	0	0.010	98.196
23YW0902-87	91	2.284	0.000	52.021	44.903	0.006	0.054	0.021	0.000	0	0.063	0	0.033	99.385
23YW0902-109	112	2.348	0.000	52.140	45.185	0.026	0.062	0.000	0.000	0	0.000	0	0.014	99.775
23YW1701-60	65	2.742	0.000	52.317	44.889	0.089	0.054	0.000	0.031	0	0.000	0	0.069	100.191
23YW1701-65	70	3.123	0.000	51.661	45.092	0.219	0.047	0.020	0.000	0	0.038	0	0.058	100.257
23YW0902-107	110	3.320	0.000	51.520	44.981	0.000	0.018	0.033	0.000	0	0.093	0	0.002	99.966
23YW1701-64	69	3.517	0.046	51.437	44.956	0.282	0.053	0.049	0.015	0	0.014	0	0.066	100.436

 https://doi.org/10.3390/min16010114

Table A1. *Cont.*

	No.	As	Sb	S	Fe	Cu	Co	Zn	Ni	Mn	Pb	Ag	Au	Total
Py2														
23YW0902-110	113	3.578	0.000	50.134	44.467	0.000	0.033	0.016	0.000	0	0.008	0	0.000	98.235
23YW0902-88	92	4.717	0.043	51.013	44.453	0.000	0.056	0.000	0.000	0	0.045	0	0.026	100.352
23YW1701-61	66	0.433	0.008	53.473	45.370	0.013	0.047	0.000	0.035	0	0.036	0	0.000	99.413
23YW1701-78	82	0.442	0.022	53.758	45.588	0.028	0.052	0.052	0.000	0	0.043	0	0.093	100.077
23YW0902-104	107	0.473	0.022	53.061	45.198	0.013	0.059	0.000	0.004	0	0.085	0	0.000	98.916
23YW0902-112	115	0.473	0.004	53.229	45.730	0.015	0.033	0.063	0.000	0	0.015	0	0.027	99.589
23YW1701-74	78	0.501	0.013	53.479	45.733	0.037	0.058	0.000	0.014	0	0.030	0	0.068	99.933
23YW0902-92	96	0.511	0.000	53.552	45.554	0.018	0.060	0.045	0.001	0	0.049	0	0.015	99.804
23YW1701-35	50	0.548	0.014	52.267	45.160	0.000	0.042	0.000	0.015	0	0.037	0	0.048	98.130
23YW1701-52	59	0.553	0.000	53.404	45.784	0.000	0.045	0.000	0.000	0	0.000	0	0.033	99.818
Apy1														
23YW06-18	126	43.972	0	20.221	33.856	0	0.027	0	0	0	0.027	0	0	98.103
23YW06-19	127	44.507	0.025	20.917	34.240	0.007	0.025	0.001	0.025	0	0	0	0.03	99.777
23YW06-20	128	44.726	0.001	20.823	34.173	0	0.032	0.013	0.101	0.01	0.011	0.01	0.016	99.916
23YW07-1-11	12	45.793	0.166	18.948	33.430	0.015	0.027	0	0	0	0.056	0	0.019	98.454
23YW07-1-12	13	47.483	0.042	18.487	33.067	0.013	0.035	0	0.637	0.007	0.04	0	0.116	99.927
23YW07-1-13	14	46.857	0.089	19.177	33.425	0.026	0.01	0	0	0.044	0.04	0	0	99.668
23YW07-1-5	5	44.802	0.104	19.688	33.411	0.016	0.049	0	0.122	0.028	0.052	0	0	98.272
23YW07-1-7	8	47.224	0.065	18.994	33.433	0	0.062	0	0	0	0.014	0.018	0	99.810
23YW07-1-8	9	45.574	0.057	18.966	33.246	0	0.042	0.035	0.138	0.018	0.057	0	0	98.134
23YW06-13	121	45.724	0	20.471	34.063	0.001	0.055	0	0	0.004	0.048	0	0.136	100.502
23YW06-14	122	45.1	0	20.309	34.502	0	0.039	0	0	0.037	0.04	0	0	100.027
23YW06-15	123	44.684	0	20.281	33.968	0.021	0.013	0	0.04	0.022	0.049	0.016	0	99.094
23YW06-16	124	44.93	0	21.349	33.770	0	0.033	0	0.009	0.028	0.085	0	0.058	100.262
23YW01-26	155	44.738	0.029	20.677	31.472	0.078	1.094	0	3.368	0.013	0.092	0	0.03	101.591
23YW01-30	159	45.202	0	20.551	33.485	0.003	0.211	0.042	0.7	0.018	0	0	0.03	100.242
Apy2														
23YW01-24	153	41.702	0.032	23.946	34.857	0.148	0.046	0.055	0	0.036	0	0	0	100.822
23YW0902-83	87	42.536	0.015	20.978	34.303	0.000	0.160	0.000	0.033	0	0.017	0	0.017	98.059
23YW0902-84	88	42.273	0.097	21.296	34.866	0.000	0.020	0.079	0.000	0	0.048	0	0.000	98.678
23YW0902-85	89	41.337	0.070	22.505	35.140	0.000	0.084	0.000	0.000	0	0.000	0	0.000	99.137
23YW1701-56	61	42.423	0.024	22.103	35.108	0.000	0.012	0.000	0.005	0	0.106	0	0.075	99.856
23YW09-2-3	45	44.291	0.043	20.868	34.455	0.025	0.041	0	0	0.012	0	0	0.04	99.775
23YW09-2-4	46	44.313	0.012	21.029	34.545	0.005	0.055	0	0	0.033	0.041	0	0.049	100.082
23YW09-1-2	89	44.318	0.069	21.105	34.346	0.013	0.018	0	0	0.006	0	0.013	0.007	99.895
23YW09-1-3	90	43.966	0.028	21.528	34.100	0.001	0.077	0.065	0.003	0	0	0	0.005	99.773
23YW09-1-4	91	43.695	0.007	21.886	34.347	0	0.014	0	0.046	0.017	0.006	0	0	100.018
23YW09-1-6	93	43.169	0.049	20.696	34.087	0	0	0.018	0.175	0.012	0	0	0.012	98.218
23YW09-1-9	96	44.382	0.041	20.246	33.600	0.01	0.029	0	0.055	0	0.041	0.012	0.054	98.469
23YW09-1-1	88	43.667	0.014	20.841	33.891	0	0.054	0	0	0	0.027	0.016	0.063	98.573
23YW09-1-12	99	42.908	0.052	20.971	34.111	0.008	0.05	0	0.083	0.025	0.074	0	0	98.283
23YW09-1-13	100	44.317	0.045	21.005	34.220	0	0.053	0	0.018	0.043	0	0.007	0.014	99.723
23YW09-1-14	101	44.518	0.011	21.170	33.926	0.014	0.059	0	0.092	0	0.114	0	0.054	99.959
23YW09-2-1	42	45.133	0.01	21.026	34.477	0	0.032	0.011	0	0	0.052	0	0.009	100.749
23YW09-2-1	43	44.874	0.021	20.941	34.556	0.019	0.018	0.024	0.141	0	0	0	0.049	100.643
23YW09-2-2	44	45.833	0	21.160	34.106	0.019	0.062	0	0.35	0.015	0.021	0.002	0	101.568
23YW09-2-13	55	44.603	0.063	21.291	34.125	0.014	0.029	0.023	0.402	0	0	0	0	100.550
23YW09-2-17	59	44.041	0.002	21.326	34.506	0	0.038	0.03	0.018	0	0.039	0	0.049	100.050
23YW09-2-24	66	42.176	0.009	23.253	35.462	0	0.055	0	0	0.014	0	0.007	0.049	101.025

References

1. Goldfarb, R.J.; Groves, D.I. Orogenic gold: Common or evolving fluid and metal sources through time. *Lithos* **2015**, *233*, 2–26. [CrossRef]
2. Qiu, K.-F.; Taylor, R.D.; Song, Y.-H.; Yu, H.-C.; Song, K.-R.; Li, N. Geologic and geochemical insights into the formation of the taiyangshan porphyry copper–molybdenum deposit, western qinling orogenic belt, China. *Gondwana Res.* **2016**, *35*, 40–58. [CrossRef]
3. Yang, L.-Q.; Deng, J.; Goldfarb, R.J.; Zhang, J.; Gao, B.-F.; Wang, Z.-L. 40Ar/39Ar geochronological constraints on the formation of the Dayingezhuang gold deposit: New implications for timing and duration of hydrothermal activity in the Jiaodong gold province, China. *Gondwana Res.* **2014**, *25*, 1469–1483. [CrossRef]
4. Deng, J.; Wang, Q.; Li, G. Tectonic evolution, superimposed orogeny, and composite metallogenic system in China. *Gondwana Res.* **2017**, *50*, 216–266. [CrossRef]
5. Goldfarb, R.; Qiu, K.; Deng, J.; Chen, Y.; Yang, L. Orogenic gold deposits of China. In *Mineral Deposits of China*; Chang, Z., Goldfarb, R.J., Eds.; Society of Economic Geologists: Littleton, CO, USA, 2019; pp. 236–324.
6. Luo, B.; Zhang, H.; Xu, W.; Yang, H.; Zhao, J.; Guo, L.; Zhang, L.; Tao, L.; Pan, F.; Gao, Z. The magmatic plumbing system for Mesozoic high-Mg andesites, garnet-bearing dacites and porphyries, rhyolites and leucogranites from West Qinling, central China. *J. Petrol.* **2018**, *59*, 447–482. [CrossRef]
7. Gou, Z.-Y.; Yu, H.-C.; Qiu, K.-F.; Geng, J.-Z.; Wu, M.-Q.; Wang, Y.-G.; Yu, M.-H.; Li, J. Petrogenesis of ore-hosting diorite in the Zaorendao gold deposit at the Tongren-Xiahe-Hezuo polymetallic district, West Qinling, China. *Minerals* **2019**, *9*, 76. [CrossRef]

https://doi.org/10.3390/min16010114

8. Kong, C.S.; Shen, J.; Santosh, M.; Li, S.; Cao, W.; Peng, Z.; Liu, H.; Li, J.; Li, J.; Bao, L. Age and genesis of the Gangcha gold deposit, western Qinling orogen, China. *Geol. J.* **2018**, *53*, 1871–1885. [CrossRef]
9. Qiu, K.F.; Deng, J.; He, D.Y.; Rosenbaum, G.; Zheng, X.; Williams-Jones, A.E.; Yu, H.C.; Balen, D. Evidence of vertical slab tearing in the Late Triassic Qinling Orogen (central China) from multiproxy geochemical and isotopic imaging. *J. Geophys. Res. Solid Earth* **2023**, *128*, e2022JB025514. [CrossRef]
10. Yu, H.C.; Qiu, K.F.; Deng, J.; Zhu, R.; Mathieu, L.; Sai, S.X.; Sha, W.J. Exhuming and preserving epizonal orogenic au-sb deposits in rapidly uplifting orogenic settings. *Tectonics* **2022**, *41*, e2021TC007165. [CrossRef]
11. Qiu, K.-F.; Long, Z.-Y.; Romer, R.L.; Halama, R.; Williams-Jones, A.; Yu, H.-C.; Li, S.-S.; Wu, M.-Q.; Deng, J. The mantle source of ree-rich alkaline silicate magmas can be enriched by continent-derived sediment subduction. *Nat. Commun.* **2025**, *16*, 11191. [CrossRef]
12. Guo, X.; Yan, Z.; Wang, Z.; Wang, T.; Wu, F.; Fu, C.; Hou, K.; Li, J. Geological characteristics and associated magmatic ages of the Xiekeng skarn-type Cu-Au deposit in the West Qinling terrane. *Acta Petrol. Sin.* **2011**, *27*, 3811–3822.
13. Hu, X.; Ding, Z.; Gong, Y.; Li, K.; He, M. Sediment-hosted disseminated gold deposits in orogenic belts: An example from the giant Jiagantan gold deposit in the West Qinling orogen, China. *Ore Geol. Rev.* **2022**, *146*, 104950. [CrossRef]
14. Luo, B.; Zhang, H.; Xiao, Z. Petrogenesis and tectonic implications of the Early Indosinian Meiwu pluton in West Qinling, central China. *Earth Sci. Front.* **2012**, *19*, 199–213. [CrossRef]
15. Qiu, K.-F.; Yu, H.-C.; Hetherington, C.; Huang, Y.-Q.; Yang, T.; Deng, J. Tourmaline composition and boron isotope signature as a tracer of magmatic-hydrothermal processes. *Am. Mineral.* **2021**, *106*, 1033–1044. [CrossRef]
16. Zhu, L.; Zhang, G.; Yang, T.; Wang, F.; Gong, H. Geochronology, petrogenesis and tectonic implications of the Zhongchuan granitic pluton in the Western Qinling metallogenic belt, China. *Geol. J.* **2013**, *48*, 310–334. [CrossRef]
17. Ping, W.; Xuanxue, M.; Xuehui, Y.; XiongFei, H.; Yi, D.; XiaoWei, L. Geochemistry, chronology and geological significance of the granitoids in Xiahe, West Qinling. *Acta Petrol. Sin.* **2013**, *29*, 3981–3992.
18. Dong, C.-Q.; Yu, H.-C.; Liu, Z.-R.; Li, Q.-Z.; Fu, J.-N.; Hao, M.-Z.; Wang, J.-Y.; Fan, C.; Zhang, X.; Gao, Y.-B. Geochemical Charac-teristics of Gold-Bearing Minerals and Their Implications for Gold Mineralization in the Maquan Gold Deposit, Western Qinling Orogen. *Earth Sci. Front.* **2025**, 1–33. [CrossRef]
19. Sui, J.-X.; Li, J.-W.; Wen, G.; Jin, X.-Y. The Dewulu reduced Au-Cu skarn deposit in the Xiahe-Hezuo district, West Qinling orogen, China: Implications for an intrusion-related gold system. *Ore Geol. Rev.* **2017**, *80*, 1230–1244. [CrossRef]
20. Huang, Y.-Q.; Qiu, K.-F.; Yu, H.C.; Jin, D.J.; He, D.Y.; Xiao, C.H.; Wang, Y.X. Petrogenesis of ore-hosting porphyry in the Gelouang gold deposit, West Qinling and its geological implications. *Acta Petrol. Sin.* **2020**, *36*, 1567–1585. [CrossRef]
21. Sui, J.-X.; Li, J.-W.; Jin, X.-Y.; Vasconcelos, P.; Zhu, R. 40Ar/39Ar and U-Pb constraints on the age of the Zaozigou gold deposit, Xiahe-Hezuo district, West Qinling orogen, China: Relation to early Triassic reduced intrusions emplaced during slab rollback. *Ore Geol. Rev.* **2018**, *101*, 885–899. [CrossRef]
22. Geng, J.-Z.; Qiu, K.-F.; Gou, Z.-Y.; Yu, H.-C. Tectonic regime switchover of Triassic Western Qinling Orogen: Constraints from LA-ICP-MS zircon U–Pb geochronology and Lu–Hf isotope of Dangchuan intrusive complex in Gansu, China. *Geochemistry* **2017**, *77*, 637–651. [CrossRef]
23. Fu, J.-N.; Pirajno, F.; Yang, F.; Shivute, E.; Sun, Y.-Z.; Ai, N.; Qiu, K.-F. Integration of zircon and apatite U–Pb geochronology and geochemical mapping of the Wude basalts (Emeishan large igneous province): A tool for a better understanding of the tectonothermal and geodynamic evolution of the Emeishan LIP. *Geosci. Front.* **2021**, *12*, 573–585. [CrossRef]
24. Liu, Y.; Gao, S.; Hu, Z.; Gao, C.; Zong, K.; Wang, D. Continental and oceanic crust recycling-induced melt–peridotite interactions in the Trans-North China Orogen: U–Pb dating, Hf isotopes and trace elements in zircons from mantle xenoliths. *J. Petrol.* **2010**, *51*, 537–571. [CrossRef]
25. Du, B.; Shen, J.; Santosh, M.; Liu, H.; Liu, J.; Wang, S.; Xu, K. Genesis of the Gangcha gold deposit, West Qinling Orogen, China: Constraints from Rb-Sr geochronology, in-situ sulfur isotopes and trace element geochemistry of pyrite. *Ore Geol. Rev.* **2021**, *138*, 104350. [CrossRef]
26. Poitrasson, F.; Chenery, S.; Shepherd, T.J. Electron microprobe and LA-ICP-MS study of monazite hydrothermal alteration: Implications for U-Th-Pb geochronology and nuclear ceramics. *Geochim. Et Cosmochim. Acta* **2000**, *64*, 3283–3297. [CrossRef]
27. Hu, F.; Liu, S.; Ducea, M.N.; Chapman, J.B.; Wu, F.; Kusky, T. Early Mesozoic magmatism and tectonic evolution of the Qinling Orogen: Implications for oblique continental collision. *Gondwana Res.* **2020**, *88*, 296–332. [CrossRef]
28. Xue, X.; Qiu, K.; He, D.; Huang, Y.; Zhu, R.; Yu, H.; Jiang, G. Petrogenesis of the quartz diorite porphyry in the zaozigou, west qinling and its geological implications. *Acta Petrol. Sin.* **2023**, *39*, 463–483. [CrossRef]
29. Yu, H.-C.; Qiu, K.-F.; Wang, J.; Zhang, P.-C.; Smeraglia, L.; Tassara, S.; Ma, J.-D.; Xue, X.-F.; Li, D.-P.; Deng, J. Competency contrast-driven extension veining and sequential orogenic au-sb mineralization: Insights from the zaorendao deposit, China. *Geol. Soc. Am. Bull.* **2025**. [CrossRef]

30. Zhang, L.; Yu, H.-C.; Fougerouse, D.; Huang, Y.-Q.; Fan, C.; Tamer, M.T.; Wang, G.-Y.; Li, C.; Qiu, K.-F. Coupled phase sep-aration and fluid cooling trigger gold precipitation in the xiakanmucang reduced intrusion-related gold deposit, China. *Geol. Soc. Am. Bull.* **2025**. [CrossRef]
31. He, D.-Y.; Qiu, K.-F.; Yu, H.-C.; Jowitt, S.M.; Zheng, X.; Mazumder, R.; Deng, J. Lithospheric architecture and evolution of the Qinling Orogen of Central China and associated controls on metallogeny. *Earth-Sci. Rev.* **2025**, *264*, 105092. [CrossRef]
32. He, D.-Y.; Qiu, K.-F.; Santosh, M.; Yu, H.-C.; Long, Z.-Y.; Wang, J.-Y. Inhomogeneous crust-mantle interaction and Triassic tectonic escape of a Proterozoic microplate: A tale of the Bikou Terrane. *Lithos* **2021**, *396*, 106227. [CrossRef]
33. Yu, H.-C.; Qiu, K.-F.; Nassif, M.T.; Geng, J.-Z.; Sai, S.-X.; Duo, D.-W.; Huang, Y.-Q.; Wang, J. Early orogenic gold mineralization event in the West Qinling related to closure of the Paleo-Tethys Ocean–Constraints from the Ludousou gold deposit, central China. *Ore Geol. Rev.* **2020**, *117*, 103217. [CrossRef]
34. Pang, Y.; Yan, L.; Liu, Y.; Tang, L.; Zhu, R.; Liu, G. Seismic Wave Finite-Difference Forward Modeling for Orogenic Gold Deposits. *Minerals* **2022**, *12*, 1465. [CrossRef]
35. Chen, B.; Zhang, Y.; Li, K. Genesis of the Jiagantan gold deposit in West Qinling, China: Constraints from Re-Os dating, in-situ trace elements and S isotope of gold bearing minerals. *Acta Petrol. Sin.* **2024**, *40*, 1767–1783. [CrossRef]
36. Qiu, K.-F.; Yu, H.-C.; Deng, J.; McIntire, D.; Gou, Z.-Y.; Geng, J.-Z.; Chang, Z.-S.; Zhu, R.; Li, K.-N.; Goldfarb, R. The giant Zaozigou Au-Sb deposit in West Qinling, China: Magmatic-or metamorphic-hydrothermal origin? *Miner. Depos.* **2020**, *55*, 345–362. [CrossRef]
37. Li, N.; Chen, Y.-J.; Santosh, M.; Pirajno, F. Compositional polarity of Triassic granitoids in the Qinling Orogen, China: Implication for termination of the northernmost paleo-Tethys. *Gondwana Res.* **2015**, *27*, 244–257. [CrossRef]
38. Xu, X.; Chen, J.; Gao, T.; Li, P.; Li, T. Granitoid magmatism and tectonic evolution in northern edge of the Western Qinling terrane, NW China. *Acta Petrol. Sin.* **2014**, *30*, 371–389.
39. Jin, X.-y.; Li, J.-w.; Hofstra, A.H.; Sui, J.-x. Magmatic-hydrothermal origin of the early Triassic Laodou lode gold deposit in the Xiahe-Hezuo district, West Qinling orogen, China: Implications for gold metallogeny. *Miner. Depos.* **2017**, *52*, 883–902. [CrossRef]
40. Jin, W.; Zhang, Q.; He, D.; Jia, X. Shrimp dating of adakites in western Qinling and their implications. *Acta Petrol. Sin.* **2005**, *21*, 959–966.
41. Dong, Y.; Yang, Z.; Liu, X.; Sun, S.; Li, W.; Cheng, B.; Zhang, F.; Zhang, X.; He, D.; Zhang, G. Mesozoic intracontinental orogeny in the Qinling Mountains, central China. *Gondwana Res.* **2016**, *30*, 144–158. [CrossRef]
42. Deng, J.; Wang, Q. Gold mineralization in China: Metallogenic provinces, deposit types and tectonic framework. *Gondwana Res.* **2016**, *36*, 219–274. [CrossRef]
43. Deng, J.; Wang, C.; Bagas, L.; Santosh, M.; Yao, E. Crustal architecture and metallogenesis in the south-eastern North China Craton. *Earth-Sci. Rev.* **2018**, *182*, 251–272. [CrossRef]
44. Dong, Y.; Santosh, M. Tectonic architecture and multiple orogeny of the Qinling Orogenic Belt, Central China. *Gondwana Res.* **2016**, *29*, 1–40. [CrossRef]
45. Yang, L.-Q.; Deng, J.; Qiu, K.-F.; Ji, X.-Z.; Santosh, M.; Song, K.-R.; Song, Y.-H.; Geng, J.-Z.; Zhang, C.; Hua, B. Magma mixing and crust–mantle interaction in the Triassic monzogranites of Bikou Terrane, central China: Constraints from petrology, geochemistry, and zircon U–Pb–Hf isotopic systematics. *J. Asian Earth Sci.* **2015**, *98*, 320–341. [CrossRef]
46. Boatner, L.A. Synthesis, structure, and properties of monazite, pretulite, and xenotime. *Rev. Mineral. Geochem.* **2002**, *48*, 87–121. [CrossRef]
47. Dong, Y.-P.; Zhou, M.-F.; Zhang, G.-W.; Zhou, D.-W.; Liu, L.; Zhang, Q. The Grenvillian Songshugou ophiolite in the Qinling Mountains, Central China: Implications for the tectonic evolution of the Qinling orogenic belt. *J. Asian Earth Sci.* **2008**, *32*, 325–335. [CrossRef]
48. Yang, L.-Q.; Deng, J.; Dilek, Y.; Qiu, K.-F.; Ji, X.-Z.; Li, N.; Taylor, R.D.; Yu, J.-Y. Structure, geochronology, and petrogenesis of the Late Triassic Puziba granitoid dikes in the Mianlue suture zone, Qinling orogen, China. *Bulletin* **2015**, *127*, 1831–1854. [CrossRef]
49. Yang, L.; Song, S.; Allen, M.B.; Su, L.; Dong, J.; Wang, C. Oceanic accretionary belt in the West Qinling Orogen: Links between the Qinling and Qilian orogens, China. *Gondwana Res.* **2018**, *64*, 137–162. [CrossRef]
50. Liu, L.; Liao, X.; Wang, Y.; Wang, C.; Santosh, M.; Yang, M.; Zhang, C.; Chen, D. Early Paleozoic tectonic evolution of the North Qinling Orogenic Belt in Central China: Insights on continental deep subduction and multiphase exhumation. *Earth-Sci. Rev.* **2016**, *159*, 58–81. [CrossRef]
51. Yang, L.-Q.; Deng, J.; Li, N.; Zhang, C.; Ji, X.-Z.; Yu, J.-Y. Isotopic characteristics of gold deposits in the Yangshan Gold Belt, West Qinling, central China: Implications for fluid and metal sources and ore genesis. *J. Geochem. Explor.* **2016**, *168*, 103–118. [CrossRef]
52. Deng, J.; Wang, Q.; Li, G.; Santosh, M. Cenozoic tectono-magmatic and metallogenic processes in the Sanjiang region, southwestern China. *Earth-Sci. Rev.* **2014**, *138*, 268–299. [CrossRef]
53. Li, S.; Jahn, B.-m.; Zhao, S.; Dai, L.; Li, X.; Suo, Y.; Guo, L.; Wang, Y.; Liu, X.; Lan, H. Triassic southeastward subduction of North China Block to South China Block: Insights from new geological, geophysical and geochemical data. *Earth-Sci. Rev.* **2017**, *166*, 270–285. [CrossRef]

 https://doi.org/10.3390/min16010114

54. Mao, J.; Goldfarb, R.J.; Zhang, Z.; Xu, W.; Qiu, Y.; Deng, J. Gold deposits in the Xiaoqinling–Xiong'ershan region, Qinling Mountains, central China. *Miner. Depos.* **2002**, *37*, 306–325. [CrossRef]
55. Dong, Y.; Zhang, G.; Neubauer, F.; Liu, X.; Genser, J.; Hauzenberger, C. Tectonic evolution of the Qinling orogen, China: Review and synthesis. *J. Asian Earth Sci.* **2011**, *41*, 213–237. [CrossRef]
56. Schandl, E.S.; Gorton, M.P. A textural and geochemical guide to the identification of hydrothermal monazite: Criteria for selection of samples for dating epigenetic hydrothermal ore deposits. *Econ. Geol.* **2004**, *99*, 1027–1035. [CrossRef]
57. Xiao, C.-H.; Chen, Z.-L.; Han, F.-B.; Wei, L.-X.; Ma, L.-T. Provenance and tectonic significance of Lower Triassic clastic sedimentary rocks of the West Qinling Orogenic Belt, Central China. *J. Asian Earth Sci.* **2022**, *224*, 105026. [CrossRef]
58. Yu, H.-C.; Qiu, K.-F.; Sai, S.-X.; McIntire, D.C.; Pirajno, F.; Duo, D.-W.; Miggins, D.P.; Wang, J.; Jia, R.-Y.; Wu, M.-Q. Paleo-tethys late triassic orogenic gold mineralization recorded by the Yidi'nan gold deposit, West Qinling, China. *Ore Geol. Rev.* **2020**, *116*, 103211. [CrossRef]
59. Yu, H.-C.; Guo, C.-A.; Qiu, K.-F.; McIntire, D.; Jiang, G.-P.; Gou, Z.-Y.; Geng, J.-Z.; Pang, Y.; Zhu, R.; Li, N.-B. Geochronological and geochemical constraints on the formation of the giant Zaozigou Au-Sb deposit, West Qinling, China. *Minerals* **2019**, *9*, 37. [CrossRef]
60. Yu, H.-C.; Qiu, K.-F.; Hetherington, C.J.; Chew, D.; Huang, Y.-Q.; He, D.-Y.; Geng, J.-Z.; Xian, H.-Y. Apatite as an alternative petrochronometer to trace the evolution of magmatic systems containing metamict zircon. *Contrib. Mineral. Petrol.* **2021**, *176*, 68. [CrossRef]
61. Qiu, K.-F.; Deng, J.; Yu, H.-C.; Rasbury, T.; Tang, Y.-W.; Zhu, R.; Zhang, P.-C.; Goldfarb, R. The Zaozigou orogenic gold-antimony deposit, West Qinling Orogen, China: Structural controls on multiple mineralization events. *Geol. Soc. Am. Bull.* **2024**, *136*, 4218–4232. [CrossRef]
62. Fan, C.; Yu, H.-C.; Tamer, M.T.; Zhang, L.; Wang, J.; Liu, P.-X.; Xue, X.-F.; Li, C.; Wang, Y.-X. Ore genesis of the Nanban gold deposit, West Qinling Orogen: Insights from monazite geochronology and geochemistry of sulfides. *Ore Geol. Rev.* **2025**, *179*, 106531. [CrossRef]
63. Wang, J.; Yu, H.-C.; Petrella, L.; Duo, D.-W.; Wang, L.; Yang, F.; Zhou, H.-Y.; Li, C.; Qiu, K.-F. Fluid evolution and genesis of the Yidinan granitoid-hosted orogenic gold deposit (China). *Geol. Soc. Am. Bull.* **2025**, *137*, 1945–1963. [CrossRef]
64. Zhang, L.; Qiu, K.; Hou, Z.; Pirajno, F.; Shivute, E.; Cai, Y. Fluid-rock reactions of the Triassic Taiyangshan porphyry Cu-Mo deposit (West Qinling, China) constrained by QEMSCAN and iron isotope. *Ore Geol. Rev.* **2021**, *132*, 104068. [CrossRef]
65. Qiu, K.; Yu, H.; Wu, M.; Geng, J.; Ge, X.; Gou, Z.; Taylor, R.D. Discrete Zr and REE mineralization of the Baerzhe rare-metal deposit, China. *Am. Mineral.* **2019**, *104*, 1487–1502. [CrossRef]
66. Zou, Y.-Z.; Li, S.-S.; Zhang, H.-F.; Tamer, M.T.; Bishikwabo, G.K.; Shivute, E.T.; Dong, W.-M.; Vatuva, A.; Zhang, C.-H. Petrogenesis, magmatic and metamorphic evolution of Salem granite: Implications for the Damara Orogeny. *Ore Geol. Rev.* **2025**, *179*, 106512. [CrossRef]
67. Tamer, M.T.; Ketcham, R.A. How many vs which: On confined track selection criteria for apatite fission track analysis. *Chem. Geol.* **2023**, *634*, 121584. [CrossRef]
68. Tamer, M.T.; Chung, L.; Ketcham, R.A.; Gleadow, A.J. The need for fission-track data transparency and sharing. *Geochronology* **2025**, *7*, 45–58. [CrossRef]
69. Qiu, K.-F.; Yang, L.-Q. Genetic feature of monazite and its U-Th-Pb dating: Critical considerations on the tectonic evolution of Sanjiang Tethys. *Acta Petrol. Sin.* **2011**, *27*, 2721–2732.
70. Kretschmar, U.; Scott, S. Phase relations involving arsenopyrite in the system Fe-As-S and their application. *Can. Miner.* **1976**, *14*, 364–386.
71. Hou, Z.; Qiu, K.; Zhou, T.; Cai, Y. An advanced image processing technique for backscatter-electron data by scanning electron microscopy for microscale rock exploration. *J. Earth Sci.* **2024**, *35*, 301–305. [CrossRef]
72. Yang, L.; Wang, Q.; Fougerouse, D.; Xian, H.; Li, H.; Wang, R.; Rickard, W.D.; Groves, D.I.; Deng, J. Release and re-enrichment of invisible gold in arsenian pyrite promoted by coupled dissolution-reprecipitation reactions. *Commun. Earth Environ.* **2024**, *5*, 241. [CrossRef]
73. Wu, Y.-F.; Li, J.-W.; Evans, K.; Vasconcelos, P.M.; Thiede, D.S.; Fougerouse, D.; Rempel, K. Late Jurassic to Early Cretaceous age of the Daqiao gold deposit, West Qinling Orogen, China: Implications for regional metallogeny. *Miner. Depos.* **2019**, *54*, 631–644. [CrossRef]
74. Ke, C.; Wang, X.; Yang, Y.; Tian, Y.; Li, J.; Nie, Z.; Lü, X.; Wang, S.; Gong, M. Petrogenesis of dykes and its relationship to gold mineralization in the western Qinling belt: Constraints from zircon U-Pb age, geochemistry and Nd-Hf-S isotopes of Liba gold deposit. *Miner. Depos.* **2020**, *39*, 42–62.
75. Yang, S.-S.; Liu, J.-J.; Zhang, F.; Yuan, F.; Zhai, D.-G.; Li, Y.-J.; Liao, Y.; Cao, Q. Petrogenesis and geodynamic setting of the Triassic granitoid plutons in West Qinling, China: Insights from LA-ICP-MS zircon U–Pb ages, Lu–Hf isotope signatures and geochemical characteristics of the Zhongchuan pluton. *Int. Geol. Rev.* **2017**, *59*, 1908–1928. [CrossRef]
76. Farsang, S.; Zajacz, Z. Sulfur species and gold transport in arc magmatic fluids. *Nat. Geosci.* **2025**, *18*, 98–104. [CrossRef]

 https://doi.org/10.3390/min16010114

77. Honglin, Z.; Aiying, W.; Jiahui, Z.; Chuandong, X. Geological characteristics and Genesis of Jiulong giant crystal calcite deposit in Luquan county, Central Yunnan province. *Contrib. Geol. Miner. Resour. Res.* **2024**, *39*, 80–86. [CrossRef]
78. Liang, Y.; Shu, L.; Ma, P.; Zhang, C.; Ma, Y.; Khan, M.; Shen, C. Gold source and ore-forming process of the Linglong gold deposit, Jiaodong gold province, China: Evidence from textures, mineral chemical compositions and sulfur isotopes of pyrite. *Ore Geol. Rev.* **2023**, *159*, 105523. [CrossRef]
79. Huang, H.-G.; Li, Z.; Liu, X.-L.; Zhang, C.-G.; Liu, L.-Q. Geological characteristics and genesis of Chafang gold deposit in Fanshi County, Shanxi Province. *Contrib. Geol. Miner. Resour. Res.* **2024**, *39*, 169–176. [CrossRef]
80. Zhu, Y.; An, F.; Tan, J. Geochemistry of hydrothermal gold deposits: A review. *Geosci. Front.* **2011**, *2*, 367–374. [CrossRef]
81. Lang, X.-C.; Han, X.-B.; Zhi, Y.-D.; Li, P.-L.; Wang, Y.; Wu, Q.-H. Geological characteristics and ore-forming mechanism of Au deposits in Kanggur area, East Tianshan mountainous district, Xinjiang Autonomous Region. *Contrib. Geol. Miner. Resour. Res.* **2024**, *39*, 177–185. [CrossRef]
82. Løge, I.A.; Anabaraonye, B.U.; Bovet, N.; Fosbøl, P.L. Crystal nucleation and growth: Supersaturation and crystal resilience determine stickability. *Cryst. Growth Des.* **2023**, *23*, 2619–2627. [CrossRef]
83. Deng, J.; Qiu, K.-F.; Wang, Q.-F.; Goldfarb, R.; Yang, L.-Q.; Zi, J.-W.; Geng, J.-Z.; Ma, Y. In situ dating of hydrothermal monazite and implications for the geodynamic controls on ore formation in the Jiaodong gold province, eastern China. *Econ. Geol.* **2020**, *115*, 671–685. [CrossRef]
84. Qiu, K.-F.; Goldfarb, R.J.; Deng, J.; Yu, H.-C.; Gou, Z.-Y.; Ding, Z.-J.; Wang, Z.-K.; Li, D.-P. Gold deposits of the jiaodong peninsula, eastern china. In *Geology of the World's Major Gold Deposits and Provinces*; Sillitoe, R.H., Goldfarb, R.J., Robert, F., Simmons, S.F., Eds.; Society of Economic Geologists: Littleton, CO, USA, 2020.
85. Ding, W.-C.; Yang, Z.-P.; Hua, B.; Liang, X.-W.; Tan, C.-X.; Wang, P. Geological and stable isotope geochemical charac-teristics of Tawuerbieke Au deposit in Xinjiang Autonomous Region. *Contrib. Geol. Miner. Resour. Res.* **2024**, *39*, 9–18. [CrossRef]
86. Deng, J.; Wang, Q.; Santosh, M.; Liu, X.; Liang, Y.; Yang, L.; Zhao, R.; Yang, L. Remobilization of metasomatized mantle lithosphere: A new model for the Jiaodong gold province, eastern China. *Miner. Depos.* **2020**, *55*, 257–274. [CrossRef]
87. Román, N.; Reich, M.; Leisen, M.; Morata, D.; Barra, F.; Deditius, A.P. Geochemical and micro-textural fingerprints of boiling in pyrite. *Geochim. Cosmochim. Acta* **2019**, *246*, 60–85. [CrossRef]
88. Qiu, K.-F.; Romer, R.L.; Long, Z.-Y.; Yu, H.-C.; Turner, S.; Wan, R.-Q.; Li, X.-Q.; Gao, Z.-Y.; Deng, J. Potassium isotopes as a tracer of hydrothermal alteration in ore systems. *Geochim. Cosmochim. Acta* **2024**, *368*, 185–196. [CrossRef]
89. Cui, T.; Yu, H.-C.; Jowitt, S.M.; Li, Z.-S.; Wei, X.-L.; Zou, S.-H.; Bo, J.-W.; Pei, L.; Li, S.-S.; Nassif, M.T. Chlorite composition records the formation of increased oxygen fugacity-induced disseminated gold mineralization within the Guocheng gold deposit, China. *Geol. Soc. Am. Bull.* **2025**, *137*, 4786–4800. [CrossRef]
90. Deng, J.; Wang, Q.; Zhang, L.; Xue, S.; Liu, X.; Yang, L.; Yang, L.; Qiu, K.; Liang, Y. Metallogenetic model of Jiaodong-type gold deposits, eastern China. *Sci. China Earth Sci.* **2023**, *66*, 2287–2310. [CrossRef]
91. Deng, J.; Wang, Q.; Liu, X.; Zhang, L.; Yang, L.; Yang, L.; Qiu, K.; Guo, L.; Liang, Y.; Ma, Y. The formation of the Jiaodong gold province. *Acta Geol. Sin.-Engl. Ed.* **2022**, *96*, 1801–1820. [CrossRef]
92. Pokrovski, G.S.; Kouzmanov, K.; Andri, S. Sulfur in hydrothermal fluids. *arXiv* **2025**, arXiv:2506.05845. [CrossRef]

 https://doi.org/10.3390/min16010114

minerals

MDPI

Article

Petrogenesis of Rhyolitic Porphyry Hosting the Newly Discovered Dengshang Mo Deposit, Northern Hebei Province

Jia-Hui Zhou [1], Nan Ju [1,2,*], Qun-Feng Miao [3], Zhuo-Er Teng [1], Xiao-Dong Wang [3], Xiao-Xia Li [1], Ming-Lu Li [1] and Shi-Ming Liu [4]

1 State Key Laboratory of Geological Processes and Mineral Resources, Frontiers Science Center for Deep-Time Digital Earth, School of Earth Sciences and Resources, China University of Geosciences, Beijing 100083, China; zhoujiahuisyx@163.com (J.-H.Z.); 2101240001@email.cugb.edu.cn (Z.-E.T.); 2101240023@email.cugb.edu.cn (X.-X.L.); 1546313001a@gmail.com (M.-L.L.)

2 Northeast Geological S&T Innovation Center of China Geological Survey, Shenyang Geological Survey Center, Shenyang 110000, China

3 The Fourth Geological Team of Hebei Bureau of Geology and Mineral Resources, Chengde 067000, China; 15232377184@163.com (Q.-F.M.); watergoon@163.com (X.-D.W.)

4 Key Laboratory of Resource Exploration Research of Hebei Province, Hebei University of Engineering, Handan 056038, China; liushiming@hebeu.edu.cn

* Correspondence: junan_cgs@163.com; Tel.: +86-18104025767

Abstract

The Dengshang Mo deposit is a recently recognized porphyry-type system within the Yanliao Mo metallogenic belt of northern Hebei Province. However, its ore-hosting rhyolitic porphyry emplacement age and petrogenesis remain insufficiently understood. This study integrates petrography, zircon U–Pb geochronology, Lu–Hf isotope analysis, zircon trace element geochemistry, and whole-rock major- and trace element data to investigate the petrogenesis of the Dengshang rhyolitic porphyry and its genetic relationship with Mo mineralization. The ore-hosting porphyry is predominantly composed of quartz and plagioclase phenocrysts. LA-ICP-MS zircon U–Pb dating yields an emplacement age of 168.3 ± 1.2 Ma, indicating that the rhyolitic porphyry was emplaced during a Middle Jurassic magmatic episode. Petrological and geochemical characteristics classify the Dengshang rhyolitic porphyry as an I-type granite. Zircon $\varepsilon_{Hf}(t)$ values range from −0.93 to −7.29, corresponding to two-stage model ages ($T_{DM}{}^2$) of 1.27–1.67 Ga, which suggests derivation from partial melting of the Mesoproterozoic lower crust. Zircon trace elements display significant positive Ce anomalies (δCe = 10.14–332.85), and calculated oxygen fugacities (ΔFMQ = −0.65 to +2.77; median = +0.51) indicate moderately oxidized magmatic conditions conducive to Mo enrichment. These results collectively imply that the Dengshang rhyolitic porphyry was emplaced at ~168 Ma associated with paleo-Pacific plate subduction. This geodynamic setting triggered partial melting of Mesoproterozoic lower crust, producing oxidized magmas that experienced fractional crystallization prior to shallow emplacement. Our findings elucidate the petrogenesis of the Dengshang rhyolitic porphyry and its control on Mo mineralization, and contribute new insight for understanding porphyry Mo genesis within the complex tectonic evolution of the Yanliao Mo Belt.

Keywords: zircon U–Pb geochronology; Lu–Hf isotopes; oxygen fugacity; Dengshang Mo deposit; Yanliao Mo Belt

1. Introduction

Porphyry deposits are among the most important metal resources worldwide, providing approximately 95% of molybdenum, three-quarters of global copper, substantial quantities of lead, zinc, gold and silver, and thus hold great economic importance [1–4]. The mineralization of porphyry deposits is generally closely linked to magmatic activity [5,6]. Therefore, understanding of the petrogenesis of ore-related magmas is fundamental to advancing mineral exploration and resource development [7,8].

The Yanliao Mo Belt is located along the northern North China block (NCB), and represents one of the most important Mo metallogenic provinces in northern China (Figure 1). Over 30 Mo deposits have been identified in this region, including several large and medium-sized deposits such as Sadaigoumen, Lanjiagou, and Taipingcun, which together constitute the third largest Mo metallogenic cluster in China [9–12]. Previous research reveals that the belt has experienced a multistage tectonic evolution, including the closure of the Paleo-Asian Ocean, the collision between the Siberian Craton and the NCB, the closure of the Mongol–Okhotsk Ocean, the subduction of the paleo-Pacific plate, and subsequent lithospheric thinning. These geodynamic processes triggered multiple episodes of magmatism and metallogenesis in the region [13–16]. Studies on the Sadaigoumen deposit have revealed that around 230 Ma, collisional tectonics associated with the closure of the Paleo-Asian Ocean may also have initiated porphyry Mo mineralization in this region [11,17]. Within the Yanliao Mo Belt, Mo mineralization was primarily concentrated during the Yanshanian period [18]. This mineralization is commonly linked to an extensional tectonic environment resulting from westward subduction of the paleo-Pacific plate [19]. During this period, partial melting of ancient lower crustal material gave rise to the ore-forming magmatism [9,11,12,20–23]. These findings imply that porphyry Mo mineralization in the Yanliao Mo Belt may not be restricted to a single tectonic regime. Its petrogenesis is more complex, involving the superposition and interaction of multiple tectonic events.

The Dengshang Mo deposit, located in northern Hebei Province within the Yanliao Mo Belt, is a newly discovered deposit that has been only sparsely studied. Early exploration in the Dengshang area (1970s–1980s) focused primarily on the pyrite mineralization developed in the upper part of the cryptoexplosive breccia. Most drillholes were terminated once only weak molybdenum mineralization was encountered, resulting in poor definition of the orebody and failure to identify the mineralized parent intrusion. It was not until recent systematic deep drilling (to depths of up to 1200 m) that a concealed porphyry-type Mo system was revealed, confirming that the major mineralization is hosted within the rhyolitic porphyry [24]. Its magmatic timing, petrogenetic characteristics, and genetic link to Mo mineralization remain unclear. In most Mo deposits of the Yanliao Mo Belt, mineralization is primarily hosted in granitic intrusions; however, the Dengshang Mo mineralization is mainly hosted in rhyolitic porphyry. This unusual host rock association may reflect a unique tectonomagmatic setting for mineralization. Therefore, a comprehensive investigation into the geochronology and petrogenesis of the Dengshang rhyolitic porphyry is essential for clarifying the genesis of the deposit and improving our understanding of Mo metallogenesis in the Yanliao Mo Belt. We conducted integrated whole-rock geochemistry, zircon U–Pb dating, trace element analysis, and Lu–Hf isotopic studies of the Dengshang rhyolitic porphyry, in combination with published geochemical data from other representative porphyry Mo deposits in the region. This study specifically seeks to (1) determine the petrogenesis of the ore-hosting rhyolitic porphyry at Dengshang, and (2) examine the links between magmatic evolution and regional tectonics, so as to provide novel insights and constraints that help elucidate the metallogenic evolution of porphyry Mo deposits in the Yanliao Mo Belt.

https://doi.org/10.3390/min16030249

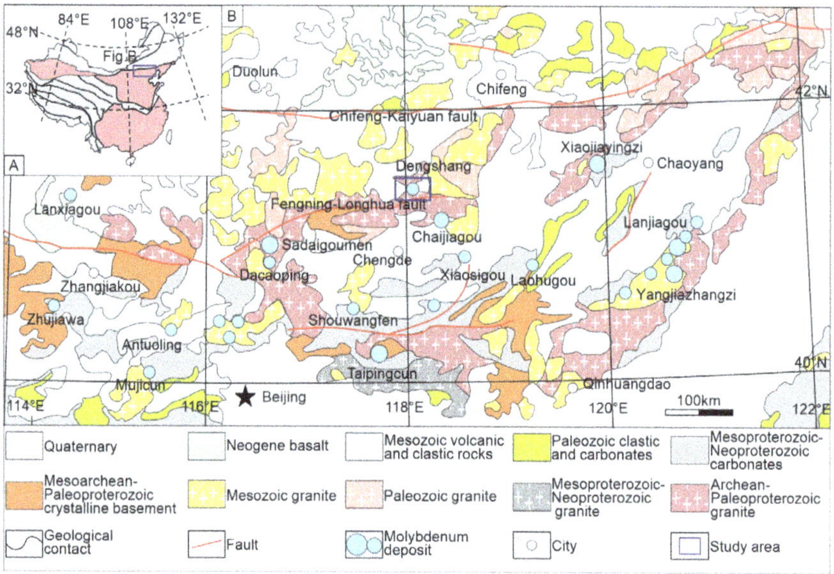

Figure 1. Map showing the geographic setting of the study area (**A**) and geological sketch map of the Yanliao Mo Belt (**B**) [25].

2. Geological Background

2.1. Regional Geology

The NCB was assembled through the amalgamation of its Eastern and Western blocks along a Paleoproterozoic orogenic belt and had largely achieved crustal stability by ~1.8 Ga (Luliang Orogeny) [15,26]. The Yanliao aulacogen developed on its northeastern margin during the Mesoproterozoic. Between the Paleozoic and Late Permian, the northern NCB experienced subduction and collision associated with the Paleo-Asian Ocean, culminating in the formation of the North China–Mongolia composite plate. During the Middle to Late Triassic, intracontinental extension was indicated by metamorphic core complexes and alkaline magmatism [27–31]. During the Middle Jurassic, the westward subduction of the paleo-Pacific plate primarily governed the tectonic evolution of the Yanliao region, coupled with the ongoing closure of the Mongol–Okhotsk Ocean to its north (Figure 1) [32–35]. These processes induced intense deformation and widespread magmatism. Regional EW and NE-trending faults strongly controlled deposit distribution and mineralization.

Stratigraphically, the region comprises Archean basement rocks overlain by Proterozoic to Paleozoic sedimentary successions, including the marine–continental sequences of the Changcheng and Jixian groups. These supracrustal sequences, together with multiple intrusive phases, provided favorable host rocks and structural traps for mineralization. Mesozoic magmatism was particularly prolific, with early Yanshanian activity characterized by intermediate to felsic intrusions (monzogranite, high Sr/Y granite), and late Yanshanian activity featuring mafic–intermediate volcanism, A-type and highly fractionated I-type granites, and alkaline complexes [36–38]. These magmas provided the essential material and thermal energy for mineralization in the belt.

2.2. Deposit Geology

The Dengshang Mo deposit is situated in Chengde city [39], Hebei Province. Tectonically, it lies north of the NCB, separated by the Chifeng–Kaiyuan Fault, within the Fengning–Longhua fault. The deposit has an estimated Mo resource of 5108.6 t with an

 https://doi.org/10.3390/min16030249

average grade of 0.108%. The mining area are mainly Archean hornblende gneiss of the Dantazi Group, Fenghuangzui Formation, and Quaternary sediments (Figure 2). Magmatic rocks intrude into the gneiss as dikes, trending NE and NW. They are dominated by Jurassic rhyolitic porphyry accompanied by cryptoexplosive breccia, and subordinately by intermediate dikes of diorite, with individual dikes extending 40–70 m in length and 1.5–3 m in thickness. The rhyolitic porphyry displays a porphyritic texture, with phenocrysts of quartz and plagioclase and the matrix is cryptocrystalline quartz, feldspar, no significant vitreous phase is detected (Figure 3A,B). Developed within the deposit are several NE-trending faults. The ore bodies are mainly hosted by rhyolitic porphyry at depth, occurring as stratiform and vein at elevations of −100 to −600 m, representing concealed mineralization. Minor disseminated molybdenite mineralization is observed in the cryptoexplosive breccia and gneiss, but does not form economic orebodies. The ores are characterized by stockwork and veinlet-style. Stockwork ores within the rhyolitic porphyry contain abundant quartz–molybdenite veinlets, in which molybdenite commonly occurs as fine flakes or aggregates (Figure 3C,D). The cryptoexplosive breccia consists mainly of rhyolitic porphyry clasts and lithic fragments, with pyrite disseminated both within the clasts and the altered matrix (Figure 3E). Pyrite occurs as fine disseminations, and quartz grains are typically surrounded by intense sericitization, indicating strong hydrothermal alteration (Figure 3F) [40]. Typical hydrothermal alteration zoning is developed in the deposit, and the alteration is predominantly distributed around the rhyolitic porphyry intrusion and its peripheral areas. The alteration zoning progresses outward from an inner zone dominated by potassic alteration (K-feldspar, sericite), which is primarily developed within the rhyolitic porphyry intrusion; the outer zone is characterized by propylitization (chlorite, carbonate), which is mostly distributed in the cryptoexplosive breccia and gneiss at the intrusion periphery [24]. In addition, silicification and sericitization are developed, accompanied by pyritization, magnetitization, and local Pb–Zn mineralization.

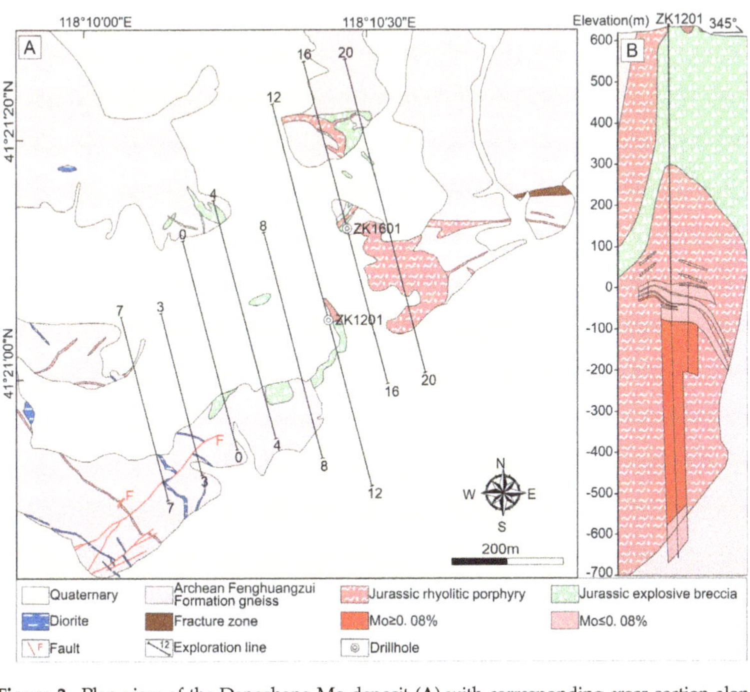

Figure 2. Plan view of the Dengshang Mo deposit (**A**) with corresponding cross-section along line 12 (**B**) [24].

 https://doi.org/10.3390/min16030249

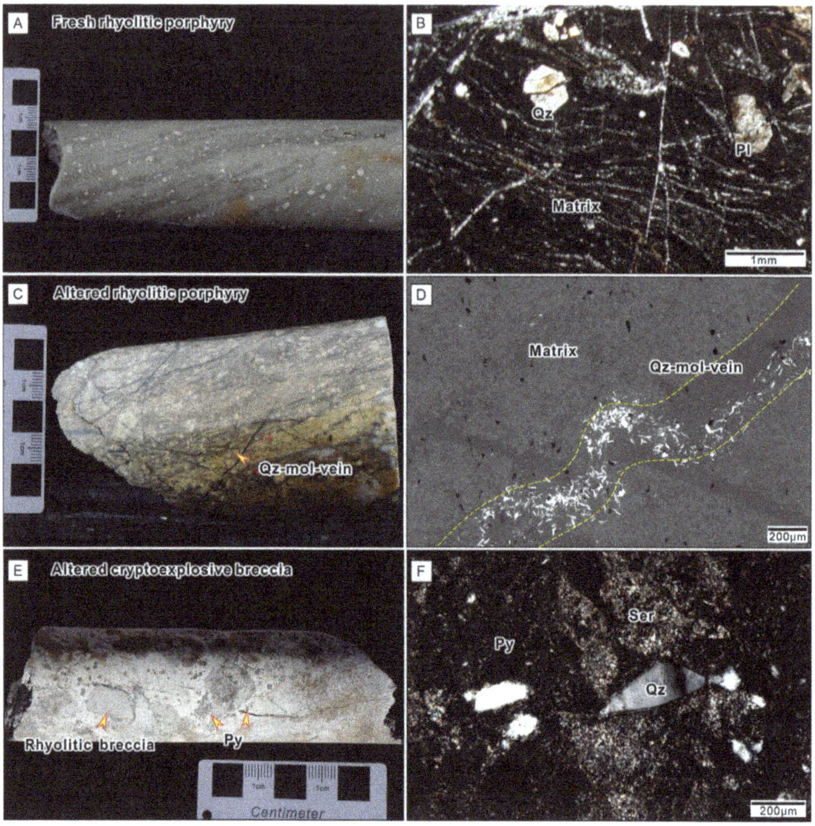

Figure 3. Core samples of fresh rhyolitic porphyry and photomicrograph under cross polarized light (**A**,**B**), showing quartz and plagioclase phenocrysts; veinlet-style rhyolitic porphyry ore and a back scattered electron (BSE) image obtained by scanning electron microscopy (SEM) of quartz–molybdenite veinlets (**C**,**D**); altered cryptoexplosive breccia and photomicrograph under cross polarized light (**E**,**F**), showing sericitic alteration and disseminated pyrite. Pl = plagioclase, Qz = quartz, Py = pyrite, and Mol = molybdenite, Ser = sericite.

3. Materials and Methods

3.1. Materials

The rhyolitic porphyry represents the main ore-hosting intrusion of the Dengshang Mo deposit, and it is predominantly exposed within the central sector of the mining area (Figure 2B). Eight fresh rhyolitic porphyry core samples (24DS106–24DS113) were collected for this study. They were obtained from drill hole ZK1201. Sample 24DS106 was used for zircon U–Pb dating, Lu–Hf isotope analysis, and zircon trace element analysis. The remaining seven samples (24DS107–24DS113) were used for whole-rock major and trace element geochemical analyses. The rocks exhibit a gray to grayish-white color with prominent flow banding and a porphyritic texture (Figure 3A). Phenocrysts constitute approximately 5%–10% of the rock and are composed essentially of quartz and plagioclase, with grain sizes ranging from 0.5 to 5 mm. The groundmass is predominantly cryptocrystalline and dominated by quartz and feldspar (Figure 3B).

https://doi.org/10.3390/min16030249

3.2. Methods

3.2.1. Whole-Rock Major and Trace Element Analyses

These analyses were determined at Langfang Geological Exploration Technology Co., Ltd. (Langfang, China) prior to analysis. Loss on ignition (LOI) was measured gravimetrically after heating to 1000 °C. For major elements, glass beads prepared from ~0.5 g of sample powder fused with lithium borate were analyzed by X-ray fluorescence (XRF) [41]. Trace elements, including rare earth elements (REEs), were quantified by inductively coupled plasma-mass spectrometry (ICP-MS) following high-pressure digestion of ~50 mg of powder in an HF–HNO_3 mixture at 150 °C. Analytical precision, based on replicate measurements, was better than 2% for major oxides and within 5% for trace elements. Accuracy, monitored using certified reference materials (GBW-07123), was within 5% of the certified value [42].

3.2.2. Zircon LA-ICP-MS U–Pb Dating and Trace Element Analyses

This analysis was performed at Beijing GeoAnalysis Co., Ltd. (Beijing, China) using LA-ICP-MS. Sample preparation included embedding zircon grains in epoxy, polished, and cleaned ultrasonically in ultrapure water followed by methanol rinsing. We used a 30 μm laser spot, 6 Hz repetition rate, and energy density of ~5 J/cm^2. Before measurement, each analysis location was pre-cleaned with ~5 laser pulses (~0.3 μm depth) to remove surface contamination. The zircon standard 91500 served as the primary reference material, and Plesovice zircon was used for quality control. Trace element contents were calibrated against NIST 610 glass, with ^{29}Si as the internal standard. During the analytical session, standard 91500 yielded a concordant age of 1061.5 ± 3.2 Ma (2σ), consistent with its reference value. Unknown age zircon Plesovice returned a weighted mean $^{206}Pb/^{238}U$ age of 337.5 ± 1.5 Ma (2σ), matching the recommended age within uncertainty [43]. Data reduction was conducted using the ICPMASDataCal 11.8 software [44] and age calculations were performed with Isoplot 3.0 [45].

3.2.3. In Situ Zircon Lu–Hf Isotope Analyses

Lu–Hf isotopic compositions were analyzed in situ on zircon domains concurrent with U–Pb dating, using a RESO 193 nm laser system (Australian Scientific Instruments, Canberra, Australia) linked to a Neptune Plus MC-ICP-MS (Thermo Fisher Instruments, Waltham, MA, USA). Operating conditions included laser spot diameter of 40 μm and helium as the carrier gas. The measured $^{176}Hf/^{177}Hf$ ratio for the Plešovice reference zircon was 0.282480 ± 0.000016 (2σ), matching its certified values [43].

For $\varepsilon_{Hf}(t)$ calculation, present-day chondrite ratios $^{176}Hf/^{177}Hf$ = 0.282772 and $^{176}Lu/^{177}Hf$ = 0.0332 were adopted, along with depleted mantle values of 0.28325 and 0.0384, respectively. Two-stage model ages ($T_{DM}{}^2$) were computed assuming $^{176}Lu/^{177}Hf$ = 0.015 for the average continental crust [43].

4. Results

4.1. Zircon U–Pb Ages and Trace Element Analyses

Cathodoluminescence (CL) images show that zircons from the rhyolitic porphyry sample 24DS106 of the Dengshang Mo deposit are mostly pristine and prismatic and exhibit euhedral to subhedral. The majority of crystals are 100 to 200 μm long, length/width between 1.5:1 and 4:1. They are generally well-preserved and display distinct oscillatory zoning, typical of magmatic zircons. In contrast, some grains exhibit inherited characteristics, such as weak or absent zoning (Figure 4A). Zircons from the rhyolitic porphyry are characterized by Th contents of 131.56–751.75 × 10^{-6} and U contents of 149.44–563.59 × 10^{-6}, and Th/U ratios between 0.62 and 2.49. Twelve analysis spots produced concordant $^{206}Pb/^{238}U$ ages between

 https://doi.org/10.3390/min16030249

162.4 and 175.0 Ma, which yield a weighted mean of 168.3 ± 1.2 Ma (MSWD = 0.31, n = 12; Supplementary Table S1, Figure 4B).

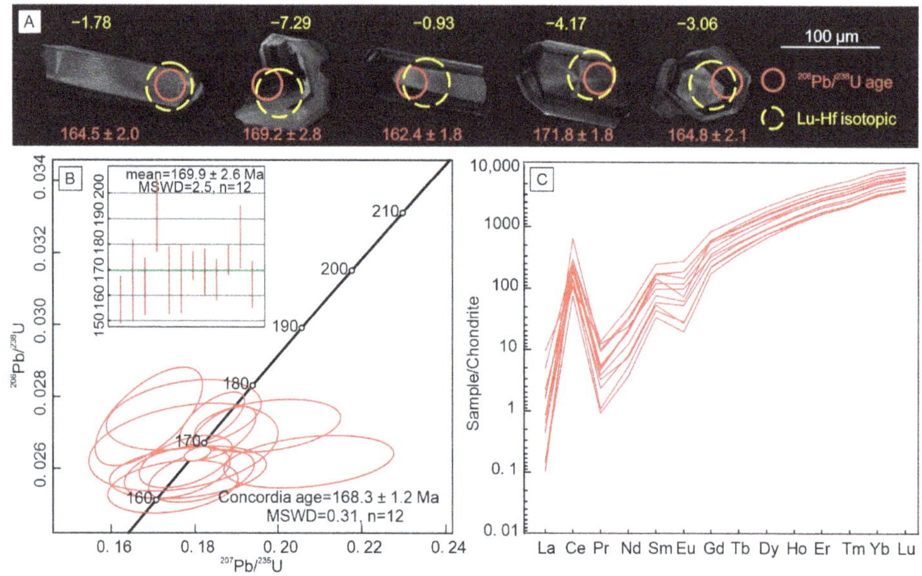

Figure 4. Representative CL images of zircons (**A**) and LA-ICP-MS U–Pb concordia diagrams for the fresh rhyolitic porphyry (**B**) and zircon chondrite-normalized REE patterns of fresh rhyolitic porphyry (**C**).

Zircons from the rhyolitic porphyry (sample 24DS106) exhibit significant positive Ce and negative Eu anomalies in their chondrite-normalized REE patterns (Figure 4C). The calculated δCe values vary widely between 10.14 and 332.85, with an average of 146.41, where δEu anomalies, between 0.25 and 0.60, with an average of 0.43. The zircon crystallization temperatures (T) range from 674.08 to 821.02 °C, with a mean value of 743.06 °C. Oxygen fugacity was calculated using an algorithm in MATLAB R2019b software [46–48]. The logarithmic oxygen fugacity (log fO_2) ranges from −18.48 to −13.28, with an average of −15.60. Relative oxygen fugacity values (ΔFMQ) vary between −0.65 and +2.77, averaging +0.51, median = +0.4 (Supplementary Table S2).

4.2. Whole-Rock Major and Trace Element Geochemistry

Major and trace element concentrations were measured for seven rhyolitic porphyry samples (24DS107-24DS113); the complete dataset is listed in Supplementary Table S3. The rhyolitic porphyry has high SiO_2 contents ratios of 73.07–74.88 wt.% and Al_2O_3 with ratios of 11.91–13.27 wt.%, with low TiO_2 (0.041–0.071 wt.%) and MgO (0.09–0.19 wt.%). In the TAS classification diagram, the samples plot within the rhyolite–granite field (Figure 5A). On the SiO_2–(Na_2O + K_2O–CaO) diagram, the samples display calc-alkaline to alkali–calcic affinities (Figure 5B). Alumina saturation indices (A/CNK = 0.97–1.09; A/NK = 1.02–1.15) fall within the metaluminous to weakly peraluminous spectrum (Figure 5C). Furthermore, all data fall within the fractionated granite (FG) field on the FeO^T/MgO vs. (Zr + Nb + Ce + Y) discrimination diagram (Figure 5D), indicating the highly evolved nature of the Dengshang rhyolitic porphyry.

https://doi.org/10.3390/min16030249

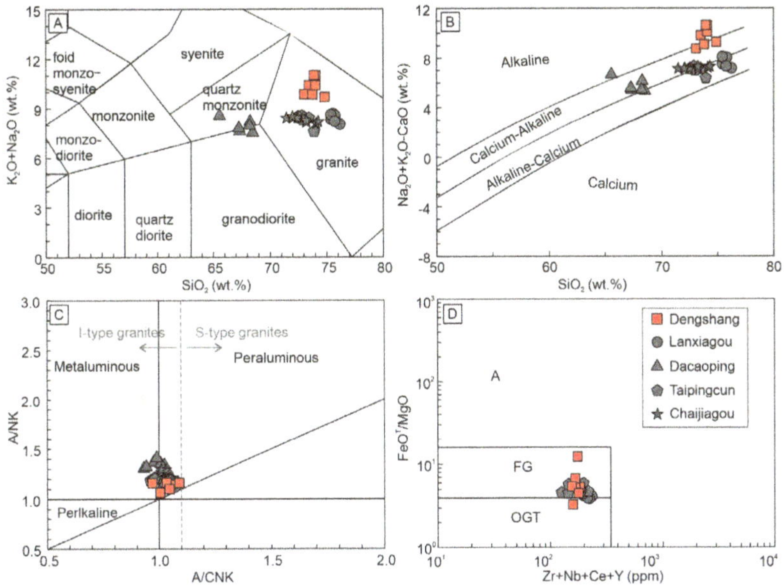

Figure 5. Geochemical classification diagrams of the Middle Jurassic–Early Cretaceous ore-hosting magmatic rocks in the Yanliao Mo Belt and Dengshang rhyolitic porphyry. Total alkali ($Na_2O + K_2O$) vs. SiO_2 (TAS) diagram (**A**) [49]. Plots of $Na_2O + K_2O + CaO$ vs. SiO_2 (**B**). A/NK vs. A/CNK diagram (**C**) [50]. FeO^T/MgO vs. Zr + Nb + Ce + Y diagram (**D**). FG = fractionated I-, S-, and M-type granites; OGT = unfractionated I-, S-, and M-type granites; A: A-type granites [51] (Data form [52–55]).

Geochemical analyses reveal pronounced light to heavy rare earth element (REE) fractionation in the rhyolitic porphyry, with total REE contents (ΣREE) ranging from 93.0×10^{-6} to 136.8×10^{-6}, LREE/HREE between 9.3 and 11.62, and $(La/Yb)_N$ between 9.7 and 18.9. The rocks display moderate negative Eu anomalies (δEu = 0.47–0.63) (Figure 6A). Primitive mantle-normalized trace element spectra exhibit pronounced spikes at Ba and Rb (LILEs) coupled with troughs at Sr, Nb, and Ta (HFSEs) (Figure 6B).

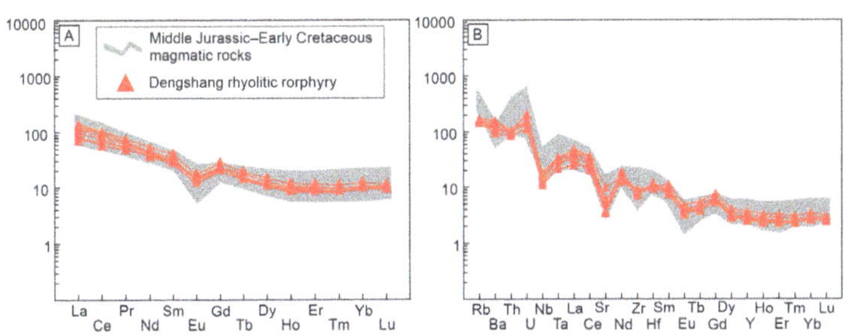

Figure 6. Chondrite-normalized REE patterns (**A**) and primitive mantle-normalized trace element spider diagrams (**B**) for the Middle Jurassic–Early Cretaceous ore-hosting magmatic rocks in the Yanliao Mo Belt and Dengshang rhyolitic porphyry (chondrite and primitive mantle normalizing values are from [56,57], data form [52–55]).

4.3. Lu–Hf Isotope

$T_{DM}{}^2$ were computed assuming a mean continental crustal $^{176}Lu/^{177}Hf$ ratio of 0.015 [58–60]. Supplementary Table S4 reports the full Lu-Hf isotopic dataset. Analyses of ten zircon grains yield $^{176}Hf/^{177}Hf$ ratios between 0.282467 and 0.282787, $^{176}Lu/^{177}Hf$

https://doi.org/10.3390/min16030249

values between 0.001793 and 0.004259. The resulting $\varepsilon_{Hf}(t)$ values vary from −7.29 to −0.93, all values lie beneath the chondritic uniform reservoir (CHUR) reference curve. The corresponding to T_{DM}^2 of 1.27–1.67 Ga (Figure 7).

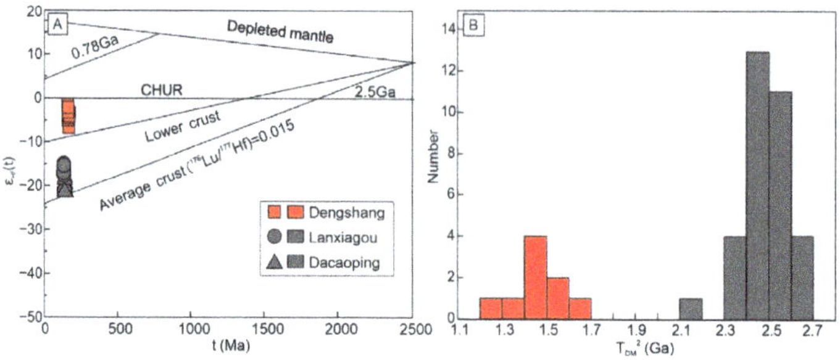

Figure 7. Plots showing $\varepsilon_{Hf}(t)$ values against zircon U–Pb ages (Ma) for the Dengshang rhyolitic porphyry and other Middle Jurassic–Early Cretaceous ore-hosting magmatic in the Yanliao Mo Belt (**A**) [61]. Histograms of corresponding T_{DM}^2 (**B**) (data form [52,62]).

5. Discussion

5.1. Emplacement Age of the Rhyolitic Porphyry and Its Constraints on Mo Mineralization

Zircon U–Pb dating constrains the crystallization age of the ore-hosting rhyolitic porphyry at the Dengshang Mo deposit to 168 ± 2.5 Ma. This age places its emplacement within a Middle Jurassic magmatic event. It has been demonstrated that the Yanliao Mo Belt experienced three major Mo mineralization events at 220–250 Ma, 180–200 Ma, and 130–160 Ma (Figure 8) [23]. In northern Hebei Province, magmatic activity is documented during only the 250–220 Ma and 160–130 Ma stages. The 250–220 Ma magmatism is mainly developed in the Zhangbei–Fengning area of northern Hebei Province, including the Sadaigoumen (240–257 Ma) [11,17,63,64]. The 185–180 Ma magmatism is mainly developed in western Liaoning, including the Lanjiagou (185~189 Ma) and Yangjiazhangzi (183~188 Ma) [12,65]. The 160–130 Ma magmatism is concentrated in the southern Zhangjiakou–Xinglong area, including the Lanxiagou (134.9 Ma), Dacaoping (142.4 Ma), Taipingcun (161.8 Ma), and Chaijiagou (163.9 Ma) [12,22,53–55] (Figure 1). These data suggest that Mo mineralization in the belt displays both temporal and spatial heterogeneity.

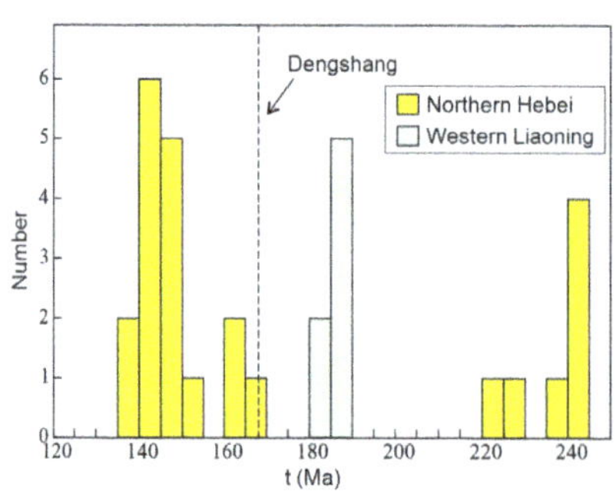

Figure 8. Histogram showing the age distribution of ore-hosting intrusions in the Yanliao Mo Belt [23].

In this study, zircon U–Pb dating by LA-ICP-MS confirms an emplacement age of 168 Ma for the Dengshang rhyolitic porphyry (Figure 4B). This age is broadly coeval with that of the Taipingcun intrusion (161.8 Ma) and the Chaijiagou intrusion (163.9 Ma) [54,55], suggesting a genetic link to the third large-scale Mo metallogenic event (160–130 Ma).

 https://doi.org/10.3390/min16030249

5.2. Magma Source and Petrogenetic Evolution

The Dengshang rhyolitic porphyry shows elevated K_2O contents (9.49–10.97 wt.%), high SiO_2 contents (73.07–74.88 wt.%), very low MgO (0.09–0.19 wt.%) and exhibits metaluminous to weakly peraluminous characteristics (A/CNK < 1.1, A/NK > 1.0) (Figure 5B). The rocks are devoid of typical alkaline mafic minerals and strongly peraluminous phases, and their geochemical affinities resemble those of I-type granites [66,67]. The rhyolitic porphyry shows high Sr/Y ratios and low Cr contents (8.9–13.4 ppm) (Table S3), consistent with the geochemical profile of the 160–130 Ma porphyry Mo deposits ore-hosting intrusions in the Yanliao Mo Belt (Figure 5).

The trace element patterns are marked by LREE enrichment and heavy HREE depletion, with significant Nb depletion, indicative of typical continental crustal signatures. Weak negative Eu anomalies (δEu = 0.47–0.63) are consistent with plagioclase fractionation or a plagioclase-bearing source residue [68]. The relatively flat HREE patterns imply negligible residual garnet [69]. In addition, FeO^T/MgO ratios combined with Zr + Nb + Ce + Y contents classify it as a fractionated granite (FG) in the discrimination diagram (Figure 5D), confirming that the magma underwent extensive fractional crystallization, and the depletion in HFSE (Nb, Ta) and enrichments in LILE (Ba, Rb) collectively point to partial melting of lower crustal material (Figure 6) [66,70,71].

The magma source is further elucidated by zircon Lu–Hf isotopes. Negative $\varepsilon_{Hf}(t)$ values (−0.93 to −7.29) reflect a major contribution from ancient crustal materials [72,73]. The data plot below the CHUR line, indicating a source in reworked ancient continental lower crust [74,75]. $T_{DM}{}^2$ range from 1.27 to 1.67 Ga, clustering mainly between 1.4 and 1.6 Ga (Figure 7), these data indicate that the magma derivation primarily from partial melting of Mesoproterozoic lower crust. The consistently high SiO_2 contents (73.07–74.88 wt.%) coupled with low Mg# values (11.63–35.71) provide additional support for this model, which are characteristic of crustal-derived felsic magma. (Supplementary Table S3, Figure 5A) [76–78].

During the Middle Jurassic–Early Cretaceous, the Mo ore-hosting magmatic rocks in northern Hebei Province yield zircon Lu–Hf compositions characterized by $\varepsilon_{Hf}(t)$ values of −10 to −25, and $T_{DM}{}^2$ of 2.7–2.1 Ga (Figure 7) [52,62]. Compared with these intrusions, zircons from the Dengshang rhyolitic porphyry exhibit elevated $\varepsilon_{Hf}(t)$ values and younger $T_{DM}{}^2$ ages. This indicates that Middle Jurassic–Early Cretaceous magmatism in northern Hebei derived from Paleoproterozoic–Mesoproterozoic lower crustal melting.

5.3. Tectonic Setting and Metallogenesis

The emplacement of the Dengshang rhyolitic porphyry falls within the Middle Jurassic magmatic pulse, as constrained by zircon U–Pb geochronology (Figure 4B). This age is distinctly younger than the Late Permian to Middle Triassic amalgamation history involving the Paleo-Asian Ocean and the collision of the Siberian and NCB [33,79–82]. Magmatic products associated with this tectonic regime are typically characterized by alkaline complexes, mafic–ultramafic rocks, and minor A-type and I-type granites [33]. In contrast, the Dengshang porphyry is a metaluminous to weakly peraluminous I-type granite that experienced significant fractional crystallization. It exhibits a pronounced arc-like geochemical signature marked by enrichment in LILE Rb, Ba and depletion in HFSE Nb, Ta (Figure 6) [83,84]. The Dengshang rhyolitic porphyry and other regional ore-hosting intrusions consistently plot within the volcanic-arc field on tectonic discrimination diagrams (Figure 9A,B), pointing to a subduction-related magma source. This interpretation is reinforced by their Th, Ta, and Yb contents, which cluster in the active continental margin fields (Figure 9C,D), thereby arguing against a collisional or intraplate setting.

Accretionary complexes along the eastern Asian margin testify to sustained westward subduction of the paleo-Pacific plate beneath Eurasia from Jurassic to Early Cretaceous

https://doi.org/10.3390/min16030249

time. Porphyry Mo mineralization dated at 180–135 Ma is attributed to a continental magmatic arc regime, where subduction-driven magmatism and back-arc extension prevailed [9,29,33,85,86]. During the Middle Jurassic, the Mongol–Okhotsk oceanic tectonic regime exerted negligible influence on the eastern NCB, since no evidence supports the subduction of this oceanic plate beneath the region at that time [33]. Independent evidence for a Jurassic subduction setting comes from contemporaneous igneous rocks, such as the ~167 Ma Nianziyu appinite. Its high water content and isotopic features, including elevated zircon $\delta^{18}O$ values, indicate metasomatism by subduction-derived fluids [33]. Together, these features indicate that the Dengshang rhyolitic porphyry formed in an active continental margin setting controlled by the subduction of the paleo-Pacific plate.

Within this subduction-related tectonic setting, a series of intermediate to felsic magmas were generated. Zircons from the Dengshang rhyolitic porphyry exhibit chondrite-normalized REE patterns characterized by significant positive Ce anomalies, weak negative Eu anomalies, depletion in LREE, and relative enrichment in HREE (Figure 4C). All analyzed zircons contain Ta > 0.2×10^{-6} and display Yb/Gd ratios < 20 (Table S1), indicating that ilmenite fractionation exerted little control on magma evolution [87]. The calculated zircon crystallization temperatures range from 674.08 to 821.02 °C, with log fO_2 values between −18.48 and −13.28. The relative oxygen fugacity (ΔFMQ) varies from −0.65 to +2.77, with a median value of +0.4, clustering around or slightly above the FMQ buffer (Figure 10, Table S2). These oxidation states indicate the Dengshang rhyolitic porphyry was derived from a moderately oxidized magma. Previous studies have shown that the oxygen fugacity (fO_2) exerts a primary control on Mo valence states in magmas. Under high-fO_2 conditions, Mo primarily occurs as Mo^{6+} in the form of molybdate ions ($MoO_4{}^{2-}$) [47,48,88]. This highly soluble species prevents Mo from substituting for Ti in crystallizing Fe-/Ti-bearing minerals or combining with reduced sulfur to form MoS_2, thereby promoting its enrichment in the residual melt as a highly incompatible element and subsequently migrates and precipitates along structurally favorable zones [89–91]. Comparison with other Mo deposits, such as the Lanxiagou deposit (log fO_2 = −16.07~−13.86) [52], shows that similarly high fO_2 characteristics are common among ore-hosting intrusions. This further confirms that high oxygen fugacity is a fundamental geochemical condition for porphyry Mo mineralization [92].

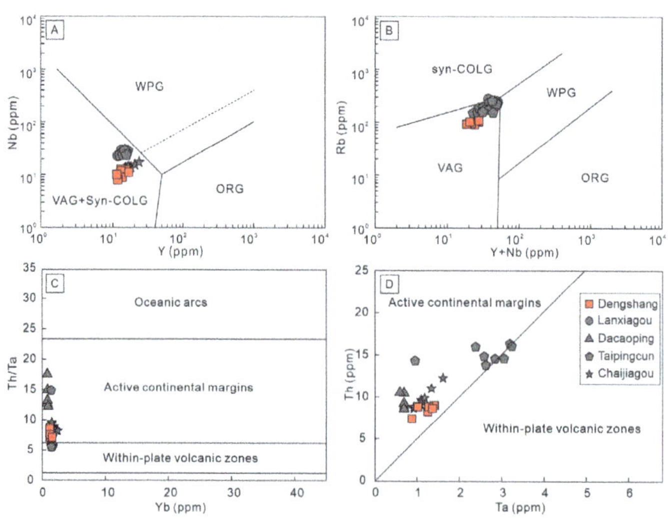

Figure 9. Tectonic discrimination diagrams for Middle Jurassic–Early Cretaceous ore-hosting magmatic rocks in the Yanliao Mo Belt and Dengshang Rhyolitic Porphyry. Nb vs. Y (**A**). Rb vs. Y + Nb

https://doi.org/10.3390/min16030249

diagram (**B**) [93]. Th/Ta vs. Yb (**C**) [34,94]. Th vs. Ta (**D**) [34,93]. The dashed line marks the upper compositional limit for ORG derived from anomalous ridge segments. Syn-COLG: syn-collisional granites; VAG: volcanic arc granites; WPG: within plate granites; ORG: ocean ridge granites (date form [52–55]).

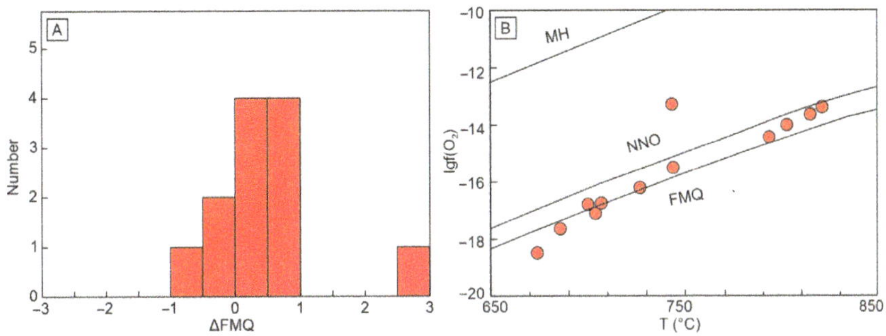

Figure 10. Oxygen fugacity discrimination diagrams of the Dengshang rhyolitic porphyry. Histogram of relative oxygen fugacity ΔFMQ (**A**). $lgf(O_2)$ vs. T diagram (**B**). MH: magnetite–hematite buffer; NNO: nickel–nickel oxide buffer; FMQ: fayalite–magnetite–quartz buffer [48].

Integrated with Dengshang rhyolitic porphyry highly fractionated I-type granite affinity, predominantly crustal magma sources, and subduction-related volcanic arc signature, this study proposes that Middle Jurassic westward subduction of the paleo-Pacific plate controlled the northern NCB. Fluids released from the subducting slab supplied volatiles and heat to the lower crust, triggering partial melting of the Paleoproterozoic–Mesoproterozoic lower crust and generating oxidized felsic magmas. Accompanied by intense hydrothermal activity and extraction–enrichment of molybdenum, these processes ultimately led to the formation of the Dengshang porphyry Mo deposit within the Yanliao Mo belt, making the porphyry a magmatic host for Mo mineralization in this metallogenic episode (Figure 11).

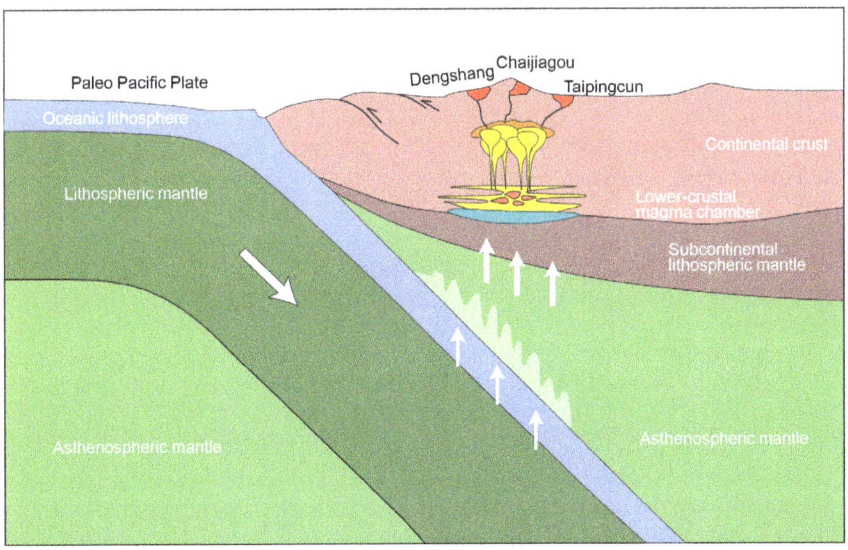

Figure 11. Genetic model showing Jurassic magma and associated with Mo mineralization modified from [95,96].

https://doi.org/10.3390/min16030249

6. Conclusions

(1) Petrographic observations reveal that the rhyolitic porphyry exhibits a porphyritic texture characterized predominantly by quartz and plagioclase phenocrysts. Zircon U–Pb analyses of the Dengshang rhyolitic porphyry constrain its crystallization age to 168.3 ± 1.2 Ma. This result indicates that the porphyry was emplaced during the Middle Jurassic, contemporaneous with the subduction of the paleo-Pacific plate, and related to the third major Mo metallogenic event (160–130 Ma) in the Yanliao Mo Belt.

(2) The Dengshang rhyolitic porphyry is metaluminous to weakly peraluminous and belongs to the calc-alkaline to alkali-calcic series. Its geochemical features are consistent with I-type granites. Combined geochemical and isotopic evidence indicates that the magma originated predominantly from partial melting of Mesoproterozoic lower crust.

(3) Tectonic discrimination diagrams indicate that the porphyry formed in an active continental margin setting, linked to paleo-Pacific subduction. The magma was moderately oxidized (ΔFMQ = −0.65 to +2.77), with zircon crystallization temperatures ranging from 674.08 to 821.02 °C. Such conditions are conducive to Mo enrichment in the melt and its subsequent transfer into hydrothermal systems, highlighting the significant metallogenic potential of the Dengshang porphyry.

(4) This study clarifies the petrogenesis of the ore-hosting intrusion in a newly discovered Dengshang Mo deposit and demonstrates a genetic link between Middle Jurassic subduction, oxidized crust-derived magmatism, and Mo mineralization. The results indicate that magmatic oxidation state is a key control on the fertility of porphyry Mo systems in continental margin settings and suggest that moderately oxidized, I-type felsic intrusions are favorable targets for regional exploration in the Yanliao Belt.

Supplementary Materials: The following supporting information can be downloaded at: https://www.mdpi.com/article/10.3390/min16030249/s1, Table S1. LA-ICP-MS zircon U–Pb age of rhyolitic porphyry samples 24DS106 in the Dengshang Mo deposit; Table S2. Trace element concentration, and calculated Ti-in-zircon temperatures (T), logarithmic oxygen fugacity ($logfO_2$), and ΔFMQ of zircons from the rhyolitic porphyry samples 24DS106; Table S3. Major (wt%) and trace elements (ppm) analytical results for the rhyolitic porphyry samples 24DS107-24DS113. Table S4. LA-ICP-MS zircon Lu–Hf isotopic of rhyolitic porphyry samples 24DS106.

Author Contributions: Conceptualization, J.-H.Z. and N.J.; methodology and investigation, Q.-F.M., X.-D.W. and S.-M.L.; data curation, J.-H.Z., X.-X.L., M.-L.L. and Z.-E.T.; funding acquisition, N.J. and Q.-F.M.; writing, J.-H.Z. and N.J. All authors have read and agreed to the published version of the manuscript.

Funding: This research was funded by National Science and Technology Major Project of Deep Earth Exploration: "Multi-scale Prospecting Prediction and Exploration of Gold Deposits in the Eastern Segment of the Northern Margin of North China" (No. 2025ZD1006400), The Funding Project Of Northeast Geological S&T Innovation Center of China Geological Survey (No. QCJJ2023-06), Hebei Bureau Of Geology And Mineral Resources Exploration (No. 13000025P003294103391), China University of Geosciences Beijing College Students' Innovative Entrepreneurial Training Plan Program (No. X202511415018), the Fundamental Research Funds for the Central Universities, China (No. 2652024008), and the Frontiers Science Center for Deep-time Digital Earth (No. 2652023001).

Data Availability Statement: All data supporting this study are contained within the article and its Supplementary Materials.

Acknowledgments: Sincere thanks go to Hao-Cheng Yu and Shan-Shan Li for their field guidance and constructive discussions. And to Jia-Dong Ma and Xian-Fa Xue for their technical support with figures and manuscript preparation. Constructive feedback from the anonymous reviewers and journal editors is also sincerely acknowledged.

Conflicts of Interest: The authors declare no conflicts of interest.

References

1. Qiu, K.F.; Song, K.R.; Song, Y.H. Magmatic-hydrothermal fluid evolution of the Wenquan porphyry molybdenum deposit in the north margin of the West Qinling, China. *Acta Petrol. Sin.* **2015**, *31*, 3391–3404.
2. Sillitoe, R.H. Porphyry copper systems. *Econ. Geol.* **2010**, *105*, 3–41. [CrossRef]
3. Qiu, K.F.; Li, N.; Taylor, R.D.; Song, Y.H.; Song, K.R.; Han, W.Z.; Zhang, D.F. Timing and duration of metallogeny of the Wenquan deposit in the West Qinling, and its constrain on a proposed classification for porphyry molybdenum deposits. *Acta Petrol. Sin.* **2014**, *30*, 2631–2643.
4. Fu, Q.X.; Cen, R.; Hong, B.; Xing, Z.F. Analysis of metallogenic regularity and resource potential of Mo deposits in Hainan Province. *Contrib. Geol. Miner. Resour. Res.* **2025**, *40*, 189–196.
5. Ma, Z.Y.; Zhang, Y.; Li, J.; Ma, Q. Discussion on metallogenic model of Qingshuihedonggou Mo deposit in East Kunlun area. *Contrib. Geol. Miner. Resour. Res.* **2024**, *39*, 293–300.
6. He, D.Y.; Qiu, K.F.; Yu, H.C.; Jowitt, S.M.; Zheng, X.; Mazumder, R.; Deng, J. Lithospheric architecture and evolution of the Qinling Orogen of Central China and associated controls on metallogeny. *Earth-Sci. Rev.* **2025**, *264*, 105092.
7. Qiu, K.F.; Taylor, R.D.; Song, Y.H.; Yu, H.C.; Song, K.R.; Li, N. Geologic and geochemical insights into the formation of the Taiyangshan porphyry copper–molybdenum deposit, Western Qinling Orogenic Belt, China. *Gondwana Res.* **2016**, *35*, 40–58. [CrossRef]
8. Richards, J. Tectono-magmatic precursors for porphyry Cu-(Mo-Au) deposit formation. *Econ. Geol.* **2003**, *98*, 1515–1533. [CrossRef]
9. Zeng, Q.D.; Liu, J.M.; Qin, K.Z.; Fan, H.R.; Chu, S.X.; Wang, Y.B.; Zhou, L.L. Types, characteristics, and time–space distribution of molybdenum deposits in China. *Int. Geol. Rev.* **2013**, *55*, 1311–1358. [CrossRef]
10. Sun, Z.J.; Wang, Z.Q.; Yu, H.; Song, W.Y.; Wang, C.Y.; Liu, G.H.; Tang, L. Geochemistry, zircon U–Pb, and molybdenite Re–Os geochronology of the Mo deposit in Fengning region, Hebei Province, China. *Geol. J.* **2018**, *54*, 3619–3642. [CrossRef]
11. Jiang, S.H.; Liang, Q.L.; Bagas, L. Re–Os ages for molybdenum mineralization in the Fengning region of northern Hebei Province, China: New constraints on the timing of mineralization and geodynamic setting. *J. Asian Earth Sci.* **2014**, *79*, 873–883. [CrossRef]
12. Zeng, Q.D.; Chu, S.X.; Liu, J.M.; Sun, S.K.; Chen, W.J. Mineralization, alteration, structure, and Re–Os age of the Lanjiagou porphyry Mo deposit, North China Craton. *Int. Geol. Rev.* **2012**, *54*, 1145–1160. [CrossRef]
13. Chen, Y.J.; Zhai, M.G.; Jiang, S.Y. Significant achievements and open issues in study of orogenesis and metallogenesis surrounding the North China continent. *Acta Petrol. Sin.* **2009**, *25*, 2695–2726.
14. Davis, G.A.; Zheng, Y.D.; Wang, C.; Darby, B.J.; Zhang, C.H.; Gehrels, G. Mesozoic tectonic evolution of the Yanshan fold and thrust belt, with emphasis on Hebei and Liaoning provinces, northern China. *Paleoz. Mesoz. Tecton. Evol. Cent. Asia Cont. Assem. Intracont. Deform.* **2001**, *194*, 171–191.
15. Dai, J.Z.; Mao, J.W.; Yang, F.Q.; Ye, H.C.; Zhao, C.S.; Xie, G.Q.; Zhang, C.Q. Geological characteristics and geodynamic background of molybdenum (copper) deposits along Yanshan-Liaoning metallogenic belt on the northern margin of North China block. *Miner. Depos.* **2006**, *25*, 598.
16. Gao, P.; Zhang, Y.J.; Liu, Z.P.; Wang, P.F.; Li, J.Y.; Li, S.B. Polymetallic metallogenic law and mineral resources prediction in Shanghuangqi Banner area in North Hebei province. *Contrib. Geol. Miner. Resour. Res.* **2024**, *39*, 35–44.
17. Chen, P.W.; Zeng, Q.D.; Zhou, T.C. Petrogenesis and Mo prospecting significance of Sadaigoumen granites on the northern margin of the North China Craton. *J. Geochem. Explor.* **2020**, *214*, 106536. [CrossRef]
18. Goldfarb, R.J.; Mao, J.W.; Qiu, K.F.; Goryachev, N. The great Yanshanian metallogenic event of eastern Asia: Consequences from one hundred million years of plate margin geodynamics. *Gondwana Res.* **2021**, *100*, 223–250. [CrossRef]
19. Deng, J.; Qiu, K.F.; Wang, Q.F.; Goldfarb, R.; Yang, L.Q.; Zi, J.W.; Geng, J.Z.; Ma, Y. In situ dating of hydrothermal monazite and implications for the geodynamic controls on ore formation in the Jiaodong gold province, eastern China. *Econ. Geol.* **2020**, *115*, 671–685. [CrossRef]
20. Feng, H.; Wu, C.Z.; Zheng, Y.C.; Gu, L.X.; Jiang, S.Y.; Sun, H.T.; Gao, L. Geochronology, Geochemistry and Petrogenic & Metallogenic Settings of the Yangmadian Porphyry Molybdenum Deposit in the Yanshan- Liaoning Metallogenic Belt, North China. *J. Jilin Univ. (Earth Sci. Ed.)* **2012**, *42*, 1711–1729.
21. Li, M.; Zhang, X.; Han, L.; Gong, E.-P.; Wang, G.-G. The metallogenic setting of the Jiangjiatun Mo deposit, North China: Constraints from a combined zircon U–Pb and molybdenite Re–Os isotopic study. *Minerals* **2019**, *9*, 723. [CrossRef]
22. Duan, H.; Qin, Z.; Lin, X.; Zhang, B.; Liu, X.; Zhang, X.; Guo, P.; Han, F.; Qin, L.; Dai, J. Zircon U-Pb ages of intrusive bodies in Dacaoping molybdenum ore district, Fengning County, Hebei Province. *Miner. Depos.* **2007**, *26*, 634.
23. Liu, Q.F.; Zhao, J.L.; Li, X.L.; Bai, M.; Xiao, S.Y.; Zhang, K.; Cao, L.; Jiang, C.W.; Fu, L.B. Geochemical characteristics of orc-forming intrusions of porphyry molybdenum deposits in Yanshan-Liaoning metallogenic belt. *Resour. Environ. Eng.* **2025**, *39*, 248–257.
24. Prospecting Report of the Dengshang Mo–Polymetallic Deposit, Chengde County, Hebei Province. *Intern. Geol. Rep.* **2025**.

 https://doi.org/10.3390/min16030249

25. Wu, G.; Li, X.Z.; Xu, L.Q.; Wang, G.R.; Liu, J.; Zhang, T.; Quan, Z.X.; Wu, H.; Li, T.G.; Zeng, Q.T. Age, geochemistry, and Sr–Nd–Hf–Pb isotopes of the Caosiyao porphyry Mo deposit in Inner Mongolia, China. *Ore Geol. Rev.* **2017**, *81*, 706–727. [CrossRef]
26. Tang, L.; Santosh, M. Neoarchean-Paleoproterozoic terrane assembly and Wilson cycle in the North China Craton: An overview from the central segment of the Trans-North China Orogen. *Earth-Sci. Rev.* **2018**, *182*, 1–27. [CrossRef]
27. Zhai, M.G. Cratonization and the Ancient North China Continent: A summary and review. *Sci. China Earth Sci.* **2011**, *54*, 1110–1120. [CrossRef]
28. Zhu, R.X.; Fan, H.R.; Li, J.W.; Meng, Q.R.; Li, S.R.; Zeng, Q.D. Decratonic gold deposits. *Sci. China Earth Sci.* **2015**, *58*, 1523–1537. [CrossRef]
29. Deng, J.F.; Zhao, G.C.; Su, S.G.; Liu, C.; Chen, Y.H.; Li, F.N.; Zhao, X.G. Structure overlap and tectonic setting of Yanshan orogenic belt in Yanshan era. *Geotecton. Metallog.* **2005**, *29*, 157–165.
30. Dong, C.Y.; Wang, Q.F.; Gregory, D.D.; Li, H.J.; Weng, W.J.; Yang, L.; Deng, J. Control of crustal deformation on orogenic Au mineralization in Himalaya: A case study from Buzhu. *J. Struct. Geol.* **2024**, *188*, 105269. [CrossRef]
31. Zhu, R.X.; Chen, L.; Wu, F.; Liu, J. Timing, scale and mechanism of the destruction of the North China Craton. *Sci. China Earth Sci.* **2011**, *54*, 789–797. [CrossRef]
32. Zhang, T.; Qiu, H.N.; Wu, Y.; Zhang, D.H. Petrogenesis of Middle Jurassic mafic dikes and granites in the eastern Hebei district, North China Craton, China: Implications for westward subduction of the Paleo-Pacific plate. *J. Asian Earth Sci.* **2024**, *272*, 106248. [CrossRef]
33. Fan, W.B.; Jiang, N.; Xu, X.Y.; Hu, J.; Zong, K.Q. Petrogenesis of the middle Jurassic appinite and coeval granitoids in the Eastern Hebei area of North China Craton. *Lithos* **2017**, *278–281*, 331–346. [CrossRef]
34. He, X.F.; Santosh, M.; Ganguly, S. Mesozoic felsic volcanic rocks from the North China craton: Intraplate magmatism associated with craton destruction. *Geol. Soc. Am. Bull.* **2017**, *129*, 947–969. [CrossRef]
35. Li, S.; Wilde, S.A.; He, Z.; Jiang, X.; Liu, R.; Zhao, L. Triassic sedimentation and postaccretionary crustal evolution along the Solonker suture zone in Inner Mongolia, China. *Tectonics* **2014**, *33*, 960–981. [CrossRef]
36. Zhang, S.H.; Zhao, Y.; Liu, X.C.; Liu, D.Y.; Chen, F.; Xie, L.W.; Chen, H.H. Late Paleozoic to Early Mesozoic mafic–ultramafic complexes from the northern North China Block: Constraints on the composition and evolution of the lithospheric mantle. *Lithos* **2009**, *110*, 229–246. [CrossRef]
37. Zhang, S.H.; Zhao, Y.; Davis, G.A.; Ye, H.; Wu, F. Temporal and spatial variations of Mesozoic magmatism and deformation in the North China Craton: Implications for lithospheric thinning and decratonization. *Earth-Sci. Rev.* **2014**, *131*, 49–87.
38. Li, S.; Chung, S.L.; Wilde, S.A.; Jahn, B.M.; Xiao, W.J.; Wang, T.; Guo, Q.Q. Early-Middle Triassic high Sr/Y granitoids in the southern Central Asian Orogenic Belt: Implications for ocean closure in accretionary orogens. *J. Geophys. Res. Solid Earth* **2017**, *122*, 2291–2309.
39. Wang, J. Geological characteristics and genesis discussed of the Dengshang Molybdenum and Pyrite Deposit. *World Nonferrous Met.* **2017**, *8*, 60+62.
40. Liu, Y.H.; Mao, J.W.; Hu, J.; Ma, J.D.; Wang, L.; Xu, D.M.; Zhao, Z.X.; Bishikwabo, G.K.; Gamboa-Herrera, J.; Fan, C. Element migration and hydrothermal alteration of the Tuwaishan orogenic gold deposit in West Hainan Island, South China. *Ore Geol. Rev.* **2025**, *180*, 106568. [CrossRef]
41. Li, S.S.; Qiu, K.F.; Hernández-Uribe, D.; Gao, Y.X.; Santosh, M.; Huang, H.C.; Zheng, Z.Y.; Zhang, Z.L.; Gao, S.C. Water Recycling in the Deep Earth: Insights From Integrated μ-XRF, THz-TDS Spectroscopy, TG, and DCS of High-Pressure Granulite. *J. Geophys. Res. Solid Earth* **2023**, *128*, e2022JB025915. [CrossRef]
42. Zhao, H.Q.; Lyu, Q.B.; Sun, H.B.; Zhang, B.K.; Zhu, Y. Uncertainty evaluation for the determination of 47 elements in rock by inductively coupled plasma mass spectrometry. *Chem. Anal. Meterage* **2020**, *29*, 115–120.
43. Sláma, J.; Košler, J.; Condon, D.J.; Crowley, J.L.; Gerdes, A.; Hanchar, J.M.; Horstwood, M.S.; Morris, G.A.; Nasdala, L.; Norberg, N. Plešovice zircon—A new natural reference material for U–Pb and Hf isotopic microanalysis. *Chem. Geol.* **2008**, *249*, 1–35. [CrossRef]
44. Lin, J.; Liu, Y.S.; Yang, Y.H.; Hu, Z.C. Calibration and correction of LA-ICP-MS and LA-MC-ICP-MS analyses for element contents and isotopic ratios. *Solid Earth Sci.* **2016**, *1*, 5–27. [CrossRef]
45. Ludwig, K.R. *ISOPLOT 3.0: A Geochronological Toolkit for Microsoft Excel*; Berkeley Geochronology Center Special Publication: Berkeley, CA, USA, 2003.
46. Smythe, D.J.; Brenan, J.M. Magmatic oxygen fugacity estimated using zircon-melt partitioning of cerium. *Earth Planet. Sci. Lett.* **2016**, *453*, 260–266. [CrossRef]
47. Yu, H.C.; Guo, C.A.; Qiu, K.F.; McIntire, D.; Jiang, G.P.; Gou, Z.Y.; Geng, J.Z.; Pang, Y.; Zhu, R.; Li, N.B. Geochronological and geochemical constraints on the formation of the giant Zaozigou Au-Sb deposit, West Qinling, China. *Minerals* **2019**, *9*, 37. [CrossRef]

 https://doi.org/10.3390/min16030249

48. Yu, H.C.; Qiu, K.F.; Pang, Y.; Zhou, J.L.; Geng, J.Z.; Gou, Z.Y.; Wang, Y. Constraints of magmatic oxidation state on mineralization in the Wenquan porphyry molybdenum deposit, West Qinling, China. *Earth Sci. Front.* **2019**, *26*, 255.
49. Middlemost, E.A. Naming materials in the magma/igneous rock system. *Earth-Sci. Rev.* **1994**, *37*, 215–224. [CrossRef]
50. Frost, B.R.; Barnes, C.G.; Collins, W.J.; Arculus, R.J.; Ellis, D.J.; Frost, C.D. A geochemical classification for granitic rocks. *J. Petrol.* **2001**, *42*, 2033–2048. [CrossRef]
51. Whalen, J.B.; Currie, K.L.; Chappell, B.W. A-type granites: Geochemical characteristics, discrimination and petrogenesis. *Contrib. Mineral. Petrol.* **1987**, *95*, 407–419. [CrossRef]
52. Ren, Y.W.; Wang, H.; Zheng, Y.; Meng, K.K.; Li, H.B.; Zhang, C.; Liang, J.L.; Dong, G.C. The petrogenesis of granite and its significance on mineralization in Lanxiagou Mo deposit of northern Hebei. *ACTA Petrol. Mineral.* **2025**, *44*, 556–572.
53. Guo, Z.; Xiao, C.D.; Wang, Z.L. Geological Characteristics and Genetic Discussion of the Dacaoping Molybdenum Deposit, Northern Hebei. *Contrib. Geol. Miner. Resour. Res.* **2011**, *26*, 28–33.
54. Sun, J.L. Geological, Geochemical Characteristics and Genesis of Taipingcun Mo Deposit in XinglongCounty, Hebei Province. Master's Thesis, Jilin University, Changchun, China, 2017.
55. Miao, G.; Dong, G.C.; Chen, Z.; Zhao, H.R.; Ren, L.; Quan, R.; Xu, Y.M.; Liu, X.Y. Origin of the granite porphyry and their geological significances in the Chaijiagou molybdenum deposit, northern Hebei. *China Min. Mag.* **2016**, *25*, 306–313.
56. Sun, S.S.; McDonough, W.F. Chemical and isotopic systematics of oceanic basalts: Implications for mantle composition and processes. *Geol. Soc. Lond. Spec. Publ.* **1989**, *42*, 313–345. [CrossRef]
57. Boynton, W.V. Cosmochemistry of the rare earth elements: Meteorite studies. In *Developments in Geochemistry*; Elsevier: Amsterdam, The Netherlands, 1984; Volume 2, pp. 63–114.
58. Qiu, K.F.; Deng, J.; Taylor, R.D.; Song, K.R.; Song, Y.H.; Li, Q.Z.; Goldfarb, R.J. Paleozoic magmatism and porphyry Cu-mineralization in an evolving tectonic setting in the North Qilian Orogenic Belt, NW China. *J. Asian Earth Sci.* **2016**, *122*, 20–40. [CrossRef]
59. Qiu, K.F.; Marsh, E.; Yu, H.C.; Pfaff, K.; Gulbransen, C.; Gou, Z.Y.; Li, N. Fluid and metal sources of the Wenquan porphyry molybdenum deposit, Western Qinling, NW China. *Ore Geol. Rev.* **2017**, *86*, 459–473. [CrossRef]
60. Griffin, W.L.; Wang, X.; Jackson, S.E.; Pearson, N.J.; O'Reilly, S.Y.; Xu, X.S.; Zhou, X.M. Zircon chemistry and magma mixing, SE China: In-situ analysis of Hf isotopes, Tonglu and Pingtan igneous complexes. *Lithos* **2002**, *61*, 237–269. [CrossRef]
61. Qiu, K.F.; Deng, J. Petrogenesis of granitoids in the Dewulu skarn copper deposit: Implications for the evolution of the Paleotethys ocean and mineralization in Western Qinling, China. *Ore Geol. Rev.* **2017**, *90*, 1078–1098. [CrossRef]
62. Sun, Z.J.; Wang, Z.Q.; Yu, H.; Yu, X.H.; Liu, G.H.; Wang, C.Y. Geochronology, geochemistry, and Hf Isotope of the granites from the Mo deposits in Fengning region, China: Implications for tectonic evolution and mineralization of the North China Craton. *Arab. J. Geosci.* **2019**, *12*, 434. [CrossRef]
63. Zhang, L.L.; Jiang, S.H.; Li, H.M.; Wu, D.; Kang, H. Metallogenic and Petrogenic Geochronology and Geochemical Features of the Ore-related Granite in the Sadaigoumen Mo Deposit, Fengning, Hebei Province. *Acta Geosci. Sin.* **2019**, *40*, 708–724. [CrossRef]
64. Wu, H.; Wu, G.; Tao, H.; Zhang, M.Y.; Wang, G.R.; Chen, J.Q.; Yang, N.N. Geochronology, geochemistry and Hf isotope of granitic complex in Dasuji porphyry Mo deposit of Inner Mongolia and their geological implications. *Miner. Depos.* **2018**, *2*, 311–338.
65. Xu, X.C.; Zhang, X.X.; Zheng, C.Q.; Cui, F.H.; Gao, Y.; Gao, F. Geochemistry and chronology characteristics of the intrusive rocks and its relationship with mineralization in Yangjiazhangzi area, the western Liaoning Province. *J. Jilin Univ.* **2015**, *45*, 804–819.
66. Chappell, B. Aluminium saturation in I-and S-type granites and the characterization of fractionated haplogranites. *Lithos* **1999**, *46*, 535–551. [CrossRef]
67. Qiu, K.F.; Deng, J.; Li, S.S.; Jowitt, S.; Hetherington, C.J.; Balen, D. Roles and perspectives of A-and I-type magmas in rare earth element and gold mineralization. *Bulletin* **2024**, *136*, 1238–1250. [CrossRef]
68. Xu, F.; Shen, H.X.; Yang, J.L.; Lei, S.; Xing, N.Z. Chronology, geochemical characteristics and ore-forming potential of Caishan Au-bearing intrusion in Dongzhi county, Anhui province. *Contrib. Geol. Miner. Resour. Res.* **2024**, *39*, 493–503.
69. Moyen, J.F. High Sr/Y and La/Yb ratios: The meaning of the "adakitic signature". *Lithos* **2009**, *112*, 556–574. [CrossRef]
70. Clemens, J.D.; Stevens, G.; Farina, F. The enigmatic sources of I-type granites: The peritectic connexion. *Lithos* **2011**, *126*, 174–181. [CrossRef]
71. Ma, J.D.; Wu, M.Q.; Diao, X.; He, D.Y.; Cui, T.; Long, Z.Y.; Qiu, K.F. Na-metasomatism and hematization implication of Zr and Nb-Be-REE mineralization: Case study of the Baerzhe deposit in Inner Mongolia. *Acta Petrol. Sin.* **2023**, *39*, 432–444. [CrossRef]
72. Geng, J.Z.; Qiu, K.F.; Gou, Z.Y.; Yu, H.C. Tectonic regime switchover of Triassic Western Qinling Orogen: Constraints from LA-ICP-MS zircon U–Pb geochronology and Lu–Hf isotope of Dangchuan intrusive complex in Gansu, China. *Geochemistry* **2017**, *77*, 637–651. [CrossRef]
73. Janne, B.T.; Francis, A. The Lu-Hf isotope geochemistry of chondrites and the evolution of the mantle-crust system. *Earth Planet. Sci. Lett.* **1997**, *148*, 243–258.
74. Yang, J.H.; Wu, F.Y.; Wilde, S.A.; Xie, L.W.; Yang, Y.H.; Liu, X.M. Tracing magma mixing in granite genesis: In situ U–Pb dating and Hf-isotope analysis of zircons. *Contrib. Mineral. Petrol.* **2007**, *153*, 177–190. [CrossRef]

 https://doi.org/10.3390/min16030249

75. Zheng, J.P.; Griffin, W.L.; O'Reilly, S.Y.; Lu, F.X.; Yu, C.M.; Zhang, M.; Li, H.M. U–Pb and Hf-isotope analysis of zircons in mafic xenoliths from Fuxian kimberlites: Evolution of the lower crust beneath the North China craton. *Contrib. Mineral. Petrol.* **2004**, *148*, 79–103. [CrossRef]
76. Blatter, D.L.; Sisson, T.W.; Hankins, W.B. Crystallization of oxidized, moderately hydrous arc basalt at mid-to lower-crustal pressures: Implications for andesite genesis. *Contrib. Mineral. Petrol.* **2013**, *166*, 861–886. [CrossRef]
77. Rapp, R.P.; Watson, E.B. Dehydration melting of metabasalt at 8–32 kbar: Implications for continental growth and crust-mantle recycling. *J. Petrol.* **1995**, *36*, 891–931. [CrossRef]
78. Xue, X.F.; Qiu, K.F.; He, D.Y.; Huang, Y.Q.; Zhu, R.; Yu, H.C.; Jiang, G.P. Petrogenesis of the quartz diorite porphyry in the Zaozigou, West Qinling and its geological implications. *Acta Petrol. Sin.* **2023**, *39*, 463–483. [CrossRef]
79. Chen, B.; Jahn, B.M.; Wilde, S.; Xu, B. Two contrasting Paleozoic magmatic belts in northern Inner Mongolia, China: Petrogenesis and tectonic implications. *Tectonophysics* **2000**, *328*, 157–182. [CrossRef]
80. Dobretsov, N.L.; Berzin, N.A.; Buslov, M.M. Opening and tectonic evolution of the Paleo-Asian Ocean. *Int. Geol. Rev.* **1995**, *37*, 335–360. [CrossRef]
81. Xiao, W.J.; Windley, B.F.; Huang, B.C.; Han, C.M.; Yuan, C.; Chen, H.L.; Sun, M.; Sun, S.; Li, J.L. End-Permian to mid-Triassic termination of the accretionary processes of the southern Altaids: Implications for the geodynamic evolution, Phanerozoic continental growth, and metallogeny of Central Asia. *Int. J. Earth Sci.* **2009**, *98*, 1189–1217. [CrossRef]
82. Zeng, Q.D.; Liu, J.M.; Xiao, W.J.; Chu, S.X.; Wang, Y.B.; Duan, X.X.; Sun, Y.; Zhou, L.L. Mineralizing types, geological characteristics and geodynamic background of Triassic molybdenum deposits in the northern and southern margins of North China Craton. *Acta Petrol. Sin.* **2012**, *28*, 357–371.
83. Qiu, K.F.; Long, Z.Y.; Romer, R.L.; Halama, R.; Williams-Jones, A.; Yu, H.C.; Li, S.S.; Wu, M.Q.; Deng, J. The mantle source of REE-rich alkaline silicate magmas can be enriched by continent-derived sediment subduction. *Nat. Commun.* **2025**, *16*, 11191. [CrossRef] [PubMed]
84. Zhang, J.Y.; Qiu, K.F.; Yin, R.; Long, Z.Y.; Feng, Y.C.; Yu, H.C.; Gao, Z.Y.; Deng, J. Lithospheric mantle as a metal storage reservoir for orogenic gold deposits in active continental margins: Evidence from Hg isotopes. *Geology* **2024**, *52*, 423–428. [CrossRef]
85. Guo, F.; Fan, W.M.; Li, X.Y.; Li, C.W. Geochemistry of Mesozoic mafic volcanic rocks from the Yanshan belt in the northern margin of the North China Block: Relations with post-collisional lithospheric extension. *Geol. Soc. Lond. Spec. Publ.* **2007**, *280*, 101–129. [CrossRef]
86. Liu, Y.H.; Mao, J.W.; Miggins, D.P.; Qiu, K.F.; Hu, J.; Wang, L.; Xu, D.M.; Goldfarb, R.J. 40Ar/39Ar geochronology constraints on formation of the Tuwaishan orogenic gold deposit, Hainan Island, China. *Ore Geol. Rev.* **2020**, *120*, 103438. [CrossRef]
87. Loader, M.A.; Wilkinson, J.J.; Armstrong, R.N. The effect of titanite crystallisation on Eu and Ce anomalies in zircon and its implications for the assessment of porphyry Cu deposit fertility. *Earth Planet. Sci. Lett.* **2017**, *472*, 107–119. [CrossRef]
88. Shen, P.; Hattori, K.; Pan, H.; Jackson, S.; Seitmuratova, E. Oxidation condition and metal fertility of granitic magmas: Zircon trace-element data from porphyry Cu deposits in the Central Asian orogenic belt. *Econ. Geol.* **2015**, *110*, 1861–1878. [CrossRef]
89. Mengason, M.J.; Candela, P.A.; Piccoli, P.M. Molybdenum, tungsten and manganese partitioning in the system pyrrhotite–Fe–S–O melt–rhyolite melt: Impact of sulfide segregation on arc magma evolution. *Geochim. Cosmochim. Acta* **2011**, *75*, 7018–7030. [CrossRef]
90. Yu, H.C.; Qiu, K.F.; Wang, J.; Zhang, P.C.; Smeraglia, L.; Tassara, S.; Ma, J.D.; Xue, X.F.; Li, D.P.; Deng, J. Competency contrast-driven extension veining and sequential orogenic au-sb mineralization: Insights from the zaorendao deposit, China. *Geol. Soc. Am. Bull.* **2025**. [CrossRef]
91. Wang, J.; Yu, H.C.; Petrella, L.; Duo, D.W.; Wang, L.; Yang, F.; Zhou, H.Y.; Li, C.; Qiu, K.F. Fluid evolution and genesis of the Yidinan granitoid-hosted orogenic gold deposit (China). *Geol. Soc. Am. Bull.* **2025**, *137*, 1945–1963. [CrossRef]
92. Sun, W.D.; Li, C.Y.; Ling, M.X.; Ding, X.; Yang, X.Y.; Liang, H.Y.; Zhang, H.; Fan, W.M. The geochemical behavior of molybdnum and mineralization. *Acta Petrol. Sin.* **2015**, *31*, 1807–1817.
93. Pearce, J.A.; Harris, N.B.; Tindle, A.G. Trace element discrimination diagrams for the tectonic interpretation of granitic rocks. *J. Petrol.* **1984**, *25*, 956–983. [CrossRef]
94. Pearce, J.A.; Peate, D.W. Tectonic implications of the composition of volcanic arc magmas. *Annu. Rev. Earth Planet. Sci.* **1995**, *23*, 251–286. [CrossRef]
95. Chen, H.Y.; Wu, C. Metallogenesis and major challenges of porphyry copper systems above subduction zones. *Sci. China Earth Sci.* **2020**, *63*, 899–918. [CrossRef]
96. Wilkinson, J.J. Triggers for the formation of porphyry ore deposits in magmatic arcs. *Nat. Geosci.* **2013**, *6*, 917–925. [CrossRef]

https://doi.org/10.3390/min16030249

minerals

MDPI

Article

Whole-Rock Element Analyses Constraining the Magmatic Evolution and Metallogenesis of the Jiaojia Fault Zone, Jiaodong Gold Province

Jiabao Jia [1,2,3], Yueliang Hu [3], Lin Gao [2,3], Yulu Lv [2], Junjie Wang [2], Xiaomei Yang [2], Yan Liu [2], Xiaoliang Shi [4], Jing Lv [4], Yanbo Xu [3,5], Mengmeng Zhang [6,7] and Wu Li [1,*]

1 School of Resources and Geosciences, China University of Mining and Technology, Xuzhou 221116, China; jiajiabao@sd-gold.com
2 Center of Technology, Shandong Gold Geologic Minerals Survey Co., Ltd., Yantai 264000, China; gaolin@sd-gold.com (L.G.); wangjunjie@sd-gold.com (J.W.)
3 Institute of Geological Research, Shandong Gold Group Co., Ltd., Jinan 250101, China; huyueliang@sd-gold.com (Y.H.); xuyb@sd-gold.com (Y.X.)
4 Investment and Development Department, Shandong Gold Geologic Minerals Survey Co., Ltd., Yantai 264000, China; shixiaoliang@sd-gold.com (X.S.); lvjing@sd-gold.com (J.L.)
5 Inner Mongolia Shandong Gold Geological and Mineral Exploration Co., Ltd., Chifeng 024005, China
6 School of Gemology, China University of Geosciences, Beijing 100083, China; mmzhang0615@163.com
7 Research Center of Genetic Mineralogy, School of Earth Sciences and Resources, China University of Geosciences, Beijing 100083, China
* Correspondence: liwu@cumt.edu.cn

Abstract

The Jiaodong Peninsula constitutes a world-class gold province in eastern China, containing more than 5000 t of identified gold resources. The Jiaojia gold deposit is one of the largest deposits within this gold province, and mineralization is primarily distributed along the northern segment of the Jiaojia Fault. The structural characteristics and mineralization processes of the northern segment have been extensively documented. In contrast, the ore-forming mechanisms of the southern Jiaojia Fault remain poorly constrained, hindering further exploration targeting. We chose several gold deposits and one drill core along the Jiaojia Fault, then present whole-rock major and trace elements data to evaluate magmatic affinities and their ore-forming potential. The results show that the lithological differences in plutonic and stratigraphic units suggest that variations in petrogenesis may have exerted a fundamental control on mineralization styles. Almost all samples are characterized by enrichment in light rare earth elements, relative enrichment in Europium, and pronounced depletion in heavy rare earth elements. Alteration characteristics indicate the northern segment is dominated by advanced argillic alteration, whereas phyllic alteration is more prevalent in the southern segment. The rare earth elements discrimination plot clearly suggests differentiation from the northern and southern fault segments. Consequently, we propose that the northern segment records synorogenic arc magmatism, while the southern segment experienced both synorogenic and a subsequent intraplate extensional transitional stage.

Keywords: metallogenesis; whole-rock geochemistry; Jiaojia Fault zone; Jiaodong gold province; tectonic evolution

1. Introduction

Since the widespread extension and metallogenic event during the Early Cretaceous occurred [1,2], the giant Jiaodong gold province formed. The metasomatized lithospheric mantle in the western Jiaodong Peninsula erupted, and is controlled by the Jiaojia Fault system [3]. Gold deposits occur along northeast (NE) trending brittle-ductile to brittle faults cutting across both terranes [4,5]. The spatial distribution of these deposits is closely coupled with fault structures, driven by regional thermal doming extension [6]. More than 85% of the overall resource is hosted to the west in the Jiaobei Uplift of the Jiaobei Terrane [7,8]. Major gold deposits, including the Jiaojia, Xincheng, Wang'ershan, and Sizhuang mines, are concentrated on the northern side of the Jiaojia Fault. The geology and genesis of this world-class gold district have been systematically reviewed, highlighting its immense scale and complex structural controls [9]. Advanced three-dimensional predictive modeling, geostatistical structural analysis, and machine learning based on geometric and geodynamic variables have been successfully applied to target deep-seated gold prospectivity in this northern concentration area [3,10,11]. In contrast, gold occurrences are sparse on the southern side, where the fault geometry remains debated, particularly with respect to whether the southern segment shares the same NE-trending orientation as the northern segment, although deep exploration in other southern fault segments of the region has begun to reveal their hidden metallogenic potential [6].

The crust plays a fundamental role in gold (Au) mineralization, particularly in orogenic gold systems [12–14]. Granitic and metamorphic rocks involved in mineralization commonly contain biotite, which can enrich iron (Fe) in ore-forming fluids [15,16]. Interactions between mantle-derived fluids and Fe-rich crustal rocks promotes the formation of gold-sulfur (Au–S) complexes and subsequent gold precipitation [3,17]. Microanalytical studies on pyrite trace elements and S-Fe isotopes have further constrained the origin, multi-stage evolution, and precipitation mechanisms of these ore-forming fluids within the Jiaojia belt [18]. Accessory minerals in granites, classified using various approaches [19,20], act as major repositories for trace and minor elements, including fluorine (F), chlorine (Cl), strontium (Sr), thorium (Th), and rare earth elements (REEs) [21–23]. These geochemical signatures preserve key information on magmatic and hydrothermal processes associated with mineralization. For the northern segment of the Jiaojia Fault (N-JJF), gold mineralization is widely interpreted to result from fluid–rock interactions accompanied by systematic wall-rock alteration zonation [3,24], and extensive studies have investigated the relative contributions of ancient and juvenile crustal components to the Jiaodong gold province [2,25–28]. In contrast, the mineralization style of the southern Jiaojia Fault remains poorly constrained.

In light of these unresolved questions, this study investigates the geochemical characteristics of the plutons and metamorphic belts along the Jiaojia gold belt. By tracing these geochemical and petrological signatures, we aim to unravel the complex fluid–rock interactions and structural–mineralization coupling processes along the Jiaojia Fault. This study provides crucial insights into regional fluid evolution and helps constrain the debated fault geometry and mineralization styles of the southern segment.

2. Geological Background

The Jiaojia Fault is a major irregular structure, approximately 60 km long and 50–500 m wide. It features an S-shaped northeastward trend (striking 25–75°) and dips to the northwest at 20–50°, opposite to the Sanshandao Fault. The structural complexity is further highlighted by its footwall, which develops branching faults arranged parallel to the main structure or forming inverted-V arrays. As a critical ore-controlling structure, it hosts over

 https://doi.org/10.3390/min16040350

20 altered-rock type gold deposits, including significant occurrences like the Jiaojia, Lingbei, and Anshi mines [3,6,29].

The Jiaodong Peninsula is bounded by the crustal-scale Tan-Lu fault system to the west and the circum-Pacific subduction system to the east, which has been active for the past ~200 Myr through subduction of the Mesozoic Izanagi and Cenozoic Pacific plates [30–32]. The region comprises two distinct tectonic units: the Jiaobei Terrane in the northwest and the Sulu ultrahigh-pressure (UHP) metamorphic belt in the southeast, separated by the Wulian–Yantai Fault [3,33,34] (Figure 1A).

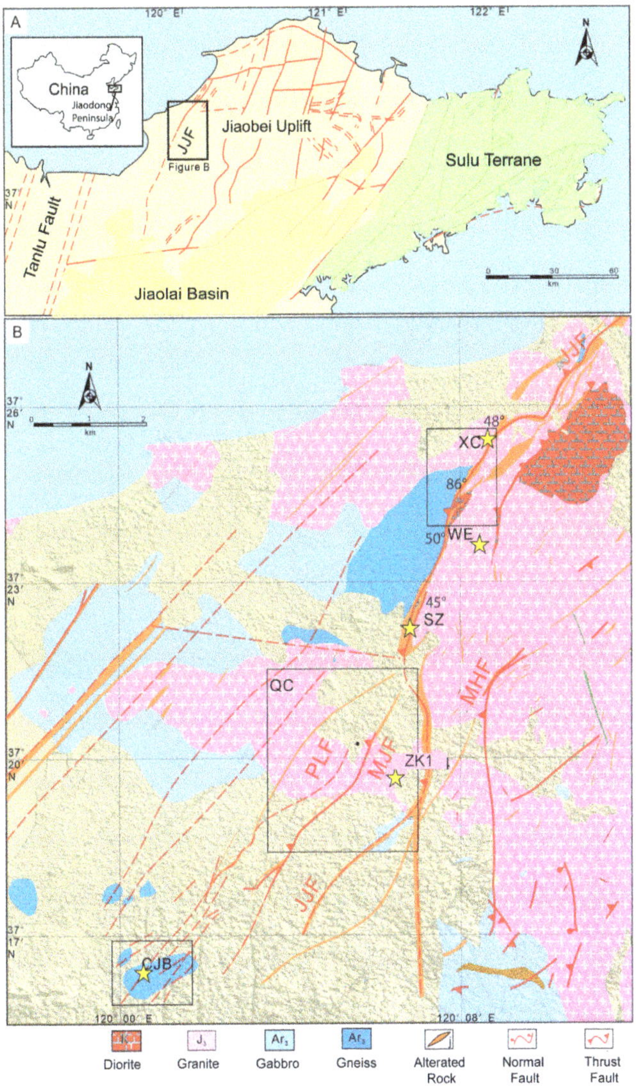

Figure 1. Geological setting of the study area (**A**) and geological sketch map of Jiaojia Fault study area (**B**). [Modified from internal materials; due to secret policy, images may appear distorted.] JJF: Jiaojia Fault, MJF: Miaojia Fault, PLF: Pinglidian Fault, MHF: Mahuigou Fault, WE: Wang'ershan gold deposit, XC: Xincheng gold deposit, SZ: Sizhuang gold deposit, CJB: Caojiabu gold deposit, ZK1: drill-hole of study area, QC: Qianchen area. Yellow star depicts gold deposits. The dashed red line depicts the inference of fault locations.

https://doi.org/10.3390/min16040350

The Jiaobei Terrane, including the Jiaolai Basin and Jiaobei Uplift, is composed of a Precambrian crystalline basement overlain and intruded by Mesozoic sedimentary and igneous rocks [35,36]. The Precambrian basement experienced high-grade amphibolite- to granulite-facies metamorphism at ~1.9–1.8 Ga [37,38]. Mesozoic intrusive rocks in the Jiaobei Uplift include the Linglong (166–149 Ma), Guojialing (130–128 Ma), and Aishan (118–115 Ma) granitoids [26,39,40]. Magma underplating induced partial melting of the lower crust, generating voluminous felsic magmas that subsequently interacted with ascending mafic magmas to form the Guojialing granodiorites at ~130 Ma [26,36,39,41,42]. Widespread Mesozoic granitoids intruding the Precambrian basement include Late Triassic alkaline and Late Jurassic to Early Cretaceous calc-alkaline intrusions [43–45]. The Late Triassic (ca. 230–200 Ma) alkaline granitoids are mainly exposed in the southeastern Sulu Terrane, and were generated by partial melting of ancient continental crust during exhumation of UHP slices following collision between the two Precambrian blocks [34]. The Late Jurassic (ca. 160–142 Ma) Linglong granitoids are extensively distributed across both terranes and are geochemically interpreted to have formed by partial melting of thickened Archean mafic lower crust with minimal mantle input [36]. Cretaceous mafic to intermediate dikes are widespread throughout the Jiaodong region [42,46], and contemporaneous volcanic rocks are mainly preserved within the Jiaolai pull-apart basin south of the major gold deposits in both terranes [47].

The basin fill comprises fluvial–lacustrine clastic rocks, volcanic rocks, and red beds interlayered with mafic to ultramafic lavas, deposited between 126 and 100 Ma [40,48]. Most gold resources in the Jiaodong province are concentrated in the Jiaojia, Xincheng, Wang'ershan, and Sizhuang gold districts within the Jiaobei Uplift (Figure 1B), as well as in the Muru gold district of the Sulu Terrane. Gold mineralization in all districts is structurally controlled by NE striking faults. Ore bodies occur predominantly as disseminated mineralization and stockwork veinlets within intensely altered and cataclastic footwall fault rocks, and as large auriferous quartz veins associated with brittle fault zones.

3. Materials and Methods

3.1. Materials

The hand sample and samples for REE analysis were collected from the mining area, including the northern side Wang'ershan mine (WE), Xincheng mine (XC), Sizhuang mine (SZ) and the southern side Caojiabu mine (CJ). REE samples were collected from each mine deposit near the rock transition and are from near to far apart with certain depths. In this study, we used a systematic spatial sampling strategy, which traversed from the central mineralized veins outward into the unaltered host rocks. In addition, we collected samples from the drill core ZK1 (Figure 2) for whole-rock analysis, which was located at the southern segment of the Jiaojia Fault. After petrographic examination, a total of 125 least-altered whole-rock samples were selected for whole-rock major and trace element analyses (Table S1).

3.2. Methods

3.2.1. Whole-Rock Major and Trace Element Analyses

The samples were crushed and milled in a ball mill to a grain size of 200 mesh prior to geochemical analysis. Major and trace element compositions were analyzed at Chengxin Geological Laboratory Ltd. (Langfang, China) using X-ray fluorescence spectrometry (XRF, PANalytical PW2424/2400, PANalytical B.V., Almelo, The Netherlands) [49] and inductively coupled plasma mass spectrometry (ICP-MS, Agilent Technologies, Santa Clara, CA, USA). Analytical uncertainties were greater than $\pm 1\%$ for major elements and $\pm 5\%$ for rare earth elements.

 https://doi.org/10.3390/min16040350

For major element analysis, calcined or ignited sample powders were mixed with lithium borate flux (50% $Li_2B_4O_7$–50% $LiBO_2$), fused in an automatic fluxer at approximately 1000 °C, and cast into homogeneous glass disks. Major element concentrations were then determined using a PANalytical PW2424 X-ray fluorescence spectrometer (PANalytical B.V., Almelo, The Netherlands). For trace element analysis, samples were mixed with lithium metaborate–tetraborate flux, fused at 1025 °C, cooled, and dissolved in a mixed acid solution of HNO_3, HCl, and HF. The solution was subsequently analyzed using an Agilent 7900 ICP-MS (Agilent Technologies, Santa Clara, CA, USA).

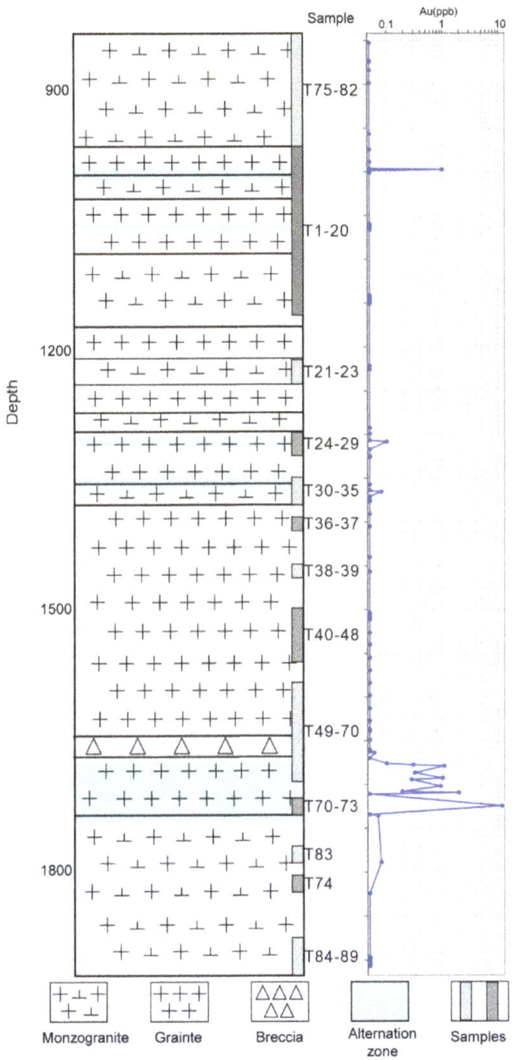

Figure 2. Sample strategy of ZK1 and its gold concentration. Blue color lines depict gold grade.

3.2.2. Gold Concentration Analysis

For Au concentrations, we applied the analytical method established in atomic absorption spectrophotometer (Thermo Scientific iCE 3500 atomic absorption spectrometer (Thermo Fisher Scientific, Cambridge, United Kingdom) at the Yantai Shandong gold

https://doi.org/10.3390/min16040350

Geological & Mining Testing Technology Co., Ltd. (Yantai, China). With the addition of mixed ^{191}Ir-^{99}Ru-^{194}Pt-^{105}Pd and ^{185}Re and ^{109}Ag spike solution, about 2 g of sample powder was dissolved by reverse aqua regia (a mixture of 5 mL of 14 M HNO_3 and 2.5 mL of 9 M HCl) in pre-cleaned Carius tubes at 240 °C for 4 days. After removing the matrix and interferences by Eichrom 50W-X8 cation resin, Pt concentrations were determined by isotope dilution methods. Gold concentrations were obtained by internal standardization of Au intensity to Pt via the high-sensitivity sector field ICP-MS (Element XR HR-ICP-MS, Thermo Fisher Scientific, Bremen, Germany).

4. Results

4.1. Field Observation and Petrography of Typical Gold Deposit

The orebodies of N-JJF in the Xincheng gold deposit are mainly hosted by the pyrite-sericite-quartz alteration zone (Figure 3A), the same as the majority of gold mineralization in the N-JJF [11]. Almost all the orebodies are parallel to the main fault. Pyrite presents as veinlets inside quartzite (Figure 3B). On the contrary, the orebody of the Caojiabu gold deposit at the S-JJF consists of the granitoid, and is almost parallel to the lineation inside wall-rock (Figure 3C). At the shallow part of the pluton, the orebody shows siderite inside granitoid with calcite veins and erosion sulfide minerals (Figure 3D).

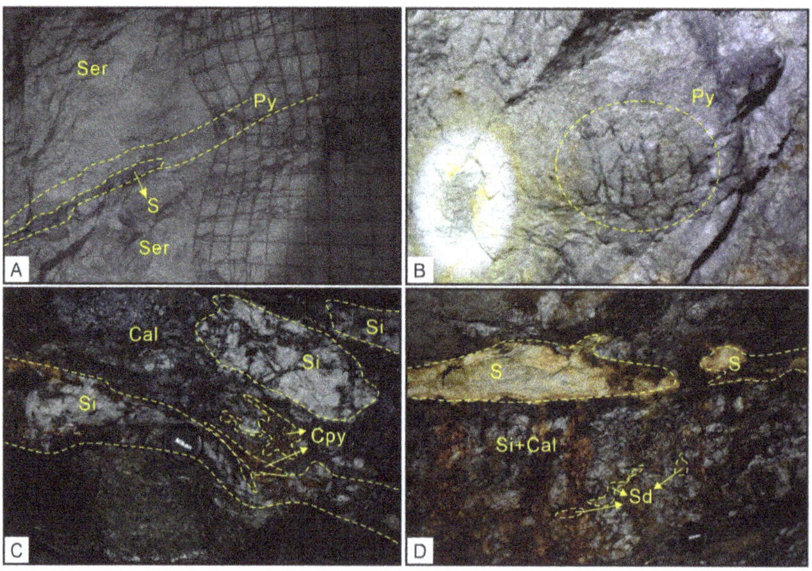

Figure 3. Field photos of Xincheng gold deposit (**A**,**B**) and field photos of Caojiabu gold deposit (**C**,**D**). Py: pyrite, Ser: sericite, Cal: calcite, Sd: siderite, S: sulfide, Si: silicide, Cpy: chalcopyrite.

Macroscopic petrographic observations further substantiate the distinct lithofacies and mineralization styles between the northern and southern segments (Figure 4). In the northern segment, representative samples from the Xincheng deposit (e.g., SD-XC-02 and SD-XC-05; Figure 4A,B) display classical "altered-rock type" disseminated mineralization. These samples are characterized by intense wall-rock alteration where fine-grained pyrite veinlets and disseminated pyrite (Py) intricately overprint the granitic host rocks, accompanied by quartz (Qtz) and remnant K-feldspar (Kfs).

 https://doi.org/10.3390/min16040350

Figure 4. Hand specimen of Xincheng gold deposit (**A**,**B**) and Caojiabu gold deposit (**C**,**D**). Kfs: K-feldspar, Py: pyrite, Qtz: quartz, Cpy: copper pyrite, Gn: galena, Sd: siderite.

In contrast, samples from the southern segment, such as those from the Caojiabu deposit (e.g., SD-CJ-01 and SD-CJ-04; Figure 4C,D) exhibit a notably different macroscopic style. These ores are predominantly characterized by massive, structurally controlled quartz veins hosting a complex polymetallic sulfide assemblage. Unlike the simple pyrite-dominated northern samples, the southern samples are markedly enriched in base metal sulfides, including chalcopyrite (Cpy), galena (Gn), and siderite (Sd) veinlets.

The modal mineralogy of selected samples from the Xincheng gold deposit exhibits a typical granitic texture and a massive structure (Figure 5A). Its primary rock-forming minerals are potassium feldspar, plagioclase, and quartz. Potassium feldspar commonly shows Carlsbad twinning, plagioclase exhibits polysynthetic twinning, and quartz displays undulatory extinction. The rock is strongly altered, primarily through the development of pyritization, sericitization, and carbonatization. Pyrite is euhedral and frequently occurs in cubic forms.

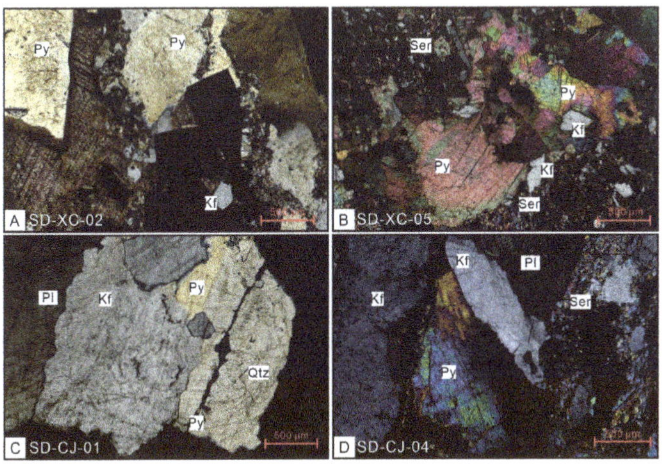

Figure 5. Pictures of mineralogical characteristics of Xincheng gold deposit (**A**,**B**) and Caojiabu gold deposit (**C**,**D**). Py: pyrite, Qtz: quartz, Ser: sericite, Kf: K-field spar, Pl: feldspar.

https://doi.org/10.3390/min16040350

Compared with sample SD-XC-01, the sericitization is less pronounced, whereas the carbonatization is more extensively developed. In addition, the sample contains a considerable amount of sulfide minerals, including pyrite, chalcopyrite, galena, and sphalerite. Based on both hand specimen observations and microscopic analysis, it has been designated as a pyrite-sericitized granitic rock. In addition, the gold grade and sericite content are higher than sample SD-XC-02 (Figure 5B).

The rock of the Caojiabu gold deposit with high gold content exhibits a typical granitic texture and blocky structure (Figure 5C), with quartz as the predominant rock-forming mineral accompanied by minor amounts of potassium feldspar and plagioclase. The quartz displays undulatory extinction and distinct pressure-shadow microstructures. The degree of alteration is relatively weak, primarily characterized by the development of fine pyrite veins, some of which crosscut earlier quartz. Based on combined hand specimen observations and microscopic analyses, the rock is classified as pyritized high-silica granite.

4.2. *Whole-Rock Major and Trace Elements*

As for the whole-rock major elements discrimination diagram (Figure 6), data of samples have wide range cover the granite, granodiorite and monzonite, which belongs to Linglong suite. The ZK1 group is characterized by SiO_2 contents of 60–72 wt.% and $Na_2O + K_2O$ contents of 5–9 wt.%, with data points forming a continuous compositional belt spanning diorite, granodiorite, quartz monzonite, and minor monzonite and syenite fields, indicating a continuous transition from intermediate to intermediate-acidic compositions. As for the data of drill-hole ZK1, the one plotting in gabbroic diorite may indicate that sample is a plutonic vein rather than wall rock.

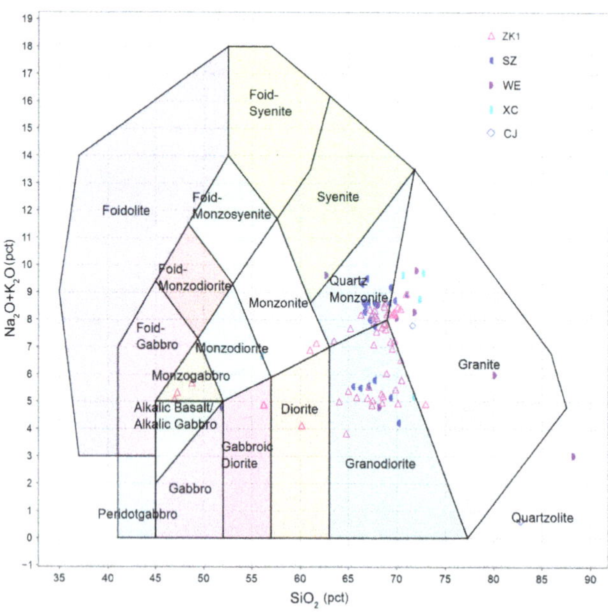

Figure 6. Whole-rock major elements rock type discrimination diagram [50].

The SZ group exhibits relatively uniform SiO_2 contents (65–72 wt.%) and variable $Na_2O + K_2O$ contents (4–10 wt.%), with data points mainly clustering in the granodiorite and quartz monzonite fields, reflecting restricted silica variability but pronounced alkalinity differentiation. The WE group displays the broadest compositional range in SiO_2, with $Na_2O + K_2O$ contents predominantly ranging from 6 to 9 wt.%; its data points extend from

https://doi.org/10.3390/min16040350

quartz monzonite and granite into the quartzolite field, indicating a distinct evolutionary trend toward high silica and high alkalinity. The XC group shows concentrated SiO_2 contents of 70–75 wt.% and $Na_2O + K_2O$ contents of 5–10 wt.%, with data points largely plotting in the granite and quartz monzonite fields, typical of high-silica, intermediate-potassium to high-potassium acidic rocks. The CJ group is represented by a single data point with ~72 wt.% SiO_2 and ~8 wt.% $Na_2O + K_2O$, located at the boundary of the granite field, and thus serves as a representative acidic lithology in the study area. The chemical- and mineralogical-based classifications of drill cores are self-consistent. The intrusions show slight differences in composition (Figure 6).

It is worth noting that several samples plot distinctly away from the main diorite–granodiorite–granite cluster in the classification diagram. These outliers are a direct consequence of the systematic spatial sampling strategy employed. Specifically, samples plotting in the extreme high-silica fields (e.g., the quartzolite field) correspond to massive pyrite-bearing quartz veins from the fault core, whereas other adjacent outliers reflect intensely phyllic-altered wall rocks (pyrite-sericite-quartz altered granites). These compositional deviations perfectly map the intensity of the localized hydrothermal fluid–rock interactions.

In the primitive mantle-normalized trace element spider diagrams (Figure 7), the SZ group, ZK1 group, WE group, CJ group, and XC group all exhibit distinct enrichment in large-ion lithophile elements (LILE, e.g., Rb, Ba, Th, U, Pb) and relative depletion in high-field-strength elements (HFSE, e.g., Nb, Ta, Ti). Among these, the ZK1 and SZ groups display the most similar curve patterns, characterized by pronounced positive Pb anomalies and negative Nb–Ta anomalies. In contrast, the WE and XC groups show higher degrees of LILE enrichment, particularly in Pb, La, and Ce. Represented by a single data point, the CJ group plots within the median range of all suites, sharing the common geochemical signatures of LILE enrichment and HFSE depletion.

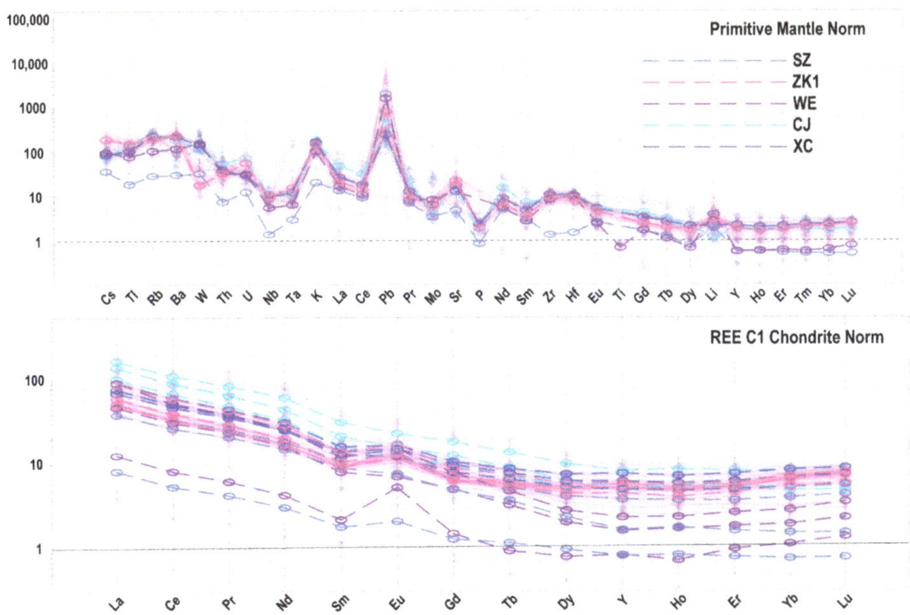

Figure 7. Primitive mantle norm spider diagram, and REE C1 chondrite norm spider diagram. Primitive mantle data and REE C1 chondrite data are based on [51]. The semi-transparent line represents the raw data, while the dashed lines indicate the fitted curves corresponding to the 50th and 75th percentiles for each sample group.

https://doi.org/10.3390/min16040350

In the chondrite-normalized REE diagrams, all groups show a right-sloping distribution pattern marked by relatively enriched light rare earth elements (LREE, La–Nd) and relatively flat heavy rare earth elements (HREE, Tb–Lu). Notably, the samples generally exhibit absent to slightly positive Eu anomalies. The ZK1, SZ, WE, and XC groups display largely absent or negligible Eu anomalies, whereas the CJ group exhibits a more distinct positive Eu anomaly.

Most samples from the WE, along with some from other sectors, can be classified as having formed in anorogenic and rift-related settings. However, given the unresolved petrogenesis of A-type granites, it is reasonable to infer that the Wang'ershan samples were likely less influenced by plate tectonic processes. In contrast, most samples from the southern segment of the Jiaojia area are associated with compressional orogenic events related to plate convergence (Figure 8).

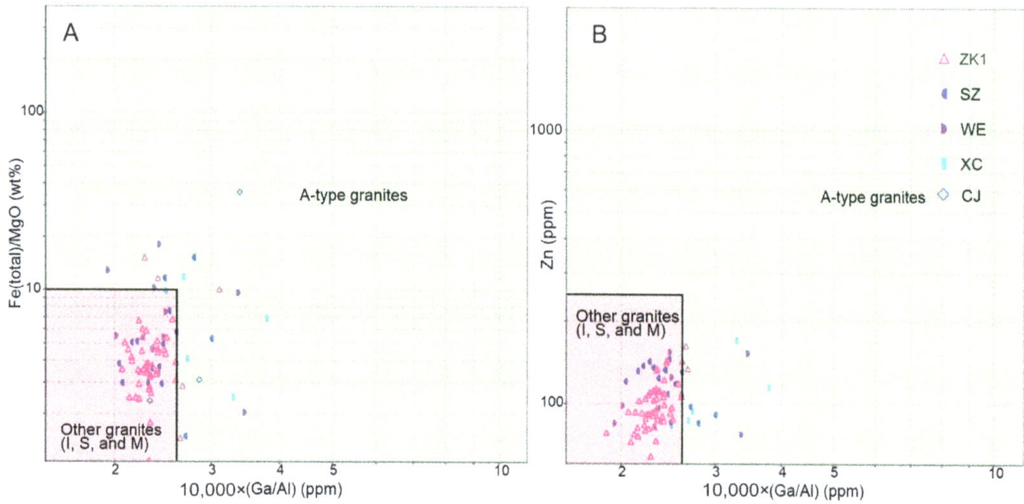

Figure 8. (**A**) The A and I-S-M type granite differentiation diagram based on Ga/Al versus Fe/MgO. (**B**) A and I-S-M type granite differentiation diagram based on Ga/Al versus Zn. The source of the boundary lines or classification fields are based on [52]. ZK1: drill core 1, SZ: Sizhuang mine, WE: Wang'ershan mine, XC: Xincheng mine, CJ: Caojiabu mine, Fe: iron, MgO: magnesium oxide, Ga: gallium, Al: aluminum, Zn: zinc, A-type granite: anorogenic granite, I-type granite: igneous-type granite, S-type granite: sedimentary-type granite, M-type granite: mantle-derived granite.

In the Zn vs. Ga/Al diagram, drill hole ZK1 shows all samples are orogenic-type gold mineralization at greater depths. Notably, high-gold-grade samples 63 to 71 plot within the transitional zone between A-type granites and orogenic gold deposits. Similarly, in the diagram, some samples from the XC and CJ also fall within the compositional field of A-type granites, indicating a comparable magmatic affinity in parts of this ore district.

Based on the Nb vs. Th discrimination diagram [53], all samples from the Wang'ershan, Xincheng, and Sizhuang areas, as well as those from drill hole ZK1, plot within the field of continental arc granites. Furthermore, in the La/10–Nb/8–Y/15 tectonic discrimination diagram, most samples fall within the island arc and calc-alkaline rock fields. Notably, some samples from drill hole ZK1 are classified in the intraplate extensional tectonic setting, indicating a later stage of magmatic activity following continental collision (Figure 9).

https://doi.org/10.3390/min16040350

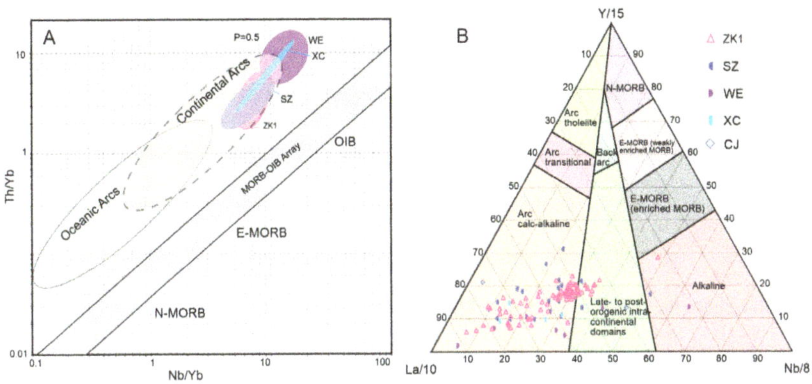

Figure 9. (**A**) Nb vs. Th discrimination diagram. The source of the boundary lines or classification fields are based on [53]. (**B**) La-Nb-Y discrimination diagram. The source of the boundary lines or classification fields are based on [54]. ZK1: drill core 1, SZ: Sizhuang mine, WE: Wang'ershan mine, XC: Xincheng mine, CJ: Caojiabu mine, Nb: niobium, Th: thorium, Yb: ytterbium, Y: yttrium, La: lanthanum, OIB: ocean island basalt, E-MORB: enriched mid-ocean ridge basalt, N-MORB: normal mid-ocean ridge basalt.

Hydrothermal alteration exhibits distinct spatial zonation, typically characterized by an inner zone rich in chlorite, pyrite, and carbonate minerals adjacent to the ore body, transitioning outward to assemblages dominated by sericite, chlorite, and carbonate (Figure 10). The pronounced deviation of samples from the Wang'ershan (WE) and Caojiabu (CJ) deposits toward the pyrophyllite node in the alteration box plot (Figure 10B) indicates a localized but critical shift in the hydrothermal regime. Notably, while other regional samples display a normal, continuous alteration trend—gradually vectoring toward the muscovite node indicative of a rock-buffered, progressive phyllic overprint—the WE and CJ samples exhibit a direct and abrupt transition to pyrophyllite-type alteration.

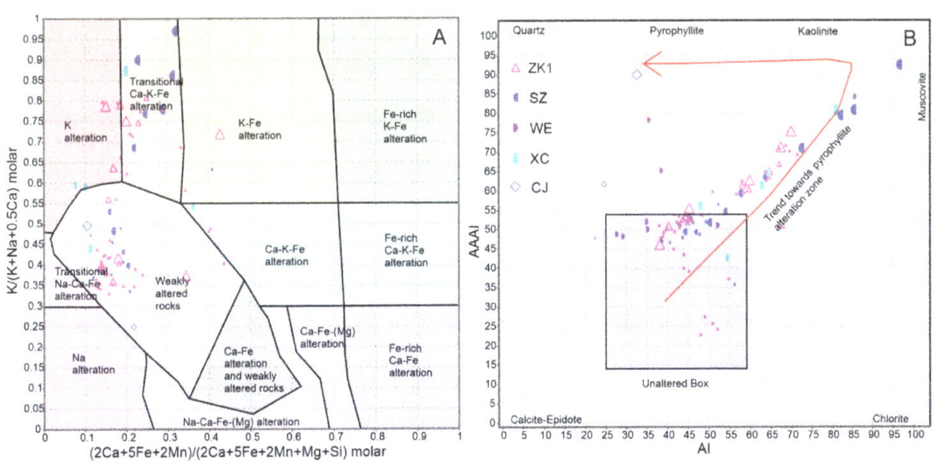

Figure 10. Iron oxide copper gold deposit alteration discrimination diagram (**A**). The source of the boundary lines or classification fields are based on [55]. (**B**) AI vs. AAAI discrimination diagram. The source of the boundary lines or classification fields are based on [56]. The size of each samples depicts the enrichment of gold grade. AI: Ishikawa alteration index, AAAI: advanced argillic alteration index. ZK1: drill core 1, SZ: Sizhuang mine, WE: Wang'ershan mine, XC: Xincheng mine, CJ: Caojiabu mine, K: potassium, Na: sodium, Ca: calcium, Mn: manganese, Mg: magnesium, Si: silicon.

 https://doi.org/10.3390/min16040350

5. Discussion

5.1. Geological Background of Jiaojia Fault System

Unlike the primary low-angle Jiaojia Fault (JJF), steeper secondary faults like the Pinglidian (PLF) and Miaojia (MJF) faults act as localized structural splays that controlled the extensional pathways and emplacement of intraplate extensional transitional magmatism, notably the ZK1 and CJ groups [57]. The relationship between this structural framework and gold mineralization is further evidenced by local petrography, where fault-driven fluid–rock interactions broke down primary mafic minerals into sericite and sulfides, liberating iron to buffer mineralizing fluids and trigger the precipitation of gold [58].

This macroscopic divergence of samples—from disseminated pyrite-altered granites in the north to polymetallic-rich massive quartz veins in the south—provides physical evidence for the different hydrothermal fluid compositions and structural evolutionary pathways across the Jiaojia Fault flanks.

The whole-rock major element distribution characteristics indicate that the continuous zonal distribution of the ZK1 group corroborates the progressive fractional differentiation from intermediate to intermediate-acidic magmas, whereas the clustered distribution of the SZ group reflects moderate differentiation dominated by magmatic fractional crystallization. Both units are characterized by intermediate-to-high silica and intermediate potassium contents, which jointly indicate a tectonic setting of continental arc or collisional orogenic belt, and most likely represent I-type or transitional granites formed by the mixing of mantle-derived magmas and crustal components. The high silica and high alkalinity signatures of the WE group, together with its extension toward the quartzolite field, reflect intense magmatic differentiation and may also be superimposed by late-stage hydrothermal alteration; its petrological affinity is consistent with highly evolved granites formed in intraplate extensional or rift setting. The high-silica granite geochemistry of the XC group is consistent with the genetic attributes of A-type granites or crustal anatectic S-type granites in an extensional regime. The data of the CJ group constrain the origin of regional A-type acidic rocks. Overall, the studied samples define a magmatic differentiation sequence from early intermediate-acidic magmas (ZK1, SZ) to late highly evolved acidic magmas (WE, XC).

As for the alteration, mechanistically, the contrast highlights the spatial variations in fluid/rock ratios and H^+ metasomatism under highly acidic conditions [59]. The continuous trend in most samples suggests a rock-buffered environment where fluid acidity was gradually neutralized by wall-rock feldspars. In striking contrast, the direct transition to pyrophyllite in the WE and CJ samples indicates a fluid-dominated, unbuffered system. Within these specific structural conduits, high-flux, volatile-rich acidic fluids—potentially generated by the disproportionation of magmatic volatiles or fluid phase separation [60,61]—rapidly overwhelmed the wall rock's buffering capacity. These unbuffered fluids leached base cations (K^+, Na^+, Ca^{2+}), bypassing or completely destroying intermediate sericite/muscovite assemblages, and leaving behind highly immobile aluminum and silica to stabilize pyrophyllite. Therefore, this sharp pyrophyllite trend precisely maps the highly focused, high-flux acidic fluid channeling within the core structural dilations of the Jiaojia Fault zone.

In the Rb/30–Hf–Ta*3 tectonic discrimination diagram (Figure 11A), samples of the ZK1 group are mainly distributed in the late and intraplate extensional field, with some points extending to the boundary between the volcanic arc and synorogenic fields. Samples of the SZ group and XC group are concentrated in the synorogenic field, with a few points proximal to the volcanic arc boundary. The WE group exhibits the widest distribution range, spanning the synorogenic, late and intraplate extensional, and within plate fields. The CJ

https://doi.org/10.3390/min16040350

group plots at the transitional position between the synorogenic and late and intraplate extensional fields.

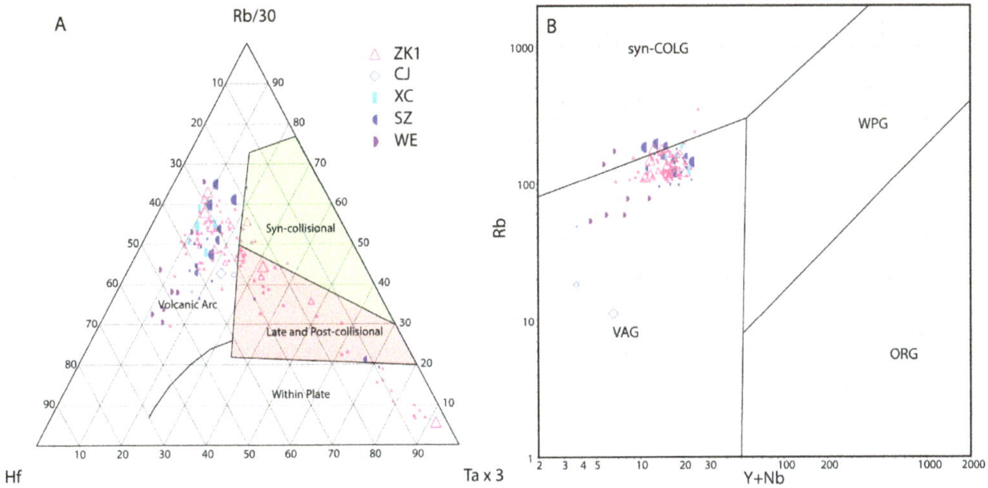

Figure 11. Hf-Ta-Rb discrimination diagram (**A**). The source of the boundary lines or classification fields are based on [62]. (**B**) Yb-Ta discrimination diagram. The source of the boundary lines or classification fields are based on [63]. The size of each sample depicts the enrichment of gold grade. ZK1: drill core 1, SZ: Sizhuang mine, WE: Wang'ershan mine, XC: Xincheng mine, CJ: Caojiabu mine, syn-COLG: syn-collisional granite, WPG: within-plate granite, ORG: ocean-ridge granite, VAG: volcanic-arc granite, Rb: rubidium, Hf: hafnium, Ta: tantalum, Yb: ytterbium.

According to the Rb vs. (Y + Nb) discrimination diagram (Figure 11B) [63], the vast majority of the analyzed samples plot securely within the volcanic arc granite field. This geochemical affinity is consistent with the regional tectonic framework of the Jiaodong Peninsula during the late Mesozoic, which was dominated by the subduction of the Paleo-Pacific plate beneath the North China Craton, forming the regional Linglong and Guojialing magmatic suites [64]. Notably, a small subset of samples deviates upward, crossing the boundary into the synorogenic field. However, these samples do not exhibit significant variations in Y + Nb; rather, their shift is driven exclusively by pronounced rubidium (Rb) enrichment. As pointed out by recent studies, Rb is highly mobile during fluid–rock interactions. Considering the intense phyllic alteration (e.g., extensive sericite and muscovite development) observed petrographically in the Jiaojia Fault zone, this Rb enrichment represents a geochemical artifact caused by fault-driven hydrothermal overprinting [65].

The weak Eu anomalies and relatively homogeneous rare earth element (REE) distribution patterns of the ZK1 and SZ groups suggest a low degree of magmatic differentiation characterized by insignificant plagioclase fractional crystallization. Consistent with the two-stage magmatic evolution of the southern segment, these geochemical features effectively preserve the signatures of the early synorogenic arc magmatism, which predates the subsequent post-orogenic extension. In contrast, the WE and XC groups are characterized by stronger enrichment of large-ion lithophile elements (LILEs), and high silica and high alkalinity features, which reflect intense plagioclase fractional crystallization and a higher degree of magmatic differentiation. These geochemical characteristics indicate that the WE and XC groups belong to highly evolved granites formed in an intraplate extensional environment. The transitional geochemical characteristics of the CJ group imply an intermediate evolutionary stage between the early low-differentiation magmas and the late highly evolved magmas.

https://doi.org/10.3390/min16040350

The pervasively observed LILE enrichment, high-field-strength element (HFSE) depletion, and negative Nb–Ta anomalies are typical geochemical indicators of arc magmas in subduction zones. These features indicate that the formation of the samples is closely related to fluid metasomatism during plate subduction and was affected by crustal contamination or residual slab components.

Geographically, the SZ and XC groups are concentrated in the synorogenic volcanic arc field, which are interpreted as early magmatic products formed in a compressional subduction-collision setting. The ZK1 and CJ groups are distributed in the intraplate extensional transitional field, recording magmatic activities during the intraplate extensional transition stage. Additionally, some samples of the WE group extend toward the within-plate field, which indicates a tectonic background of intraplate extensional collapse.

5.2. Tectonic Setting and Metallogenesis

The ore-forming processes across the Jiaodong gold province exhibit remarkable consistency with the results we find in this research. The homogeneous nature of the ore-forming fluids, as evidenced by fluid inclusion studies [66–70], and the absence of significant mineralogical zoning suggest that fluid source and precipitation mechanisms are unlikely to be primary causes for the observed heterogeneity in gold endowment. The similarities we found in wall rock assemblages, structural settings along the major fault systems, and geochemical characteristics further support this interpretation. Consequently, the heterogeneous gold distribution among different districts is unlikely to be attributable to variations in fluid sources, metal precipitation efficiency, or local structural and lithological conditions ([3] and references therein).

Furthermore, thermotectonic modeling and paleo-geothermal gradient reconstructions indicate a broadly consistent exhumation magnitude across the gold province [71]. Estimates suggest the deposits in the Jiaobei Uplift have been unroofed from depths of 6 ± 2 km [72,73]. This similar degree of post-mineralization erosion on a regional scale effectively rules out variable exhumation levels as a decisive factor for the district-scale differences in gold endowment [71]. Therefore, the fundamental cause for the heterogeneous gold distribution must be sought in processes predating or synchronous with mineralization, rather than in post-ore modifications. These differences are likely influenced by structural inheritance related to first-order tectonic architectures, namely the Tan-Lu and Wulian-Qingdao-Yantai fault systems [74].

Field observations indicate that gold ores in the Jiaobei Uplift are predominantly hosted in low-angle detachment faults, which exhibit listric fault geometries controlled by the Tan-Lu Fault ([3] and references therein). The formation of the giant Jiaodong gold province is widely attributed to an extensional regime linked to a major clockwise rotation of the Izanagi plate at approximately 125 Ma [3,4,75]. Although genetic models remain debated, slab rollback of the subducting Izanagi–Pacific plate is considered a fundamental driver for hydrothermal fluid flow and gold precipitation in the region [73,76,77]. It is proposed that substantial amounts of water, gold, and sulfur—originally derived from the subducting slab—were stored in the serpentinized mantle wedge and released during a transient geodynamic event at around 120 Ma [3,7,11,44].

The Goldfarb and Santosh [78] model for Jiaodong mineralization provides a comprehensive framework that explains several key geological relationships observed in orogenic gold systems globally. This model successfully accounts for the consistent associations of gold deposits with subduction zones and upper crust, the characteristic late-metamorphic timing of mineralization, and the temporal link between auriferous fluid release and the transition from compressional to transgressional tectonics—particularly plausible in scenarios where stalled subduction slabs induce changes in far-field stress regimes.

https://doi.org/10.3390/min16040350

Furthermore, the occurrence of gold deposits at crustal depths deeper than those predicted by supracrustal metamorphic models; lead isotope evidence indicating fluid interaction with basement rocks; and the resolution of apparently conflicting isotopic, halogen, and noble gas signatures suggest contributions from upper to lower crustal and mantle sources.

Notably, the characteristically higher CO_2 content observed in Archean ore fluids can be explained by more extensive carbonation of oceanic crust during the Archean—a consequence of the predominance of high-MgO basalts with greater susceptibility to carbonation, combined with limited CO_2 sinks in contemporary oceans [79]. According to their models, mid-ocean ridge basalts initially carbonated at spreading centers, and during seafloor spreading would subsequently undergo decarbonation when subducted at convergent margins, completing the deep carbon cycle.

In the Jiaojia Fault zone, the study area provides a representative window into the tectono-magmatic evolutionary sequence of the studied segment of the Jiaojia Fault system from early synorogenic volcanic arc magmatism (SZ and XC groups), through intraplate extensional transitional magmatism (ZK1 and CJ groups), to late-stage intraplate extensional highly evolved magmatism (WE group) (Figure 12).

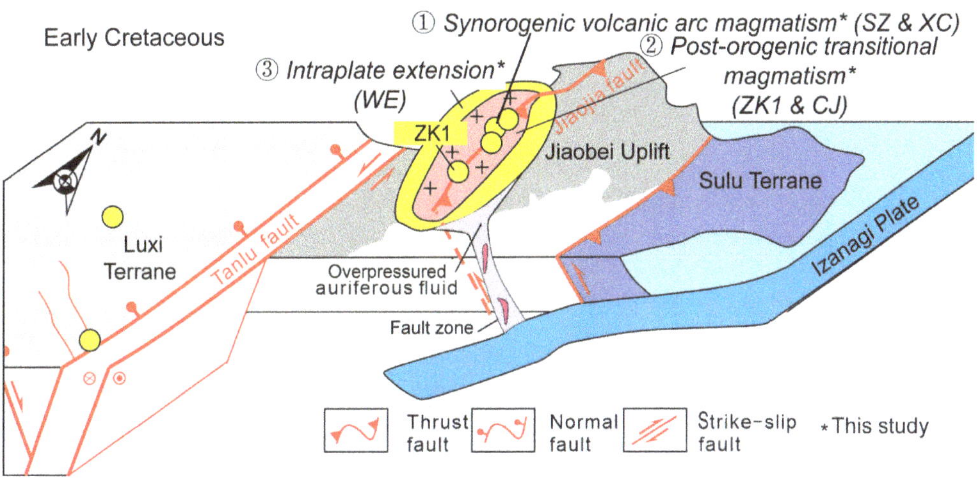

Figure 12. Magmatic evolution of the Jiaojia gold deposit belt during Early Cretaceous. Figure is redrawn and modified after [80]. Yellow circle depicts the sample site. SZ: Sizhuang mine, WE: Wang'ershan mine, XC: Xincheng mine, CJ: Caojiabu mine, ZK1: drill core 1.

This sequence is highly comparable to the Mesozoic tectonic evolution of the Jiaodong region: both record a continuous geodynamic process of "plate subduction–collisional compression, intraplate extensional transition and extensional collapse". During the arc magmatic stage, both are characterized by typical geochemical signatures including enrichment of LILE, depletion of HFSE, and negative Nb–Ta anomalies. During the extensional stage, both developed highly evolved granites with high silica and high alkalinity, and both are closely associated with mineralization.

Geodynamic and structural investigations of the North China Craton (NCC) destruction depict that regional differences in hydrothermal overprinting and metallogenesis across the Jiaojia Fault are controlled by differential structural extensions [81–83]. During the Early Cretaceous, the subduction and subsequent rollback of the Paleo-Pacific plate induced intense lithospheric thinning and widespread continental margin transtension [78,84]. This tectonic regime reactivated pre-existing structural weaknesses, primarily

 https://doi.org/10.3390/min16040350

governed by the far-field kinematic stress of the trans-lithospheric Tan-Lu strike-slip fault system [85].

Seismic and structural data have demonstrated that this crustal extension was highly heterogeneous across the Jiaodong Peninsula [86]. While the northern segment experienced extensional weakening and structural dilation that provided physical permeability for voluminous fluid migration, the southern segment experienced a relatively reduced degree of extensional strain and crustal thinning [86]. Therefore, the intraplate extensional magmatism and superimposed hydrothermal signatures observed in our southern samples are direct manifestations of this heterogeneous structural architecture and reduced extensional permeability.

Nevertheless, differences exist between the two regions in terms of geodynamic drivers, temporal framework of magmatic evolution, and dominant mineralization styles. In this study, we proposed that the tectonic evolution of the Jiaodong region was jointly controlled by the collision between the North China and Yangtze plates and subduction of the Paleo-Pacific plate, with prolonged magmatic activity and characterized by large-scale gold mineralization. In contrast, the study area was likely influenced by a single subduction system, exhibiting a more concentrated evolutionary sequence, and its highly evolved granites are more closely associated with rare metal mineralization.

We propose that the more pronounced extension in the Jiaobei Uplift enhanced crustal permeability and facilitated large-scale fluid circulation and metal precipitation through multiple fluid-cycling events under reduced lithostatic pressure. In contrast, the relatively weaker extensional regime in the Luxi Terrane may have limited such extreme fluid focusing, thereby contributing to regional differences in gold endowment. The crustal-scale faults extend to the Moho discontinuity causing extensive lateral fluid flow, which allows direct channeling of slab-derived metamorphic fluids upward through the crust, eliminating the need for implausibly large-scale lateral fluid migration.

In addition, the intermediate to high-silica and intermediate-potassic compositions of the early arc magmatic stage (SZ and XC groups) in the study area are consistent with the metallogenic settings of arc granites, such as the Linglong and Guojialing plutons in Jiaodong. If accompanied by anomalies in Cu, Mo, and other elements, these units may indicate the potential for porphyry-type Cu-Mo mineralization.

6. Conclusions

(1) Petrographic observations reveal distinct lithofacies associations between the southern and northern segments of the Jiaojia Fault, which indicate different ore-forming processes.

(2) Whole-rock major element geochemistry demonstrates that the study area is characterized by medium- to high-silica and potassium-rich compositions, primarily plotting within the diorite, granodiorite, quartz monzonite, and granite fields. Specifically, the ZK1 group displays a continuous compositional trend, reflecting a transition from intermediate to intermediate-acidic magmatism.

(3) Trace element geochemistry reveals that all samples are enriched in large ion lithophile elements and depleted in high-field-strength elements, with strongly fractionated rare earth element patterns indicative of a subduction-related magma source. Notably, samples from the ZK1 drill hole reflect late- to post-orogenic and intraplate tectonic signatures, in contrast to the volcanic arc affinities observed in the northern Jiaojia Fault area.

(4) Regarding alteration mineralogy, samples from Caojiabu and Wang'ershan exhibit a strong tendency toward pyrophyllite formation, whereas other regional samples are dominated by an alteration assemblage characterized by pronounced muscovite development.

 https://doi.org/10.3390/min16040350

(5) Metallogenically, the northern segment is characterized by a single mineralization event associated with synorogenic island arc magmatism. In contrast, the southern segment records this initial event but is overprinted by a distinct second mineralization stage. This later stage was driven by intraplate extensional transitional magmatism, accompanied by structural reworking and superimposed hydrothermal alteration.

(6) Although the Jiaojia Fault system was influenced by the subduction of the Izanagi and Paleo-Pacific plates, we propose that the observed regional differences across the Jiaodong Peninsula are primarily attributed to reduced extensional weakening in the southern Jiaojia Fault. This differential structural response is interpreted to reflect variations in far-field strain induced by the regional Tan-Lu strike-slip fault system.

Supplementary Materials: The following supporting information can be downloaded at: https://www.mdpi.com/article/10.3390/min16040350/s1, Table S1. Sample Geological and Depth Data.

Author Contributions: Conceptualization, J.J. and Y.H.; methodology, L.G., Y.L. (Yulu Lv), and J.W.; data curation, J.J., X.Y., and Y.L. (Yan Liu); investigation, X.S. and J.L.; funding acquisition, Y.X.; supervision, W.L.; writing, J.J. and M.Z. All authors have read and agreed to the published version of the manuscript.

Funding: This research was funded by the National Key Research and Development Program of China: "Research on Comprehensive Prospecting Technology and Reserve Increase Demonstration of Gold Deposits in the Covered Area Southward Extension of the Jiaojia Fault, Jiaodong" (No. 2022YFC2903605-2), and the Geological Research program of Shandong Gold Group Co., Ltd., China: "Gold Metallogenic System and Prospecting Prediction of the Jiaojia Fault Zone" (No. 23A1001).

Data Availability Statement: The whole-rock element analysis data and gold-concentration data are unavailable due to privacy policy. For further information regarding the data, please contact the Shandong Gold Group Co., Ltd., Jinan, China.

Acknowledgments: Sincere thanks go to Yanmin Dong and Lifang Ren for their technical support with methodology and manuscript preparation. Constructive feedback from the anonymous reviewers and journal editors is also sincerely acknowledged.

Conflicts of Interest: The rock samples in this study were collected with the support of Shandong Gold Geologic Minerals Survey Co., Ltd., Shandong Gold Group Co., Ltd., and Inner Mongolia Shandong Gold Geological and Mineral Exploration Co., Ltd., acting as a liaison with various mining sites. While the company facilitated logistical access and sample collection, the authors maintain full ownership of the resulting data and interpretative figures. All analyses and interpretations were conducted independently by the research team to ensure the transparency and objectivity of the scientific conclusions.

References

1. Goldfarb, R.J.; Mao, J.-W.; Qiu, K.-F.; Goryachev, N. The great Yanshanian metallogenic event of eastern Asia: Consequences from one hundred million years of plate margin geodynamics. *Gondwana Res.* **2021**, *100*, 223–250. [CrossRef]
2. Liu, J.; Ni, J.; Chen, X.; Craddock, J.P.; Zheng, Y.; Ji, L.; Hou, C. Early Cretaceous tectonics across the North Pacific: New insights from multiphase tectonic extension in Eastern Eurasia. *Earth-Sci. Rev.* **2021**, *217*, 103552. [CrossRef]
3. Deng, J.; Yang, L.-Q.; Groves, D.I.; Zhang, L.; Qiu, K.-F.; Wang, Q.-F. An integrated mineral system model for the gold deposits of the giant Jiaodong province, eastern China. *Earth-Sci. Rev.* **2020**, *208*, 103274. [CrossRef]
4. Goldfarb, R.; Qiu, K.; Deng, J.; Chen, Y.; Yang, L. Orogenic gold deposits of China. In *Mineral Deposits of China*; Society of Economic Geologists: Littleton, CO, USA, 2019.
5. Mills, S.E.; Tomkins, A.G.; Weinberg, R.F.; Fan, H.-R. Anomalously silver-rich vein-hosted mineralisation in disseminated-style gold deposits, Jiaodong gold district, China. *Ore Geol. Rev.* **2015**, *68*, 127–141. [CrossRef]

https://doi.org/10.3390/min16040350

6. Song, M.; Wang, B.; Song, Y.; Li, J.; Zheng, J.; Li, S.; Fan, J.; Yang, Z.; He, C.; Gao, M.; et al. Spatial coupling relationship between faults and gold deposits in the Jiaodong ore concentration area and the effect of thermal doming-extension on mineralisation. *Ore Geol. Rev.* **2023**, *153*, 105277. [CrossRef]
7. Qiu, K.-F.; Goldfarb, R.J.; Deng, J.; Yu, H.-C.; Gou, Z.-Y.; Ding, Z.-J.; Wang, Z.-K.; Li, D.-P. Gold deposits of the Jiaodong Peninsula, eastern China. In *Geology of the World's Major Gold Deposits and Provinces*; Society of Economic Geologists: Littleton, CO, USA, 2020.
8. Zhu, G.; Xu, J. Displacement, timing and tectonic model of the Tan-Lu fault zone. In *Structural Geology and Geomechanics*; CRC Press: Boca Raton, FL, USA, 2018; pp. 217–228.
9. Dong, L.; Song, M.; Li, J.; Wang, L.; Chi, N.; Wang, R.; Yang, Z.; Zhao, R. The geology and genesis of the Jiaojia-Xincheng gold district (∼1200 t) in the Jiaodong Peninsula, China: A review. *Ore Geol. Rev.* **2025**, *188*, 107068. [CrossRef]
10. Liu, X.; Wang, G.; Lv, J.; Peng, Y.; Zhao, X. Three-dimensional modeling and geostatistical structural analysis for ore deposit prospecting trend and wisdom mining: A case study in the Jiaoxibei gold field, China. *Front. Earth Sci.* **2024**, *12*, 1217016. [CrossRef]
11. Yu, X.-f.; Li, D.-p.; Tian, J.-x.; Yang, D.-p.; Shan, W.; Geng, K.; Xiong, Y.-x.; Chi, N.-j.; Wei, P.-f.; Liu, P.-r. Deep gold mineralization features of Jiaojia metallogenic belt, Jiaodong gold Province: Based on the breakthrough of 3000 m exploration drilling. *China Geol.* **2020**, *3*, 385–401.
12. Goldfarb, R.; Groves, D.; Gardoll, S. Rotund versus skinny orogens: Well-nourished or malnourished gold? *Geology* **2001**, *29*, 539–542. [CrossRef]
13. Pitcairn, I.K.; Teagle, D.A.; Craw, D.; Olivo, G.R.; Kerrich, R.; Brewer, T.S. Sources of metals and fluids in orogenic gold deposits: Insights from the Otago and Alpine Schists, New Zealand. *Econ. Geol.* **2006**, *101*, 1525–1546. [CrossRef]
14. Wang, Z.; Cheng, H.; Zong, K.; Geng, X.; Liu, Y.; Yang, J.; Wu, F.; Becker, H.; Foley, S.; Wang, C.Y. Metasomatized lithospheric mantle for Mesozoic giant gold deposits in the North China craton. *Geology* **2020**, *48*, 169–173. [CrossRef]
15. Fan, H.R.; Zhai, M.G.; Xie, Y.H.; Yang, J.H. Ore-forming fluids associated with granite-hosted gold mineralization at the Sanshandao deposit, Jiaodong gold province, China. *Miner. Depos.* **2003**, *38*, 739–750. [CrossRef]
16. Deng, J.; Liu, X.; Wang, Q.; Dilek, Y.; Liang, Y. Isotopic characterization and petrogenetic modeling of Early Cretaceous mafic diking—Lithospheric extension in the North China craton, eastern Asia. *Bulletin* **2017**, *129*, 1379–1407. [CrossRef]
17. Phillips, G.; Powell, R. Formation of gold deposits: A metamorphic devolatilization model. *J. Metamorph. Geol.* **2010**, *28*, 689–718. [CrossRef]
18. Cheng, J.; Yang, L.; Wang, Q.; Li, Y.; Zhang, W.; Sun, B.; Dong, X.; Zhao, S.; Li, D.; Zhao, K. Origin, evolution and precipitation mechanism of ore-forming fluids in Jiaojia belt, Jiaodong Peninsula: Insights from trace elements and SFe isotopes of pyrite. *J. Geochem. Explor.* **2026**, *283*, 107969. [CrossRef]
19. Bouzari, F.; Hart, C.J.; Bissig, T.; Barker, S. Hydrothermal alteration revealed by apatite luminescence and chemistry: A potential indicator mineral for exploring covered porphyry copper deposits. *Econ. Geol.* **2016**, *111*, 1397–1410. [CrossRef]
20. Mao, M.; Rukhlov, A.S.; Rowins, S.M.; Spence, J.; Coogan, L.A. Apatite trace element compositions: A robust new tool for mineral exploration. *Econ. Geol.* **2016**, *111*, 1187–1222. [CrossRef]
21. RØnsbo, J.G. Coupled substitutions involving REEs and Na and Si in apatites in alkaline rocks from the Ilimaussaq intrusion, South Greenland, and the petrological implications. *Am. Mineral.* **1989**, *74*, 896–901.
22. Sha, L.-K.; Chappell, B.W. Apatite chemical composition, determined by electron microprobe and laser-ablation inductively coupled plasma mass spectrometry, as a probe into granite petrogenesis. *Geochim. Cosmochim. Acta* **1999**, *63*, 3861–3881. [CrossRef]
23. Pan, Y.; Fleet, M.E. Compositions of the apatite-group minerals: Substitution mechanisms and controlling factors. *Rev. Mineral. Geochem.* **2002**, *48*, 13–49. [CrossRef]
24. Hu, H.-L.; Liu, S.-L.; Fan, H.-R.; Yang, K.; Zuo, Y.; Cai, Y. Structural networks constraints on alteration and mineralization processes in the Jiaojia gold deposit, Jiaodong Peninsula, China. *J. Earth Sci.* **2020**, *31*, 500–513. [CrossRef]
25. Hu, H.-L.; Fan, H.-R.; Lan, T.-G.; Xu, Y.; Cai, Y.-C.; Yang, K.-F.; Dai, Z.-H. New constraints for crustal sulfur contamination of gold source: Evidence from complex δ34S of pyrite in the northwestern Jiaodong gold province, China. *Precambrian Res.* **2022**, *378*, 106773. [CrossRef]
26. Li, X.-H.; Fan, H.-R.; Santosh, M.; Yang, K.-F.; Peng, H.-W.; Hollings, P.; Hu, F.-F. Mesozoic felsic dikes in the Jiaobei Terrane, southeastern North China Craton: Constraints from zircon geochronology and geochemistry, and implications for gold metallogeny. *J. Geochem. Explor.* **2019**, *201*, 40–55. [CrossRef]
27. Tian, Z.; Liu, P.; Wen, F.; Zhu, H.; Wang, W.; Liu, F. Mesoproterozoic accretionary orogenesis: Evidence from ∼ 1.4 Ga metamorphism on the southeastern margin of the North China Craton. *J. Asian Earth Sci.* **2022**, *232*, 105247. [CrossRef]
28. Xu, W.-G.; Fan, H.-R.; Yang, K.-F.; Hu, F.-F.; Cai, Y.-C. Gold mineralizing efficiency during hydrothermal alteration of the Mesozoic granitoids in the northwest Jiaodong Peninsula: Contrasting conditions between the Guojialing and Linglong plutons. *Geochemistry* **2017**, *77*, 387–398. [CrossRef]

 https://doi.org/10.3390/min16040350

29. Yang, L.-Q.; Guo, L.-N.; Wang, Z.-L.; Zhao, R.-X.; Song, M.-C.; Zheng, X.-L. Timing and mechanism of gold mineralization at the Wang'ershan gold deposit, Jiaodong Peninsula, eastern China. *Ore Geol. Rev.* **2017**, *88*, 491–510. [CrossRef]
30. Collins, W.J.; Belousova, E.A.; Kemp, A.I.; Murphy, J.B. Two contrasting Phanerozoic orogenic systems revealed by hafnium isotope data. *Nat. Geosci.* **2011**, *4*, 333–337. [CrossRef]
31. Goldfarb, R.J.; Taylor, R.D.; Collins, G.S.; Goryachev, N.A.; Orlandini, O.F. Phanerozoic continental growth and gold metallogeny of Asia. *Gondwana Res.* **2014**, *25*, 48–102. [CrossRef]
32. Sun, H.; Li, H.; Liu, L.; Chen, Q.; Yang, H.; Wu, P. Exhumation history of the Jiaodong and its adjacent areas since the Late Cretaceous: Constraints from low temperature thermochronology. *Sci. China Earth Sci.* **2017**, *60*, 531–545. [CrossRef]
33. Song, Z.; Liu, H.; Meng, F.; Yuan, X.; Feng, Q.; Zhou, D.; Romaní, J.R.V.; Yan, H. Zircon U-Pb ages and Hf isotopes of Neoproterozoic meta-igneous rocks in the Liansandao area, northern Sulu Orogen, eastern China, and the tectonic implications. *J. Earth Sci.* **2019**, *30*, 1230–1242. [CrossRef]
34. Zhao, R.; Wang, Q.; Liu, X.; Wang, W.; Pan, R. Architecture of the Sulu crustal suture between the North China Craton and Yangtze Craton: Constraints from Mesozoic granitoids. *Lithos* **2016**, *266*, 348–361. [CrossRef]
35. Jahn, B.-m.; Liu, D.; Wan, Y.; Song, B.; Wu, J. Archean crustal evolution of the Jiaodong Peninsula, China, as revealed by zircon SHRIMP geochronology, elemental and Nd-isotope geochemistry. *Am. J. Sci.* **2008**, *308*, 232–269. [CrossRef]
36. Yang, K.-F.; Fan, H.-R.; Santosh, M.; Hu, F.-F.; Wilde, S.A.; Lan, T.-G.; Lu, L.-N.; Liu, Y.-S. Reactivation of the Archean lower crust: Implications for zircon geochronology, elemental and Sr–Nd–Hf isotopic geochemistry of late Mesozoic granitoids from northwestern Jiaodong Terrane, the North China Craton. *Lithos* **2012**, *146*, 112–127. [CrossRef]
37. Zhai, M.-G.; Santosh, M.; Zhang, L. Precambrian geology and tectonic evolution of the North China Craton. *Gondwana Res.* **2011**, *20*, 1–5. [CrossRef]
38. Zhai, M.-G.; Santosh, M. Metallogeny of the North China Craton: Link with secular changes in the evolving Earth. *Gondwana Res.* **2013**, *24*, 275–297. [CrossRef]
39. Yang, Q.; Santosh, M.; Shen, J.; Li, S. Juvenile vs. recycled crust in NE China: Zircon U–Pb geochronology, Hf isotope and an integrated model for Mesozoic gold mineralization in the Jiaodong Peninsula. *Gondwana Res.* **2014**, *25*, 1445–1468. [CrossRef]
40. Zhang, L.; Weinberg, R.F.; Yang, L.-Q.; Groves, D.I.; Sai, S.-X.; Matchan, E.; Phillips, D.; Kohn, B.P.; Miggins, D.P.; Liu, Y. Mesozoic orogenic gold mineralization in the Jiaodong Peninsula, China: A focused event at 120 ± 2 Ma during cooling of pregold granite intrusions. *Econ. Geol.* **2020**, *115*, 415–441. [CrossRef]
41. Jiang, G.; Tang, X.; Rao, S.; Gao, P.; Zhang, L.; Zhao, P.; Hu, S. High-quality heat flow determination from the crystalline basement of the south-east margin of North China Craton. *J. Asian Earth Sci.* **2016**, *118*, 1–10. [CrossRef]
42. Yang, J.-H.; Chung, S.-L.; Zhai, M.-G.; Zhou, X.-H. Geochemical and Sr–Nd–Pb isotopic compositions of mafic dikes from the Jiaodong Peninsula, China: Evidence for vein-plus-peridotite melting in the lithospheric mantle. *Lithos* **2004**, *73*, 145–160. [CrossRef]
43. Dou, J.-Z.; Zhang, H.-F.; Tong, Y.; Wang, F.; Chen, F.-K.; Li, S.-R. Application of geothermo-barometers to Mesozoic granitoids in the Jiaodong Peninsula, eastern China: Criteria for selecting methods of pressure estimation and implications for crustal exhumation. *J. Asian Earth Sci.* **2018**, *160*, 271–286. [CrossRef]
44. Fan, H.-R.; Zhai, M.; Yang, K.-F.; Hu, F.-F. Late Mesozoic gold mineralization in the North China craton. In *Main Tectonic Events and Metallogeny of the North China Craton*; Springer: Berlin/Heidelberg, Germany, 2016; pp. 511–525.
45. Goss, S.C.; Wilde, S.A.; Wu, F.; Yang, J. The age, isotopic signature and significance of the youngest Mesozoic granitoids in the Jiaodong Terrane, Shandong Province, North China Craton. *Lithos* **2010**, *120*, 309–326. [CrossRef]
46. Liang, Y.; Liu, X.; Qin, C.; Li, Y.; Chen, J.; Jiang, J. Petrogenesis of early cretaceous mafic dikes in southeastern Jiaolai basin, Jiaodong Peninsula, China. *Int. Geol. Rev.* **2017**, *59*, 131–150. [CrossRef]
47. Li, Y.; He, H.; Deng, C.; Pan, Y.; Ji, Q.; Wang, C.; Zheng, D.; Zhu, R. 40Ar/39Ar dating results from the Shijiatun Formation, Jiaolai Basin: New age constraints on the Cretaceous terrestrial volcanic-sedimentary sequence of China. *Cretac. Res.* **2018**, *86*, 251–260. [CrossRef]
48. Zhao, R.; Wang, Q.; Deng, J.; Santosh, M.; Liu, X.; Cheng, H. Late Mesozoic magmatism and sedimentation in the Jiaodong Peninsula: New constraints on lithospheric thinning of the North China Craton. *Lithos* **2018**, *322*, 312–324. [CrossRef]
49. Li, S.S.; Qiu, K.F.; Hernández-Uribe, D.; Gao, Y.X.; Santosh, M.; Huang, H.C.; Zheng, Z.Y.; Zhang, Z.L.; Gao, S.C. Water Recycling in the Deep Earth: Insights From Integrated μ-XRF, THz-TDS Spectroscopy, TG, and DCS of High-Pressure Granulite. *J. Geophys. Res. Solid Earth* **2023**, *128*, e2022JB025915. [CrossRef]
50. Middlemost, E.A. Naming materials in the magma/igneous rock system. *Earth-Sci. Rev.* **1994**, *37*, 215–224. [CrossRef]
51. Sun, S.-S.; McDonough, W.F. Chemical and isotopic systematics of oceanic basalts: Implications for mantle composition and processes. *Geol. Soc. Lond. Spec. Publ.* **1989**, *42*, 313–345. [CrossRef]
52. Whalen, J.B.; Currie, K.L.; Chappell, B.W. A-type granites: Geochemical characteristics, discrimination and petrogenesis. *Contrib. Mineral. Petrol.* **1987**, *95*, 407–419. [CrossRef]

 https://doi.org/10.3390/min16040350

53. Pearce, J.A. Geochemical fingerprinting of oceanic basalts with applications to ophiolite classification and the search for Archean oceanic crust. *Lithos* **2008**, *100*, 14–48. [CrossRef]
54. Cabanis, B.; Lecolle, M. The La/10-Y/15-Nb/8 diagram: A tool for distinguishing volcanic series and discovering crustal mixing and/or contamination. *Comptes Rendus l'Academie Sci. Ser. 2 Sci. Terre* **1989**, *309*, 20.
55. Montreuil, J.-F.; Corriveau, L.; Grunsky, E.C. Compositional data analysis of hydrothermal alteration in IOCG systems, Great Bear magmatic zone, Canada: To each alteration type its own geochemical signature. *Geochem. Explor. Environ. Anal.* **2013**, *13*, 229–247. [CrossRef]
56. Large, R.R.; Gemmell, J.B.; Paulick, H.; Huston, D.L. The alteration box plot: A simple approach to understanding the relationship between alteration mineralogy and lithogeochemistry associated with volcanic-hosted massive sulfide deposits. *Econ. Geol.* **2001**, *96*, 957–971. [CrossRef]
57. Ma, Z.; Qiao, Y.; Jing, M. Prospecting potential evaluation based on geological factor analysis—A case study of Jiaojiacheng ore belt. *Front. Earth Sci.* **2024**, *12*, 1464429. [CrossRef]
58. Li, R.; Albert, N.N.; Yun, M.; Meng, Y.; Du, H. Geological and geochemical characteristics of the Archean basement-hosted gold deposit in Pinglidian, Jiaodong Peninsula, eastern China: Constraints on auriferous quartz-vein exploration. *Minerals* **2019**, *9*, 62. [CrossRef]
59. Hemley, J.J.; Jones, W. Chemical aspects of hydrothermal alteration with emphasis on hydrogen metasomatism. *Econ. Geol.* **1964**, *59*, 538–569. [CrossRef]
60. Hedenquist, J.W.; Lowenstern, J.B. The role of magmas in the formation of hydrothermal ore deposits. *Nature* **1994**, *370*, 519–527. [CrossRef]
61. Sillitoe, R.H. Porphyry copper systems. *Econ. Geol.* **2010**, *105*, 3–41. [CrossRef]
62. Harris, N.B.; Pearce, J.A.; Tindle, A.G. Geochemical characteristics of collision-zone magmatism. *Geol. Soc. Lond. Spec. Publ.* **1986**, *19*, 67–81. [CrossRef]
63. Pearce, J.A.; Harris, N.B.; Tindle, A.G. Trace element discrimination diagrams for the tectonic interpretation of granitic rocks. *J. Petrol.* **1984**, *25*, 956–983. [CrossRef]
64. Yang, L.-Q.; Dilek, Y.; Wang, Z.-L.; Weinberg, R.F.; Liu, Y. Late Jurassic, high Ba–Sr Linglong granites in the Jiaodong Peninsula, east China: Lower crustal melting products in the eastern north China craton. *Geol. Mag.* **2018**, *155*, 1040–1062. [CrossRef]
65. McCuaig, T.C.; Kerrich, R. *P—T—t*—Deformation—Fluid characteristics of lode gold deposits: Evidence from alteration systematics. *Ore Geol. Rev.* **1998**, *12*, 381–453. [CrossRef]
66. Groves, D.I.; Santosh, M.; Goldfarb, R.J.; Zhang, L. Structural geometry of orogenic gold deposits: Implications for exploration of world-class and giant deposits. *Geosci. Front.* **2018**, *9*, 1163–1177. [CrossRef]
67. Hannington, M.; Harðardóttir, V.; Garbe-Schönberg, D.; Brown, K.L. Gold enrichment in active geothermal systems by accumulating colloidal suspensions. *Nat. Geosci.* **2016**, *9*, 299–302. [CrossRef]
68. Henley, R.W.; King, P.L.; Wykes, J.L.; Renggli, C.J.; Brink, F.J.; Clark, D.A.; Troitzsch, U. Porphyry copper deposit formation by sub-volcanic sulphur dioxide flux and chemisorption. *Nat. Geosci.* **2015**, *8*, 210–215. [CrossRef]
69. Richards, J.P.; Mumin, A.H. Magmatic-hydrothermal processes within an evolving Earth: Iron oxide-copper-gold and porphyry Cu $\pm$ Mo $\pm$ Au deposits. *Geology* **2013**, *41*, 767–770. [CrossRef]
70. Wilkinson, J.J. Triggers for the formation of porphyry ore deposits in magmatic arcs. *Nat. Geosci.* **2013**, *6*, 917–925. [CrossRef]
71. Qiu, K.-F.; Deng, J.; Sai, S.X.; Yu, H.C.; Tamer, M.T.; Ding, Z.J.; Yu, X.F.; Goldfarb, R. Low-Temperature Thermochronology for Defining the Tectonic Controls on Heterogeneous Gold Endowment Across the Jiaodong Peninsula, Eastern China. *Tectonics* **2023**, *42*, e2022TC007669. [CrossRef]
72. Liu, X.; Fan, H.-R.; Evans, N.J.; Yang, K.-F.; Danišík, M.; McInnes, B.I.; Qin, K.-Z.; Yu, X.-F. Exhumation history of the Sanshandao Au deposit, Jiaodong: Constraints from structural analysis and (U-Th)/He thermochronology. *Sci. Rep.* **2017**, *7*, 7787. [CrossRef] [PubMed]
73. Zhang, L.; Groves, D.I.; Yang, L.-Q.; Wang, G.-W.; Liu, X.-D.; Li, D.-P.; Song, Y.-X.; Shan, W.; Sun, S.-C.; Wang, Z.-K. Relative roles of formation and preservation on gold endowment along the Sanshandao gold belt in the Jiaodong gold province, China: Importance for province-to district-scale gold exploration. *Miner. Depos.* **2020**, *55*, 325–344. [CrossRef]
74. Hao, T.Y.; Suh, M.; Wang, Q.S.; Choi, S.; JIiang, W.W.; Song, H.B.; Yan, X.W.; Liu, J.H.; Yao, C.L. A study on the extension of fault zones in Yellow Sea and its adjacent areas based on gravity data. *Chin. J. Geophys.* **2002**, *45*, 393–410. [CrossRef]
75. Li, J.-W.; Bi, S.-J.; Selby, D.; Chen, L.; Vasconcelos, P.; Thiede, D.; Zhou, M.-F.; Zhao, X.-F.; Li, Z.-K.; Qiu, H.-N. Giant Mesozoic gold provinces related to the destruction of the North China craton. *Earth Planet. Sci. Lett.* **2012**, *349*, 26–37. [CrossRef]
76. Engebretson, D.C.; Cox, A.; Gordon, R.G. *Relative Motions Between Oceanic and Continental Plates in the Pacific Basin*; Geological Society of America: Boulder, CO, USA, 1985; Volume 206.
77. Sager, W.W. Cretaceous paleomagnetic apparent polar wander path for the Pacific plate calculated from Deep Sea Drilling Project and Ocean Drilling Program basalt cores. *Phys. Earth Planet. Inter.* **2006**, *156*, 329–349. [CrossRef]

78. Goldfarb, R.J.; Santosh, M. The dilemma of the Jiaodong gold deposits: Are they unique? *Geosci. Front.* **2014**, *5*, 139–153. [CrossRef]
79. Santosh, M.; Tsunogae, T.; Ohyama, H.; Sato, K.; Li, J.; Liu, S. Carbonic metamorphism at ultrahigh-temperatures: Evidence from North China Craton. *Earth Planet. Sci. Lett.* **2008**, *266*, 149–165. [CrossRef]
80. Groves, D.I.; Santosh, M. The giant Jiaodong gold province: The key to a unified model for orogenic gold deposits? *Geosci. Front.* **2016**, *7*, 409–417. [CrossRef]
81. Zhu, R.; Xu, Y.; Zhu, G.; Zhang, H.; Xia, Q.; Zheng, T. Destruction of the North China craton. *Sci. China Earth Sci.* **2012**, *55*, 1565–1587. [CrossRef]
82. Zhu, R.; Fan, H.; Li, J.; Meng, Q.; Li, S.; Zeng, Q. Decratonic gold deposits. *Sci. China Earth Sci.* **2015**, *58*, 1523–1537. [CrossRef]
83. Deng, J.; Wang, Q. Gold mineralization in China: Metallogenic provinces, deposit types and tectonic framework. *Gondwana Res.* **2016**, *36*, 219–274. [CrossRef]
84. Li, J.-W.; Vasconcelos, P.M.; Zhang, J.; Zhou, M.-F.; Zhang, X.-J.; Yang, F.-H. 40Ar/39Ar constraints on a temporal link between gold mineralization, magmatism, and continental margin transtension in the Jiaodong gold province, eastern China. *J. Geol.* **2003**, *111*, 741–751. [CrossRef]
85. Deng, J.; Liu, X.; Wang, Q.; Pan, R. Origin of the Jiaodong-type Xinli gold deposit, Jiaodong Peninsula, China: Constraints from fluid inclusion and C–D–O–S–Sr isotope compositions. *Ore Geol. Rev.* **2015**, *65*, 674–686. [CrossRef]
86. Yu, G.; Xu, T.; Ai, Y.; Chen, L.; Yang, J. Significance of crustal extension and magmatism to gold deposits beneath Jiaodong Peninsula, eastern North China Craton: Seismic evidence from receiver function imaging with a dense array. *Tectonophysics* **2020**, *789*, 228532. [CrossRef]

 https://doi.org/10.3390/min16040350

minerals

MDPI

Article

Trace-Element Systematics and Multivariate Insights into Gold Fertility of Arsenopyrite from the Um Rus Orogenic Gold Deposit, Central Eastern Desert, Egypt

ElSayed A. Saber [1,*], H. M. Hamouda [2,*], A. S. Hamid [3] and Ahmed A. El-Sheikh [1]

1 Department of Geology, Faculty of Science, Sohag University, Sohag 82524, Egypt; ahmed_abdelrashid@science.sohag.edu.eg

2 Department of Management Information Systems, College of Business and Economics, Qassim University, Buraidah 52571, Saudi Arabia

3 Department of Physics, College of Science, Qassim University, Buraidah 51452, Saudi Arabia; asabrahiem@qu.edu.sa

* Correspondence: elsayedahmed@science.sohag.edu.eg (E.A.S.); h.hamouda@qu.edu.sa (H.M.H.)

Abstract

Arsenopyrite (FeAsS) is an important sulfide that holds gold in orogenic systems. Its arsenic content is often used as a proxy for gold fertility. However, arsenopyrite from the Um Rus gold deposit in Egypt's Central Eastern Desert shows a complicated gold distribution that makes simple Au-As correlations hard to make. Integrated electron microprobe analysis (EMPA), laser ablation ICP-MS, and principal component analysis (PCA) reveal three unique textural and geochemical domains. Fine-grained arsenopyrite inclusions within pyrite aggregates (28–31 at% As) are devoid of detectable gold; PCA elucidates 84% of their variance through Fe–S versus Co-As substitution (PC1: 61.8%) and Pb-decoupled variability (PC2: 22.2%), suggesting crystallization from a Co-rich, Au-poor fluid. On the other hand, coarse oscillatory-zoned arsenopyrite can hold up to 6154 ppm of invisible gold. This is because of a moderate Au-As substitution ($R = 0.41063$, $p = 0.08074$) that was overprinted by a separate Au-Ag-Sb-Te hydrothermal pulse (Au–Ag: $R = 0.97762$; Au–Sb: $R = 0.97608$). PCA finds four parts (72.8% variance): Ag-Cu-As associations (PC1: 25.1%), Te versus Bi-Au signatures (PC2: 17.8%), Fe–S stoichiometry (PC3: 17.1%), and an Au versus Pb-decoupled event (PC4: 12.9%). This shows that minerals formed in more than one stage. Irregular As-rich overgrowths, containing ≤950 ppm gold and lacking significant Au–As correlation ($R = -0.14011$, $p = 0.56726$), show PCA (74.3% variance) that highlights S-As contrasts (PC1: 25.2%), Co-Ni enrichment (PC2: 18.8%), Cu-Fe-Ni associations (PC3: 16.2%), and a late Au-decoupled event (PC4: 14.2%), indicating barren recrystallization. These results show that just adding arsenic is not a good way to tell if gold is fertile. The highest amounts of invisible gold, on the other hand, are found in oscillatory-zoned domains with Ag-Sb-Te signatures. This research highlights the importance of combining PCA, geochemistry, and microtextures to differentiate auriferous from barren arsenopyrite, thereby enhancing exploration methodologies for structurally intricate orogenic gold systems.

Keywords: Um Rus gold deposit; statistical model; Central Eastern Desert; Egypt; arsenopyrite; electron microprobe analysis (EMPA); laser ablation ICP-MS; principal component analysis (PCA)

1. Introduction

Arsenopyrite (FeAsS) is a significant gold-bearing sulfide in orogenic and hydrothermal gold systems, capable of containing both invisible (lattice-bound) gold and acting as a structural control for visible (free) gold precipitation [1–6]. The fact that it can take in gold through coupled substitution mechanisms, like $Au^{+} + As^{3-} \leftrightarrow Fe^{2+} + S^{2-}$, has led many people to believe that a lot of arsenic in arsenopyrite is a good sign that it is rich in gold [7–10]. The "arsenic fence" model is now a key part of exploration strategies. It uses As anomalies in sulfides to show where gold might be found [11–14]. Nevertheless, increasing evidence indicates that the association between As and Au in arsenopyrite is not universal. In numerous deposits, the distribution of gold is heterogeneous at the microscale, with enrichment concentrated in specific textural domains such as oscillatory zones, fractures, or recrystallized rims, rather than exhibiting a uniform correlation with arsenic content [15–17]. Additionally, late-stage fluid events, deformation, and element partitioning during recrystallization can separate Au from As, resulting in domains that are rich in As but low in gold [18]. These complexities reveal the inadequacies of univariate exploration models and emphasize the necessity for integrated textural and multivariate geochemical methodologies to elucidate the authentic paragenetic history of gold mineralization. Orogenic gold deposits are one of the most important sources of the global gold production and have been extensively studied over the past three decades [19–24]. These deposits are typically located within the convergent margin settings and are structurally controlled, metamorphic devolatilization, and multistage mineralization processes. Gold deposition is commonly associated with the transcrustal fluid systems and deformation-pathway permeability within major shear zones, leading to complex temporal and spatial variations in metal distribution. Principal Component Analysis (PCA) is a powerful multivariate tool for making sense of complicated geochemical datasets. It is commonly used in mineralogy to find elemental associations and paragenetic sequences [25–27]. PCA is different from traditional bivariate plots because it breaks down high-dimensional data into principal components. This can show hidden patterns that univariate methods miss, like fluid-driven zoning or alteration events, primary growth vs. overgrowth, or lattice substitution vs. micro inclusion enrichment. In arsenopyrite research, PCA has been employed to differentiate gold-bearing from barren domains and correlate trace element signatures with mineralization stages, thereby enhancing the comprehension of ore-forming processes [28]. The Um Rus gold deposit, situated in the Central Eastern Desert of Egypt within the Neoproterozoic Arabian Nubian Shield, is a structurally controlled orogenic system that emerged during late Pan-African tectonism. Gold mineralization takes place in auriferous quartz–carbonate veins and disseminated sulfides, with arsenopyrite and pyrite being the most common minerals and chalcopyrite, pyrrhotite, sphalerite, and galena being less common [29]. Mineralization is closely linked to a mixed contact zone between tonalite and granodiorite intrusions and gabbroic rocks. In this zone, NE–SW trending faults and shear zones acted as paths for hydrothermal fluids [30,31]. The interaction of magmatism, deformation, and hydrothermal activity created an optimal thermal and structural setting for ore deposition, rendering Um Rus a significant site for investigating the relationships among intrusion-related processes, structural controls, and gold enrichment [32,33]. Using high-resolution electron microprobe analysis (EMPA) and principal component analysis (PCA), this study looks at the trace element chemistry of arsenopyrite from the Um Rus deposit to figure out how gold gets into different types of textures. Two distinct generations are identified: (1) fine-grained arsenopyrite (50–250 μm), encapsulated within pyrite aggregates and lacking detectable gold, and (2) coarse-grained arsenopyrite (up to 4.5 mm), characterized by oscillatory zoning and irregular As-rich overgrowths. These different textures make a natural lab for testing models of how trace elements get into rocks and how

https://doi.org/10.3390/min16050439

fertile gold is in orogenic systems. To tackle the intricacy of gold distribution, we present the subsequent essential questions: (1) How is the zoning of trace elements in arsenopyrite related to the amount and spread of gold? (2) What do the multivariate patterns that PCA shows tell us about how metals are added to paragenetic evolution and how they are added? (3) Is the traditional Au-As correlation applicable to all textural domains, or does it fail under certain circumstances? This study aims to describe zoning patterns, find geochemical fingerprints of gold enrichment, and improve exploration models for orogenic gold systems by combining microtextural observations with in situ geochemistry and multivariate analysis. These results show that simple Au-As correlations do not always work and that we need to understand mineralization in a more complex, process-based way.

2. Geological Setting

The Um Rus gold mine is in Egypt's Central Eastern Desert (Figure 1A). The area is geologically complex because of the interaction of magmatic, volcaniclastic, and hydrothermal processes. A composite intrusive system lies beneath the area. It is made up of a younger granite pluton that is shaped like a dome and intrudes into a semicircular younger gabbro body (Figure 1B). This body then cuts through the volcaniclastic country rocks of the Wadi Mubarak Belt [29,34]. The Um Rus younger granite pluton is the main part of this intrusive system. It is a round body about 2.7 km wide that is mostly made of granodiorite with some tonalitic phases (Figure 2A). It has a lot of apophyses, offshoots, and xenoliths (Figure 2B) that came from the gabbroic wall rocks around it [35]. Radiometric dating indicates that this granitoid phase was emplaced 643 $\pm$ 9 Ma [23], identifying it as an early, relatively undifferentiated component of the Pan-African younger granite suite [36]. The younger gabbro unit surrounds the granite core and makes up a semicircular intrusion that is about 8 km wide (Figure 2C). It has inward zoning with clear magmatic layering, going from olivine gabbros and troctolites at the bottom to more evolved gabbroic units with anorthositic streaks (Figure 2D). The outside is fine-grained and cold, while the inside is medium- to coarse-grained. A major part of the Um Rus area is the creation of a hybrid contact zone between the granodiorite and gabbro, which is about 200 m wide on average (Figure 2E). This zone shows how magma from different sources can mix and mingle. On the granitic side, it goes from quartz diorite and diorite to meladiorite (altered gabbro) on the gabbroic side. Field evidence and geochronological data indicate that the granodiorite intruded a partially molten gabbro body, resulting in a coherent intrusive complex with emplacement ages ranging from 573 to 615 Ma [36,37]. The intrusive complex is cut by many dykes, which range in composition from acidic (dacite, rhyolite (Figure 2F) to intermediate (porphyrite) and mafic to ultramafic (lamprophyres Figure 2G). These dykes, which are mostly found in the eastern part of the pluton, are between 0.5 and 4 m wide and up to 2 km long. Based on cross-cutting relationships, felsic dykes that run from N-S to NNE are older than mafic and lamprophyric dykes that run from N-S to NNW. Several fault systems cut through the area, with the main trend going from NE to SW. These faults, especially the ones near the southeastern granodiorite contact, dip gently (about 40 °E) and go on for more than 4 km. They have fault zones that are several meters wide and are often marked by intense fracturing. They are very important for finding hydrothermal alteration and gold-bearing quartz veins and mineralized lodes (Figure 2H,I).

https://doi.org/10.3390/min16050439

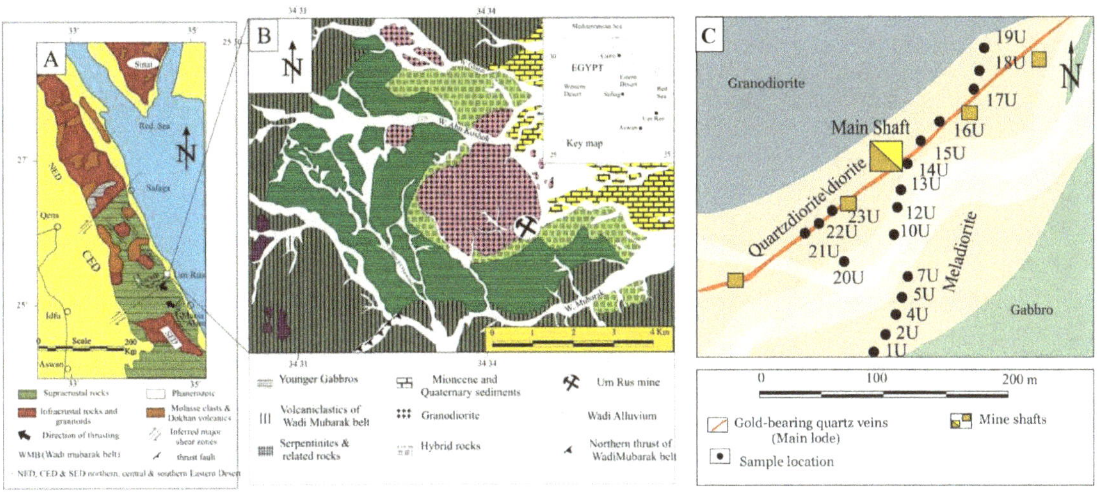

Figure 1. (**A**) Location of Um Rus area in a geological map of Egyptian basement rocks complied from [38,39]. (**B**) Geological map of the Um Rus gold mine area showing tonalite–granodiorite intrusion in the Gabbro rocks, modified after [29,30]. (**C**) A simplified geological map of Um Rus gold mine area showing the main gold-bearing quartz vein and the location of samples for mineralogical studies.

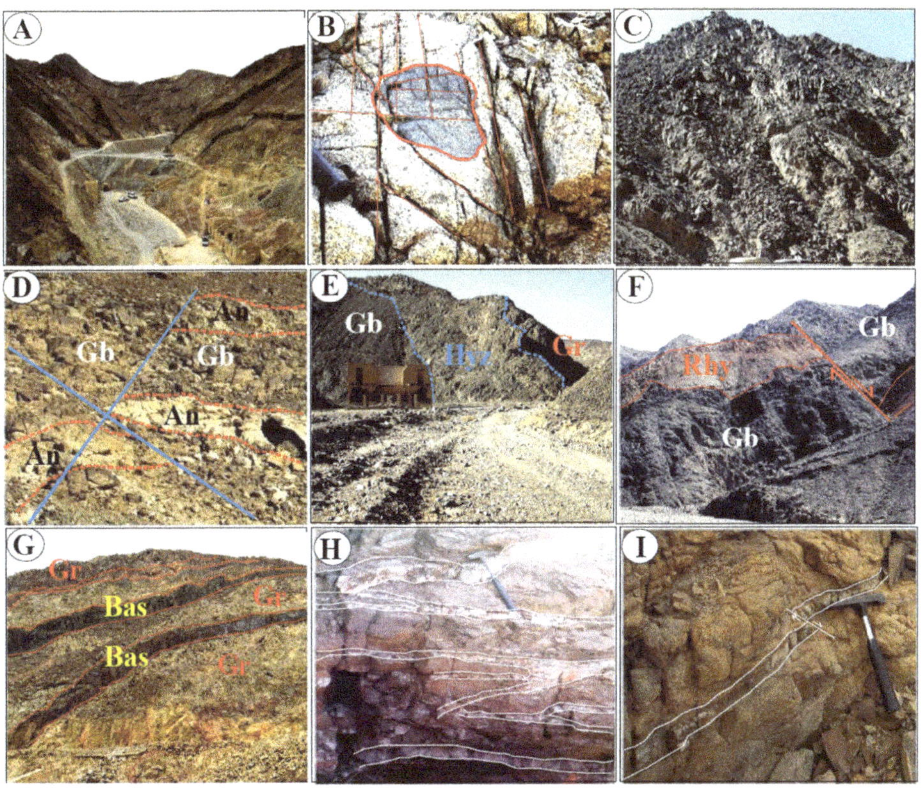

Figure 2. (**A**) Open adit of the Um Rus gold mine, Central Eastern Desert, Egypt, exposing the hybrid intrusive contact zone between tonalite–granodiorite and gabbroic units, with mineralized quartz–

 https://doi.org/10.3390/min16050439

carbonate, highlighting structural and lithological controls on orogenic gold mineralization. (**B**) Xenoliths of older country rocks incorporated within the tonalite and granodiorite intrusions at the Um Rus area. These xenoliths were engulfed during magma emplacement and preserved as rounded to angular fragments within the intrusive bodies. Their occurrence provides evidence for magma–country rock interaction and highlights the hybrid nature of the intrusive contact zone that played a key role in localizing gold mineralization. (**C**) General view of gabbroic rocks at the Um Rus area forming a terrain of relatively moderate relief, these gabbros form the country rocks into which the younger granodiorite intruded, giving rise to a hybrid contact zone that exerted significant control on hydrothermal alteration and gold mineralization. (**D**) Anorthosite lenticular body concordant with the NNE-SSW banding trend of the meta-gabbro mass at the Um Rus gold mine, Central Eastern Desert, Egypt, highlighting its leucocratic texture and structural alignment with gold-mineralized quartz–carbonate veins. (**E**) Hybrid intrusive contact zone of gabbro, tonalite, and granodiorite rocks near the Um Rus gold mine, Central Eastern Desert, Egypt, showcasing transitional quartz diorite, xenoliths, and gold-mineralized quartz–carbonate veins controlled by shear-related fracturing. (**F**) NE–SW trending acidic dikes crosscutting the gabbroic rocks at Um Rus. These felsic dikes, composed mainly of dacite to rhyolite, are steeply dipping and represent late-stage intrusive phases that truncate the magmatic banding of the gabbro. Their orientation reflects the regional NE–SW structural fabric, which also played a significant role in controlling fluid circulation and gold-bearing quartz–carbonate vein emplacement. (**G**) NE-SW parallel basic dikes cutting through tonalite–granodiorite rocks at the Um Rus gold mine, Central Eastern Desert, Egypt, highlighting mafic intrusions aligned with regional shear zones and associated with gold-mineralized quartz–carbonate veins. (**H**) NEE-SWW trending discontinuous quartz veins cutting tonalite–granodiorite rocks near the Um Rus gold mine, Central Eastern Desert, Egypt, showcasing boudinaged and brecciated veins with sulfide mineralization and hydrothermal alteration, controlled by regional shear-related fracturing. (**I**) Sericitization of tonalite–granodiorite rocks in the alteration zone near the Um Rus gold mine, Central Eastern Desert, Egypt, displaying fine-grained sericite replacing feldspars in halos around gold-mineralized quartz–carbonate veins, associated with regional shear zones. Abbreviations; Gb = gabbro, Gr = granite, Hyz = Hybrid intrusive contact, Rhy = rhyolite, Bas = Basalt.

Arsenopyrite in the Um Rus orogenic gold deposit is restricted to the quartzdiorite–diorite side of the hybrid contact zone, where it occurs within phyllic to argillic alteration halos and gold-bearing quartz veins. These veins are structurally controlled by NE–SW to NNE–SSW shear fractures and faults related to reactivation of the Najd Fault System (D3), which acted as conduits for hydrothermal fluids. This spatial confinement, distinct from earlier magmatic sulfides (pyrrhotite-I $\pm$ pentlandite) in the meladiorite, indicates that the trace-element systematics of the analyzed arsenopyrite reflect the fertile hydrothermal events responsible for gold deposition within structurally prepared traps.

3. Materials and Methods

3.1. Study Area and Sample Collection

The samples that were studied were taken from the Um Rus gold deposit, which is in the Central Eastern Desert of Egypt at about 25°27′52″ N and 34°34′54″ E, about 54 km northwest of Marsa Alam city (Figure 1A). You can get to the site by taking a 10 km desert road that branches off the coastal highway near Marsa Mubarak harbor on the Red Sea. Total 18 representative sulfide-bearing samples (labels (1U–23U) were collected from quartz veins, alteration zones, and hybrid rocks across the mineralized zone for mineralogical and geochemical investigations (Figure 1C). The collected samples represent different textural and paragenetic stages of arsenopyrite formation.

https://doi.org/10.3390/min16050439

3.2. Petrographic Analysis

We used reflected light microscopy and scanning electron microscopy (SEM) to look at sub-microscopic features and get high-resolution back-scattered electron (BSE) images that show compositional zoning within sulfide minerals based on differences in atomic number (Z-contrast). The Institute of Mineralogy and Petrology at the University of Graz in Austria used a JEOL JSM-6310 SEM (JEOL Ltd., Tokyo, Japan) to do the tests. The tool has an Energy Dispersive X-ray Spectroscopy (EDS) system and a MICROSPEC Wavelength Dispersive Spectroscopy (WDS) system that can be used for both qualitative and semi-quantitative elemental analysis.

3.3. Electron Probe Microanalysis (EPMA)

Using a JEOL JXA-8800 electron probe microanalyzer (JEOL Ltd., Tokyo, Japan) at the Center for Cooperative Research at Kanazawa University in Japan, we got quantitative elemental analyses and maps of the distribution of arsenopyrite and other sulfide minerals. The analytical conditions consisted of an accelerating voltage of 25 kV and a probe current of 20 nA. EPMA was employed to quantify both major and minor elements and to facilitate internal standardization for LA-ICP-MS trace element analysis.

3.4. Laser Ablation Inductively Coupled Plasma Mass Spectrometry (LA-ICP-MS)

LA-ICP-MS was used at the Incubation Business Laboratory Center at Kanazawa University in Japan to find the concentrations of trace elements like gold (Au), cobalt (Co), nickel (Ni), copper (Cu), silver (Ag), antimony (Sb), and bismuth (Bi). A 193 nm ArF excimer laser (GeoLas Q-plus) (Coherent Inc., Santa Clara, CA, USA) and an Agilent 7500s ICP-MS (Coherent Inc., Santa Clara, CA, USA) were used to do the analyses. The ablation parameters were a spot size of 30 μm for fine-grained sulfides and 50 μm for coarse grains, a repetition rate of 5 Hz, an energy density of 8 J/cm^2 per pulse, and an ablation time of 50 s. NIST SRM 612 and 614 glasses were used to calibrate the system. The Co content measured by EPMA was used as an internal standard, and Si, Al, and Ca were used as extra reference elements.

The following are the detection limits (DL) and average concentrations for some trace elements in NIST standards.

The complete raw geochemical dataset, including all EPMA and LA-ICP-MS analyses used for the statistical evaluations, is provided in Table S1 of the Supplementary Materials.

3.5. Principal Component Analysis (PCA)

Principal Component Analysis (PCA) was utilized on the EPMA dataset to discern trace element correlations and paragenetic sequences within arsenopyrite. The dataset comprised analyses of 12 elements (S, Fe, Co, Ni, Cu, Zn, As, Ag, Te, Au, Pb, Bi) from both coarse-grained and fine-grained arsenopyrite. We used Minitab 18 to standardize the data by taking the mean and dividing it by the standard deviation. We kept the parts that had eigenvalues greater than 1 (the Kaiser criterion).

4. Results

4.1. Mineralogical and Textural Observations

Arsenopyrite at Um Rus is only found in the quartz diorite and where it meets gabbroic rocks. It is mostly found in NNE-SSW and N-S quartz veins and alteration zones within the granodiorite pluton and hybrid zone. Petrographic analysis and Backscattered Electron (BSE) imaging showed that there are two different types of arsenopyrite. They are different in size, texture, host environment, and how they relate to gold mineralization. Fine-grained arsenopyrite (50–250 μm) can be found as subhedral to anhedral inclusions in older pyrite

 https://doi.org/10.3390/min16050439

aggregates, usually in sericitic and carbonatized alteration haloes (Figure 3A,B). These crystals often occur as isolated or in groups along microfractures, forming a fine-grained dispersed texture. In BSE images, these grains do not have any visible gold and show weak compositional zoning, with brighter edges suggesting slight As enrichment.

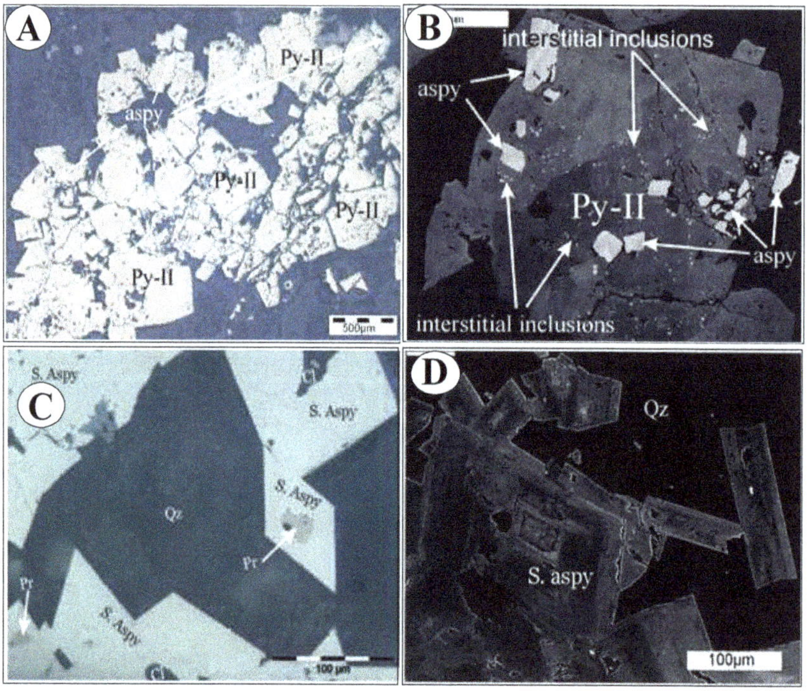

Figure 3. (**A**) Photomicrograph showing the monomineralic aggregates of pyrite-II (Py-II) with arsenopyrite inclusions. (**B**) Back-scattered electron (BSE) image in pyrite-II (Py-II) showing interstitial inclusions of arsenopyrite. (**C**) Photomicrograph showing small arsenopyrite type (S. Aspy) forming rhombic euhedral crystals with inclusions of pyrrhotite (pr), and chlorite (Cl). P.P.L. Reflected light. (**D**) High-contrast (BSE) image showing broad zoning in small arsenopyrite (S. Aspy) due to slight variation in As-content.

Sometimes, fine-grained arsenopyrite has tiny bits of galena inside it or growing on it. Chlorite, ilmenite, and pyrrhotite are also often found in the As-poor core zones (Figure 3C). Back-scattered electron (BSE) images show that the internal cores are wide and have less As, while the outer rims are thinner and have more As (Figure 3D).

On the other hand, coarse-grained arsenopyrite makes larger, well-formed euhedral to subhedral crystals that can be up to 4.5 mm long (Figure 4A). These crystals can also be twinned and deformed, and they have complex textures like oscillatory zoning and patches with a lot of As (Figure 4B). Oscillatory-zoned domains have zones that change rhythmically (0.5–10 μm wide) based on changes in the amount of As and S they contain. BSE images show bands of light and dark that change rhythmically in composition. The brighter zones are arsenic-rich domains, and the darker bands are relatively As-poor (Figure 4B). These main growth zones are covered by uneven patches of As-rich material that tend to form along fractures, grain boundaries, and deformed edges. These patches are often seen with gold, especially when they come into contact with carbonate inclusions or pyrite (Figure 4C,D). Pyrite inclusions and sulfides that go with them, like pyrrhotite, chalcopyrite, and galena, are common.

https://doi.org/10.3390/min16050439

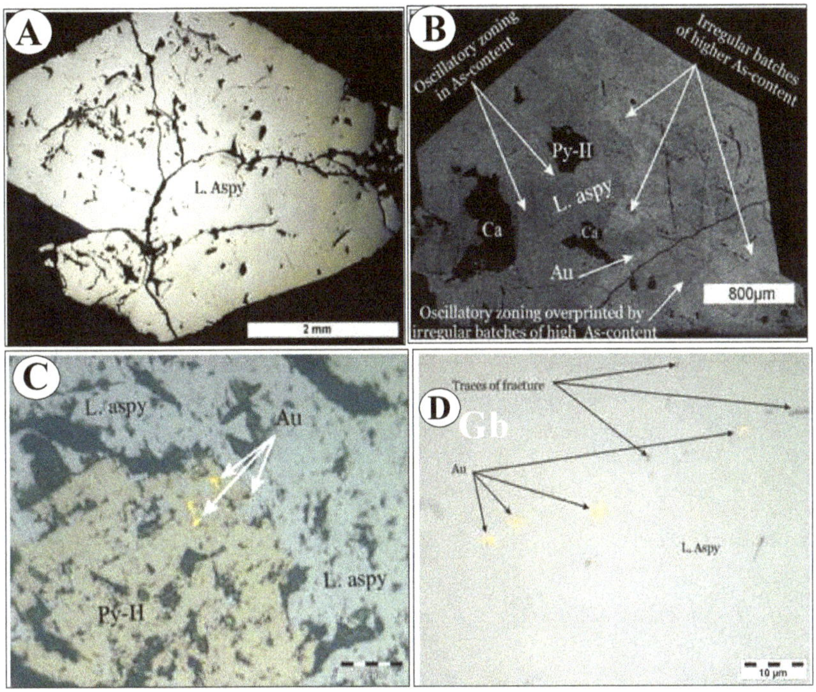

Figure 4. (**A**) S large grain of late arsenopyrite (L. Aspy) from the Um Rus deposit. The crystal is anhedral to subhedral and displays prominent brittle deformation features, including multiple microfractures and veinlets that dissect the grain. (**B**) Backscattered electron (BSE) image of late arsenopyrite (L. aspy) from the Um Rus deposit, showing oscillatory zoning in As content overprinted by irregular patches of higher As concentrations. The arsenopyrite is associated with pyrite-II (Py-II), carbonate (Ca), and native gold (Au). (**C**) Photomicrograph under reflected plane-polarized light showing gold (Au) at the contact between large arsenopyrite (L. Aspy) and a relict inclusion of pyrite-II (Py-II). (**D**) Photomicrograph under reflected plane-polarized light showing gold (Au) occurrences in large arsenopyrite (L. Aspy) near and along fractures.

4.2. Geochemical Results

According to the texture of the identified arsenopyrite types, the geochemical characteristics were investigated using EPMA and LA-ICP-MS analyses. Distinct compositional differences are observed between fine-grained, oscillatory-zoned, and As-rich overgrowth domains, particularly in terms of trace element distribution and gold content.

4.2.1. EPMA and LA-ICP-MS Geochemistry

Electron microprobe (EPMA) analyses reveal that fine-grained arsenopyrite displays significant arsenic variability, ranging from 28.23 to 31.07 atomic percent (at%), while iron (Fe) and sulfur (S) contents fluctuate between 28.5–31.8 at% and 30.1–33.4 at%, respectively. Even with this change in As, neither EPMA nor LA-ICP-MS found any gold (<100 ppm) in multiple grains. These crystals are instead rich in cobalt (up to 3.37 wt%), especially in areas with little As near the outer zones. This suggests that Co replaces Fe in the lattice when the crystals first form, possibly from a fluid that is rich in Co and poor in Au.

Figure 5 shows how arsenic (As) changes with sulfur (S) and iron (Fe) in fine-grained arsenopyrite from the Um Rus deposit. The data show a consistent negative correlation between As and S. This means that when As levels go up, S levels go down in the crystal lattice. On the other hand, it does not seem to have a consistent relationship with Fe, which

https://doi.org/10.3390/min16050439

suggests that iron stays pretty stable across the range of compositions of the fine-grained arsenopyrite. This geochemical behavior suggests that the mechanisms of substitution in these arsenopyrite grains are mainly governed by As-S exchange rather than As-Fe interactions. Such patterns are in line with the fact that arsenopyrite is fine-grained and usually forms quickly when conditions are right for compositional variability.

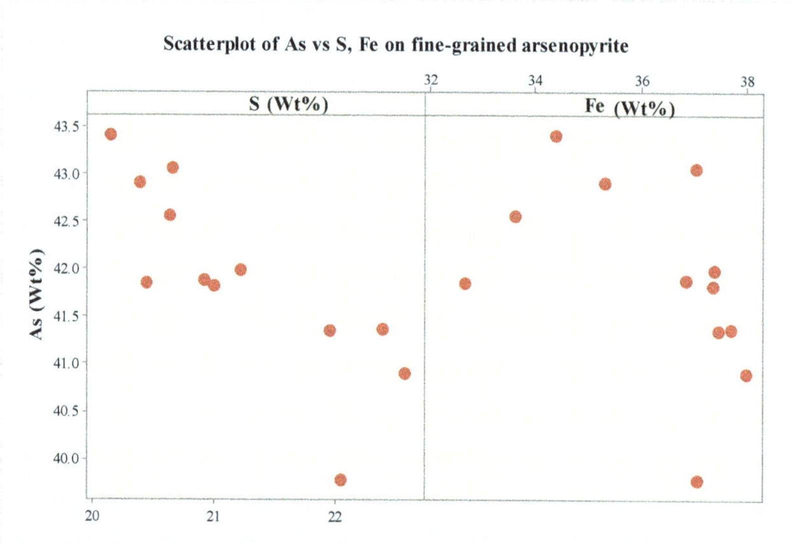

Figure 5. Scatterplots showing the relationship between arsenic (As) and sulfur (S) and between As and iron (Fe) in fine-grained arsenopyrite from the studied samples. The plots illustrate the inverse As–S trend and the corresponding positive association between As and Fe, reflecting compositional substitution mechanisms within the arsenopyrite lattice.

Coarse-grained arsenopyrite with oscillatory-zoned domains has As (28.31–30.13 wt%), Au (1926–6154 ppm), and other traces of Ag (up to 0.16 at %), Bi (up to 0.18 at %), Cu (up to 0.8 at %), Pb (up to 0.08 at %), and Te (up to 0.07 at %). The irregular As-rich patches contain As (28.47–31.48 wt%), Au ($\leq$950 ppm), and other important differences, such as being rich in other trace elements, especially Co (up to 0.85 at %) and Ni (up to 0.1 at %).

Electron microprobe (EMPA) analyses of arsenopyrite from the Um Rus deposit demonstrate a weak-to-moderate positive correlation between gold and arsenic in oscillatory-zoned domains (R = 0.41063, p = 0.08074, n = 19), suggesting that gold incorporation is partially, but not predominantly, influenced by arsenic content (Figure 6). This trend indicates that some gold may be integrated within the arsenopyrite lattice-possibly via coupled substitution mechanisms like $Au^{+} + As^{3-} \leftrightarrow Fe^{2+} + S^{2-}$ yet the moderate strength and limited statistical significance of the correlation suggest that lattice substitution alone cannot explain the observed gold enrichment. Instead, a large amount of gold is probably found in micro- or nanoparticulate forms, which is what petrographic studies have found when they looked at visible gold in fractures and along grain boundaries.

https://doi.org/10.3390/min16050439

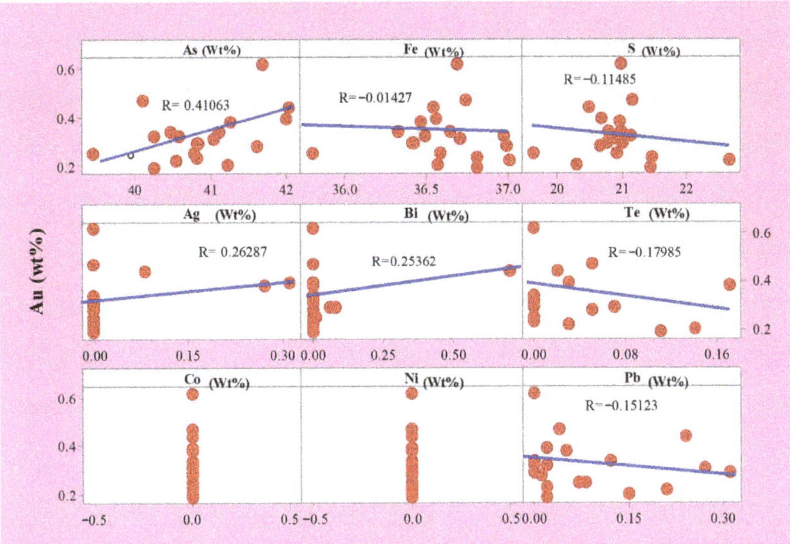

Figure 6. Scatter plots showing correlations between gold (Au) concentrations and other elements (Ag, As, Bi, Co, Fe, Ni, Pb, S, Te), with R-values indicating the strength of linear relationships: Au-Ag (R = 0.26387), Au-As (R = −0.41063), Au-Bi (R = 0.25362), Au-Co (R not shown), Au-Fe (R = −0.01427), Au-Ni (R not shown), Au-Pb (R = −0.15123), Au-S (R = −0.11485), and Au-Te (R = −0.17985). based on Electron microprobe (EMPA) analyses from oscillatory-zoned domains.

LA-ICP-MS analyses furnish substantial evidence for a unique, gold-rich hydrothermal pulse during the development of oscillatory zones. Gold has a very strong positive relationship with silver (R = 0.97762, $p < 0.001$, $n = 6$) and with antimony (R = 0.97608, $p < 0.001$, $n = 6$; however, the LA-ICP-MS correlations are statistically important; the small sample size ($n \approx 6$) needs careful interpretation so that the results here can be considered exploratory and supportive rather than definitive). This means that they are almost perfectly co-enriched (Figure 7). This extraordinary association indicates that gold was introduced as part of a chalcophile-rich fluid phase, potentially resulting in nanoscale inclusions of Ag–Au alloys, tellurides, or Sb-bearing sulfosalts such as petzite, hessite, or jamesonite. The strong link between Au and Ni (R = 0.87426, $p = 0.02272$) backs up the idea that a metal-rich fluid was involved (Figure 7), even though Ni is not usually found with gold. This could be due to fluid mixing, localized sulfide saturation, or the presence of Ni-rich microphases in fracture networks. The Ag-Sb association is very strong (R = 0.9612, $p = 0.00223$, Figure 7), which shows that these elements were moved and deposited together. There is also a moderate, almost significant correlation between silver and Ni (R = 0.79636, $p = 0.05798$) and between Sb and Ni (R = 0.76226, $p = 0.07806$). This means that Ni was part of the same hydrothermal event as Au-Ag-Sb, but it was not as tightly bound to the system. The relationships between Ag and Bi (R = 0.46468, $p = 0.35314$) and Sb and Bi (R = 0.44896, $p = 0.37181$) are weak and not significant. This means that Bi acts on its own and may have been added during a separate, minor alteration event. In stark contrast, the irregular As-rich patches, despite their elevated arsenic content (up to 31.48 at%) and enrichment in Co, Ni, and Bi, exhibit no coherent association between gold and other trace elements (Figure 8). LA-ICP-MS analyses show that there are very small and not statistically significant correlations between Au and Ag (R = 0.36282, $p = 0.54839$), Au and Sb (R = 0.70583, $p = 0.18284$), Au and Bi (R = 0.33038, $p = 0.58713$), and Au and Ni (R = −0.13679, $p = 0.82637$; $n = 5$ Figure 9). This lack of correlation, even in areas with a lot of arsenic, shows that

https://doi.org/10.3390/min16050439

having a lot of arsenic does not mean that gold will be found. These patches seem to be post-crystallization overprints that were probably made during fluid-assisted recrystallization or deformation. This is because the conditions favored the movement of As and Co–Ni but not gold, which could have been caused by lattice strain, charge imbalance, or changes in the fluid's redox state. The different ways that gold behaves in different textural domains show that primary oscillatory growth and secondary As-rich alteration are not linked. The highest concentrations of gold are not found in the areas with the most arsenic, but in areas with moderate zoning where gold is closely linked to silver, antimony, and tellurium. These elements are signs of a specific hydrothermal pulse that is rich in gold. On the other hand, the uneven As-rich patches are geochemically different but lack gold, which shows that just having more arsenic is not a good way to tell if there is gold. These results contradict the traditional "arsenic fence" model and underscore the significance of textural context and multielement signatures in the interpretation of gold mineralization. The multistage growth of the arsenopyrite and trace-element systematics documented in this work are in agreement with the progressive deformation-controlled and the mineralized fluid evolution typical of orogenic gold systems [20,22,23,40]. The strong association of Au with Ag–Sb–Te in oscillatory-zoned domains supports metal-rich hydrothermal fluids episodic influxes, that is a characteristic feature widely known in the orogenic systems as revealing changes in fluid composition, pressure, and redox conditions.

There was not just one way that gold was spread out at Um Rus; it changed over time in many ways. As-rich growth with little Au replacement, then a separate Te-Ag-Sb-Au-rich fluid event, and finally As-Co-Ni overprinting that made the gold content harder to see but did not add to it.

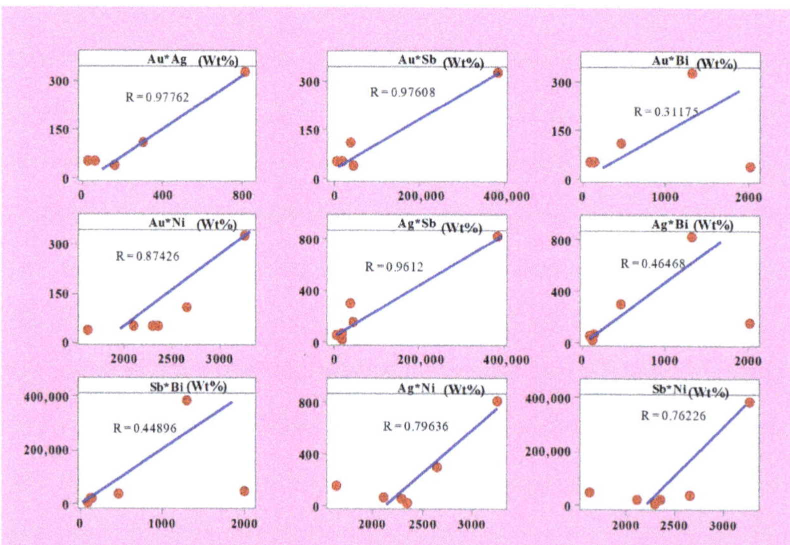

Figure 7. Matrix of scatterplots illustrating pairwise correlations among gold (Au), silver (Ag), antimony (Sb), bismuth (Bi), and nickel (Ni) coarse-grained arsenopyrite with oscillatory-zoned domains. Notably strong positive correlations occur for Au–Ag (R = 0.97762), Au–Sb (R = 0.97608), Au–Ni (R = 0.87426), and Ag–Sb (R = 0.9612), suggesting shared geochemical controls on element incorporation. Moderate correlations are observed for Ag–Ni (R = 0.79636) and Sb–Ni (R = 0.76226), while Au–Bi (R = 0.31175), Sb–Bi (R = 0.44896), and Ag–Bi (R = 0.46456) display weaker relationships, indicating more variable incorporation mechanisms. based on LA-ICP-MS data from oscillatory-zoned domains. The small LA-ICP-MS sample ($n \approx 6$) makes these correlations exploratory.

https://doi.org/10.3390/min16050439

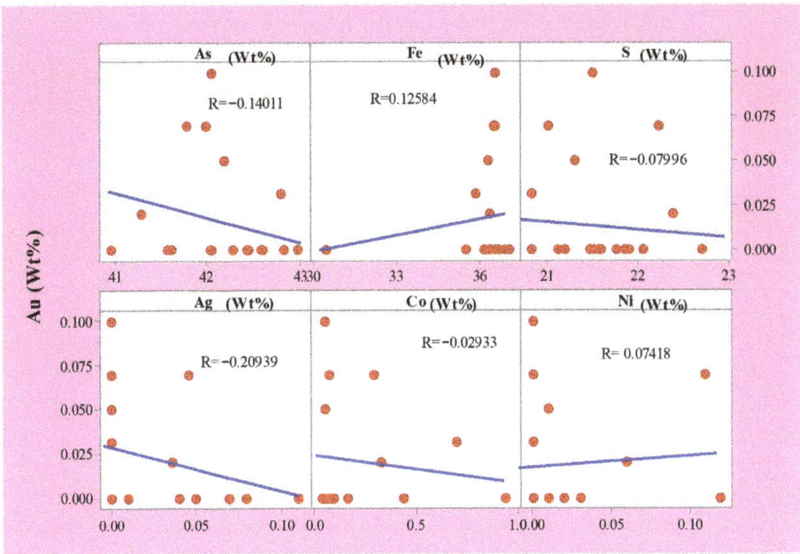

Figure 8. Scatter plots showing correlations between gold (Au) concentrations and other elements (As, Fe, S, Ag, Co, Ni), with R-values indicating the strength of linear relationships: Au-As (R = −0.14011), Au-Fe (R = 0.12584), Au-S (R = −0.07996), Au-Ag (R = −0.20939), Au-Co (R = 0.02933), and Au-Ni (R = 0.07418). based on Electron microprobe (EMPA) analyses from As rich irregular patches.

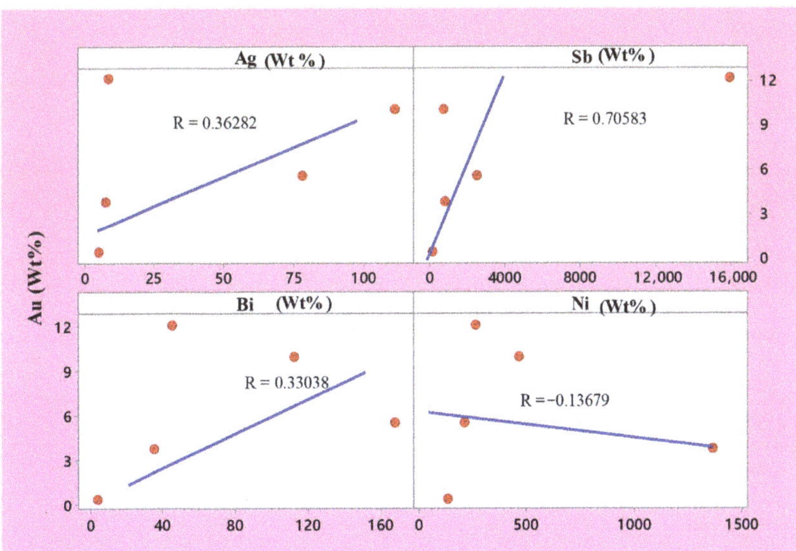

Figure 9. Scatter plots illustrating the correlations between gold (Au) and trace elements (Ag, Sb, Bi, Ni) in the irregular As-rich patches arsenopyrite from the Um Rus deposit. The plots show (**top left**) Au vs. Ag with R = 0.36282, (**top right**) Au vs. Sb with R = 0.70583, (**bottom left**) Au vs. Bi with R = 0.33038, and (**bottom right**) Au vs. Ni with R = −0.13679, based on LA-ICPMS data from As-rich irregular patches.

https://doi.org/10.3390/min16050439

4.2.2. Principal Component Analysis (PCA)

The Principal Component Analysis (PCA) performed on the EMPA dataset, which includes 50 analyses of 12 elements (S, Fe, Co, Ni, Cu, Zn, As, Ag, Te, Au, Pb, Bi) from both fine-grained and coarse-grained arsenopyrite samples, showed clear patterns in how trace elements are spread out and how minerals form.

Fine-Grained Arsenopyrite

Eigenanalysis of the correlation matrix shows that the first two principal components (PC1 and PC2) explain 84.0% of the total variance, with PC1 accounting for 61.8% and PC2 for 22.2%. PC3 explains 13.3%, while PC4 (2.6%) and PC5 (0.1%) contribute negligibly to Table 1. PC1has strong positive loadings for Fe (0.526) and S (0.511), and strong negative loadings for Co (−0.514) and As (−0.437). PC2 is dominated by a strong negative loading for Pb (−0.860), moderate positive loading for S (0.299), and a negative contribution from As (−0.301; Table 2). These loadings indicate that PC1 primarily separates Fe-S-rich compositions from As-Co-rich compositions, whereas PC2 reflects Pb-related variability (Figure 10).

Coarse-grained arsenopyrite

PCA of oscillatory-zoned arsenopyrite

Principal Component Analysis (PCA) of oscillatory-zoned arsenopyrite utilizing S, Fe, Cu, Zn, As, Ag, Te, Au, Pb, and Bi indicates that the initial four principal components (PCs) account for 72.8% of the total variance (Table 3). PC1 explains 25.1% of the variance and is strongly linked to Ag (0.512), Cu (0.477), and As (0.371). It also has negative contributions from S (−0.277) and Fe (−0.201; Table 4). PC2 (17.8% variance) has mostly negative loadings from Bi (−0.490), As (−0.455), and Au (−0.382), but a positive Te association (0.351) makes up for them. PC3 (17.1%) shows negative correlations with Fe (−0.636) and S (−0.580). On the other hand, PC4 (12.9%) is mostly affected by Pb (−0.730) and positive Au loading (0.397). These loadings show that PC3 mostly shows the intrinsic Fe-S-As composition of arsenopyrite. PC1 and PC2 show how hydrothermal controls affect the incorporation of trace elements, and PC4 shows the separate Au component (Table 4, Figure 11).

Table 1. Eigenvalues, proportion of variance explained, and cumulative variance for the first five principal components from eigenanalysis of the correlation matrix for fine-grained arsenopyrite.

Element	Co	Ni	Cu	Ag	Sb	Au	Bi
Av. Conc.	0.749	1.049	1.314	0.405	0.820	0.443	0.563
Av. DL	0.059	0.229	0.649	0.079	0.143	0.068	0.053

Table 2. Principal component loadings of major variables (S, Fe, Co, As, Pb) on the first four components derived from eigenanalysis of the correlation matrix for fine-grained arsenopyrite.

Variable	PC1	PC2	PC3	PC4
S	0.511	0.299	−0.187	−0.729
Fe	0.526	−0.187	0.396	−0.096
Co	−0.514	0.213	−0.409	−0.405
As	−0.437	−0.301	0.642	−0.523
Pb	0.088	−0.86	−0.477	−0.15

https://doi.org/10.3390/min16050439

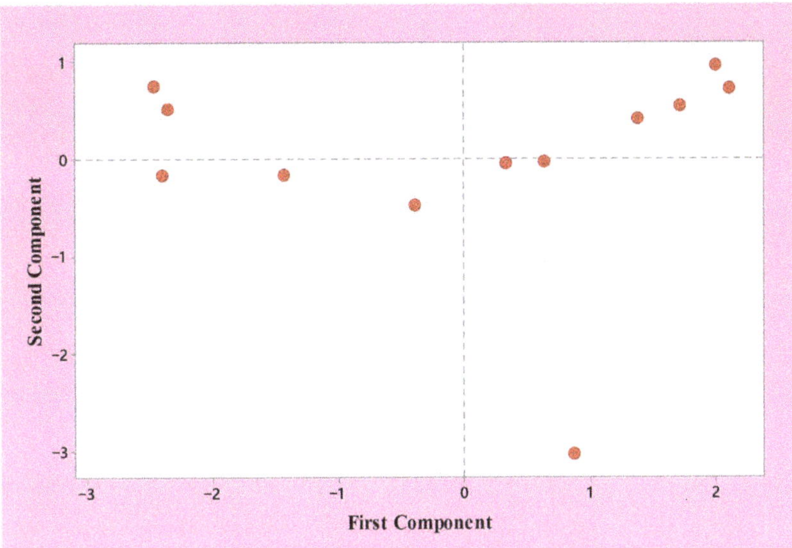

Figure 10. Score plot of the first two principal components (PC1 and PC2) for fine-grained arsenopyrite, showing the distribution of samples in multivariate space based on S, Fe, Co, As, and Pb concentrations.

Table 3. Eigenvalues, proportion of variance explained, and cumulative variance for the first ten principal components derived from eigenanalysis of the correlation matrix for S, Fe, Cu, Zn, As, Ag, Te, Au, Pb, and Bi of oscillatory-zoned domains.

Component	Eigenvalue	Proportion	Cumulative
PC1	2.5077	0.251	0.251
PC2	1.785	0.178	0.429
PC3	1.7055	0.171	0.6
PC4	1.2855	0.129	0.728
PC5	0.7888	0.079	0.807
PC6	0.656	0.066	0.873
PC7	0.3998	0.04	0.913
PC8	0.3886	0.039	0.952
PC9	0.3327	0.033	0.985
PC10	0.1504	0.015	1

PCA of Irregular As-rich Overgrowths

Principal Component Analysis (PCA) of S, Fe, Co, Ni, Cu, As, Ag, and Au in irregular As-rich overgrowths shows that the first four components account for 74.3% of the total variance (PC1 = 25.2%, PC2 = 18.8%, PC3 = 16.2%, PC4 = 14.2% Table 5).

PC1 is dominated by high positive loading of S (0.618) and negative loadings for As (−0.581) and Ag (−0.390), indicating a strong S–As compositional contrast. PC2 is characterized by strong positive loadings for Co (0.656) and Ni (0.496), moderate positive loading for As (0.293), and negative loading for Fe (−0.337). PC3 shows high positive loadings for Cu (0.641), Fe (0.582), and Ni (0.344), suggesting a Fe–Cu–Ni association. PC4 is strongly influenced by Au (0.781; Table 6) with moderate negative loading from Ag

 https://doi.org/10.3390/min16050439

(−0.400), indicating that gold enrichment is largely decoupled from other compositional trends in these domains.

These loadings indicate PC1 reflects the fundamental S–As compositional contrast in As-rich overgrowths, PC2 and PC3 capture hydrothermal controls through Co–Ni and Fe–Cu–Ni associations, respectively, while PC4 isolates the distinct behavior of Au, highlighting its decoupling from other trace elements (Figure 12).

Based on petrographic observations, textural relationships, and in situ geochemical data, the paragenetic sequence of mineralization is summarized in Table 7.

Table 4. Principal component loadings of major and trace elements (S, Fe, Cu, Zn, As, Ag, Te, Au, Pb, Bi) on the first four components obtained from eigenanalysis of the correlation matrix of oscillatory-zoned domains.

Variable	PC1	PC2	PC3	PC4
S	−0.277	−0.099	−0.58	−0.16
Fe	−0.201	−0.203	−0.636	−0.061
Cu	0.477	0.178	−0.118	−0.024
Zn	−0.243	−0.378	0.013	0.333
As	0.371	−0.455	−0.208	0.028
Ag	0.512	0.01	−0.195	0.07
Te	0.302	0.351	−0.304	−0.218
Au	0.258	−0.382	0.009	0.397
Pb	−0.027	−0.241	0.159	−0.73
Bi	0.195	−0.49	0.214	−0.339

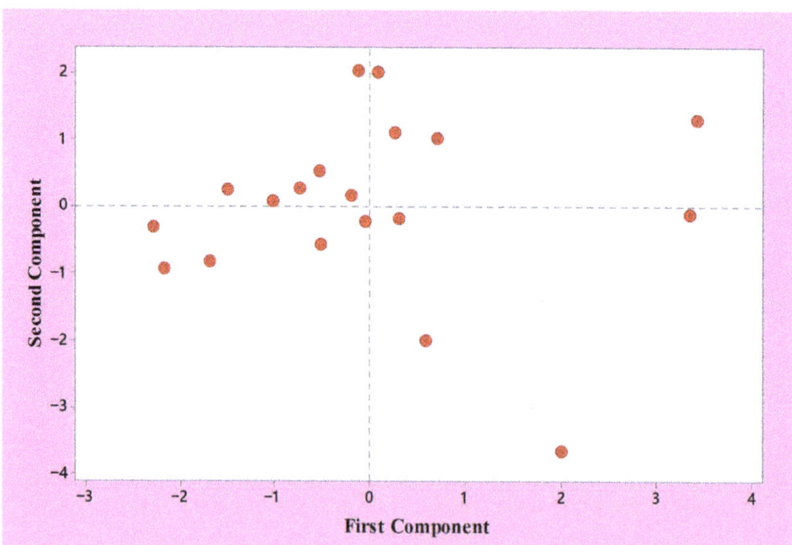

Figure 11. Score plot of the first two principal components (PC1 vs. PC2) for oscillatory-zoned arsenopyrite from Um Rus.

https://doi.org/10.3390/min16050439

Table 5. Eigenvalues, proportion of variance explained, and cumulative variance for the first eight principal components derived from eigenanalysis of the correlation matrix for S, Fe, Co, Ni, Cu, As, Ag, and Au, the irregular As-rich patches.

Component	Eigenvalue	Proportion	Cumulative
PC1	2.0144	0.252	0.252
PC2	1.5048	0.188	0.44
PC3	1.2929	0.162	0.602
PC4	1.1343	0.142	0.743
PC5	0.8294	0.104	0.847
PC6	0.6877	0.086	0.933
PC7	0.4344	0.054	0.987
PC8	0.1022	0.013	1

Table 6. Principal component loadings of major and trace elements (S, Fe, Co, Ni, Cu, As, Ag, Au) on the first four components obtained from eigenanalysis of the correlation matrix of the irregular As-rich patches.

Variable	PC1	PC2	PC3	PC4
S	0.618	−0.135	0.051	−0.339
Fe	0.014	−0.337	0.582	0.133
Co	0.035	0.656	−0.025	0.066
Ni	0.303	0.496	0.344	−0.285
Cu	−0.135	0.178	0.641	−0.1
As	−0.581	0.293	−0.011	−0.037
Ag	−0.39	−0.273	0.248	−0.4
Au	0.128	0.021	0.259	0.781

Table 7. Summarized the paragenetic sequence highlighting the transition from early magmatic sulfides to hydrothermal arsenopyrite associated with gold deposition.

Stage	Mineral Assemblage	Texture/Features	Interpretation
Stage I (Pre-ore)	Pyrrhotite-I ± Pentlandite	Disseminated, magmatic textures	Early magmatic sulfide stage (meladiorite-hosted)
Stage II (Early hydrothermal)	Arsenopyrite-I ± Pyrite	Coarse-grained, oscillatory zoning	Initial hydrothermal fluid influx
Stage III (Main ore stage)	Arsenopyrite-II + Au ± Ag ± Sb ± Ni	Overgrowths, inclusion-rich, trace-element enrichment	Gold mineralization stage linked to fluid evolution
Stage IV (Late hydrothermal)	Quartz ± Carbonates ± Minor sulfides	Vein filling, crosscutting earlier phases	Late fluid circulation and remobilization

https://doi.org/10.3390/min16050439

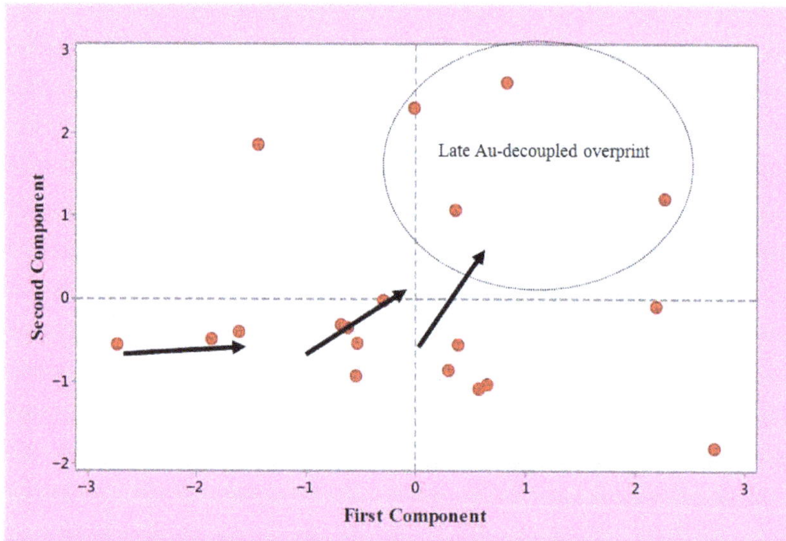

Figure 12. PCA score plot for irregular As-rich overgrowths in coarse-grained arsenopyrite from the Um Rus deposit. Samples cluster primarily along PC1 (S–Fe–Co–Ni–Cu association) with moderate spread toward PC2, reflecting varibility in Co–Ni enrichment and partial decoupling of Au from As. The distribution pattern suggests a later-stage chemical overprint distinct from the compositional trends seen in oscillatory-zoned domains (Short black arrows pointing to Late Au-decoupled overprint).

5. Discussion

5.1. Compositional Zoning and Paragenesis Reflected in PCA Results

The principal component analysis (PCA) of trace element data from different textural generations of arsenopyrite at the Um Rus deposit reveals a complex paragenetic history involving multiple fluid pulses, substitution mechanisms, and overprinting events. The geochemical controls on gold distribution vary significantly between fine-grained, and Coarse-grained arsenopyrite includes both the oscillatory-zoned domains, and irregular As-rich overgrowths—highlighting that gold enrichment is not uniform across arsenopyrite types, nor is it reliably predicted by arsenic content alone.

In fine-grained arsenopyrite, which occurs as euhedral inclusions within pyrite aggregates, PCA identifies two dominant components explaining 84.0% of compositional variance. The first component reflects a strong inverse relationship between Fe and S (negative loadings) and Co and As (positive loadings), indicating that major geochemical variation is governed by Co-for-Fe substitution, likely coupled with As-for-S or charge-balanced mechanisms during growth from a Co-rich, Au-poor fluid [4,41]. This anti-correlation suggests crystal-chemical constraints under non-ideal crystallization conditions, possibly linked to fluctuating fO_2 or As activity [8]. The second component separates As from Co, highlighting episodic As enrichment independent of Co, potentially during transient fluid pulses (Figure 13). Notably, no detectable gold was found in this phase, despite As variability, confirming that As-rich domains do not necessarily host gold—especially when formed in closed or isolated microenvironments like pyrite aggregates. Fluid inclusion analysis of quartz associated with gold-bearing arsenopyrite reveals three main types—(1) single $CO_2 \pm CH_4$, (2) plane-arranged $CO_2 \pm CH_4$, and (3) aqueous H_2O–NaCl $\pm$ KCl inclusions—with homogenization temperatures of 311–373 °C and pressures of 0.3–1.5 kbar,

 https://doi.org/10.3390/min16050439

indicating trapping from boiling, heterogeneous CO_2–H_2O fluids and providing constraints on the chemical environment and physical conditions of gold mineralization [29].

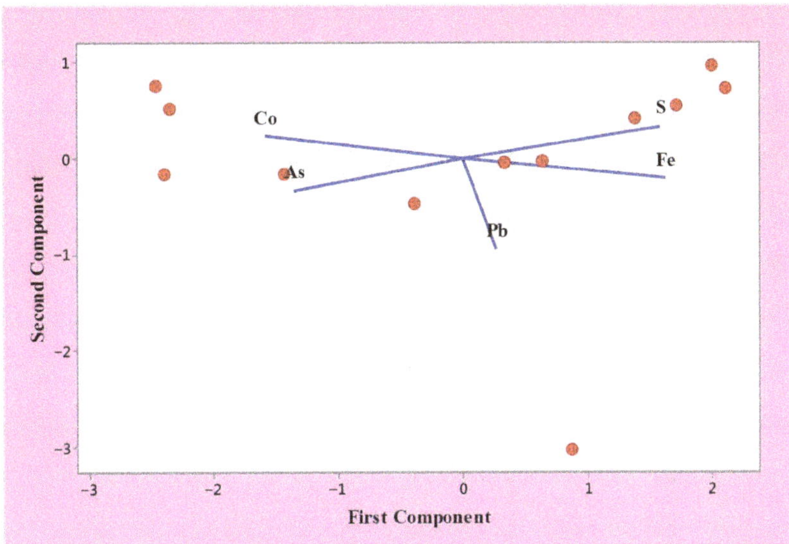

Figure 13. PCA biplot of PC1 versus PC2 for fine-grained arsenopyrite, illustrating both sample distribution and variable loadings.

The negative loading of Pb on PC2 indicates that Pb enrichment is largely decoupled from the primary Fe–S–As–Co substitution trends. This may result from Pb-bearing micro-inclusions, localized enrichment during late-stage fluid infiltration, or heterogeneity acquired during crystal growth. This separation into two dominant components highlights that the composition of fine-grained arsenopyrite is shaped by both primary substitution processes and secondary trace element incorporation, laying the foundation for understanding its barren nature relative to gold.

In contrast, the oscillatory-zoned domains within coarse-grained arsenopyrite exhibit a more complex mineralization history, as revealed by PCA. The first principal component (PC1) is dominated by positive loadings for Ag, Cu, and As, indicating an early As-rich growth stage where minor Au was structurally incorporated into the arsenopyrite lattice, likely through limited substitution for Fe or As [6]. However, the weak Au loading on PC1 suggests this stage contributed only modestly to the total gold content. The second principal component (PC2) features positive Te loadings, a moderate positive association with Cu, and negative associations with Au and Bi. This pattern, corroborated by LA-ICP-MS evidence of a strong Au–Ag correlation and elevated Te, points to a discrete Te–Ag–Sb–Au-rich hydrothermal event (Figure 14) that overprinted the earlier As-rich domains [42–44]. This fluid influx likely enriched Au through precipitation along growth bands rather than widespread lattice substitution. Subsequent variation captured in PC3 and PC4 separates the Fe–S stoichiometric framework from trace-metal associations, indicating later chemical re-equilibration. PC3, defined by strong negative Fe and S correlations, may represent Fe- and S-depleted zones enriched in other trace metals, consistent with compositional oscillations linked to redox changes at the crystal–fluid interface [45]. Finally, PC4, showing a strong negative Pb association with a positive Au loading, suggests that Pb was introduced during a late mineralizing stage where Au enrichment occurred, possibly from a distinct, Pb-poor hydrothermal pulse. These patterns collectively support

https://doi.org/10.3390/min16050439

the interpretation that oscillatory zoning records periodic changes in fluid chemistry, temperature, and redox state during arsenopyrite growth, likely reflecting a dynamic hydrothermal environment with multiple metal inputs [46,47].

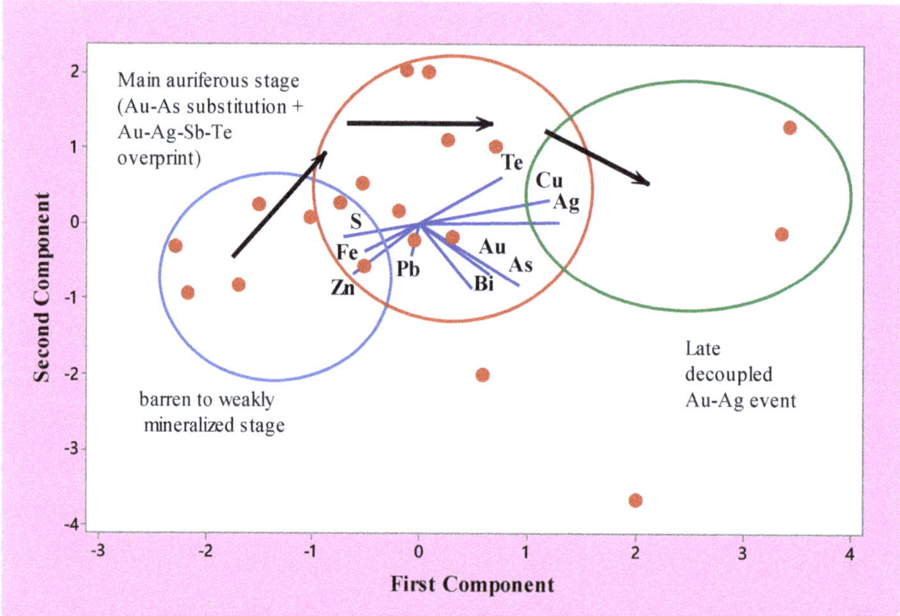

Figure 14. Biplot of PC1 vs. PC2 for oscillatory-zoned arsenopyrite from Um Rus, showing both sample distribution and variable loadings. The plot illustrates the association of Au with Ag, Te, and Cu in certain growth zones, supporting a multi-stage gold enrichment model involving early As-rich growth and later Te–Ag–Sb–Au hydrothermal overprinting.

In contrast, the irregular As-rich patches—which overprint oscillatory zones along fractures and deformed margins—exhibit a fundamentally different geochemical signature. PC1 shows a strong inverse relationship between S and As, suggesting that As enrichment in these zones is decoupled from the Fe–S structural framework [7,48]. This may reflect non-stoichiometric growth, lattice strain, or local sulfur loss during fluid-assisted recrystallization. The negative loading on further indicates that this fluid phase was poor in precious metals, despite high As [49].

PC2 is dominated by Co and Ni, identifying a Co–Ni-rich hydrothermal influx, likely associated with deep-circulating fluids interacting with mafic sources [45]. The lack of Au loading confirms that this event did not enhance gold content, even though it modified As distribution. PC3 highlights a Cu–Fe–Ni association, suggesting deposition of base metal sulfides (e.g., chalcopyrite) along microfractures—possibly due to late sulfide saturation [50]. Finally, PC4 is defined by an extremely strong positive loading on Au, with negative contributions from S and Ag. This indicates that visible gold was introduced late, likely via focused fluid flow along fractures, where pressure drop or phase separation triggered physical precipitation [51,52]. This pattern implies that gold introduction in these domains was decoupled from As content, likely reflecting a late-stage overprint unrelated to the primary arsenopyrite growth chemistry (Figure 15).

https://doi.org/10.3390/min16050439

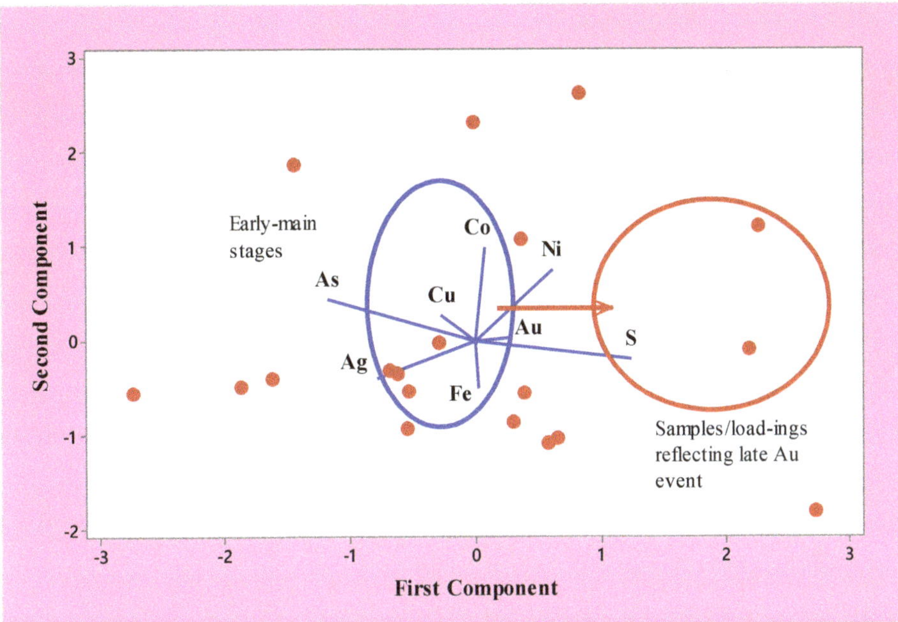

Figure 15. PCA biplot for irregular As-rich overgrowths showing loading vectors for S, Fe, Co, Ni, Cu, As, Ag, and Au. Au is isolated along PC4, indicating minimal correlation with the dominant S–Fe–Co–Ni–Cu assemblage. This vector pattern supports the interpretation that Au was introduced during a late, structurally controlled event, rather than through direct substitution into As-rich growth phases.

5.2. Mechanisms of Gold Incorporation Across Arsenopyrite Textures

The processes that control how gold gets into Um Rus arsenopyrite are very different in different textural domains. This goes against the common Au-As correlation model, which says that high arsenic content is a direct sign of gold fertility.

In fine-grained arsenopyrite, found in pyrite aggregates, the lack of detectable gold despite broad As zoning indicates that early substitution mechanisms failed to incorporate substantial gold. The PCA results show that the first component shows a strong Fe-S-As-Co framework. This means that structural stability was more important than precious metal enrichment at this stage. This is different from the oscillatory-zoned domains in coarse-grained arsenopyrite, where a positive Au–As correlation in the early stages of growth supports limited lattice substitution, which may happen through the coupled mechanism $Au^{+} + As^{3-} \leftrightarrow Fe^{2+} + S^{2-}$ [41]. But the low amount of Au on PC1 suggests that this part of the total gold budget is small. The primary mechanism of gold incorporation in these domains transitions to a late-stage process, as indicated by robust Au-Ag-Sb correlations from LA-ICP-MS, associated with the fourth principal component. This suggests that a hydrothermal pulse deposited gold as micro-inclusions or electrum along growth bands and microfractures, rather than through extensive lattice substitution [41,53]. In the As-rich overgrowths, the inverse Au-As relationship suggests that gold enrichment is independent of arsenic content, probably because the lattice became saturated or re-equilibrated during late-stage fluid events. The prominent Co/Ni association in PC2 of these overgrowths, indicative of subsequent fluid infiltration enriched in base metals, signifies remobilization, further obscuring prior gold signatures and highlighting the influence of secondary fluid pulses on gold distribution [54]. These observations collectively substantiate a four-stage

https://doi.org/10.3390/min16050439

mineralization model that clarifies the intricate paragenesis of gold at Um Rus: (1) primary oscillatory growth of As-rich stage arsenopyrite [55] with limited Au substitution, as indicated by PC1; (2) a Te-Ag-Sb-Au-rich hydrothermal fluid influx altering As and Au distribution [56], evident in the negative Au loading on PC2; (3) post-zoning growth overprinted by Co-Ni-As-rich fluids creating irregular patches [57], geochemically distinct yet Au-poor in an invisible form, as shown by PC2; and (4) a late fracture-controlled deposition of visible gold, physically concentrated in As-rich patches but geochemically decoupled, as reflected in the positive Au loading on PC4 [29]. This paragenetic framework shows that not all As-rich arsenopyrite contains gold [29,58,59]. As summarized in Table 7, the transition from early barren arsenopyrite to Au-rich overgrowths reflects evolving fluid conditions, progressive element enrichment, and the multistage paragenetic evolution of mineralization at the Um Rus deposit. The highest amounts of invisible gold are found in moderately As-rich, oscillatory-zoned domains where strong Ag-Sb-Te co-enrichment (backed up by LA-ICP-MS correlations) happens at the same time as structural continuity. On the other hand, the irregular As-rich patches, which have visible gold, have much less invisible gold ($\leq$950 ppm) and formed in conditions that were not good for stable Au incorporation, such as lattice instability or redox changes [60]. These results contest the prevalent "arsenic fertility" model, which posits that arsenic anomalies in sulfides directly signify gold potential [10,61]. At Um Rus, arsenic enrichment alone is not a good way to find minerals, because the best signs of high-grade mineralization are oscillatory zoning, co-enrichment in Te, Ag, and Sb, and structural continuity that kept fluid pathways open. Combining microtextural analysis, high-resolution geochemistry (such as EMPA and LA-ICP-MS), and multivariate statistics like PCA creates a strong framework for telling the difference between primary gold-forming events and secondary overprints. This method improves targeting in orogenic gold systems by taking into account both textural context and chemical composition. It is a better way to find auriferous zones than using simple geochemical proxies.

5.3. Implications of PCA Findings and Exploration Significance

Principal Component Analysis (PCA) is a very useful tool for figuring out the complicated geochemical signals in the Um Rus gold deposit. It does this by breaking down multidimensional trace element data into parts that can be understood. This multivariate analysis shows that the distribution of gold is not determined by a single geochemical factor but rather by several fluid pulses that occur at different times, each leaving a distinct multivariate signature. In the oscillatory-zoned domains of coarse-grained arsenopyrite, PC1 captures an early As-Ag-Cu-rich growth phase, while PC4 isolates a late Au–Pb-decoupled event, with gold loading positively and Pb strongly negatively. This change from an As-dominated crystallization stage to an Au-focused pulse shows that the chemistry of the fluid has changed. The late event brought most of the invisible gold through a hydrothermal influx rich in precious metals. PCA of fine-grained arsenopyrite, on the other hand, shows a Fe-S versus Co-As framework with no Au connection. This confirms that early crystallization happened in a gold-poor, structurally isolated environment. These patterns reveal a significant flaw in conventional exploration models: arsenic concentration alone is an unreliable indicator of gold fertility [62–64]. The highest gold concentrations are located not in the most arsenic-rich areas, but in moderately arsenic-rich, oscillatory-zoned regions, where gold is closely associated with silver, antimony, and tellurium, which are signs of a certain type of auriferous fluid event [64]. The very strong Au–Ag and Au–Sb correlations from LA-ICP-MS back this up. When combined with PCA, they make a more accurate geochemical fingerprint that is more useful than As anomalies for diagnosis. The irregular As-rich patches, despite their geochemical extremity, are geochemically barren in

 https://doi.org/10.3390/min16050439

terms of lattice-bound gold, representing post-growth overprints that obscure rather than enhance the primary mineralization signal [65]. The exploration ramifications for Um Rus and analogous orogenic gold systems are substantial. When targeting, texture continuity and multielement co-enrichment should come first, especially Ag, Sb, and Te over bulk As content [64–68]. It is important to have a unified plan that includes high-resolution textural analysis (like BSE and SEM imaging), in situ geochemistry (like EMPA and LA-ICP-MS), and multivariate statistics (like PCA). This method makes it possible to tell the difference between primary gold-forming events and secondary, barren overprints. This lets you find areas with preserved fluid signatures. By changing the focus from bulk composition to a process-based interpretation, exploration efficiency can be greatly improved. This will guide drilling and sampling toward structurally sound, chemically coherent areas that are the core of the mineral system.

6. Conclusions

This study shows that the distribution of gold in arsenopyrite at the Um Rus deposit is not just related to the amount of arsenic present. Instead, it is shaped by a complex, multistage paragenetic history that includes different fluid pulses, changes in texture, and structural controls.

Two dietetic generation of arsenopyrite are recognized: (1) fine-grained inclusions within pyrite that show As zoning (28.23–31.07 at%) but no detectable gold, which suggests that crystallization started with Fe–S in a gold-poor environment; and (2) coarse oscillatory-zoned with irregular As-rich overgrowths that host high concentrations of invisible gold (up to 6154 ppm).The highest concentrations of lattice-bound gold occur in moderately As-rich oscillatory zones characterized by co-enrichment in Ag, Sb, and Te, reflecting a discrete precious-metal-rich hydrothermal pulse. In contrast, the irregular As-rich overgrowths are relatively poor in invisible gold ($\leq$950 ppm) and represent later recrystallization events that largely decoupled gold from arsenic.

These findings show that the traditional "arsenic fence" model is not universally applicable. Arsenic enrichment by itself is an unreliable indicator of gold fertility. Instead, the most auriferous domains are best identified by specific textural features (oscillatory zoning) combined with multi-element signatures involving Ag–Sb–Te. The results highlight the value of integrating high-resolution microtextural analysis (BSE imaging and SEM), in situ geochemical techniques (EMPA and LA-ICP-MS), and multivariate statistics (PCA) to distinguish primary gold-bearing assemblages from barren or overprinted domains. This integrated approach provides a more robust framework for understanding gold mineralization processes in structurally complex orogenic systems and should improve exploration targeting at Um Rus and similar deposits.

These findings have direct implications for ore processing, as they indicate that gold recovery efficiency will depend on targeting the specific sulfide populations hosting the metal. While this study did not include detailed liberation or ex-traction tests, identifying the primary host phases and their element associations and refractory ore treatment strategies. Future work integrating these mineralogical insights with metallurgical characterization will enable optimized processing of gold from complex arsenopyrite-bearing ores.

Supplementary Materials: The following supporting information can be downloaded at: https://www.mdpi.com/article/10.3390/min16050439/s1, Table S1: Raw EPMA and LA-ICP-MS trace element data of arsenopyrite from the Um Rus deposit.

Author Contributions: Conceptualization, E.A.S.; Methodology, E.A.S.; Formal analysis, E.A.S.; Resources, A.A.E.-S.; Data curation, A.S.H.; Writing—original draft, E.A.S.; Writing—review & editing, A.A.E.-S.; Visualization, A.A.E.-S.; Supervision, E.A.S. and H.M.H.; Project administration,

https://doi.org/10.3390/min16050439

E.A.S.; Funding acquisition, H.M.H. and A.S.H. All authors have read and agreed to the published version of the manuscript.

Funding: This research was funded by the Deanship of Graduate Studies and Scientific Research at Qassim University grant number [QU-APC-2026]. The APC was funded by the Deanship of Graduate Studies and Scientific Research at Qassim University.

Data Availability Statement: All data generated or analyzed during this study are included in this published article.

Acknowledgments: The Researcher would like to thank the Deanship of Graduate Studies and Scientific Research at Qassim University for financial support (QU-APC-2026). During the preparation of this manuscript, the authors used Gemini 1.5 Pro to refine the English language and improve clarity and readability. This tool was used solely for linguistic enhancement and grammatical correction. The authors thoroughly reviewed, edited, and verified the output and take full responsibility for the scientific content.

Conflicts of Interest: The author declares that there are no relevant financial or non-financial competing interests to report. The research was conducted in the absence of any commercial or financial relationships that could be construed as a potential conflict of interest.

References

1. Cook, N.J.; Ciobanu, C.L.; Mao, J. Textural control on gold distribution in As-free pyrite from the Dongping, Huangtuliang and Hougou gold deposits, North China Craton (Hebei Province, China). *Chem. Geol.* **2009**, *264*, 101–121. [CrossRef]
2. Kreuzer, O.P. Intrusion-hosted mineralization in the Charters Towers Goldfield, North Queensland: New isotopic and fluid inclusion constraints on the timing and origin of the auriferous veins. *Econ. Geol.* **2005**, *100*, 1583–1603. [CrossRef]
3. Kravtsova, R.; Tauson, V.; Nikitenko, E. Modes of Au, Pt, and Pd occurrence in arsenopyrite from the Natalkinskoe deposit, NE Russia. *Geochem. Int.* **2015**, *53*, 964–972. [CrossRef]
4. Naumov, E.; Kalinin, Y.; Palyanova, G.; Kryuchkova, L.; Voitenko, V.; Abramova, V.; Pirajno, F. Combined study of Au-bearing arsenopyrite of orogenic gold deposits (NE Asia): High resolution 3D X-ray computed tomography, LA-ICP-MS, and EMPA data. *Geosci. Front.* **2025**, *16*, 101953. [CrossRef]
5. Taylor, R.D.; Hofstra, A.H. *Critical Minerals in Orogenic (Gold) and Coeur D'Alene-Type Mineral Systems of the United States*; US Geological Survey: Reston, VA, USA, 2025.
6. Zhang, H.; Zhu, Y. Textural, trace elemental and sulfur isotopic signatures of arsenopyrite and pyrite from the Mandongshan gold deposit (west Junggar, NW China): Implications for the conditions of gold mineralization. *Ore Geol. Rev.* **2021**, *129*, 103938. [CrossRef]
7. Cardenas-Vera, A.; MacDonald, M.; Lentz, D.R.; Thorne, K.G. Trace element characteristics of pyrite and arsenopyrite from the Golden Ridge gold deposit, New Brunswick, Canada: Implications for ore genesis. *Minerals* **2023**, *13*, 954. [CrossRef]
8. Xing, Y.; Brugger, J.; Tomkins, A.; Shvarov, Y. Arsenic evolution as a tool for understanding formation of pyritic gold ores. *Geology* **2019**, *47*, 335–338. [CrossRef]
9. Wu, Y.-F.; Fougerouse, D.; Evans, K.; Reddy, S.M.; Saxey, D.W.; Guagliardo, P.; Li, J.-W. Gold, arsenic, and copper zoning in pyrite: A record of fluid chemistry and growth kinetics. *Geology* **2019**, *47*, 641–644. [CrossRef]
10. Pokrovski, G.; Escoda, C.; Blanchard, M.; Testemale, D.; Hazemann, J.-L.; Gouy, S.; Kokh, M.; Boiron, M.-C.; de Parseval, F.; Aigouy, T.; et al. An arsenic-driven pump for invisible gold in hydrothermal systems. *Geochem. Perspect. Lett.* **2021**, *17*, 39–44. [CrossRef]
11. Behera, R.C.; Singh, S.; Srivastava, S.; Kumar, K.V.; Reddy, N.R.; Barla, A.; Pandey, R. Trace elemental systematics of auriferous sulfides within the dolerites of northern Cuddapah Basin, southern India: Implications for invisible gold mineralization. *Ore Geol. Rev.* **2025**, *180*, 106569. [CrossRef]
12. Barbosa, L. Examination of Micro-Scale Sulfide Distribution for the Gold Mineralization in the Lappberget Deposit, Garpenberg Mine, Sweden: Towards a Geometallurgical Approach. Master's Thesis, Luleå University of Technology, Luleå, Sweden, 2021.
13. Chi, N.; Han, Z.; Liu, C.; Zhang, W.; Zhang, Y.; Shan, W.; Li, Z.; Li, M.; Wang, X.; Sun, Y. Metallogenic mechanism of the Houge'zhuang gold deposit, Jiaodong, China: Evidence from fluid inclusion, in situ trace element, and sulfur isotope compositions. *Front. Earth Sci.* **2022**, *10*, 1009715. [CrossRef]
14. Konopelko, D.L.; Zhdanova, V.S.; Stepanov, S.Y.; Sidorova, E.S.; Petrov, S.V.; Kozin, A.K.; Aliyev, E.S.; Saltanov, V.A.; Kalinin, M.A.; Korneev, A.V.; et al. Mineralization Styles in the Orogenic (Quartz Vein) Gold Deposits of the Eastern Kazakhstan Gold Belt: Implications for Regional Prospecting. *Minerals* **2025**, *15*, 885. [CrossRef]

 https://doi.org/10.3390/min16050439

15. Hazarika, P.; Mishra, B.; Pruseth, K.L. Trace-element geochemistry of pyrite and arsenopyrite: Ore genetic implications for late Archean orogenic gold deposits in southern India. *Mineral. Mag.* **2017**, *81*, 661–678. [CrossRef]
16. Li, W.; Cook, N.J.; Xie, G.-Q.; Mao, J.-W.; Ciobanu, C.L.; Li, J.-W.; Zhang, Z.-Y. Textures and trace element signatures of pyrite and arsenopyrite from the Gutaishan Au–Sb deposit, South China. *Miner. Depos.* **2019**, *54*, 591–610. [CrossRef]
17. Tauson, V.; Lipko, S.; Kravtsova, R.; Smagunov, N.; Belozerova, O.; Voronova, I. Distribution of "invisible" noble metals between pyrite and arsenopyrite exemplified by minerals coexisting in orogenic Au deposits of North-Eastern Russia. *Minerals* **2019**, *9*, 660. [CrossRef]
18. Wu, Y.-F.; Evans, K.; Hu, S.-Y.; Fougerouse, D.; Zhou, M.-F.; Fisher, L.A.; Guagliardo, P.; Li, J.-W. Decoupling of Au and As during rapid pyrite crystallization. *Geology* **2021**, *49*, 827–831. [CrossRef]
19. Groves, D.I.; Goldfarb, R.; Gebre-Mariam, M.; Hagemann, S.; Robert, F. Orogenic gold deposits: A proposed classification in the context of their crustal distribution and relationship to other gold deposit types. *Ore Geol. Rev.* **1998**, *13*, 7–27. [CrossRef]
20. Goldfarb, R.; Groves, D.; Gardoll, S. Orogenic gold and geologic time: A global synthesis. *Ore Geol. Rev.* **2001**, *18*, 1–75. [CrossRef]
21. Goldfarb, R.J.; Baker, T.; Dubé, B.; Groves, D.I.; Hart, C.J.; Gosselin, P.; Hedenquist, J.W.; Thompson, J.F.H.; Richards, J.P. Distribution, character, and genesis of gold deposits in metamorphic terran. In *One Hundredth Anniversary Volume*; Society of Economic Geologists: Littleton, CO, USA, 2005.
22. Goldfarb, R.J.; Groves, D.I. Orogenic gold: Common or evolving fluid and metal sources through time. *Lithos* **2015**, *233*, 2–26. [CrossRef]
23. Groves, D.I.; Santosh, M.; Goldfarb, R.J.; Zhang, L. Structural geometry of orogenic gold deposits: Implications for exploration of world-class and giant deposits. *Geosci. Front.* **2018**, *9*, 1163–1177. [CrossRef]
24. Goldfarb, R.J.; Pitcairn, I. Orogenic gold: Is a genetic association with magmatism realistic? *Miner. Depos.* **2023**, *58*, 5–35. [CrossRef]
25. Dmitrijeva, M.; Cook, N.J.; Ehrig, K.; Ciobanu, C.L.; Metcalfe, A.V.; Kamenetsky, M.; Kamenetsky, V.S.; Gilbert, S. Multivariate statistical analysis of trace elements in pyrite: Prediction, bias and artefacts in defining mineral signatures. *Minerals* **2020**, *10*, 61. [CrossRef]
26. Winderbaum, L.; Ciobanu, C.L.; Cook, N.J.; Paul, M.; Metcalfe, A.; Gilbert, S. Multivariate analysis of an LA-ICP-MS trace element dataset for pyrite. *Math. Geosci.* **2012**, *44*, 823–842. [CrossRef]
27. Raič, S.; Molnár, F.; Cook, N.; O'BRien, H.; Lahaye, Y. Application of lithogeochemical and pyrite trace element data for the determination of vectors to ore in the Raja Au-Co prospect, northern Finland. *Solid Earth Discuss.* **2022**, *13*, 271–299. [CrossRef]
28. Thiruvengadam, S.; Murphy, M.E.; Tan, J.S.; Watling, R.J.; Stewart, J.I.; Miller, K. Multivariate modelling of the trace element chemistry of arsenopyrite from gold deposits using higher-dimensional algebras. *Math. Geosci.* **2020**, *52*, 865–928. [CrossRef]
29. Zoheir, B.; Goldfarb, R.; Holzheid, A.; Helmy, H.; El Sheikh, A. Geochemical and geochronological characteristics of the Um Rus granite intrusion and associated gold deposit, Eastern Desert, Egypt. *Geosci. Front.* **2020**, *11*, 325–345. [CrossRef]
30. Anwar, M.; Abu El-Leil, I.; Salem, S. Lithological and alteration mapping at the Um El-Rus Area, Central Eastern Desert, Egypt, using remote sensing techniques. *J. Indian Soc. Remote Sens.* **2023**, *51*, 829–848. [CrossRef]
31. Abd El Monsef, M. An Overview for Gold Mineralization in the Egyptian Segment of the Northern Continuation of the East African Orogen. In *Gold Deposits in Egypt: Geology, Settings, Types, Genesis and Spatiotemporal Distribution*; Springer: Cham, Switzerland, 2025; pp. 85–97.
32. El Tokhi, M.; El Muslem, A. Fluid inclusions in the gold-bearing quartz veins at Um Rus area, Eastern Desert, Egypt. *Chin. J. Geochem.* **2002**, *21*, 131–139. [CrossRef]
33. Harraz, H.; El-Dahhar, M. Nature and composition of gold-forming fluids at Umm Rus area, Eastern Desert, Egypt: Evidence from fluid inclusions in vein materials. *J. Afr. Earth Sci.* **1993**, *16*, 341–353. [CrossRef]
34. Amer, R.; Kusky, T.; El Mezayen, A. Remote sensing detection of gold related alteration zones in Um Rus area, Central Eastern Desert of Egypt. *Adv. Space Res.* **2012**, *49*, 121–134. [CrossRef]
35. El-Leil, I.A.; Sakr, S.; Abd El-Salam, L.M.; Azzam, H.M. Factors control of gold mineralization and associated ore metals, Um Rus area, Central Eastern Desert, Egypt. *J. Am. Sci.* **2016**, *12*, 110–125.
36. Akaad, M.; Noweir, A.M.; Abu El Ela, A.M.; El-Bahariya, G.A. The Muweilih conglomerate; Qift-Quseir road region, and the problem of the Atud formation. In *Proceedings of the Egyptian Geological Survey Centennial Conference*; The Geological Survey Egypt: Cairo, Egypt, 1996; pp. 23–45.
37. Kamel, O.; Elmahallawi, M.; Hilmy, H. Mineralogy of the Umm Rus goldbearing quartz veins and surrounding alteration zones. *Egypt. Mineral.* **1992**, *4*, e86.
38. El Gaby, S.; List, F.K.; Tehrani, R. The basement complex of the Eastern Desert and Sinai. In *The Geology of Egypt*; Routledge: Oxfordshire, UK, 2017; pp. 175–184.
39. Shalaby, A.; Stüwe, K.; Fritz, H.; Makroum, F. The El Mayah molasse basin in the Eastern Desert of Egypt. *J. Afr. Earth Sci.* **2006**, *45*, 1–15. [CrossRef]

https://doi.org/10.3390/min16050439

40. Goldfarb, R.J. Gold Deposit Types and Settings. In *Gold Deposits in Egypt: Geology, Settings, Types, Genesis and Spatiotemporal Distribution*; Springer: Cham, Switzerland, 2025; pp. 21–34.
41. Ai Zamruddin, N.N.S.; Abidin, N.S.Z.; Endut, Z.; Makoundi, C.; Lok, L.K.; Ismail, M.S. Trace Element Analysis of Pyrite and Arsenopyrite Using the LA-ICPMS Technique in Pulai, Central Belt of Peninsular Malaysia. *Minerals* **2023**, *13*, 1026. [CrossRef]
42. Cardenas-Vera, A.; Lentz, D.R.; McFarlane, C.R.; Thorne, K.G. Assessment of pyrite and arsenopyrite compositions, in situ S isotopes, and bulk Pb isotopes from the Cape Spencer gold deposit, New Brunswick, Canada. *Ore Geol. Rev.* **2025**, *178*, 106492. [CrossRef]
43. Cook, N.J.; Ciobanu, C.L.; Meria, D.; Silcock, D.; Wade, B. Arsenopyrite-pyrite association in an orogenic gold ore: Tracing mineralization history from textures and trace elements. *Econ. Geol.* **2013**, *108*, 1273–1283. [CrossRef]
44. Augustin, J.; Gaboury, D. Multi-stage and multi-sourced fluid and gold in the formation of orogenic gold deposits in the world-class Mana district of Burkina Faso–Revealed by LA-ICP-MS analysis of pyrites and arsenopyrites. *Ore Geol. Rev.* **2019**, *104*, 495–521. [CrossRef]
45. Kovalchuk, E.; Tagirov, B.R.; Vikentyev, I.V.; Chareev, D.A.; Tyukova, E.E.; Nikolsky, M.S.; Borisovsky, S.E.; Bortnikov, N.S. "Invisible" gold in synthetic and natural arsenopyrite crystals, Vorontsovka Deposit, Northern Urals. *Geol. Ore Depos.* **2019**, *61*, 447–468. [CrossRef]
46. Gourcerol, B.; Kontak, D.; Petrus, J.; Thurston, P. Application of LA ICP-MS analysis of arsenopyrite to gold metallogeny of the Meguma Terrane, Nova Scotia, Canada. *Gondwana Res.* **2020**, *81*, 265–290. [CrossRef]
47. Zhang, Z.; Xie, G.; Olin, P. Texture, in-situ geochemical, and S isotopic analyses of pyrite and arsenopyrite from the Longshan Sb-Au deposit, southern China: Implications for the genesis of intrusion-related Sb-Au deposit. *Ore Geol. Rev.* **2022**, *143*, 104781. [CrossRef]
48. Yang, L.; Wang, Q.; Fougerouse, D.; Xian, H.; Li, H.; Wang, R.; Rickard, W.D.A.; Groves, D.I.; Deng, J. Release and re-enrichment of invisible gold in arsenian pyrite promoted by coupled dissolution-reprecipitation reactions. *Commun. Earth Environ.* **2024**, *5*, 241. [CrossRef]
49. Sun, S.-C.; Yang, L.-Q.; Zhang, L.; Olin, P.; Gao, X.; Li, R.-H.; Wang, J.-Y.; Li, Z.-Q.; Zhang, F.; Wen, T. In-situ trace elements on pyrite and arsenopyrite of the Zhengchong gold deposit, Jiangnan Orogen: Insights for the mineralization mechanism. *Ore Geol. Rev.* **2020**, *122*, 103486. [CrossRef]
50. Tomkins, A.G.; Frost, B.R.; Pattison, D.R. Arsenopyrite melting during metamorphism of sulfide ore deposits. *Can. Mineral.* **2006**, *44*, 1045–1062. [CrossRef]
51. Tavares Nassif, M.; Monecke, T.; Reynolds, T.J.; Kuiper, Y.D.; Goldfarb, R.J.; Piazolo, S.; Lowers, H.A. Formation of orogenic gold deposits by progressive movement of a fault-fracture mesh through the upper crustal brittle-ductile transition zone. *Sci. Rep.* **2022**, *12*, 17379. [CrossRef] [PubMed]
52. Xiao, J.; Xie, Z.; Xia, Y.; Lan, T.; Zhang, L.; Pan, L.; Hu, H.; Wang, H.; Tan, Q.; Xu, Y. Predominant fracture gold in the Jiaodong gold province, China: Constraints on gold mineralization processes. *Ore Geol. Rev.* **2024**, *165*, 105915. [CrossRef]
53. Voute, F.; Hagemann, S.G.; Evans, N.J.; Villanes, C. Sulfur isotopes, trace element, and textural analyses of pyrite, arsenopyrite and base metal sulfides associated with gold mineralization in the Pataz-Parcoy district, Peru: Implication for paragenesis, fluid source, and gold deposition mechanisms. *Miner. Depos.* **2019**, *54*, 1077–1100. [CrossRef]
54. Tang, Q.; Di, P.; Yu, M.; Bao, J.; Zhao, Y.; Liu, D.; Wang, Y. Mineralogy and geochemistry of pyrite and arsenopyrite from the Zaozigou gold deposit in West Qinling orogenic belt, central China: Implications for ore genesis. *Resour. Geol.* **2019**, *69*, 314–332. [CrossRef]
55. Zhang, H.; Cai, Y.; Sha, G.; Brugger, J.; Pring, A.; Ni, P.; Qian, G.; Luo, Z.; Zhang, Y.; Tan, W. Effects of arsenic on the distribution and mode of occurrence of gold during fluid-pyrite interaction: A case study of pyrite from the Qiucun gold deposit, China. *Am. Mineral.* **2022**, *107*, 914–929. [CrossRef]
56. Morey, A.A.; Tomkins, A.G.; Bierlein, F.P.; Weinberg, R.F.; Davidson, G.J. Bimodal distribution of gold in pyrite and arsenopyrite: Examples from the Archean Boorara and Bardoc shear systems, Yilgarn craton, Western Australia. *Econ. Geol.* **2008**, *103*, 599–614. [CrossRef]
57. Cabral, A.R.; Koglin, N. Hydrothermal fluid source constrained by Co/Ni ratios in coexisting arsenopyrite and tourmaline: The auriferous lode of Passagem, Quadrilátero Ferrífero of Minas Gerais, Brazil. *Mineral. Petrol.* **2012**, *104*, 137–145. [CrossRef]
58. Cepedal, A.; Fuertes-Fuente, M.; Martin-Izard, A.; Gonzalez-Nistal, S.; Barrero, M. Gold-bearing As-rich pyrite and arsenopyrite from the El Valle gold deposit, Asturias, northwestern Spain. *Can. Mineral.* **2008**, *46*, 233–247. [CrossRef]
59. Kovalev, K.; Kalinin, Y.; Naumov, E.; Kolesnikova, M.; Korolyuk, V. Gold-bearing arsenopyrite in eastern Kazakhstan gold-sulfide deposits. *Russ. Geol. Geophys.* **2011**, *52*, 178–192. [CrossRef]
60. Qiu, W.J.; Zhou, M.-F. Self-organizational origin of oscillatory zoning in pyrite: A coupled diffusion–reaction model and pH-dependent As adsorption feedback mechanism. *Geochim. Cosmochim. Acta* **2025**, *403*, 94–111. [CrossRef]

61. Abdus-Salam, M.; Brugger, J.; Etschmann, B.; McNab, R.; Li, W. Arsenopyrite-Pyrite Association in Mineral Replacement Experiments: A Microscopic Study of Arsenic-Controlled Iron Sulfide Formation and implications for Orogenic Gold Mineralisation. In *Proceedings of the 13th Asia Pacific Microscopy Congress 2025 (APMC13)*; ScienceOpen: Berlin, Germany, 2025.
62. Ashley, P.; Creagh, C.; Ryan, C. Invisible gold in ore and mineral concentrates from the Hillgrove gold-antimony deposits, NSW, Australia. *Miner. Depos.* **2000**, *35*, 285–301. [CrossRef]
63. Asghari, G.; Alipour, S.; Azizi, H.; Mirnejad, H. Geology and mineral chemistry of gold mineralization in Mirge-Naqshineh occurrence (Saqez, NW Iran): Implications for transportation and precipitation of gold. *Acta Geol. Sin.-Engl. Ed.* **2018**, *92*, 210–224. [CrossRef]
64. Ehrig, K.; Ciobanu, C.L.; Verdugo-Ihl, M.R.; Dmitrijeva, M.; Cook, N.J.; Slattery, A. Lifting the cloak of invisibility: Gold in pyrite from the Olympic Dam Cu-U-Au-Ag deposit, South Australia. *Am. Mineral. J. Earth Planet. Mater.* **2023**, *108*, 259–276. [CrossRef]
65. King, S.A.; Cook, N.J.; Ciobanu, C.L.; Ehrig, K.; Rodriguez, Y.T.C.; Basak, A.; Gilbert, S. Coupled microstructural EBSD and LA-ICP-MS trace element mapping of pyrite constrains the deformation history of breccia-hosted IOCG ore systems. *Minerals* **2024**, *14*, 198. [CrossRef]
66. Fougerouse, D.; Reddy, S.M.; Sumail; Brugger, J.; Thébaud, N.; Rickard, W.D.; Yang, L.; Quadir, Z.; Roberts, M.P.; Tomkins, A.G.; et al. Dislocation-mediated interfacial re-equilibration of pyrite: An alternative model to interface-coupled dissolution-reprecipitation and gold remobilisation. *Geochim. Cosmochim. Acta* **2024**, *374*, 136–145. [CrossRef]
67. de Araujo Silva, A.; Hector, S.; Patten, C.; Beranoaguirre, A.; Eiche, E.; Walter, B.F.; Kolb, J. Hydrothermal co-enrichment of Au, As, Co, and Cu in orogenic deposits: Insights from the Pohjanmaa belt. *Miner. Depos.* **2026**, *61*, 91–115. [CrossRef]
68. Kravtsova, R.G.; Makshakov, A.S.; Tauson, V.L.; Belozerova, O.Y. Distribution and speciation features of finely dispersed and "Invisible" gold in arsenopyrite and pyrite of the Natalka Deposit (Northeastern Russia). *Геология Рудных Месторождений* **2025**, *67*, 161–201. [CrossRef]

 https://doi.org/10.3390/min16050439

minerals

MDPI

Article

Genesis of the Longkou Gold Deposit in the Northeastern Jiaolai Basin: Constraints from Sericite Rb-Sr Geochronology and Pyrite Geochemistry

Jin-Shuai Zhang [1], Hao-Cheng Yu [1,2,*], Guo-Long Yan [3,*], Ming Ma [1,4,5], Tao Cui [1], Ya-Peng Li [1], Lian-Yuan Qin [6] and Chun-Ting Xu [1]

1 State Key Laboratory of Geological Processes and Mineral Resources, Frontiers Science Center for Deep-Time Digital Earth, School of Earth Sciences and Resources, China University of Geosciences, Beijing 100083, China; 15831450506@163.com (J.-S.Z.); sddkmaming@163.com (M.M.); cuitao95@163.com (T.C.); 19106382043@163.com (Y.-P.L.); chuntingxu@163.com (C.-T.X.)

2 North China Center for Geoscience Innovation, Tianjin Center, China Geological Survey, Tianjin 300170, China

3 China National Gold Group Corporation Limited, Beijing 100011, China

4 No. 1 Geological Team of Shandong Provincial Bureau of Geology and Mineral Resources, Jinan 250100, China

5 Shandong Engineering Research Center of High-Grade Iron Deposits Exploration and Exploitation, Jinan 250100, China

6 Shandong Yantai Xintai Gold Mining Industry Co., Ltd., Yantai 265147, China; lianyuanqin@126.com

* Correspondence: haochengyu@cugb.edu.cn (H.-C.Y.); w43628025763@163.com (G.-L.Y.); Tel.: +86-13361322893 (H.-C.Y.); +86-15076901657 (G.-L.Y.)

Abstract

Whether the genesis of gold deposits in the Northeastern Jiaolai Basin is consistent with that in the Northwestern Jiaodong area remains controversial. This study presents in situ Rb-Sr dating of sericite, along with in situ trace element and sulfur isotope analyses of pyrite in the Longkou gold deposit. The sericite Rb-Sr inverse isochron yields an age of 120.9 ± 2.4 Ma, indicating that gold mineralization occurred in the Early Cretaceous. Two generations of pyrite, Py1 and Py2, were identified. Py1 is anhedral and hosted in relatively low-grade, weakly altered marble wall rock. Py2 is euhedral to subhedral and hosted in relatively high-grade, strongly altered marble ore. The $\delta^{34}S$ value of Py1 is 7.38‰, whereas that of Py2 is 6.79‰. The decrease in $\delta^{34}S$ values from Py1 to Py2 reflects an increase in the oxygen fugacity of the ore-forming system. These features suggest that fluid–rock interaction led to an increase in oxygen fugacity, thereby triggering gold precipitation. The mineralization age and precipitation mechanism of the Longkou gold deposit are consistent with those of the Northwestern Jiaodong area. The Longkou gold deposit is best classified as a Jiaodong-type gold deposit.

Keywords: sericite Rb-Sr geochronology; pyrite geochemistry; fluid–rock interaction; Longkou gold deposit; northeastern Jiaolai Basin

1. Introduction

The Jiaodong Peninsula hosts the world's third-largest gold province, with total gold resources reaching nearly 6000 t. Its gold reserves and production account for one-quarter and one-third of China's total, respectively [1–5]. Gold deposits in the Northwestern Jiaodong area exhibit consistency in terms of metallogenic setting and mineralization processes. The mineralization age is concentrated in the Early Cretaceous (ca. 120 Ma), coinciding with the peak periods of lithospheric thinning in eastern China and destruction of the North China Craton [6,7]. The ore-controlling structures are NNE-trending detachment faults developed along the contact zones between Precambrian metamorphic rocks and Mesozoic granites. The hydrothermal alteration assemblages are dominated by pyritization, silicification, and sericitization. The ore-forming fluids are low-temperature, low-salinity, reduced H_2O–CO_2–NaCl±CH_4 system [1,8]. Metal precipitation is mainly driven by fluid boiling or fluid–rock interaction [9]. These characteristics differ from magmatic–hydrothermal deposits, so gold deposits in the Northwestern Jiaodong area are defined as the Jiaodong-type gold deposit.

The Longkou gold deposit is located in the Guocheng-Yazi area of the Northeastern Jiaolai Basin [10]. The genesis of this deposit remains controversial. Some researchers classify it as a Jiaodong-type gold deposit based on its similarities with the gold deposits in the Northwestern Jiaodong area. These similarities include NNE-trending structural control, alteration characteristics, mineralization styles, and ore-forming fluid features. Other scholars emphasize its spatiotemporal association with Mesozoic magmatic rocks and support a magmatic–hydrothermal origin [11–18]. Resolving this controversy is crucial for improving our understanding of the ore-forming processes in the gold district of the Northeastern Jiaolai Basin.

The Longkou gold deposit contains abundant sericite and pyrite that are closely associated with gold mineralization. In recent years, in situ Rb-Sr dating techniques have advanced rapidly, and Laser Ablation Inductively Coupled Plasma Tandem Mass Spectrometry (LA-ICP-MS/MS) enables precise determination of mineralization ages in low-temperature hydrothermal systems [19,20]. Sericite, a common K-rich mineral in hydrothermal alteration assemblages, readily incorporates Rb via isomorphous substitution for K, making it suitable for in situ Rb-Sr dating to precisely determine the mineralization age. Pyrite, as the main gold-hosting sulfide, directly participates in the ore-forming process, and its trace element and isotopic signatures can provide information on the ore-forming fluids. Moreover, pyrite exhibits complex textures and high stability, allowing such information to be effectively preserved. Therefore, in situ trace element and sulfur isotope analyses can effectively constrain the mechanisms of metal precipitation [21]. This study presents in situ Rb-Sr geochronology of sericite from the Longkou gold deposit, combined with in situ trace element and sulfur isotope analyses of pyrite, to precisely constrain its mineralization age and metal precipitation mechanism, thereby providing evidence for elucidating the gold mineralization in the Northeastern Jiaolai Basin and its genetic link with the regional metallogenic system of the Jiaodong Peninsula.

2. Regional Geologic Setting

The Jiaodong Peninsula is located in the southeastern North China Block and represents an intracontinental composite tectonic domain that transitioned from the Tethyan to the Circum-Pacific tectonic regime [3]. The Jiaodong Peninsula is bounded to the east by the subduction zone of the Paleo-Pacific Plate and to the south by the Dabie–Sulu ultrahigh-pressure metamorphic belt. To the west, it is bounded by the Tan–Lu deep sinistral strike-slip fault, and to the north by the region where the continental lithosphere is thinnest in China. (Figure 1A). This region experienced a complex geological history: Archean to Paleoproterozoic cratonization; the collisional amalgamation of the Longgang and Langlin blocks during the Paleoproterozoic (1.9 Ga); subduction and orogenesis of the Paleo-Tethys Ocean plate during the Late Paleozoic; continent–continent collision and deep subduction between the South China and North China blocks during the Triassic (220 Ma); subduction-rollback of the Paleo-Pacific Plate during the Late Mesozoic, accompanied by asthenospheric upwelling and lithospheric mantle delamination (120 Ma) [3,22–25]. The NE–SW-trending Wulian-Yantai deep sinistral strike-slip fault divides the peninsula into the Jiaobei Terrane to the west and the Sulu Terrane to the east (Figure 1B) [7,26,27]. The Jiaobei Terrane is

https://doi.org/10.3390/min16050485

mainly composed of Neoarchean to Paleoproterozoic basement rocks, including tonalite–trondhjemite–granodiorite (TTG) and amphibolite of the Jiaodong Group (2.9–2.6 Ga). It is also overlain by Paleoproterozoic metasedimentary sequences of the Fenzishan Group and Jingshan Group [28,29]. The Sulu Terrane consists of Neoproterozoic granitic gneisses, high-pressure eclogites, and Triassic granites [30,31]. Mesozoic magmatic activity is dominated by Late Jurassic to Early Cretaceous granitic intrusions and intermediate–mafic dikes. The granitic intrusions include the Linglong type (163.2–149.0 Ma), Guojialing type (132.0–125.0 Ma), Weideshan type (125.0–110.5 Ma), and Laoshan type (125.0–108.9 Ma). The intermediate–mafic dikes are represented by island-arc-type and ocean-island-type lamprophyre dikes [32]. Gold mineralization is spatially controlled by NE–SW to NNE–SSW-trending fault systems. These fault systems define several major gold districts, including the Sanshandao–Cangshang, Jiaojia–Xincheng, Zhaoyuan–Pingdu, Qixia–Penglai, Northeastern Jiaolai Basin, and Muping–Rushan districts (Figure 1B).

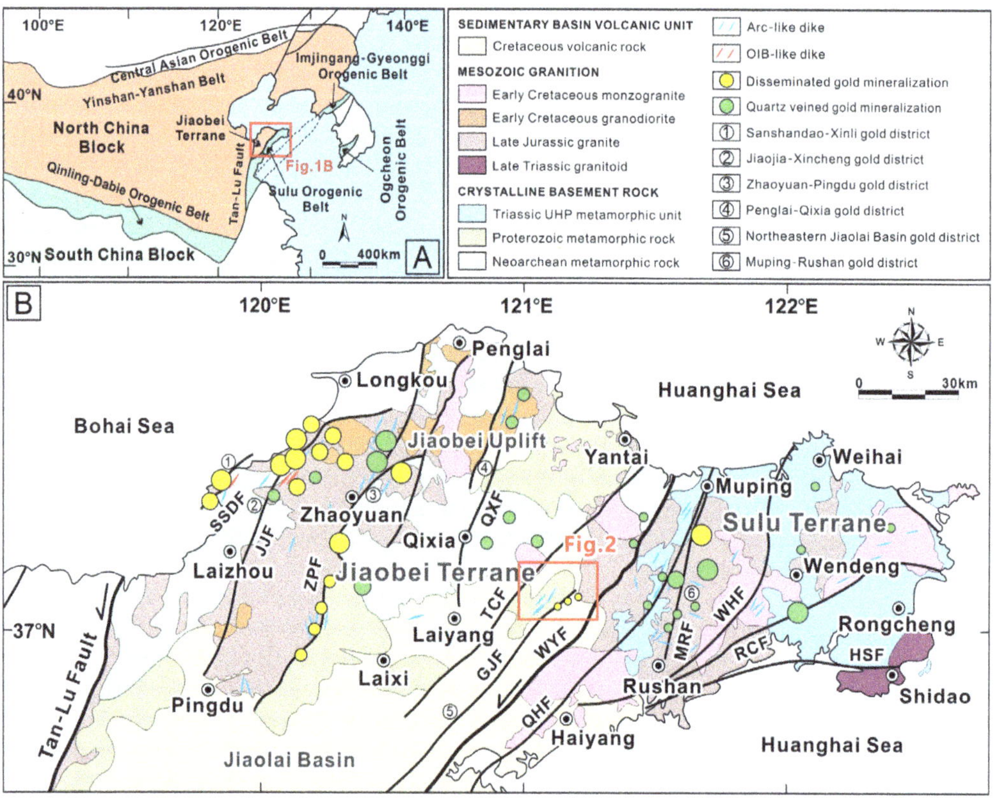

Figure 1. (**A**) Simplified geological map showing the tectonic setting of the Jiaodong gold province, China. (**B**) Geological map of the Jiaodong gold province showing the distribution of gold deposits in terms of size, modified after Deng et al. (2017) [33]. UHP: ultra high-pressure; OIB: ocean island basalt; SSDF: Sanshandao Fault; JJF: Jiaojia Fault; ZPF: Zhaoping Fault; QXF: Qixia Fault; TCF: Taocun Fault; GJF: Guoji Fault; WYF: Wulian-Yantai Fault; QHF: Qingdao-Haiyang Fault; MRF: Muru Fault; WHF: Weihai Fault; RCF: Rongcheng Fault; HSF: Haiyang-Shidao Fault. The circle size represents the amount of gold resources.

The Northeastern Jiaolai Basin gold district is located in the southern Jiaobei Terrane (36°55′00″–37°10′20″ N, 121°00′00″–121°26′00″ E). It has proven gold resources exceeding 200 t, indicating significant metallogenic potential (Figure 1B) [34,35]. The Jiaolai Basin

https://doi.org/10.3390/min16050485

is a NE–SW-trending Cretaceous extensional basin. Its basement is mainly composed of the Paleoproterozoic Jingshan Group, and it is overlain by the Cretaceous Laiyang Group and Qingshan Group (Figure 2). The Jingshan Group is subdivided into the Yetou Formation and the Douya Formation. The Yetou Formation consists of lower amphibolite-facies granulite and upper marble units, whereas the Douya Formation is composed of graphite-bearing gneiss and marble (Figure 3). The Laiyang Group is dominated by sedimentary rocks such as conglomerate and sandy conglomerate; the Qingshan Group mainly comprises intermediate–mafic to felsic volcaniclastic rocks, tuffs, and lavas. Magmatic records include the Paleoproterozoic Muniushan monzogranite (2105 ± 7 Ma) [36] and the Late Jurassic Queshan pluton (160–150 Ma) [25]. Early Cretaceous intrusions is represented by diorite porphyrite (116 ± 1 Ma) and monzonite porphyry (114 ± 2 Ma) [37]. The structural framework is dominated by NE–SW-trending faults, including the Guocheng Fault, Yazi Fault, and their secondary faults. These faults control the distribution of gold deposits such as Longkou, Liaoshang, Qianchuiliu, Xilaokou, Xijingkou, and Songjiagou (Figure 2).

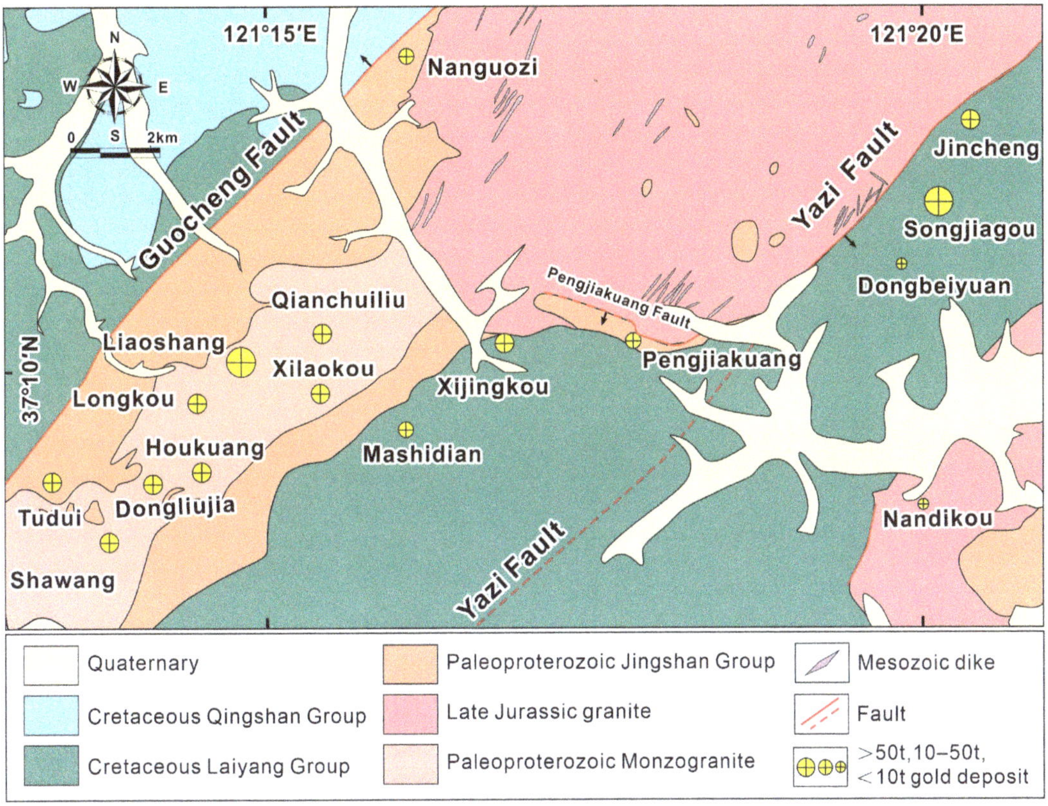

Figure 2. Simplified geologic map of the Northeastern Jiaolai Basin gold district, modified after Cui et al. (2026) and Bo et al. (2025) [35,38]. The arrows indicate the dip direction.

https://doi.org/10.3390/min16050485

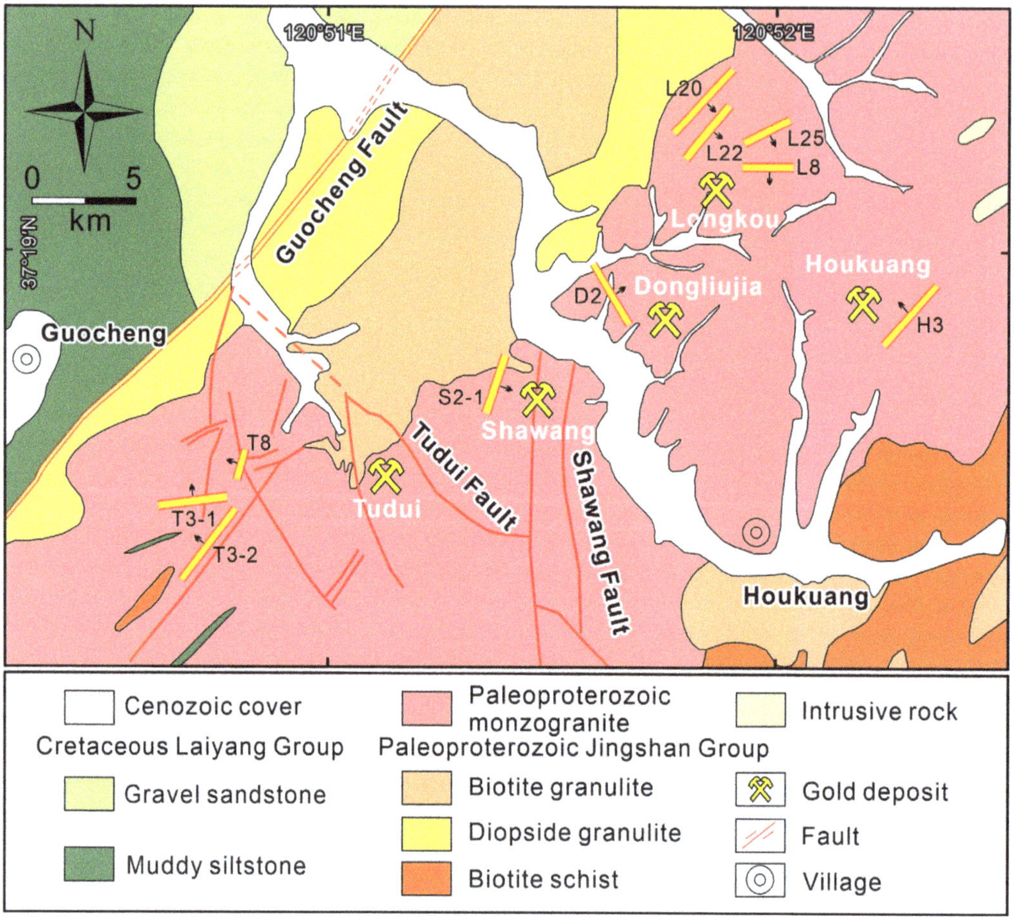

Figure 3. Simplified geological map of the Longkou gold deposit in China, modified after Cui et al. (2025) [10].

3. Deposit Geology

The Longkou gold deposit is located in the footwall of the Guocheng Fault, within secondary fault zones (Figure 2). Gold orebodies are mainly hosted in the marble unit of the Jingshan Group and the Muniushan monzogranite (Figure 3). Representative orebodies L22 and L20 are approximately 260–280 m in length, 0.90–24.42 m in width, and 1.00–3.60 m in thickness, with an average grade of 2.67 g/t. The orebodies strike NE–SW, dip SE, and have dip angles ranging from 14° to 41° (Figure 4). Gold is veinlet-disseminated. Petrographic observations indicate that ore minerals are predominantly pyrite and pyrrhotite, with minor chalcopyrite, sphalerite, and magnetite (Figure 5A–C). Gangue minerals include quartz, plagioclase, K-feldspar, muscovite, sericite, chlorite, and calcite (Figure 5D–I). Hydrothermal alteration types mainly include pyritization, sericitization, silicification, chloritization, and carbonatization. Two generations of pyrite, Py1 and Py2, were identified. Py1 is anhedral and hosted in relatively low-grade, weakly altered marble wall rock. Py2 is euhedral to subhedral, commonly fractured with visible fracture-hosted gold, and hosted in relatively high-grade, strongly altered marble ore (Figure 5C).

 https://doi.org/10.3390/min16050485

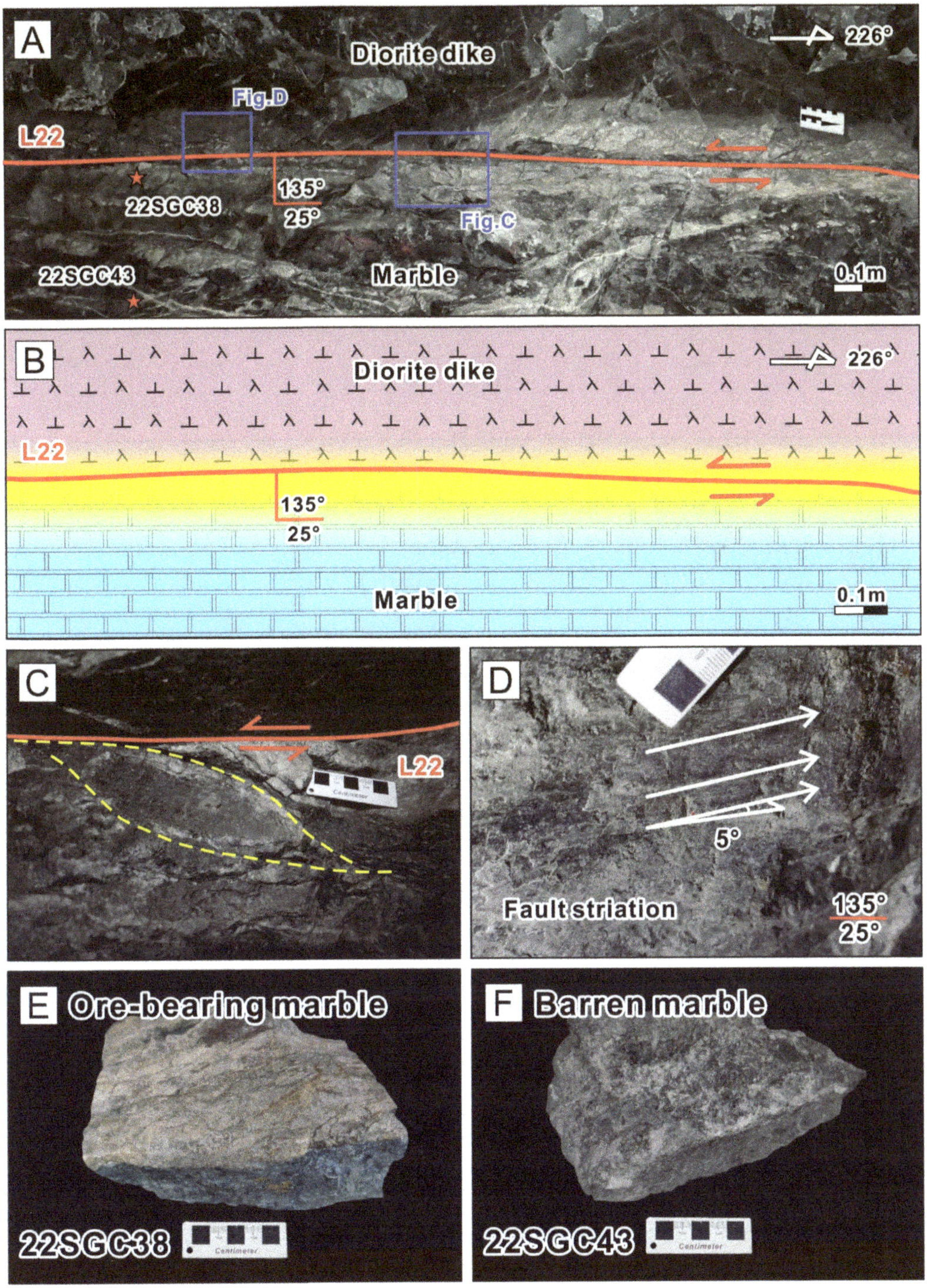

Figure 4. Field photographs and specimens of the Longkou gold deposit. (**A**,**B**) Cross-section showing the diorite dike intruding into marble (L22 orebody). (**C**) Lens-shaped orebody in marble. (**D**) Slickenside developed on the fault surface. (**E**) Ore-bearing marble. (**F**) Barren marble.

https://doi.org/10.3390/min16050485

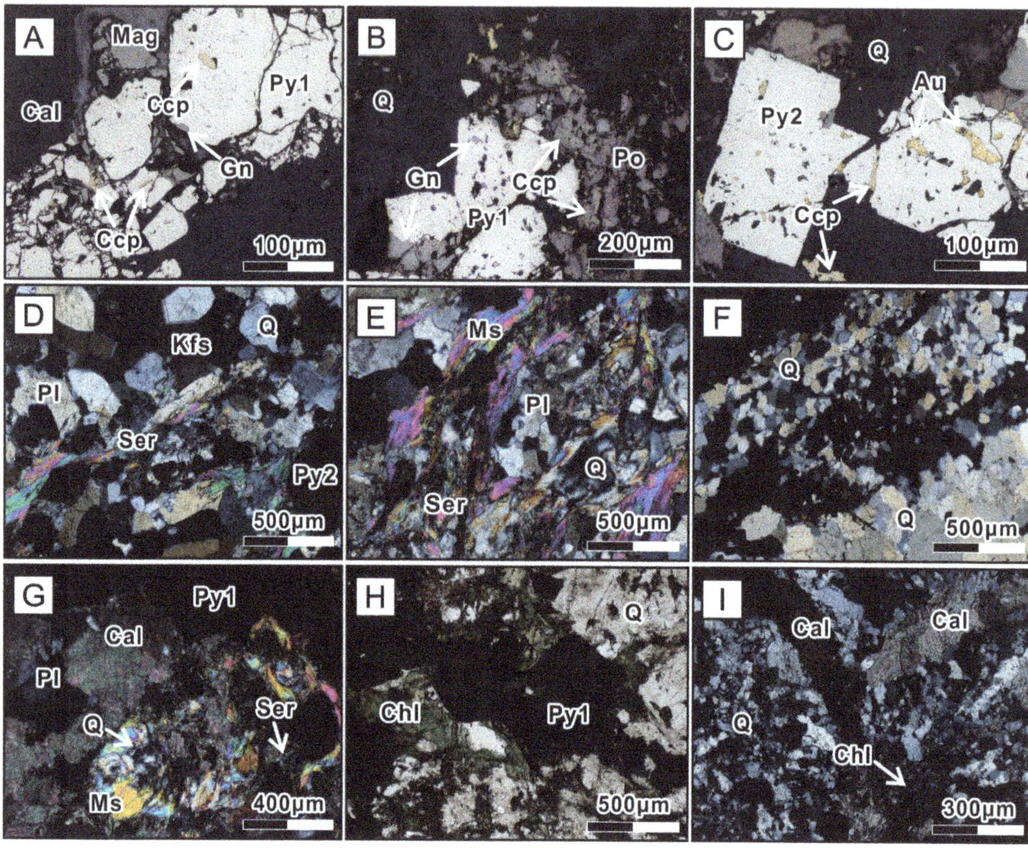

Figure 5. Representative photomicrographs showing mineral assemblages of the Longkou gold deposit. (**A–C**) Assemblage of metal minerals. (**D**) Pyrite–sericite–quartz alteration. (**E**) Sericite–quartz alteration. (**F**) Silicification. (**G**) Carbonatization. (**H**,**I**) Chloritization. Py: Pyrite; Po: Pyrrhotite; Ccp: Chalcopyrite; Gn: Galena; Mag: Magnetite; Au: Gold; Ser: Sericite; Ms: Muscovite; Pl: Plagioclase; Kfs: K-feldspar; Chl: Chlorite; Cal: Calcite; Q: Quartz.

4. Samples and Analysis Methods

4.1. Samples Strategy

Eleven samples were collected from the Longkou gold deposit for optical microscopy, in situ Rb-Sr dating of sericite, and in situ trace element and sulfur isotope analyses of pyrite.

4.2. In Situ Sericite Rb-Sr Dating

In situ Rb-Sr dating of sericite was conducted at the National Research Center for Geoanalysis, Chinese Academy of Geological Sciences, Beijing, China, using a laser ablation inductively coupled plasma tandem mass spectrometer (LA-ICP-MS; Agilent 8900 ICP-MS/MS triple quadrupole coupled with a New Wave Research 193^{TM} laser). The laser repetition rate was 5 Hz, the energy density was 3.5 J/cm^2, and the spot diameter was 67 μm for sericite ablation. Each analysis consisted of 30 s of background acquisition, 120 s of ablation signal acquisition, and 30 s of washout time. Dwell times for ^{85}Rb and Sr isotopes (^{86}Sr, ^{87}Sr, ^{88}Sr, ^{86}Sr^{16}O, ^{87}Sr^{16}O, and ^{88}Sr^{16}O) were all set to 50 ms. Mica-mg, NIST-610, and NIST-612 reference materials were used for calibration. The obtained data

https://doi.org/10.3390/min16050485

were processed using the Isoplot4 R software package [39]. Detailed analytical procedures are described in Yu et al. (2025) [19] and Huang et al. (2023) [20].

4.3. In Situ Pyrite Trace Element Analysis

In situ trace element analysis of pyrite was conducted at the National Research Center for Geoanalysis, Chinese Academy of Geological Sciences, Beijing, China, using an NWR 193^{TM} ArF excimer laser ablation system coupled with an Agilent 8900 ICP-MS/MS mass spectrometer. The laser repetition rate was 6 Hz, the energy density was 5 J/cm^2, and the spot diameters ranged from 30 to 40 μm for pyrite ablation. Each analysis consisted of 20 s of background acquisition and 50 s of ablation signal acquisition. NIST-610 and MASS-1 reference materials were used as primary standards. The obtained data were processed using the ICPMSDataCal 11.8 software package [40].

4.4. In Situ Pyrite Sulfur Isotope Analysis

In situ sulfur isotope analysis of pyrite was conducted at the National Research Center for Geoanalysis, Chinese Academy of Geological Sciences, Beijing, China, using laser ablation multi-collector inductively coupled plasma mass spectrometry (LA-MC-ICP-MS). A Neptune Plus multi-collector inductively coupled plasma mass spectrometer coupled with an NWR 193^{TM} laser ablation system was used. The laser repetition rate was 6 Hz, the energy density was 4 J/cm^2, and the spot diameter was 30 μm for pyrite ablation. JX and HN reference materials were used for standard-sample bracketing calibration to determine $\delta^{34}S$ values ($\delta^{34}S = (^{34}S/^{32}S)_{sample}/(^{34}S/^{32}S)_{VCDT} - 1$).

5. Results

5.1. In Situ Sericite Rb-Sr Geochronology

In situ sericite Rb-Sr isotope data for the Longkou gold deposit are presented in Table S1. A total of 62 analytical spots yielded $^{87}Rb/^{86}Sr$ ratios ranging from 0.0027 to 210.5805, defining a clear inverse isochron in the $^{87}Rb/^{86}Sr$ vs. $^{86}Sr/^{87}Sr$ diagram. The calculated inverse isochron age is 120.9 ± 2.4 Ma (MSWD = 1.7), with an initial $^{87}Sr/^{86}Sr$ ratio of 0.7145 ± 0.0026.

5.2. In Situ Pyrite Trace Element Composition

In situ LA-ICP-MS analyses were conducted on pyrite from the Longkou gold deposit, yielding 71 data points, which are presented in Table S2. The data show systematic variations in trace element content from Py1 to Py2: As content increase from 84 ppm to 410 ppm; Co content increase from 27 ppm to 81 ppm; Ni content decrease from 35 ppm to 22 ppm; Cu content increase from 9.1 ppm to 32 ppm; Pb content increase from 1.1 ppm to 6.3 ppm; and Zn content decrease from 2.32 ppm to 0.93 ppm. Au, Ag, and Sb contents are generally very low (<1 ppm) but also show increasing trends.

5.3. In Situ Pyrite Sulfur Isotope Composition

In situ sulfur isotope analyses were conducted on 25 pyrite samples from the Longkou gold deposit, with the results presented in Table S3. The data reveal a systematic decrease in $\delta^{34}S$ values from Py1 to Py2, with mean values of 7.38‰ for Py1 and 6.79‰ for Py2.

6. Discussion

6.1. Mineralization Age

This study presents in situ Rb-Sr dating on sericite, a hydrothermal alteration mineral, providing reliable constraints on the formation age of the deposit. The sericite used for dating was obtained from ore samples of the Longkou gold deposit. Microscopic ob-

 https://doi.org/10.3390/min16050485

servations show that sericite occurs as fine-grained scaly aggregates that are intimately intergrown with gold-bearing pyrite (Py2) (Figure 6A). The sericite grains display uniform interference colors, without microscopic evidence of alteration rims, replacement textures, or crosscutting by later veinlets. These features indicate that the sericite represents a single generation, coeval with gold-bearing pyrite during the main mineralization stage [41]. Chlorite thermometry indicates a mineralization temperature of approximately 300–350 °C [38], lower than the closure temperature of the sericite Rb-Sr system [42]. This implies that the sericite Rb-Sr system closed immediately after crystallization, thus directly recording the timing of mineralization. The in situ LA-ICP-MS/MS sericite Rb-Sr inverse isochron age of (120.9 ± 2.4) Ma (Figure 6A) precisely indicates that gold mineralization occurred during the Early Cretaceous. This age shows a significant temporal gap from the Late Jurassic (160–150 Ma) regional magmatic activity [3] (Figure 6C). The Longkou gold deposit shares the same mineralization age (ca. 120 Ma) with other gold deposits in the Northeastern Jiaolai Basin (Figure 6B), and gold deposits in the Northwestern Jiaodong area (e.g., Sanshandao, Jiaojia, Linglong) were also formed in the Early Cretaceous (ca. 120 Ma; Figure 6C). This indicates a unified regional metallogenic event that coincides with the subduction of the Paleo-Pacific Plate [43,44]. In summary, the Longkou gold deposit is coeval with gold deposits in the Northwestern Jiaodong area and is a product of Early Cretaceous plate subduction.

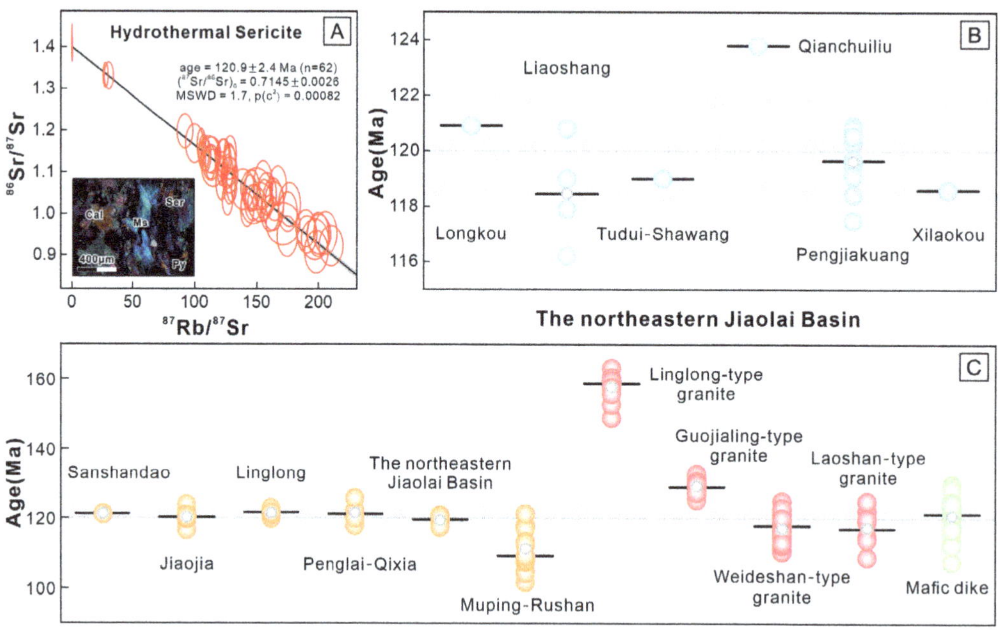

Figure 6. Sericite Rb-Sr geochronology and regional mineralization age correlation for the Longkou gold deposit. (**A**) Rb-Sr inverse isochron diagram for hydrothermal sericite. (**B**) Mineralization ages of gold deposits in the Northeastern Jiaolai Basin [16,45,46]. (**C**) Mineralization ages of gold deposits in the Jiaodong Peninsula [16,33,47,48].

6.2. Mechanism of Gold Enrichment and Precipitation

In the Longkou gold deposit, pyrite, an important gold-bearing mineral, provides critical insights into the mechanisms of gold precipitation through its sulfur isotopic composition. In this study, the $\delta^{34}S$ values of pyrite decrease from Py1 to Py2 (Figure 7). Petrographic observations reveal that pyrite is replaced by magnetite, and together with

https://doi.org/10.3390/min16050485

the coexistence of reduced pyrrhotite and oxidized magnetite (Figure 5A–C), this supports an increase in oxygen fugacity during mineralization. Elevated oxygen fugacity promotes the partitioning of the heavy isotope ^{34}S into oxidized SO_4^{2-}. This results in depletion of ^{34}S in the remaining reduced HS^- and, consequently, lower $\delta^{34}S$ values in precipitated sulfides [49–52]. Therefore, the decrease in $\delta^{34}S$ indicates an increase in oxygen fugacity within the mineralizing system.

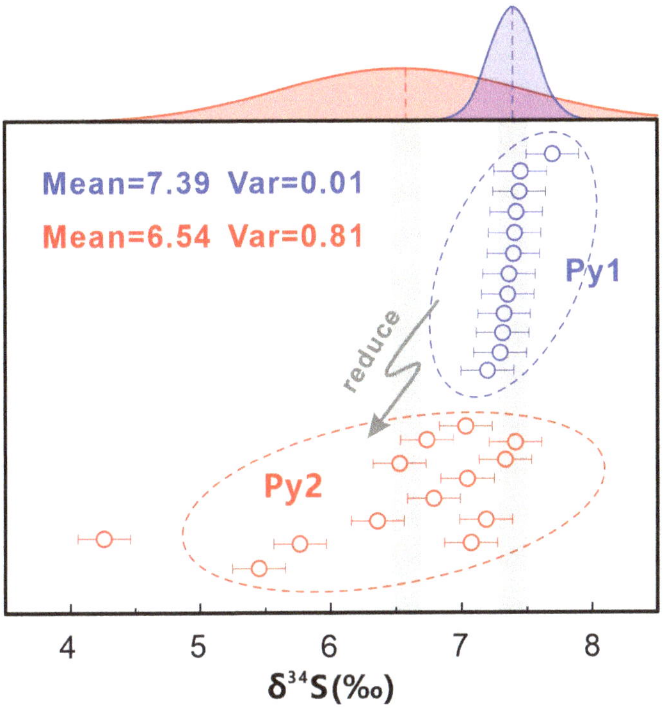

Figure 7. Sulfur isotope trend of pyrite from different generations in the Longkou gold deposit.

The Au-As solubility diagram indicates that gold occurs in solid solution in pyrite [53] (Figure 8C). The presence of As in pyrite suggests that the ore-forming fluids were near-neutral [54]. The Cu/Au ratios (>1) of pyrite suggest that Cu directly substitutes for Au in the pyrite lattice, which occurs under relatively reducing conditions [54]. Under such conditions, gold is primarily transported as $Au(HS)_2^-$ complexes [55]. Thermodynamic modeling shows that a slight increase in oxygen fugacity can cause significant gold precipitation [10]. This is because elevated oxygen fugacity leads to the oxidation of reduced H_2S to oxidized SO_4^{2-}, which reduces the H_2S content in the system and thereby destabilizes $Au(HS)_2^-$ complexes [10,56,57], ultimately triggering gold precipitation.

Py1 is hosted in relatively low-grade, weakly altered marble wall rock, indicating weak fluid–rock interaction; Py2 is hosted in relatively high-grade, strongly altered marble ore, indicating strong fluid–rock interaction. Therefore, gold precipitation in the Longkou gold deposit was primarily triggered by fluid–rock interaction. This interaction increased oxygen fugacity and destabilized $Au(HS)_2^-$ complexes, leading to gold mineralization.

https://doi.org/10.3390/min16050485

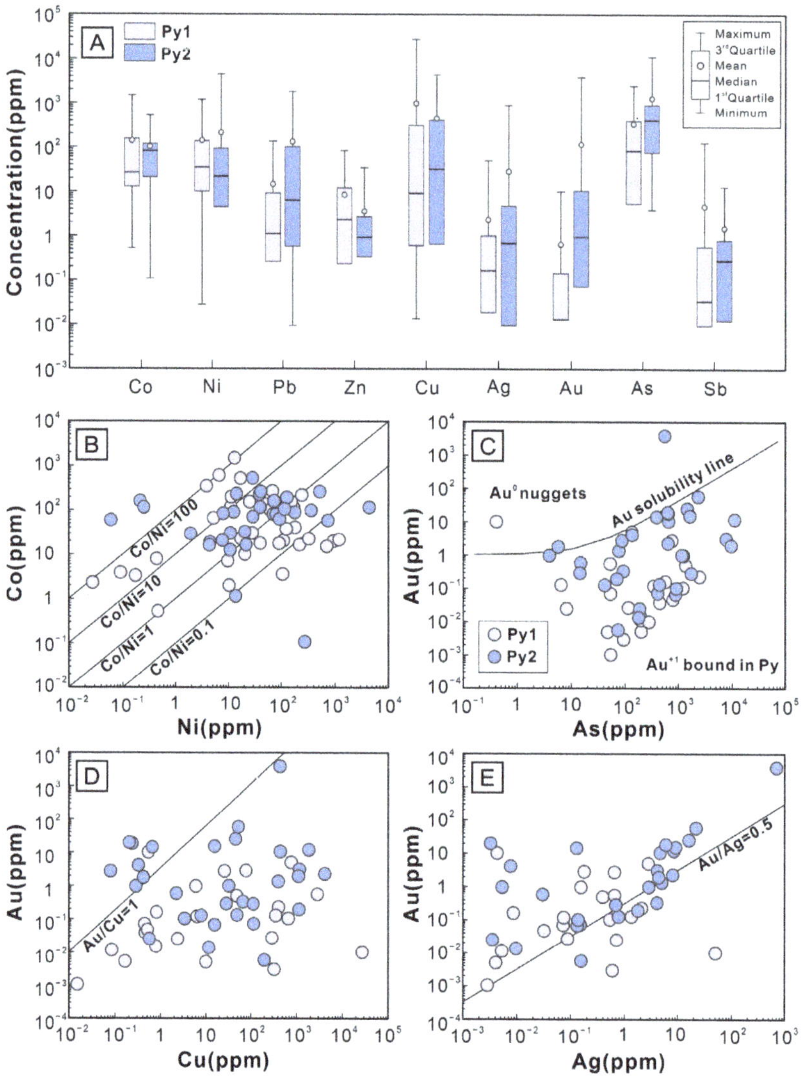

Figure 8. (**A**) Trace element content (in ppm) for the different generations of pyrite. (**B**) Co vs. Ni. (**C**) Au vs. As. (**D**) Au vs. Cu. (**E**) Au vs. Ag.

6.3. Ore Genesis

Based on the close spatial association between the Longkou gold deposit and magmatic rocks, some researchers have proposed a magmatic–hydrothermal genetic model [46,58,59]. However, the mineralization age obtained in this study, ca. 120 Ma (Figure 6B), is significantly later than the formation ages of regional granites (ca. 160–150 Ma) [3], with a gap of 30–40 Ma. Furthermore, hydrogen and oxygen isotope data fall outside the range of magmatic–hydrothermal fluids and are predominantly metamorphic fluids [60,61]. This substantial temporal gap, combined with the stable isotope evidence, indicates that the Longkou gold deposit cannot be explained by a magmatic–hydrothermal model directly linked to the regional granites.

The Longkou gold deposit exhibits consistency with gold deposits in the Northwestern Jiaodong area in terms of metallogenic geological characteristics: (1) Its mineralization

https://doi.org/10.3390/min16050485

space is strictly controlled by NE–SW- to NNE–SSW-trending fault systems, and it is characterized by a typical suite of alteration mineral assemblages, including pyritization, silicification, sericitization, K-feldspar alteration, chloritization, and carbonatization. (2) The mineralization age is concentrated in the Early Cretaceous (ca. 120 Ma; Figure 6B), synchronous with the regional-scale metallogenic event, which is widely considered to be closely related to plate subduction [27,43,62–64]. (3) The ore-forming fluids are characterized by moderate temperatures (ca. 200–400 °C), low salinities (<11 wt.% NaCl), low densities, and are classified as a typical H_2O-CO_2-NaCl±CH_4 system [65,66]. (4) The sulfur isotope $\delta^{34}S$ values range from 6‰ to 12‰, showing regional consistency (Figure 9). (5) Disseminated mineralization is extensively developed in the deposit, triggered by fluid–rock interaction. This interaction directly destabilized the gold-bearing sulfur complexes by modifying the physicochemical conditions of the system, thereby significantly reducing gold solubility and ultimately leading to extensive gold precipitation.

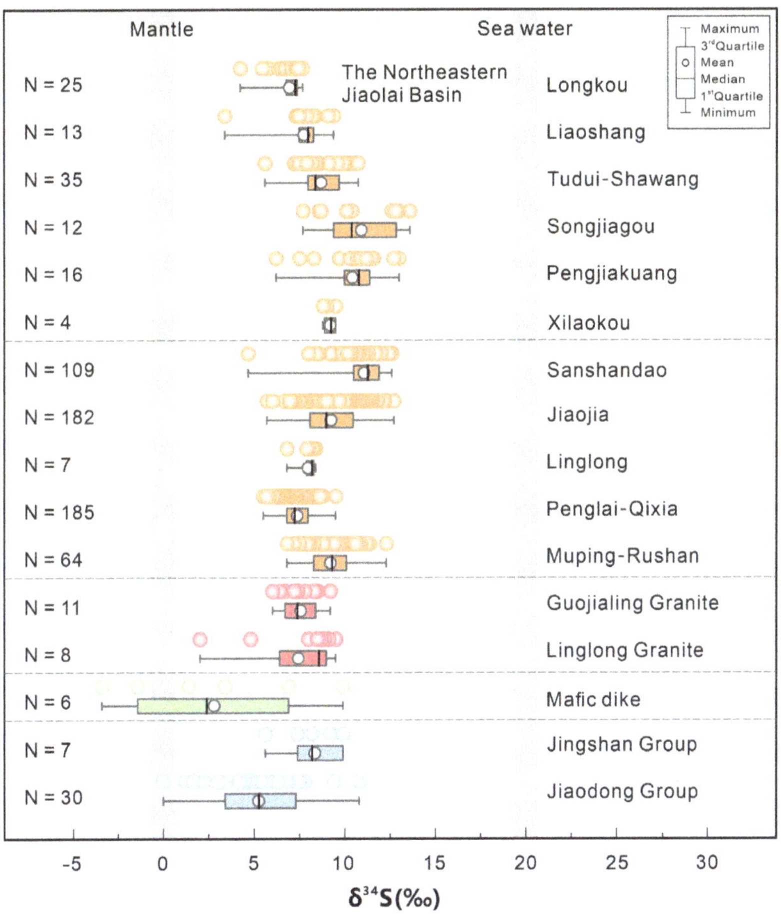

Figure 9. Sulfur isotope compositions ($\delta^{34}S$) of pyrite from the Jiaodong gold district [16,67–71], metamorphic rocks of the Jingshan and Jiaodong groups, granites, and mafic dikes [2].

In summary, the Longkou gold deposit is genetically consistent with the regional gold metallogenic system. It represents a product of Early Cretaceous hydrothermal fluids that migrated along deep NE–SW to NNE–SSW-trending faults and reacted with wall rocks, resulting in gold precipitation.

https://doi.org/10.3390/min16050485

7. Conclusions

From weakly altered wall rock to strongly altered ore, the $\delta^{34}S$ values of pyrite decrease. This reflects that fluid–rock interaction led to an increase in oxygen fugacity, thereby triggering gold precipitation. In situ sericite Rb–Sr dating yields an age of 120.9 Ma, constraining the gold mineralization to the Early Cretaceous. This age is 30–40 Ma younger than the regional Late Jurassic granites (ca. 160–150 Ma), ruling out a direct genetic link to that magmatic episode. Geological, geochronological, and geochemical evidence indicates that the Longkou gold deposit is consistent with gold deposits in the northwestern Jiaodong area in terms of mineralization age and precipitation mechanism, suggesting a common genesis.

Supplementary Materials: The following supporting information can be downloaded at: https://www.mdpi.com/article/10.3390/min16050485/s1, Table S1: In situ LA-ICP-MS/MS Rb-Sr dating of hydrothermal sericite in the Longkou deposit. Table S2: Trace element of pyrite in the different generations from Longkou deposit. Table S3: Sulfur isotopic compositions of pyrite in the different generations from Longkou deposit.

Author Contributions: Conceptualization, J.-S.Z.; methodology and investigation, H.-C.Y., G.-L.Y. and T.C.; data curation, J.-S.Z., M.M., Y.-P.L. and C.-T.X.; funding acquisition, H.-C.Y. and L.-Y.Q.; writing, J.-S.Z. All authors have read and agreed to the published version of the manuscript.

Funding: This research was supported by the National Key R&D Program of China (grant 2023YFC2906900), the Talent Cultivation Fund from North China Center for Geoscience Innovation of China Geological Survey (2024HBPJ-G03), the Joint Project of Ministry of Natural Resources and Shandong Province (2024ZRBSHZ130), the Natural Science Foundation of Shandong Province (ZR2023QD057), the Fundamental Research Funds for the Central Universities, China (No. 2652026301), the Chronological and geochemical characteristics of Guocheng gold deposit in Jiaodong and their implications for mineralization (S202511415031), the Frontiers Science Center for Deep-Time Digital Earth (grant 2652023001), Technology Research Project of No. 6 Geological Team of Shandong Provincial Bureau of Geology and Mineral Resources (LDKJ20240101), the Taishan Scholar Program of Shandong (tstp20240847).

Data Availability Statement: All data supporting this study are contained within the article and its Supplementary Materials.

Acknowledgments: Sincere thanks go to Mu-Yuan Guo and Xi-Yue Liang for their field guidance and constructive discussions. And to Qiang Gao and Yan-Xing Jian for their technical support with figures and manuscript preparation. Constructive feedback from the anonymous reviewers and journal editors is also sincerely acknowledged.

Conflicts of Interest: Author Guo-Long Yan was employed by the company China National Gold Group Corporation Limited. Author Lian-Yuan Qin was employed by the company Shandong Yantai Xintai Gold Mining Industry Co., Ltd. The remaining authors declare that the research was conducted in the absence of any commercial or financial relationships that could be construed as a potential conflict of interest.

References

1. Deng, J.; Liu, X.-F.; Wang, Q.-F.; Pan, R.-G. Origin of the Jiaodong-type Xinli gold deposit, Jiaodong Peninsula, China: Constraints from fluid inclusion and C–D–O–S–Sr isotope compositions. *Ore Geol. Rev.* **2015**, *65*, 674–686. [CrossRef]
2. Deng, J.; Wang, C.-M.; Bagas, L.; Santosh, M.; Yao, E. Crustal architecture and metallogenesis in the south-eastern North China Craton. *Earth-Sci. Rev.* **2018**, *182*, 251–272. [CrossRef]
3. Deng, J.; Wang, Q.-F.; Zhang, L.; Xue, S.-C.; Liu, X.-F.; Yang, L.; Yang, L.-Q.; Qiu, K.-F.; Liang, Y.-Y. Metallogenetic model of Jiaodong-type gold deposits, eastern China. *Sci. China Earth Sci.* **2023**, *66*, 2287–2310. [CrossRef]
4. Lv, C.-X.; Wang, J.-P.; Chen, X.-L. Discussion on the estimation method of diagenetic and metallogenic depth of Jiaodong gold deposits. *Geosci. Front.* **2024**, *38*, 910–921.

5. Hu, W.-X.; Song, M.-C.; Li, J.; Dong, L.-L.; Zhao, R.-Q.; Zhang, L.-L.; Li, J.; Bai, T.-H. Source of ore-forming materials of Jiaodong gold deposits: Constraints from the gold content of geological units related to gold mineralization. *Gold Sci. Technol.* **2024**, *32*, 781–797.
6. Deng, J.; Yang, L.-Q.; Ge, L.-S.; Wang, Q.-F.; Zhang, J.; Gao, B.-F.; Zhou, Y.-H.; Jiang, S.-Q. Research advances in the Mesozoic tectonic regimes during the formation of Jiaodong ore cluster area. *Prog. Nat. Sci.* **2006**, *16*, 777–784. [CrossRef]
7. Goldfarb, R.J.; Santosh, M. The dilemma of the Jiaodong gold deposits: Are they unique? *Geosci. Front.* **2014**, *5*, 139–153. [CrossRef]
8. Fan, H.-R.; Zhai, M.-G.; Xie, Y.-H.; Yang, J.-H. Ore-forming fluids associated with granite-hosted gold mineralization at the Sanshandao deposit, Jiaodong gold province, China. *Miner. Depos.* **2003**, *38*, 739–750. [CrossRef]
9. Yang, L.-Q.; Deng, J.; Wang, Z.-L.; Guo, L.-N.; Li, R.-H.; Groves, D.I.; Danyushevsky, L.V.; Zhang, C.; Zheng, X.-L.; Zhao, H. Relationships Between Gold and Pyrite at the Xincheng Gold Deposit, Jiaodong Peninsula, China: Implications for Gold Source and Deposition in a Brittle Epizonal Environment. *Econ. Geol.* **2016**, *111*, 105–126. [CrossRef]
10. Cui, T.; Yu, H.-C.; Jowitt, S.M.; Li, Z.-S.; Wei, X.-L.; Zou, S.-H.; Bo, J.-W.; Pei, L.; Li, S.-S.; Nassif, M.T.; et al. Chlorite composition records the formation of increased oxygen fugacity-induced disseminated gold mineralization within the Guocheng gold deposit, China. *GSA Bull.* **2025**, *137*, 4786–4800. [CrossRef]
11. Tan, J.; Wei, J.-H.; Li, Y.-J.; Fu, L.-B.; Li, H.-M.; Shi, W.-J.; Tian, N. Origin and geodynamic significance of fault-hosted massive sulfide gold deposits from the Guocheng–Liaoshang metallogenic belt, eastern Jiaodong Peninsula: Rb–Sr dating, and H–O–S–Pb isotopic constraints. *Ore Geol. Rev.* **2015**, *65*, 687–700. [CrossRef]
12. Goldfarb, R.; Qiu, K.-F.; Deng, J.; Chen, Y.-J.; Yang, L.-Q. Chapter 8 Orogenic Gold Deposits of China. In *Mineral Deposits of China*; Society of Economic Geologists: Littleton, CO, USA, 2019; pp. 263–324. [CrossRef]
13. Song, M.-C.; Song, Y.-X.; Ding, Z.-J.; Li, S.-Y. Jiaodong gold deposits: Basic characteristics and main controversies. *Gold Sci. Technol.* **2018**, *26*, 406–422. [CrossRef]
14. Wang, Q.-F.; Deng, J.; Zhao, H.-S.; Yang, L.; Ma, Q.-Y.; Li, H.-J. Research progress on orogenic gold deposits: With discussion on the orogenic gold metallogenesis in China. *Earth Sci.* **2019**, *44*, 2155–2186.
15. Zeng, G.-P.; Xiang, W.-S.; Wu, F.-F.; Li, F.-L.; Wang, J.-X.; Hu, P. Comparison between orogenic gold deposits in Northeast Africa and Jiaodong, China, and its implications for mineral exploration in Africa. *Geol. Bull. China* **2022**, *41*, 48–59.
16. Ding, Z.-J.; Sun, F.-Y.; Bo, J.-W.; Qiu, K.-F.; Liu, J.-H.; Li, G.-H.; Li, Y.; Wu, F.-P.; Zhang, Q.-B.; Chen, J.; et al. *Geology of Gold Deposits in the Northeastern Margin of the Jiaolai Basin, Shandong Province*; Geological Publishing House: Beijing, China, 2022; p. 145.
17. Song, M.-C.; Song, Y.-X.; Li, J.; Liu, H.-B.; Li, J.; Dong, L.-L.; He, C.-Y.; Wang, R.-S. Thermal uplift-extension metallogenic system of Jiaodong-type gold deposits. *Acta Petrol. Sin.* **2023**, *39*, 1241–1260.
18. Yu, H.-C.; Qiu, K.-F.; Chew, D.; Yu, C.; Ding, Z.-J.; Zhou, T.; Li, S.; Sun, K.-F. Buried Triassic rocks and vertical distribution of ores in the giant Jiaodong gold province (China) revealed by apatite xenocrysts in hydrothermal quartz veins. *Ore Geol. Rev.* **2022**, *140*, 104612. [CrossRef]
19. Yu, P.-Y.; Li, C.; Zhang, J.-H.; Zhang, H.; Ren, X.-Y.; Zuo, P.-F.; Zhang, S.-K.; Feng, S.-P.; Yu, H.-C.; Qiu, K.-F. Laser in-situ Rb-Sr geochronology: Principles, analytical methods, and geological applications. *Earth Sci. Front.* **2025**, *33*, 1–31. [CrossRef]
20. Huang, C.; Wang, H.; Shi, W.-B.; Sun, J.-F.; Hu, F.-J.; Xu, L.; Yang, Y.-H.; Wu, S.-T.; Xie, L.-W.; Yang, J.-H. In-situ LA-ICP-MS/MS determination of Rb-Sr isochron ages of mica. *Sci. China Earth Sci.* **2023**, *53*, 2648–2668.
21. Wu, H.; Zhu, X.-P.; Jia, Q.; Zhang, B.; Li, Y.; Yang, L.-C. In-situ major and trace element characteristics of pyrite from the Bada copper-gold deposit in Tibet and their environmental implications. *Acta Geol. Sin.* **2023**, *97*, 2261–2280.
22. Li, S.-Z.; Hao, D.-F.; Han, Z.-Z.; Zhao, G.-C.; Sun, M. Paleoproterozoic tectono-thermal evolution and deep processes of the Jiao-Liao massif. *Acta Geol. Sin.* **2003**, *77*, 328–340.
23. Zheng, Y.-F.; Xiao, W.-J.; Zhao, G. Introduction to tectonics of China. *Gondwana Res.* **2013**, *23*, 1189–1206. [CrossRef]
24. Zheng, Y.-F.; Zhao, Z.-F.; Chen, R.-X. Ultrahigh-pressure metamorphic rocks in the Dabie–Sulu orogenic belt: Compositional inheritance and metamorphic modification. *Geol. Soc. Lond. Spec. Publ.* **2018**, *474*, 89–132. [CrossRef]
25. Qiu, K.-F.; Goldfarb, R.J.; Deng, J.; Yu, H.-C.; Gou, Z.-Y.; Ding, Z.-J.; Wang, Z.-K.; Li, D.-P. Chapter 35: Gold Deposits of the Jiaodong Peninsula, Eastern China. In *Geology of the World's Major Gold Deposits and Provinces*; Society of Economic Geologists: Littleton, CO, USA, 2020; pp. 753–774. [CrossRef]
26. Deng, J.; Yang, L.-Q.; Groves, D.I.; Zhang, L.; Qiu, K.-F.; Wang, Q.-F. An integrated mineral system model for the gold deposits of the giant Jiaodong province, eastern China. *Earth-Sci. Rev.* **2020**, *208*, 103274. [CrossRef]
27. Qiu, K.-F.; Romer, R.L.; Long, Z.-Y.; Williams-Jones, A.E.; Yu, H.-C.; Turner, S.; Wang, Q.-F.; Li, S.-S.; Zhang, J.-Y.; Duan, H.-R.; et al. The role of an oxidized lithospheric mantle in gold mobilization. *Sci. Adv.* **2024**, *10*, eado6262. [CrossRef]
28. Zhao, G.-C.; Zhai, M.-G. Lithotectonic elements of Precambrian basement in the North China Craton: Review and tectonic implications. *Gondwana Res.* **2013**, *23*, 1207–1240. [CrossRef]
29. Deng, J.; Wang, Q.-F.; Santosh, M.; Liu, X.-F.; Liang, Y.-Y.; Yang, L.-Q.; Zhao, R.; Yang, L. Remobilization of metasomatized mantle lithosphere: A new model for the Jiaodong gold province, eastern China. *Miner. Depos.* **2019**, *55*, 257–274. [CrossRef]

30. Liu, F.L.; Liou, J.G. Zircon as the best mineral for P–T–time history of UHP metamorphism: A review on mineral inclusions and U–Pb SHRIMP ages of zircons from the Dabie–Sulu UHP rocks. *J. Asian Earth Sci.* **2011**, *40*, 1–39. [CrossRef]
31. Qiu, K.F.; Deng, J.; Sai, S.X.; Yu, H.C.; Tamer, M.T.; Ding, Z.J.; Yu, X.F.; Goldfarb, R. Low-Temperature Thermochronology for Defining the Tectonic Controls on Heterogeneous Gold Endowment Across the Jiaodong Peninsula, Eastern China. *Tectonics* **2023**, *42*, e2022TC007669. [CrossRef]
32. Song, M.-C.; Yang, L.-Q.; Fan, H.-R.; Yu, X.-F.; Ding, Z.-J.; Zhang, Y.-W.; Qiu, K.-F.; Li, J.; Zhang, L.; Wang, B.; et al. Metallogenic theory and deep exploration progress of Jiaodong gold deposits during the ten-year prospecting breakthrough strategy action. *Geol. Bull. China* **2022**, *41*, 903–935.
33. Deng, J.; Liu, X.-F.; Wang, Q.-F.; Dilek, Y.; Liang, Y.-Y. Isotopic characterization and petrogenetic modeling of Early Cretaceous mafic diking—Lithospheric extension in the North China craton, eastern Asia. *GSA Bull.* **2017**, *129*, 1379–1407. [CrossRef]
34. Tan, J.; Wei, J.-H.; He, H.-Y.; Su, F.; Li, Y.-J.; Fu, L.-B.; Zhao, S.-Q.; Xiao, G.-L.; Zhang, F.; Xu, J.-F.; et al. Noble gases in pyrites from the Guocheng-Liaoshang gold belt in the Jiaodong province: Evidence for a mantle source of gold. *Chem. Geol.* **2018**, *480*, 105–115. [CrossRef]
35. Bo, J.-W.; Ding, Z.-J.; Deng, J.; Sun, F.-Y.; Qiu, K.-F.; Song, M.-C. Metallogenic characteristics and mineralization of supergiant gold deposits (223 t) in the northeastern margin of the Jiaolai Basin, Shandong Province: A review. *China Geol.* **2025**, *8*, 806–842. [CrossRef]
36. Cheng, S.-B.; Liu, Z.-J.; Wang, Q.-F.; Feng, B.; Wei, X.-L.; Liu, B.-Z.; Qin, L.-Y.; Zhao, B.-J.; Shui, P.; Xu, L.; et al. SHRIMP zircon U–Pb dating and Hf isotope analyses of the Muniushan Monzogranite, Guocheng, Jiaobei Terrane, China: Implications for the tectonic evolution of the Jiao–Liao–Ji Belt, North China Craton. *Precambrian Res.* **2017**, *301*, 36–48. [CrossRef]
37. Tan, J.; Wei, J.-H.; Guo, L.-L.; Zhang, K.-Q.; Yao, C.-L.; Lu, J.-P.; Li, H.-M. LA-ICP-MS zircon U-Pb dating and phenocryst EPMA of dikes, Guocheng, Jiaodong Peninsula: Implications for North China Craton lithosphere evolution. *Sci. China Ser. D Earth Sci.* **2008**, *51*, 1483–1500. [CrossRef]
38. Cui, T.; Yu, H.-C.; Jowitt, S.M.; Zhang, J.-J.; Pei, L.; Fu, Q.; Zhang, J.-S.; Ding, Z.-J.; Qiu, K.-F. Genesis of the gold deposits in the Northeastern Jiaolai Basin, Jiaodong, China: Constraints from geochronology, geochemistry, and thermodynamic modeling. *Geol. Soc. Am. Bull.* **2026**. [CrossRef]
39. Vermeesch, P. IsoplotR: A free and open toolbox for geochronology. *Geosci. Front.* **2018**, *9*, 1479–1493. [CrossRef]
40. Liu, Y.-S.; Hu, Z.-C.; Gao, S.; Günther, D.; Xu, J.; Gao, C.-G.; Chen, H.-H. In situ analysis of major and trace elements of anhydrous minerals by LA-ICP-MS without applying an internal standard. *Chem. Geol.* **2008**, *257*, 34–43. [CrossRef]
41. Środoń, J. Nature of mixed-layer clays and mechanisms of their formation and alteration. *Annu. Rev. Earth Planet. Sci.* **1999**, *27*, 19–53. [CrossRef]
42. Liebmann, J.; Kirkland, C.L.; Kelsey, D.E.; Korhonen, F.J.; Rankenburg, K. Lithological fabric as a proxy for Rb–Sr isotopic complexity. *Chem. Geol.* **2022**, *608*, 121041. [CrossRef]
43. Deng, J.; Qiu, K.-F.; Wang, Q.-F.; Goldfarb, R.; Yang, L.-Q.; Zi, J.-W.; Geng, J.-Z.; Ma, Y. In Situ Dating of Hydrothermal Monazite and Implications for the Geodynamic Controls on Ore Formation in the Jiaodong Gold Province, Eastern China. *Econ. Geol.* **2020**, *115*, 671–685. [CrossRef]
44. Qiu, K.-F.; Long, Z.-Y.; Romer, R.L.; Halama, R.; Williams-Jones, A.E.; Yu, H.-C.; Li, S.-S.; Wu, M.-Q.; Deng, J. The mantle source of REE-rich alkaline silicate magmas can be enriched by continent-derived sediment subduction. *Nat. Commun.* **2025**, *16*, 11191. [CrossRef]
45. Li, J.; Zhang, L.-P.; Li, C.-Y.; Jiang, M.-Y. Rb-Sr isochron age of pyrite from the Guocheng gold deposit in Jiaodong. *Geol. China* **2020**, *47*, 894–895.
46. Han, X.-M.; Duan, L.-A.; Zhao, P.-F.; Wang, J.-T.; Guo, Y.-C. Rb-Sr isochron age of pyrite from the Qianchuiliu gold deposit in the northeastern margin of the Jiaolai Basin, Shandong Province. *Geol. China* **2024**, *51*, 366–367.
47. Zhang, X.-O.; Cawood, P.A.; Wilde, S.A.; Liu, R.-Q.; Song, H.-L.; Li, W.; Snee, L.W. Geology and timing of mineralization at the Cangshang gold deposit, north-western Jiaodong Peninsula, China. *Miner. Depos.* **2003**, *38*, 141–153. [CrossRef]
48. Zhang, L.-C.; Shen, Y.-C.; Liu, T.-B.; Zeng, Q.-D.; Li, G.-M.; Li, H.-M. 40Ar/39Ar and Rb-Sr isochron dating of the gold deposits on northern margin of the Jiaolai Basin, Shandong, China. *Sci. China Ser. D Earth Sci.* **2003**, *46*, 708–718. [CrossRef]
49. Ohmoto, H. Systematics of Sulfur and Carbon Isotopes in Hydrothermal Ore Deposits. *Econ. Geol.* **1972**, *67*, 551–578. [CrossRef]
50. Hodkiewicz, P.F.; Groves, D.I.; Davidson, G.J.; Weinberg, R.F.; Hagemann, S.G. Influence of structural setting on sulphur isotopes in Archean orogenic gold deposits, Eastern Goldfields Province, Yilgarn, Western Australia. *Miner. Depos.* **2008**, *44*, 129–150. [CrossRef]
51. LaFlamme, C.; Sugiono, D.; Thébaud, N.; Caruso, S.; Fiorentini, M.; Selvaraja, V.; Jeon, H.; Voute, F.; Martin, L. Multiple sulfur isotopes monitor fluid evolution of an Archean orogenic gold deposit. *Geochim. Cosmochim. Acta* **2018**, *222*, 436–446. [CrossRef]
52. Wu, Y.-F.; Evans, K.; Li, J.-W.; Fougerouse, D.; Large, R.R.; Guagliardo, P. Metal remobilization and ore-fluid perturbation during episodic replacement of auriferous pyrite from an epizonal orogenic gold deposit. *Geochim. Cosmochim. Acta* **2019**, *245*, 98–117. [CrossRef]

 https://doi.org/10.3390/min16050485

53. Kusebauch, C.; Gleeson, S.A.; Oelze, M. Coupled partitioning of Au and As into pyrite controls formation of giant Au deposits. *Sci. Adv.* **2019**, *5*, eaav5891. [CrossRef]
54. Cao, G.-S.; Zhang, Y.; Chen, H.-Y. Implications of trace elements in pyrite from orogenic gold deposits for metallogenic mechanism. *Acta Petrol. Sin.* **2023**, *39*, 2330–2346. [CrossRef]
55. Pokrovski, G.S.; Akinfiev, N.N.; Borisova, A.Y.; Zotov, A.V.; Kouzmanov, K. Gold speciation and transport in geological fluids: Insights from experiments and physical-chemical modelling. *Geol. Soc. Lond. Spec. Publ.* **2014**, *402*, 9–70. [CrossRef]
56. Williams-Jones, A.E.; Bowell, R.J.; Migdisov, A.A. Gold in Solution. *Elements* **2009**, *5*, 281–287. [CrossRef]
57. Liu, X.-D.; Deng, J.; Zhang, L.; Lin, S.-Y.; Zhou, M.-L.; Song, Y.-Z.; Xu, X.-L.; Lian, C.-Q. Hydrothermal alteration of the Sizhuang gold deposit in Northwestern Jiaodong. *Acta Petrol. Sin.* **2019**, *35*, 1551–1565. [CrossRef]
58. Liu, X.-Y.; Tan, J.; He, H.-Y.; Gan, J.-R. Origin of the Tudui–Shawang gold deposit, Jiaodong Peninsula, north China Craton: Constraints from fluid inclusion and H-O-He-Ar-S-Pb isotopic compositions. *Ore Geol. Rev.* **2021**, *133*, 104125. [CrossRef]
59. Xu, Y.-W.; Mao, G.-Z.; Geng, H.-Y.; He, T.-L.; Xu, Q.-L.; Meng, Y.-K.; Cao, M.-P.; Yang, F.-J.; An, P.-R.; Song, L.-G.; et al. Ore-forming materials and fluids and ore-controlling factors of the Liaoshang gold deposit in Jiaodong Peninsula, NE China. *Ore Geol. Rev.* **2023**, *154*, 105330. [CrossRef]
60. Qiu, K.-F.; Taylor, R.D.; Song, Y.-H.; Yu, H.-C.; Song, K.-R.; Li, N. Geologic and geochemical insights into the formation of the Taiyangshan porphyry copper–molybdenum deposit, Western Qinling Orogenic Belt, China. *Gondwana Res.* **2016**, *35*, 40–58. [CrossRef]
61. Chen, C.-X.; Wang, L.; Chen, Y.; Sun, L.-W. Study on fluid inclusions of Shawang gold deposit in Haiyang of Shandong. *Glob. Geol.* **2015**, *18*, 164–173. [CrossRef]
62. Qiu, K.-F.; Deng, J.; Laflamme, C.; Long, Z.-Y.; Wan, R.-Q.; Moynier, F.; Yu, H.-C.; Zhang, J.-Y.; Ding, Z.-J.; Goldfarb, R. Giant Mesozoic gold ores derived from subducted oceanic slab and overlying sediments. *Geochim. Cosmochim. Acta* **2023**, *343*, 133–141. [CrossRef]
63. He, D.-Y.; Qiu, K.-F.; Simon, A.C.; Pokrovski, G.S.; Yu, H.-C.; Connolly, J.A.D.; Li, S.-S.; Turner, S.; Wang, Q.-F.; Yang, M.-F.; et al. Mantle oxidation by sulfur drives the formation of giant gold deposits in subduction zones. *Proc. Natl. Acad. Sci. USA* **2024**, *121*, e2404731121. [CrossRef]
64. Zhang, J.-Y.; Qiu, K.-F.; Yin, R.-S.; Long, Z.-Y.; Feng, Y.-C.; Yu, H.-C.; Gao, Z.-Y.; Deng, J. Lithospheric mantle as a metal storage reservoir for orogenic gold deposits in active continental margins: Evidence from Hg isotopes. *Geology* **2024**, *52*, 423–428. [CrossRef]
65. Wu, J.-H.; Chen, Y.-L.; Zheng, C.-Y.; Li, H.; Yonezu, K.; Tang, Y.-Y.; Zong, Q. Genesis of the Longkou–Tudui gold deposit, Jiaodong Peninsula, eastern China: Constraints from zircon U-Pb dating, fluid inclusion studies and C–H–O–S stable isotopes. *Ore Geol. Rev.* **2021**, *139*, 104449. [CrossRef]
66. Li, Z.-Q.; Ding, Z.-J.; Bo, J.-W.; Wang, Z.-X.; Li, Y.; Ji, P.; Li, T.-T. Ore-forming fluid characteristics and ore genesis of the Guocheng gold deposit in the northeastern margin of the Jiaolai Basin. *Gold* **2024**, *45*, 57–63.
67. Wen, B.-J.; Fan, H.-R.; Hu, F.-F.; Liu, X.; Yang, K.-F.; Sun, Z.-F.; Sun, Z.-F. Fluid evolution and ore genesis of the giant Sanshandao gold deposit, Jiaodong gold province, China: Constrains from geology, fluid inclusions and H–O–S–He–Ar isotopic compositions. *J. Geochem. Explor.* **2016**, *171*, 96–112. [CrossRef]
68. Yang, L.-Q.; Deng, J.; Zhang, L.; Yang, W.; Xie, D.; Wang, L.; Qiu, K.-F.; Li, D.-P. Jiaodong-type gold deposits. *Acta Petrol. Sin.* **2024**, *40*, 1691–1711. [CrossRef]
69. Zhang, C.; Liu, Y.; Liu, X.-D.; Feng, J.-Q.; Huang, T.; Zhang, Q.; Wang, X.-D. Sulfur isotope geochemistry of the Xincheng gold deposit in Northwestern Jiaodong. *Acta Petrol. Sin.* **2014**, *30*, 2495–2506.
70. Wang, Y.-W.; Zhu, F.-S.; Gong, R.-T. Structural isotope geochemistry: A further study of sulfur isotopes in the Jiaodong gold deposit cluster. *Gold* **2002**, *4*, 1–16.
71. Huang, D.-Y. A study of sulfur isotopes in the metallogenic series of Jiaodong gold deposits. *Miner. Depos.* **1994**, *13*, 75–87.

 https://doi.org/10.3390/min16050485

minerals

MDPI

Article

Hyperspectral Drill Core Spectroscopy for Alteration Zoning in Orogenic Gold Deposits: Insights from the Liba Deposit, West Qinling, China

Zhe-Han Li [1], Hong-Jie Tan [1], Ding-Fan Xing [1,*], Tao Yang [2,3,4], Tong-Tong He [5], Jia-Nan Fu [1] and Kun-Feng Qiu [1]

1 State Key Laboratory of Geological Processes and Mineral Resources, Frontiers Science Center for Deep-Time Digital Earth, School of Earth Sciences and Resources, China University of Geosciences, Beijing 100083, China; 18102057420@163.com (Z.-H.L.); thjthj1209@163.com (H.-J.T.); jiananfu.geo@gmail.com (J.-N.F.); kunfengqiu@cugb.edu.cn (K.-F.Q.)

2 College of Earth and Planetary Sciences, Chengdu University of Technology, Chengdu 610059, China; yangtlzu@163.com

3 Key Laboratory of Mineral Resources Exploration, Gansu Provincial Bureau of Geology and Mineral Exploration, Lanzhou 735000, China

4 Gansu Provincial Bureau of Geology and Mineral Exploration, Lanzhou 735000, China

5 The Third Geological and Mineral Exploration Institute of Gansu Provincial Bureau of Geology and Mineral Exploration, Lanzhou 730050, China; hett2024@lzu.edu.cn

* Correspondence: xingdf@cugb.edu.cn

Abstract

Orogenic gold deposits host a substantial proportion of global gold resources, yet their internal alteration architecture and mineral assemblage variability are commonly reconstructed from random samples, limiting our ability to resolve along-profile zonation and its linkage to gold enrichment. Here we evaluate hyperspectral drill-core spectroscopy, integrated with surface spectroscopy and petrographic validation, as a rapid and spatially continuous approach to delineate alteration zoning in the Liba orogenic gold deposit (West Qinling Orogen, China). We acquired hyperspectral spectra (0.35–2.50 μm) from 255 evenly spaced surface points across two orebodies and from nine representative drill cores scanned at 1 m intervals, and organized the spectral dataset according to Au-grade domains (<0.05 g/t, 0.05–0.5 g/t, and >0.5 g/t). Spectra were quality controlled and interpreted using ENVI-based spectral library matching and The Spectral Geologist (TSG) processing workflows. Petrographic observations from 76 polished thin sections provide independent mineralogical constraints. The hyperspectral results resolve a systematic alteration progression from barren chlorite-dominated assemblages (Au-grade < 0.05 g/t) (locally with minor carbonate) through a weakly mineralized transition (Au-grade 0.05–0.5 g/t) characterized by diminishing chlorite and emergent kaolinite–carbonate with intensified sericitization, to an ore-grade (Au-grade > 0.5 g/t) assemblage dominated by kaolinite–sericite–carbonate–pyrite where biotite persists in distal and proximal zones but was progressively replaced by chlorite during subsequent hydrothermal stages; its modal abundance exhibits a clear inverse correlation with mineralization intensity. These results demonstrate that hyperspectral core scanning, coupled with targeted validation, can rapidly reconstruct a three-dimensional alteration architecture and provide practical mineralogical vectors for exploration targeting and process-oriented studies in orogenic gold systems.

Keywords: Liba deposit; alteration zoning; hyperspectral drill core spectroscopy

 https://doi.org/10.3390/min16050526

1. Introduction

Orogenic gold deposits constitute approximately one-third of the world's gold resources and represent one of the most widely distributed and economically significant deposit types globally [1,2]. Orogenic gold deposits typically form within ductile–brittle shear zones along convergent plate boundaries, occurring as veins or disseminations and exhibiting a close spatial and temporal association with orogenic events [3–6]. Understanding alteration zonation and mineral assemblages is essential for interpreting ore-forming processes and guiding mineral exploration [7]. Conventional approaches to delineating alteration zonation, encompassing field geological mapping, petrographic analysis, and geochemical characterization, furnish essential mineralogical insights. However, these techniques are predicated on discrete sampling points and therefore lack the spatial continuity necessary to resolve continuous mineralogical variations along drill cores or across alteration gradients. In particular, traditional alteration zonation studies have predominantly relied on field-based geological methods, including structural mapping, the quantification of alteration zone widths and orientations, macroscopic observations integrated with petrographic microscopy, and scanning electron microscopy–energy dispersive X-ray spectroscopy (SEM–EDS). Chinese researchers have systematically characterized alteration zoning patterns in the Liba gold deposit using field-based visual identification and established corresponding prospecting criteria [8].

Although these petrographic and geochemical methods yield high-precision data and provide valuable insights, sample collection and laboratory analysis are time-consuming and lack spatial continuity, making it difficult to capture the continuous spatial variation of alteration zones both within orebodies and in surrounding wall rocks.

Remote-sensing approaches also represent a widely used method for mapping alteration zonation in mineral deposits. For example, in the Yemaoquanxi gold field in Xinjiang, China, the integration of hyperspectral mineral-mapping results with ground geological verification enabled the successful identification of sericite, chlorite, and hematite alteration anomalies closely associated with Au–Ag–Cu polymetallic mineralization, leading to the delineation of new exploration targets [9]. In the Huaniushan Au–Ag–Pb–Zn district in Gansu Province, China, researchers utilized the CASI/SASI airborne hyperspectral system to generate lithological maps, identify mineralization-associated alteration minerals, and delineate alteration zones [10,11]. However, remote-sensing approaches face several significant limitations. Intrinsic constraints on image resolution, combined with atmospheric interference, cloud cover, and related factors, can substantially reduce identification accuracy [12,13]. Additionally, regions with dense vegetation or thick Quaternary cover are poorly suited for remote-sensing applications.

Therefore, developing an identification approach that integrates efficiency, accuracy, and spatial visualization has become a critical need for characterizing alteration zonation in the Liba deposit and similar orogenic gold systems.

Hyperspectral core scanning has demonstrated significant advantages in mineral exploration owing to its nanometer-scale spectral resolution and high sensitivity to diagnostic mineral absorption features [14–16]. This technique acquires continuous spectral data over the 0.4–2.5 μm wavelength range, enabling rapid mineral identification through recognition of diagnostic absorption features. Compared with traditional methods and remote-sensing approaches, hyperspectral techniques not only significantly enhance the efficiency and accuracy of mineral identification but also enable quantitative and spatially continuous analysis along drill cores or surface transects, providing a powerful new tool for detailed delineation of alteration zonation [2,17].

Hyperspectral core scanning acquires data across numerous narrow spectral bands (typically < 10 nm bandwidth), facilitating precise identification and discrimination of rocks,

 https://doi.org/10.3390/min16050526

diagnostic minerals, and associated alteration assemblages based on their unique spectral signatures [18–20]. This capability provides a useful basis for hyperspectral mineral identification and classification. Such high spectral resolution (≤10 nm contiguous bandwidths) enables clear discrimination of rock-forming and ore-related minerals with diagnostic absorption features [21,22]. Globally, reflectance spectroscopy has been successfully applied to characterize alteration minerals and support mineral exploration in a range of deposit types. For example, spectral techniques have been used to map hydrothermal alteration minerals and vectors toward mineralization in epithermal and porphyry systems, as well as in orogenic gold deposits [23]. In the context of orogenic gold systems, hyperspectral core analysis has proven highly effective across various global metallogenic provinces. For instance, in the central Victorian goldfields of Australia, continuous hyperspectral scanning of drill cores successfully delineated cryptic white mica and carbonate alteration halos that serve as robust vectors toward mineralized structures [24]. In South America, integrated point and imaging spectroscopy was applied to successfully characterize complex phyllosilicate and propylitic alteration assemblages in the Pedra Branca orogenic gold deposit (Northeast Brazil), demonstrating the method's capability to unravel hydrothermal evolution in higher metamorphic-grade terrains [7]. Similarly, reflectance spectroscopy has been applied to Archean greenstone-hosted gold deposits in the Carajás Mineral Province (Northern Brazil) to map decreasing chlorite abundances toward hydrothermal centers, providing a reliable technique for mapping alteration footprints [25]. In China, hyperspectral analysis of drill cores from the Zhuxi deposit in Jiangxi Province, China, was conducted to investigate the relationships between alteration minerals and mineralization, and the spectral characteristics of alteration assemblages in different orebody types and their associations with mineralization environments were examined [26]. At the Xishan rare-earth deposit in Shandong Province, China, a high-resolution hyperspectral scanner (CMS350B; Nanjing Center, China Geological Survey, Nanjing, China) was employed to investigate the spectral characteristics of alteration minerals and their applications in mineral exploration. Through systematic core scanning, principal alteration minerals, including phlogopite, chlorite, and montmorillonite, were successfully identified, and their downhole distribution patterns were characterized [27]. These studies demonstrate that spectral mineral identification can provide valuable insights into alteration zonation and fluid–rock interaction processes.

Gold production from the North China Craton margin and adjacent orogenic belts accounts for more than 70% of China's total output. The West Qinling Orogen, located along the southwestern margin of the North China Craton, represents one of China's most important orogenic gold provinces [28–30]. This belt hosts typical orogenic mineralization systems, large-scale deposits, and notably concentrated gold resources. The Liba gold deposit is one of the largest and most extensively studied orogenic gold deposits in this belt, with estimated resources exceeding several hundred tonnes. Mineralization is intimately associated with subduction–collision processes related to closure of the Paleo-Tethys Ocean. The deposit holds major scientific significance for elucidating tectonic–magmatic–hydrothermal coupling mechanisms in the Qinling Orogen and serves as a key case study for understanding orogenic gold mineralization processes throughout central and western China. Although previous studies have systematically investigated tectonic controls and fluid evolution in the Liba deposit and other orogenic gold systems, limitations remain in detailed delineation of alteration zonation.

This study focuses on the Liba orogenic gold deposit in the West Qinling Orogen to evaluate the application of reflectance spectroscopy measurements using an Analytical Spectral Devices (ASD) field spectrometer, combined with petrographic observations, to characterize the spatial distribution of vertical and lateral alteration zonation within

https://doi.org/10.3390/min16050526

orebodies and the surrounding wall rocks. The objectives of this study are to (1) evaluate the capability of spectral mineral identification for delineating alteration patterns along drill cores and (2) assess its consistency with petrographic observations.

2. Regional Geological Setting

2.1. Geological Background of the Western Qinling Region

The West Qinling metallogenic belt is located in western China, bounded by the Qilian Orogenic Belt to the north, the Songpan–Ganzi Orogenic Belt and Animaqing–Mianlue suture zone to the south, the East Qinling Orogenic Belt to the east, and the East Kunlun and Qaidam blocks to the west [31]. This belt extends approximately 600 km and represents a key segment of the Central China Orogenic Belt, which has been active from the Neoproterozoic to the present. The region has undergone multiple stages of complex tectonic evolution accompanied by voluminous magmatism [3,28]. Numerous gold deposits have been discovered, including Liba, Zhaishang, Jinshan, Maquan, Zaozigou, Liziyuan, and Shuangwang, establishing this belt as China's second-largest gold deposit (Figure 1) [32].

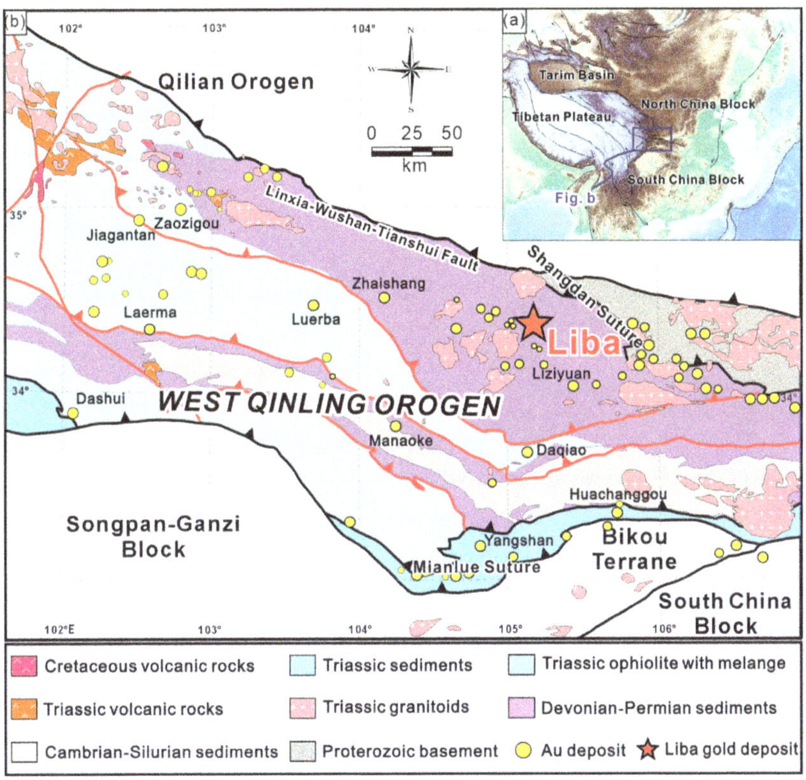

Figure 1. Geological map of the Western Qinling region. (**a**) Geographic location of the West Qinling Orogenic Belt. (**b**) Simplified geological map of the West Qinling Orogenic Belt (modified from Qiu et al. (2020)) [3].

2.2. Geological Background and Research Status of the Liba Gold Deposit

The Liba gold deposit is located in the Li–Min gold concentration area along the northern margin of the West Qinling Orogen. It is one of the largest open-pit gold deposits in the region, with proven reserves of approximately 161 tonnes at an average grade of 2.0 g/t [2].

https://doi.org/10.3390/min16050526

The Liba Group is widely exposed in the study area and hosts the main ore bodies. It is subdivided into three submembers ($D_2lb_1{}^2$–$D_2lb_3{}^2$). The lower submember ($D_2lb_1{}^2$) consists mainly of dark gray sericite–chlorite slate with minor metasandstone (~459 m thick). The middle submember ($D_2lb_2{}^2$) is composed of gray to gray-green silty sericite slate with minor metasandstone (>330 m thick). The upper submember ($D_2lb_3{}^2$) comprises dark gray to gray-black silty sericite–chlorite slate interbedded with metasandstone and spotted slate. All units show lepidoblastic to granoblastic textures and are characterized by phyllitic, spotted, and relict bedding structures (Figure 2) [33].

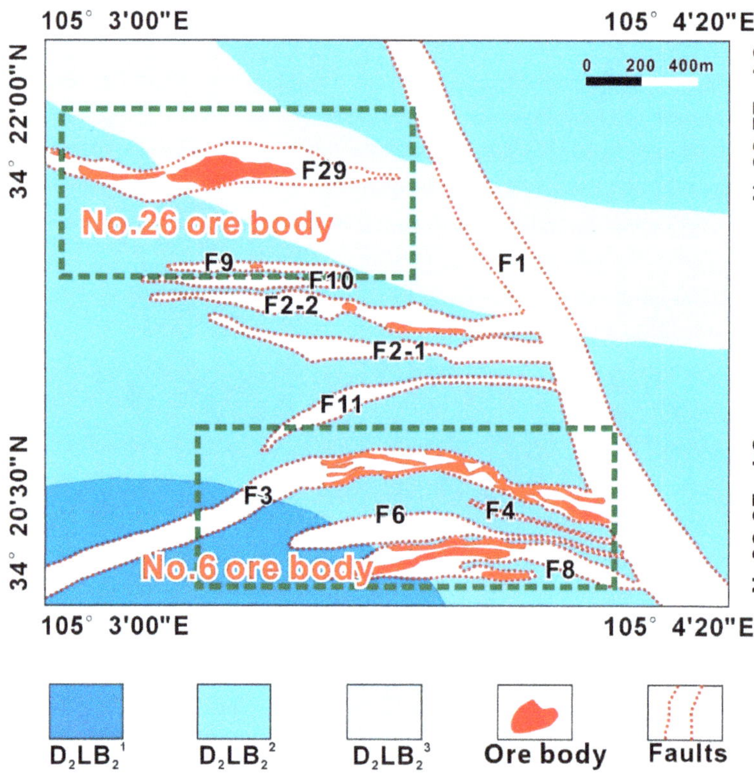

Figure 2. Geological map of the Liba gold deposit area. $D_2LB_2{}^1$: the first submember of the second member of the Middle Devonian Liba Group; $D_2LB_2{}^2$: the second submember of the second member of the Middle Devonian Liba Group; $D_2LB_2{}^3$: the third submember of the second member of the Middle Devonian Liba Group; Fn: Faults (where n represents the specific fault number) [33].

Regionally, the deposit lies along the northeastern margin of the Late Triassic Zhongchuan pluton [32], with mineralization intimately associated with collision and amalgamation of the North China and South China plates during closure of the Paleo-Tethys Ocean. The deposit is structurally controlled by a NNW-trending fault system that developed under Indosinian NE–SW regional compression, comprising the Lixian–Luoba–Suolongkou and Lixian–Taoping–Jiziba faults [34]. These faults constitute a fault network that served as the primary conduit and host for ore deposition [35].

The host rocks in the mineralized area belong to the Middle Devonian Liba Group, which comprises a greenschist-facies metamorphosed clastic-carbonate assemblage dominated by slate, metasandstone, and metasiltstone [36].

https://doi.org/10.3390/min16050526

Orebodies are primarily hosted within the thermal contact-metamorphic aureole in the exocontact zone of the Zhongchuan pluton, with their spatial distribution strictly controlled by fault structures and the pluton–host rock contact. The Zhongchuan composite pluton and associated lamprophyre, diorite, and granite porphyry dikes (210 ± 1 Ma) belong to an alkaline to high-K peraluminous series [37,38]. Zircon U–Pb ages of the pluton show strong temporal coupling with the gold mineralization event (210–200 Ma), indicating that magmatic intrusion both supplied the thermal drive for mineralization and created structurally favorable contact zones that facilitated the focusing of ore-forming fluids [39].

Wall-rock alteration exhibits pronounced zonal patterns: an inner intense silicification–pyrite zone grades outward into a sericite–chlorite zone, with a peripheral carbonate zone. Alteration assemblages display three distinct paragenetic stages: (1) an early sericite–quartz–pyrite assemblage constituting the principal gold host; (2) middle-stage chlorite derived largely from biotite breakdown; and (3) late-stage carbonate filling quartz–sulfide veins as veinlets or aggregates. Ore textures are dominated by fine-grained disseminated sulfides with locally developed brecciated fabrics. Gold occurs primarily in pyrite and arsenopyrite (As–Sb-bearing sulfides), with minor pyrrhotite. Sulfide paragenetic relationships indicate multiple episodes of sulfur and gold enrichment [2]. The close spatiotemporal coupling among structure, magmatism, and mineralization indicates that the deposit represents a typical orogenic gold system, providing useful implications for exploring down-dip and along-strike extensions of known orebodies as well as similar structurally controlled targets in the region.

3. Data

In this study, surface spectral data and core hyperspectral data were integrated to delineate the three-dimensional spatial distribution of alteration zonation in the Liba gold deposit. Core hyperspectral data were used to characterize vertical alteration zonation, whereas surface spectral data mapped lateral zonation patterns and provided independent validation of core-based mineral identifications. This integrated approach enables robust three-dimensional characterization of alteration zonation while validating the capability of hyperspectral techniques for mapping hydrothermal alteration in orogenic gold systems. To validate hyperspectral interpretations, 76 rock samples were collected from orebodies, proximal, and distal alteration zones in Pits 6 and 26, and polished thin sections were prepared for petrographic and micro-hyperspectral analysis.

3.1. Surface Spectral Data

In Pits 6 and 26 of the Liba gold deposit, surface spectral data were collected at 255 locations in total along transects extending from orebodies into distal wall rocks, comprising 150 measurement points in Pit 6 and 105 measurement points in Pit 26 (Figure 3). Surface spectral measurements were acquired using an Analytical Spectral Devices (ASD) portable spectrometer to characterize lateral alteration zonation patterns; the equipment is described in Section 4.

3.2. Core Hyperspectral Data

Nine representative drill cores from Orebodies 6 and 26 were selected for analysis. Table 1 summarizes the orientation, spatial coordinates, and orebody intersections of the analyzed cores. Each core was scanned at 1 m intervals using an ASD portable spectrometer to acquire hyperspectral data. Based on assay data, each drill core was categorized into three gold-grade classes: <0.05 g/t (barren), 0.05–0.5 g/t (weakly mineralized), and >0.5 g/t (ore-grade). The gold grades of the core samples were determined by the analytical laboratory of Xiamen Zijin Mining and Metallurgy Technology Co., Ltd. (Xiamen, China) using a

https://doi.org/10.3390/min16050526

standard Fire Assay coupled with Atomic Absorption Spectrometry (FA-AAS). Then, the corresponding spectral measurements were grouped and analyzed accordingly.

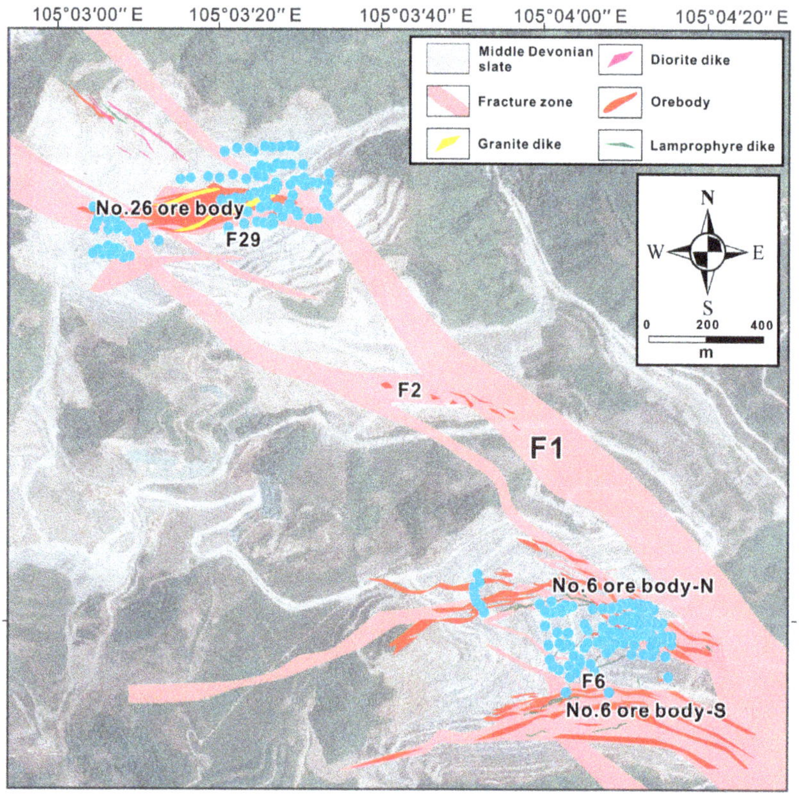

Figure 3. Spatial distribution of surface spectral data points.

Table 1. Orientation, spatial parameters, and orebody intersections of the analyzed drill cores.

Core	Depth (m)	Azimuth (°)	Tilt Angle (d°)	Main Ore Body Area (m)
zzk79-2	199	0	70	142–168
zzk135-12	354	180	78	302–342
zzk151-5	284	180	67	220–224; 237–240; 263–271
mzk15-8	197	40	76	91–106
mzk24-3	146	0	74	36–41
djg120-3	314	40	61	228–242; 252–262
mzk8-4	361	220	72	221–267
zzk95-2	355	180	69	155–221
zzk135-10	341	83	83	79–114; 167–179; 199–205; 213–226; 301–314

4. Methods

4.1. Hyperspectral Core Scanning and Spectral Data Processing

Data acquisition in this study relied primarily on hyperspectral technology. Hyperspectral core scanning is a high-resolution analytical technique that typically operates over the 0.4–2.5 μm wavelength range [20,40]. This range encompasses the visible (0.4–0.7 μm), near-infrared (0.7–1.1 μm), and shortwave infrared (1.1–2.5 μm) regions, yielding continuous spectral profiles. Extensive research on mineral spectral characteristics, combined with successful applications in regional geological mapping and mineral exploration, has

https://doi.org/10.3390/min16050526

demonstrated that hyperspectral core scanning enables efficient and accurate mineral identification through the recognition of diagnostic spectral features [41–43]. Therefore, hyperspectral core scanning was employed as the primary method for characterizing alteration zonation in the Liba gold deposit.

In this study, the ASD field spectrometer (ASD Inc., Falls Church, VA, USA) was used to acquire spectral data from surface samples and drill cores. Key instrument specifications are summarized in Table 2.

Table 2. Key technical specifications of the Analytical Spectral Devices (ASD) field spectrometer.

Parameter	Specification
Spectral range	350–2500 nm
Resolution	3 nm @ 700 nm, 6 nm @ 1400/2100 nm
Wavelength accuracy	0.5 nm
Sensitivity	75 photons/count value
Signal-to-noise ratio	Visible and Near-Infrared (VNIR) 9000:1 @ 700 nm, Short-Wave Infrared (SWIR) 1 9000:1 @ 1400 nm, SWIR 2 4000:1 @ 2100 nm
Dynamic range	7.5×10^9
Stray light	VNIR 5000:1 (0.02%), NIR 10,000:1 (0.01%)

The standardized hyperspectral core scanning procedure employed in this study included the following steps: Fresh core sections were selected from core boxes, with representative samples collected at approximately 1 m intervals along the alteration gradient. Core surfaces were cleaned prior to scanning to remove contaminants that could interfere with spectral measurements. The spectral measurements were conducted using a portable ASD spectrometer equipped with a standard High-Intensity Contact Probe. The contact probe features a built-in halogen light source, which ensures consistent illumination conditions and minimizes the interference from ambient light. Measurements were taken with a standard spot size of 10 mm in diameter. Prior to formal data acquisition, the instrument was sufficiently warmed up, and optimization and white reference calibration were performed regularly using a panel to prevent instrument drift. During white-reference measurements, the probe–target distance, field of view, and illumination geometry were kept consistent with those used for sample measurements. Each sample was measured three times consecutively, and the averaged spectrum was taken as the final reflectance curve to reduce random noise and improve the signal-to-noise ratio (SNR). Following the recommended ASD protocol, the raw digital signal (raw DN) was continuously monitored during acquisition to prevent the detector from entering saturation or low-response regions [23]. When significant changes occurred in ambient temperature or instrument response, optimization and white reference calibration were repeated.

Spectral data were systematically processed and interpreted following established hyperspectral analysis methods [7,44,45]. Raw spectral data from the ASD field spectrometer were first exported and converted to text format and then reformatted into ENVI-compatible .sli files for downstream processing. Within ENVI, all spectra underwent visual inspection and quality control, including the removal of spectra exhibiting obvious saturation, spectral discontinuities, abnormal white-reference calibration, or excessive noise; samples showing large variability among repeated measurements were re-measured; and spectral regions severely affected by strong absorption and low SNR were masked. Particular attention was given to potential detector splice-related discontinuities and low-SNR intervals around 1350–1400 nm and 1810–1940 nm, which are associated with strong H_2O/OH-related absorption features. Because the spectra were acquired using a contact probe, atmospheric water-vapor interference was considered negligible. These intervals were therefore treated as spectral quality-control regions evaluated as part of the quality-control procedure. For mineral interpretation, the 350–1000 nm range was excluded because the main diagnostic

https://doi.org/10.3390/min16050526

features of the alteration minerals considered in this study occur predominantly in the SWIR region.

Prior to spectral interpretation, raw spectral data were corrected to absolute reflectance using the white reference standard according to the following equation [46]:

$$R = (I_{sample} - I_{dark})/(I_{white} - I_{dark}) \quad (1)$$

In this equation, R represents the absolute reflectance of the sample; I_{sample} is the raw measured radiance intensity of the target core sample; I_{white} is the measured radiance intensity of the standard white reference panel (e.g., Spectralon); and I_{dark} denotes the dark current intensity (instrument background noise) measured with a closed shutter.

Following this preprocessing, spectral data were processed and analyzed using The Spectral Geologist (TSG) software (version 8.0; CSIRO, Canberra, Australia), a platform designed specifically for hyperspectral core analysis that enables the complete workflow from raw data preprocessing to mineral identification, where sample IDs were first aligned with depth information. Subsequently, SWIR diagnostic absorption features were used to invoke the built-in automated mineral identification module for initial interpretation. The automated interpretation results are based on the built-in spectral algorithm (TSA+) and reference spectral libraries and provide mineral identification outputs at the group level, species level, and relevant chemical-indicator levels, while generating mineral weighting factors for each sample to represent relative abundances.

To improve the geological plausibility and reproducibility of the identification results, the spectral library used in this study consists of two components: (1) the built-in mineral reference library in TSG, used for automated preliminary screening; and (2) an "active mineral set (active minerals/domain)" constrained by the geological background of the study area and known mineral assemblages, retaining only those mineral entries that are theoretically expected to occur and exhibit diagnostic absorption features in the SWIR/VNIR range, namely sericite, chlorite, kaolinite, calcite, and biotite. In addition, the USGS and ASTER spectral libraries were used as external references for cross-validation to improve the interpretability of mineral identification; these two public spectral libraries are widely used standard references in hyperspectral mineral identification [47–50].

During spectral matching, this study did not adopt an unconstrained full-library search; instead, automated identification was conducted within a constrained mineral domain to avoid misclassification of minerals that are unlikely to occur in the study area. This approach recognizes that for accurate mineral identification, the quality and applicability of the reference spectral library are highly dependent on the specific geological context.

Therefore, the TSG outputs are interpreted as relative mineral abundance rather than uncalibrated absolute mass percentages. For the automated identification results, manual verification was further performed by integrating the original reflectance spectra, diagnostic absorption positions, and geological context; results with low identification confidence or those inconsistent with the lithological context were corrected or discarded. Finally, the depth-registered results exported from TSG were output as .csv files, and variation curves of relative mineral abundances were plotted against drill-core depth.

4.2. Mineral Reference Spectra and Spectral Matching

The spectral data processed in TSG (Section 4.1) and ENVI (Section 4.2) share the exact same source. In our workflow, TSG was primarily utilized for systematic data interpretation, including batch spectral matching and mineral identification. Subsequently, ENVI was specifically employed for the extraction and high-quality visualization of representative spectral curves. The results derived from the two software tools are completely consistent. Specifically, the extraction and visualization of these spectral curves were conducted using

 https://doi.org/10.3390/min16050526

ENVI 5.6 (Environment for Visualizing Images; NV5 Geospatial Software, Broomfield, CO, USA). The built-in Spectral Library Viewer module was used to visualize and compare spectral signatures of alteration minerals. Reference spectral libraries and sample spectral profiles were loaded to analyze the diagnostic spectral characteristics of target minerals, supporting mineral identification and alteration zone delineation. This module enables intuitive visualization of multi-band spectral curves and has been widely applied in hyperspectral mineral mapping.

Prior to comparison with reference spectra, continuum removal was applied to the representative spectra in ENVI to normalize local absorption features to a common baseline. This treatment was used to enhance diagnostic absorption features and assist mineral identification, particularly by comparing absorption-feature position, relative depth, and spectral shape [51,52]. It should be emphasized that continuum removal was used as a spectral feature-enhancement and interpretation step rather than as a direct method for calculating absolute mineral abundances.

Standard mineral reference spectra were obtained from three spectral libraries accessed through ENVI: the ASTER library version 2.0 (NASA/JPL) [53], focused on geological materials, and USGS spectral libraries V6 and V7, with the latter representing an updated version with enhanced spectral coverage and accuracy.

In this study, five key alteration minerals closely associated with orogenic gold systems—sericite, chlorite, kaolinite, calcite, and biotite—were selected for detailed spectral analysis. Each mineral exhibits diagnostic spectral absorption features at characteristic wavelengths, enabling robust mineral discrimination [54].

Figure 4 presents reference spectral curves for the principal alteration minerals examined in this study. Figure 4a demonstrates the reference spectral curves of sericite, which exhibits a prominent absorption feature at 2200 nm. This feature arises from combination modes of Al–OH stretching and bending vibrations and serves as a diagnostic criterion for hydroxyl-bearing phyllosilicates. Figure 4b shows chlorite, which displays diagnostic doublet absorptions at 2235–2255 nm and 2330–2360 nm. The shorter-wavelength feature corresponds to Al(Mg, Fe)–OH or Mg–OH vibrations, whereas the longer-wavelength feature reflects metal–OH bonds at octahedral sites [7,23,55,56]. This doublet pattern is diagnostic for distinguishing chlorite from other phyllosilicates. Reference spectral curves of kaolinite are shown in Figure 4c. Kaolinite typically displays an asymmetric doublet near 2200 nm, comprising a principal absorption at 2209 nm and a weaker feature at 2159 nm. This doublet reflects differences in OH vibrational modes between inner and outer hydroxyl groups. In well-crystallized samples, sharp features appear at 2180 and 2190 nm, together with a 2208 nm absorption from inner-surface hydroxyls [57], constituting the most prominent diagnostic spectral signature of kaolinite-group minerals. In Figure 4d, calcite is presented and is characterized by a pronounced absorption near 2.3 μm (2300–2350 nm), which is attributed to overtone and combination modes of the CO_3^{2-} group. This feature is diagnostic of carbonate minerals. The identification of calcite in this study relies primarily on its diagnostic carbonate absorption feature, located near 2340 nm. It is worth noting that due to the extreme scarcity of dolomite in the Liba mining area, there is virtually no corresponding spectral interference. Theoretically, if dolomite were present, the incorporation of magnesium into the crystal lattice would cause a blue-shift of the main carbonate absorption feature to approximately 2320 nm. The consistent position of the absorption valleys at ~2340 nm in our samples further confirms that the carbonate minerals present are overwhelmingly calcite. The reference spectral curves of biotite are shown in Figure 4e, characterized by a diagnostic absorption at 2250 nm from OH stretching vibrations. Notably, pure biotite lacks a distinct 2000 nm absorption; when present, this

https://doi.org/10.3390/min16050526

feature indicates chloritization or chlorite admixture and serves as a proxy for alteration intensity [40,47,58,59].

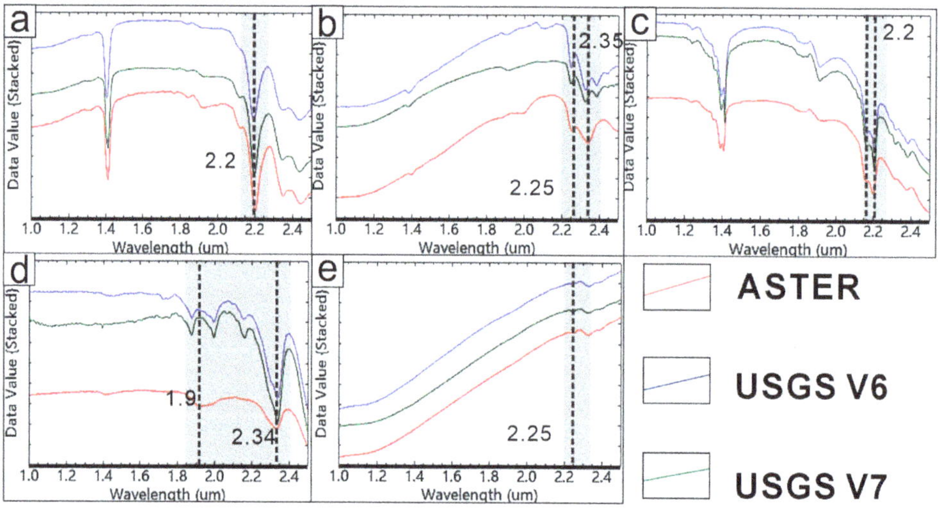

Figure 4. Reference spectral curves of principal alteration minerals: (**a**) sericite; (**b**) chlorite; (**c**) kaolinite; (**d**) calcite; (**e**) biotite. The grey shaded areas highlight the primary spectral regions used for mineral identification, and the vertical dashed lines indicate the diagnostic absorption features for specific minerals.

The SWIR absorption features of these minerals are governed primarily by fundamental, overtone, and combination modes of molecular bonds (e.g., OH, CO_3, H_2O) within the mineral lattice. Variations in chemical composition and crystal structure produce diagnostic absorption features at characteristic wavelengths and intensities, providing the theoretical basis for hyperspectral mineral identification and alteration zone mapping [60].

Therefore, building on these observations, systematic spectral data were acquired from surface outcrops of Orebodies 6 and 26 and from nine drill cores, with sampling stratified by Au-grade intervals (<0.05 g/t, 0.05–0.5 g/t, and >0.5 g/t). Spectral feature analysis was performed in ENVI 5.6 using the Spectral Library Viewer module. Eight representative spectral curves from each Au-grade interval were selected for analysis; spectra exhibiting anomalously high reflectance or from highly fragmented samples were excluded owing to poor interpretability. Spectral curves are displayed in stacked format, with curve morphology reflecting spectral-feature trends and relative absorption intensities rather than sampling depth or spatial position relative to orebodies.

To validate the accuracy of spectral-based mineral assemblage and alteration zonation identification, samples were collected from three Au-grade domains: <0.05 g/t (barren), 0.05–0.5 g/t (weakly mineralized), and >0.5 g/t (ore-grade). The mineral assemblage characteristics of each domain were verified through petrographic observation.

5. Results

Based on surface and core hyperspectral data acquired during fieldwork and laboratory processing, TSG analysis of nine drill cores revealed systematic downhole variations in mineral composition and abundance, with distinct mineral assemblages characterizing orebodies, weakly mineralized domains, and barren domains. Mineral identification and semi-quantitative analysis were conducted using the built-in interpretation workflow of TSG on ASD spectral data. The relative abundance data of minerals along the drill core

https://doi.org/10.3390/min16050526

depth generated by TSG were extracted and compiled to analyze the variation trends of major minerals (sericite, chlorite, calcite, kaolinite, and biotite) across different boreholes and alteration zones. Specifically, mineral relative abundance values were arranged according to core depth, and their variations within continuous depth intervals were summarized and visualized. The resulting plots illustrate the distribution of mineral relative abundance as a function of depth, reflecting the spatial variation of alteration characteristics in different zones. Aside from minor amounts of accessory minerals (such as trace montmorillonite and dolomite) varying among cores, the principal compositional variations involved chlorite, kaolinite, sericite, biotite, and carbonates (Figure 5).

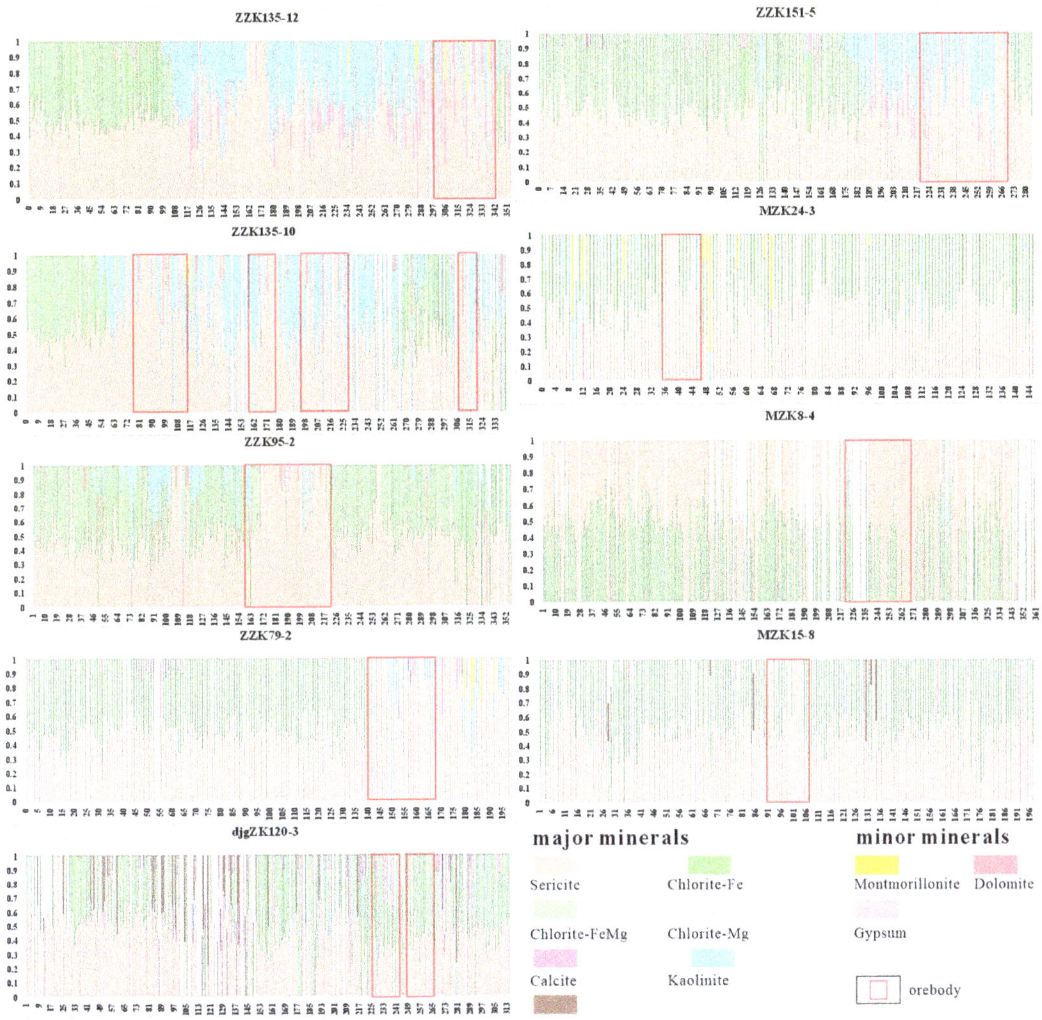

Figure 5. Changes in mineral contents as a function of drill-core depth.

TSG analysis reveals that within orebody intervals, chlorite abundance decreases markedly, whereas sericite and kaolinite increase substantially. Additionally, biotite content is relatively low and shows a negative correlation with mineralization intensity as well.

https://doi.org/10.3390/min16050526

Spectral characteristics of mineral assemblages exhibit pronounced spatial variations and can be classified into three domains: distal, proximal, and orebody zones (Figure 5).

5.1. Spectral Characteristics of the Barren Domain (Au < 0.05 g/t)

Figure 6 presents representative spectral curves from the barren domain (Au < 0.05 g/t) acquired from surface outcrops of Orebodies 6 and 26 and from nine drill cores. Spectral characteristics are remarkably consistent across all samples and reveal the following mineral assemblage: (1) A sharp Al–OH absorption at ~2200 nm together with Al-OH secondaries indicates pervasive sericitization. (2) The presence of chlorite is confirmed by the diagnostic Al(Mg, Fe)–OH absorption feature near 2250 nm, coupled with the Mg–OH absorption band at 2310–2370 nm and a broad peak near 2000 nm from molecular water combination vibrations. An additional absorption peak at 2380 nm points to the coexistence of biotite. Notably, while both chlorite and biotite share an absorption near 2250 nm, chlorite exhibits additional features at 2000 nm, attributed to combination vibrations of molecular water (H_2O), and biotite shows characteristic SWIR absorptions at 2380 nm [23]. (3) Weak absorptions near 2150 nm and 2330 nm indicate minor carbonate. Notably, diagnostic kaolinite doublet absorptions (2100–2200 nm) are absent, consistent with the distal alteration assemblage [7].

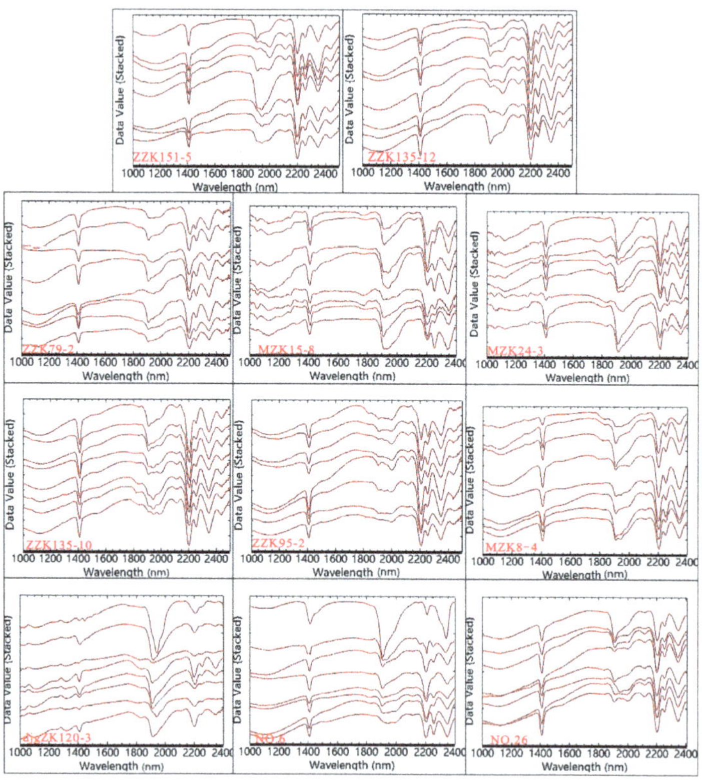

Figure 6. Spectral curves of mineral assemblages in the barren domain (Au < 0.05 g/t).

5.2. Spectral Characteristics of the Weakly Mineralized Domain (Au 0.05–0.5 g/t)

Figure 7 presents representative spectral curves from the weakly mineralized domain (Au 0.05–0.5 g/t), acquired from surface outcrops of Orebodies 6 and 26 and from nine drill cores. Spectral analysis reveals the following characteristic alteration assemblage: (1) prominent asymmetric doublet absorptions at 2100–2200 nm are diagnostic of kaolinite;

 https://doi.org/10.3390/min16050526

(2) a sharp Al–OH absorption at ~2200 nm together with Al-OH secondaries indicates intense sericitization; (3) absorptions near 2150 nm and 2330 nm confirm carbonate; (4) weak absorptions at 2250 nm and 2350 nm indicate minor chlorite (2000 nm) together with biotite (2380 nm), with abundance markedly lower than in distal zones; (5) a broad absorption near 1000 nm indicates Fe^{2+}, consistent with the pyrite identified in hand specimens. This assemblage is characteristic of proximal alteration zones.

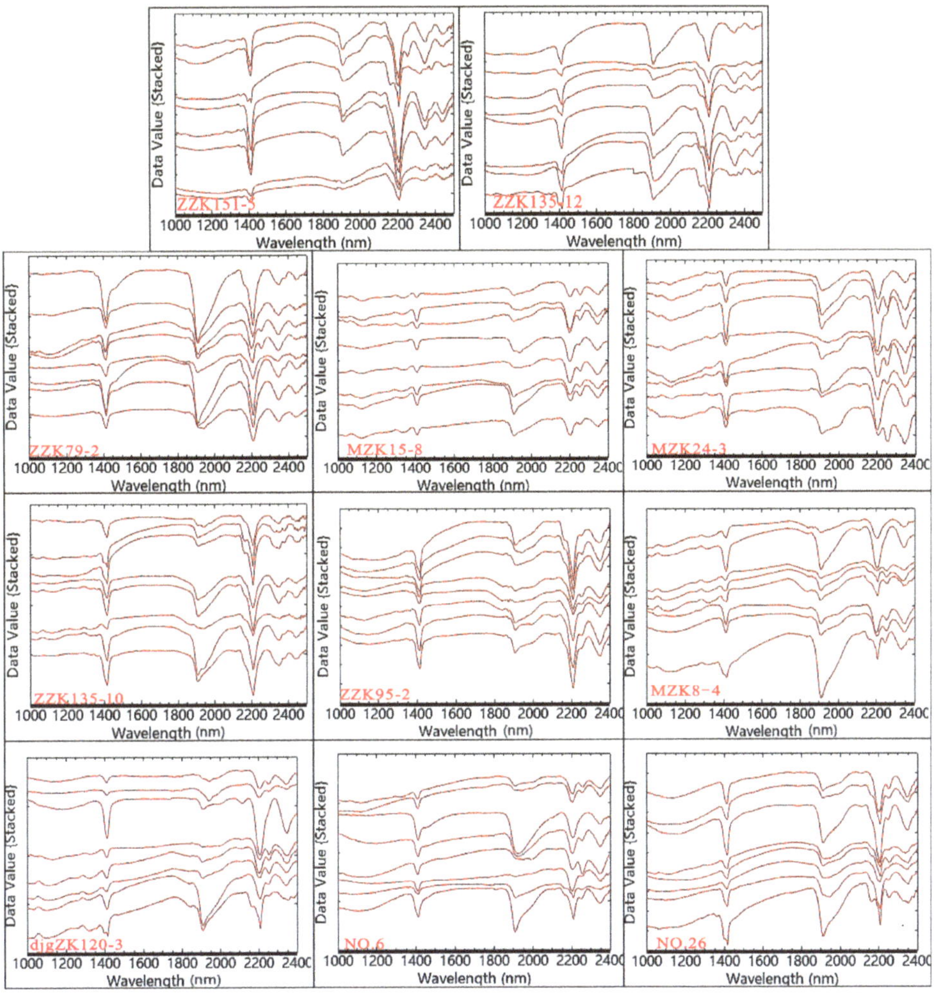

Figure 7. Spectral curves of mineral assemblages in the weakly mineralized domain (Au 0.05–0.5 g/t).

5.3. Spectral Characteristics of the Ore-Grade Domain (Au > 0.5 g/t)

Figure 8 presents representative spectral curves from the ore-grade domain (Au > 0.5 g/t) acquired from surface outcrops of Orebodies 6 and 26 and from orebody intervals in nine drill cores. Spectral analysis reveals the following characteristic orebody assemblage: (1) prominent asymmetric doublet absorptions at 2160 nm and 2200 nm (2100–2200 nm range) are diagnostic of kaolinite; (2) a sharp Al–OH absorption at ~2200 nm (full width at half maximum < 50 nm) together with Al-OH secondaries indicates intense sericitization; (3) strongly enhanced absorptions at 2150 nm and 2330 nm indicate a carbonate abundance significantly higher than in

 https://doi.org/10.3390/min16050526

proximal zones; (4) notably, chlorite-diagnostic absorptions at 2250 nm (Fe–OH) and 2350 nm (Mg–OH) are absent, consistent with chlorite depletion in the orebody. This assemblage, dominated by kaolinite–sericite–carbonate–pyrite, is characteristic of orebody alteration zones.

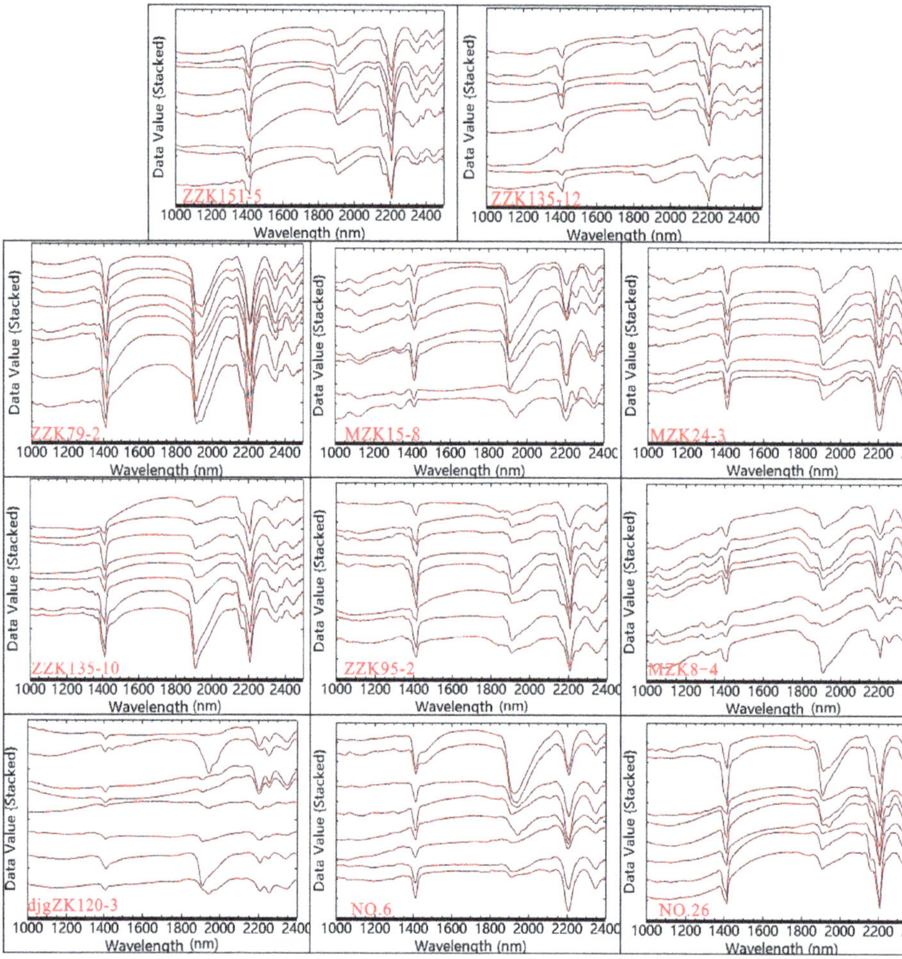

Figure 8. Spectral curves of mineral assemblages in the ore-grade domain (Au > 0.5 g/t).

Based on diagnostic spectral variations observed in surface and core samples, representative spectral models were established for each alteration domain. Figure 9a presents the distal zone spectral model, characterized by the absence of kaolinite absorptions as well as prominent chlorite and biotite features; Figure 9b presents the proximal zone model, characterized by diagnostic kaolinite absorptions, diminished chlorite and biotite features, and elevated carbonate abundance; Figure 9c presents the orebody model, characterized by prominent kaolinite absorptions, the virtual absence of chlorite and biotite, and a sharp Al–OH absorption at ~2200 nm together with Al-OH secondaries diagnostic of intense sericitization. It should be noted that while specific spectral features (e.g., asymmetric water absorption) indicative of minor smectite were observed in certain samples, smectite was not included in our predefined active mineral domain. The current study focuses strictly on the major alteration assemblage (sericite, chlorite, kaolinite, biotite, and calcite) that is

https://doi.org/10.3390/min16050526

most critical to the target mineralization. Therefore, detailed spectral characterization and continuum-removal analysis for smectite were excluded from the scope of this paper.

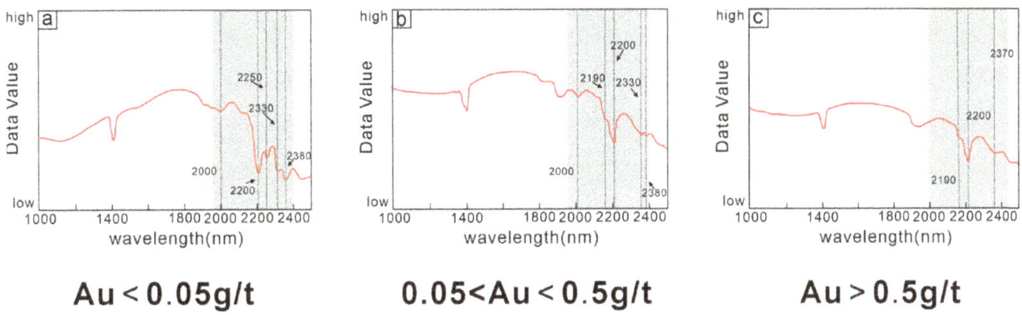

Figure 9. Spectral models for each alteration domain: (**a**) the distal zone spectral model; (**b**) the proximal zone model; (**c**) the orebody model. The grey shaded areas highlight the primary spectral regions used for mineral identification, and the vertical dashed lines indicate the diagnostic absorption features for specific minerals.

5.4. Microscopic Characteristics of the Mineral Assemblage

Petrographic observations were conducted as a supporting method to validate hyperspectral mineral identification, providing a direct and robust means of validating the capability of hyperspectral data to delineate alteration zonation and discriminate mineral assemblages within the deposit. A total of 76 representative samples were selected from different spatial domains, including distal, proximal, and ore-zone locations, and prepared as standard thin sections. Thin sections were examined using a polarizing microscope under plane-polarized light (PPL) and cross-polarized light (XPL). These samples are considered representative of the major lithological and alteration variations in the study area, thus providing a reasonable basis for validating hyperspectral results.

Petrographic results demonstrate that in barren zones (Au < 0.05 g/t), alteration assemblages are dominated by chlorite with minor sericite and clay minerals; carbonate contents are negligible, characteristic of chlorite-dominated alteration. In the weakly mineralized domain (Au 0.05–0.5 g/t), chlorite decreases markedly, whereas sericite and pyrite increase substantially, accompanied by the appearance of kaolinite and minor carbonate minerals. Notably, residual biotite is locally preserved in the distal and proximal zones of Orebody 6 and is commonly partially replaced by chlorite, indicating progressive alteration of earlier high-temperature minerals by later hydrothermal fluids. Such replacement textures and mineralogical overprinting relationships are widely interpreted as evidence for multi-stage hydrothermal processes [61]. In ore-grade zones (Au > 0.5 g/t), intense sericitization and pervasive pyritization are prominent; chlorite is near-absent, whereas kaolinite and carbonate abundances increase further relative to proximal zones.

Figure 10 presents the microscopic characteristics of mineral assemblages from domains with Au < 0.5 g/t. In Figure 10a, quartz is colorless, with low relief and low-order interference colors, commonly showing undulose extinction, whereas calcite exhibits high birefringence, rhombohedral cleavage and polysynthetic twinning. Both of these constitute the dominant mineral phases, accounting for approximately 50% and 30% modal abundance, respectively, with minor associated sericite. This mineral assemblage suggests formation under relatively low-temperature hydrothermal conditions. The absence of high-temperature silicate minerals and the presence of carbonate and fine-grained white mica indicate precipitation from relatively low-temperature fluids. Such quartz–calcite–sericite

 https://doi.org/10.3390/min16050526

assemblages are commonly associated with late- to post-mineralization hydrothermal stages in many hydrothermal systems.

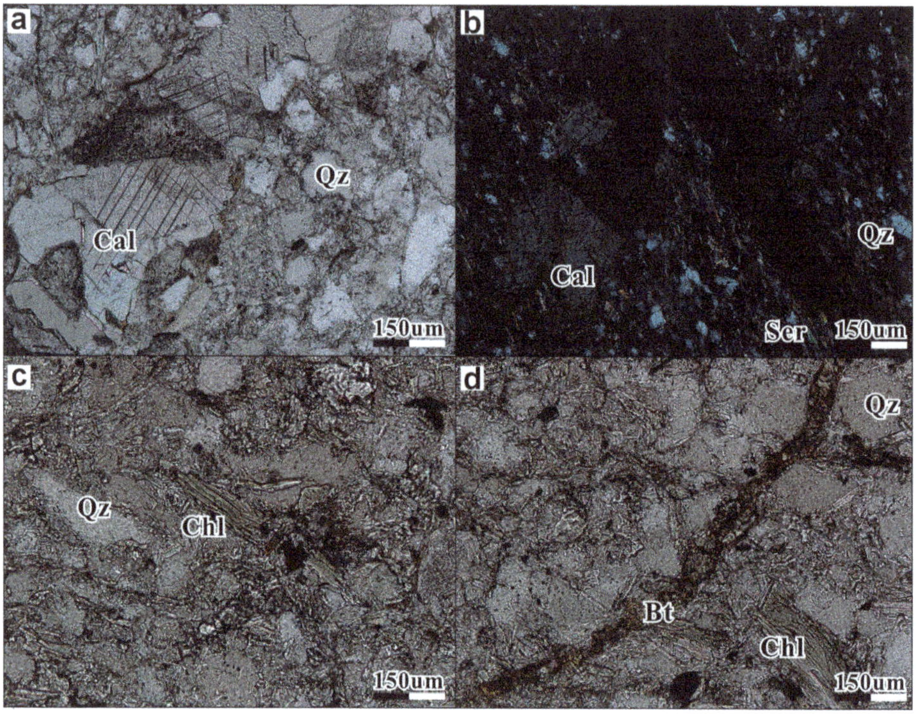

Figure 10. Microscopic photograph of rock proximal zone end in the Liba area. (**a**) Quartz (Qz) and calcite (Cal) from domains in proximal zone under plane-polarized light; (**b**) gray-green slate observed under crossed polarizers under cross-polarized light; (**c**) cataclastic slate under polarized-light microscopy without biotite; (**d**) cataclastic slate under polarized-light microscopy with biotite (Bt) and chlorite (Chl).

Figure 10b illustrates the mineral paragenesis of gray-green slate observed under crossed polarizers. Microscopic petrographic observations reveal that the major mineral constituents are quartz (30% modal abundance), calcite (20%), and sericite (30%). Sericite appears as fine-grained aggregates of white mica with low relief and low-order interference colors, while the dark bands visible in the photomicrograph are interpreted as argillaceous metamorphic layers formed during regional metamorphism. Detailed comparison with Figure 10a indicates that this sample contains markedly higher sericite content and that quartz grains exhibit poorer euhedral forms (more anhedral crystal forms). These petrographic features collectively suggest that this sample was likely collected from a location in closer spatial proximity to the orebody relative to the sample shown in Figure 10a [62].

Figure 10c,d document the mineral assemblage characteristics of slate developed within the F1 fault fracture zone under polarized-light microscopy. The major mineral constituents identified are quartz (50% modal abundance) and chlorite (30%), with quartz occurring as evenly distributed subhedral to anhedral granular grains. Chlorite occurs as greenish flakes with weak pleochroism, anomalous interference colors, and platy flakes and is typically formed by hydrothermal alteration of biotite or other ferromagnesian silicate minerals under water-rich, weakly alkaline to neutral pH conditions [63–65], indicating a relatively stable hydrothermal alteration environment characteristic of distal zones pe-

https://doi.org/10.3390/min16050526

ripheral to the orebody. Although chlorite can also form through meteoric weathering, the chlorite identified in this study is considered to be of hydrothermal origin. This is supported by the fact that chlorite is widely distributed in the deeper sections of the drill cores without showing a trend of increasing abundance toward the surface, which contradicts the typical profile of meteoric weathering. The fine veinlet-like yellowish-brown mineral phase intergrown among chlorite aggregates is identified as biotite, which is characterized by strong pleochroism and perfect basal cleavage; it constitutes less than 10% modal abundance and occurs in banded or irregular spatial distributions. These biotite occurrences are interpreted as relict mineral grains formed during primary deposition or early metamorphic processes and not completely replaced by later hydrothermal fluids [66].

Figure 11 presents representative microscopic mineral assemblages from ore-grade zones, which are used to constrain mineralogical characteristics of the orebody and to support interpretation of hyperspectral alteration signatures. Figure 11a and b show reflected-light photomicrographs under plane-polarized light of dark-colored veins from Orebodies 6 and 26, respectively. Pyrite occurs as golden-yellow anhedral grains disseminated within quartz veins, constituting ~10% modal abundance. Grains are irregularly distributed and commonly occupy intergranular pores or fractures within the quartz matrix. Arsenopyrite occurs as relatively euhedral grains formed during the main ore-forming stage. These sulfide minerals are not readily identifiable in hyperspectral data but provide direct petrographic evidence for the presence of mineralized zones. The ore mineral is arsenopyrite, which exhibits relatively euhedral crystal forms and was predominantly formed during the main ore-forming stage. Previous studies demonstrate that gold occurs predominantly as a solid solution ("invisible gold") within pyrite and arsenopyrite [36], both of which constitute the principal ore minerals in the Liba deposit. Therefore, this mineral assemblage serves as both a direct indicator of gold occurrence and a critical criterion for discriminating hydrothermal mineralization stages and predicting high-grade ore zones, while the associated alteration minerals (more sericite and carbonates, less chlorite) correspond well with hyperspectral responses observed in these zones.

Figure 11c shows the microscopic characteristics of sericite in a cataclastic slate from ore-grade zones under crossed polarizers. Sericite occurs as rust-colored anhedral grains constituting ~90% modal abundance, reflecting intense metamorphic overprint. Disseminated opaque grains, interpreted as pyrite, occur sparsely and show subhedral to anhedral crystal forms. Opaque minerals are evenly distributed throughout the matrix, constituting ~10% modal abundance.

Figure 11d shows calcite in cataclastic slate from ore-grade zones under plane-polarized light. Calcite grains are relatively euhedral, with well-developed cleavage, locally arranged in multiple intersecting sets. Intracrystalline deformation structures are prominent, with well-developed twins that exhibit bending, offset, and crosscutting microfractures. These features indicate that calcite experienced post-crystallization ductile deformation under tectonic stress and underwent multi-stage structural modification. Fine inclusions are sparsely distributed within calcite grains, occurring as isolated dots to elongate bands, with some preferentially concentrated along cleavage planes and microfractures. Locally along grain margins, irregular dissolution–replacement textures and dark mineral infillings are observed, reflecting reactivation by late-stage hydrothermal fluids and superimposed alteration. The occurrence of calcite is consistent with the carbonate-related spectral features identified in hyperspectral data.

https://doi.org/10.3390/min16050526

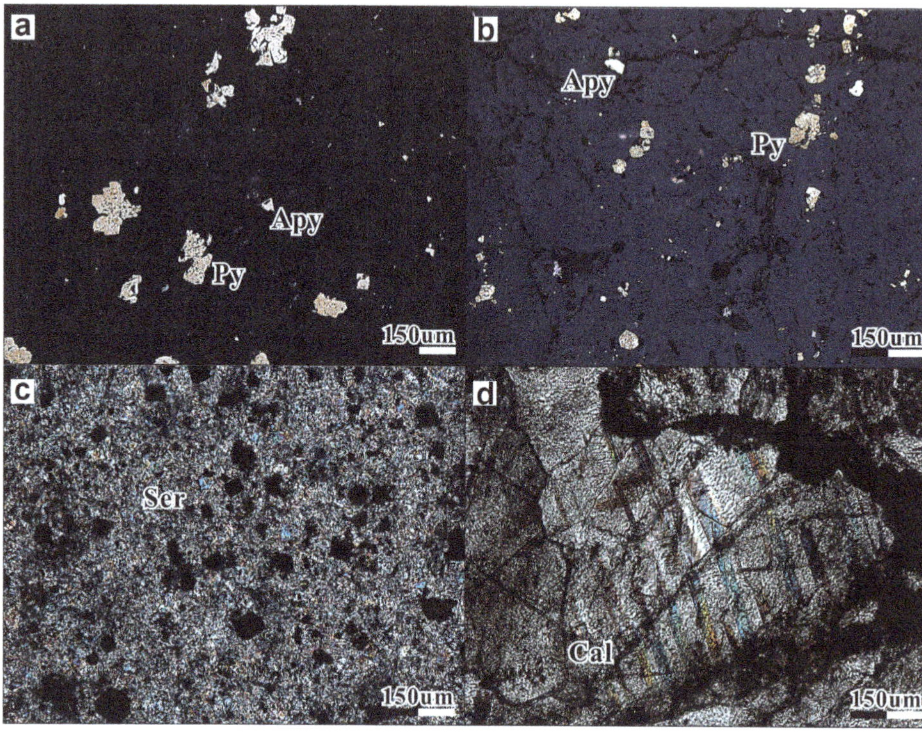

Figure 11. Microscopic photograph of ore body end in the Liba area. (**a**) Pyrite (Py) and arsenopyrite (Apy) from Orebody 6 under plane-polarized light; (**b**) pyrite and arsenopyrite from Orebody 26 under plane-polarized light; (**c**) sericite (Ser) in cataclastic slate from ore-grade zones under crossed polarizers; (**d**) calcite (Cal) in cataclastic slate from ore-grade zones under plane-polarized light.

6. Discussion

6.1. Hyperspectral Characteristics of the Surface and Core

The study demonstrates that hyperspectral data provide a reliable and efficient way of identifying mineral assemblages and alteration zones. The quantitative comparisons presented here are based on relative mineral abundances derived from hyperspectral data using the built-in interpretation algorithms in TSG (The Spectral Geologist) (Table 3). These values represent relative spectral abundances rather than absolute mineral contents but provide a consistent basis for evaluating spatial variations in mineral distribution across alteration zones.

Table 3. Average relative abundances of major minerals in different alteration zones.

	Barren	Weakly Mineralized Domain	Orebody
	Mean (%)	**Mean (%)**	**Mean (%)**
sericite	49	52	71
chlorite	38	24	12
kaolinite	2	12	8
calcite	3	6	2
biotite	1	0.3	0

It should be noted that all mineral abundance values in Table 3 are expressed as percentages. However, these percentages refer only to the five minerals emphasized in this

https://doi.org/10.3390/min16050526

study, and other minerals present in the samples, particularly those with relatively low abundances or weak spatial regularity, were not included in this calculation. Accordingly, these values are intended to compare relative variations among the selected alteration minerals rather than to represent a closed mineralogical mass balance.

These data indicate systematic variations in mineral assemblages from distal to ore zones, characterized by progressive chlorite and biotite depletion and relative enrichment of sericite. Although some minerals (e.g., kaolinite, carbonate) show local variability, this behavior is likely influenced by their relatively low modal abundances and spatially heterogeneous distribution. In particular, minerals occurring at low abundance are more sensitive to local lithological variability and sampling effects, while late-stage alteration minerals may exhibit patchy or discontinuous distributions controlled by fluid pathways and permeability contrasts. These factors can lead to deviations from idealized zonation patterns without contradicting the overall alteration trend.

In distal alteration zones, chlorite minerals (predominantly Fe-chlorite and Fe–Mg chlorite) are widespread and abundant; biotite also occurs in relatively high abundance, reflecting intense metasomatism of wall rocks by early low-temperature hydrothermal fluids [2]. Carbonatization is also moderately developed in these zones. Importantly, sericite is present across all alteration zones and maintains relatively high abundance, although its abundance shows a moderate increase toward ore zones, indicating that sericitization represents a pervasive alteration style throughout the deposit rather than being confined to proximal or ore-grade zones. This pervasive distribution of white mica is consistent with observations in other major gold systems, such as the Lavra Velha deposit in Brazil, where hyperspectral data revealed that white mica abundance and its compositional shifts (e.g., shifting to more phengitic compositions) provide a highly sensitive exploratory vector toward higher-grade mineralization [67]. Our results confirm that the relative enrichment of sericite, coupled with the systematic depletion of Fe-Mg silicates like chlorite, serves as a reliable proximal indicator for gold mineralization in the Liba deposit.

In proximal alteration zones, chlorite abundance decreases markedly, particularly Fe-chlorite, indicating that hydrothermal evolution toward orebodies resulted in progressive diminution of chloritization, likely reflecting increasing temperature, fluid acidity, and changes in fluid composition along the fluid pathways. Concurrently, kaolinite and carbonate abundances increase locally, reflecting evolving hydrothermal fluid composition and temperature conditions.

In ore-grade zones, sericite constitutes the dominant alteration mineral, attaining maximum abundance and exhibiting the most continuous distribution throughout the entire alteration profile, with pervasive fine-grained muscovitization characteristics. Chlorite is markedly diminished or absent in these zones, reflecting hydrothermal conditions characterized by elevated temperature and enhanced acidity that destabilized chlorite and promoted sericitization. This inverse relationship between sericite and chlorite perfectly mirrors the globally recognized fluid–rock interaction models for orogenic gold systems. As established by classical studies [68], the influx of CO_2- and K-rich auriferous fluids into host rocks typically drives the breakdown of precursor Fe-Mg silicates (e.g., chlorite and biotite) to form robust muscovite/sericite and carbonate assemblages. By quantitatively capturing this precise mineralogical transition through SWIR absorption depth variations, our spectral data provide objective validation of this global thermodynamic fluid model at the Liba deposit. Carbonates and kaolinite locally form co-existing assemblages within ore-grade zones, characteristic of late-stage metasomatic alteration [34].

https://doi.org/10.3390/min16050526

6.2. Hyperspectral Characteristics Under the Microscope

The systematic integration of petrographic observations with hyperspectral data provides a robust framework for validating the accuracy and feasibility of spectral-based alteration mapping. By conducting a detailed microscopic examination of representative core intervals—spanning distal, proximal, and ore-grade zones—we established a strong correspondence between microstructural characteristics and interpreted spectral signatures. This cross-validation is particularly effective in resolving the diagnostic limitations of individual techniques. Specifically, petrography compensates for the inherent constraints of SWIR spectroscopy in identifying opaque metallic sulfides like pyrite and arsenopyrite, which lack diagnostic vibrational absorption features in the SWIR range [69]. Conversely, hyperspectral analysis provides high-resolution identification of fine-grained clay minerals, such as kaolinite, which are often challenging to definitively recognize under the microscope. The excellent agreement between these two independent methods not only confirms the reliability of the vertical alteration zonation but also ensures a comprehensive interpretation of the complex hydrothermal mineral assemblages. Such cross-scale validation addresses a common challenge in hyperspectral geology: the "scaling effect" between point-based spectral measurements and bulk mineralogy. As demonstrated in recent studies characterizing the hydrothermal footprint of deposits like Crixás in Brazil, integrating multi-scale analytical data allows for a more nuanced understanding of mineral paragenesis, particularly in identifying complex overprinting relationships [70].

6.3. Feasibility

Integration of core hyperspectral scanning, microscopic hyperspectral analysis, and petrographic observations demonstrates that hyperspectral core scanning technology exhibits good reliability and practical applicability in delineating alteration zonation based on consistent mineral distribution patterns identified from hyperspectral data within orogenic gold deposits. First, core hyperspectral data clearly reveal systematic downhole variations in mineral assemblage composition throughout the deposit, demonstrating particularly high resolution in mapping the spatial distribution of principal alteration minerals, including chlorite, sericite, kaolinite, and carbonate minerals [23,71]. The observed alteration zonation pattern—characterized by peripheral chlorite-dominated zones grading through transitional assemblages to core sericite-dominated zones—is highly consistent with the typical alteration zonation models for orogenic gold deposits documented in the literature, indicating that this technique effectively captures the thermal and chemical evolutionary trajectory of hydrothermal fluid systems.

Second, integrated microscopic petrographic observations and comparison with hyperspectral results provide robust confirmation of the reliability and accuracy of hyperspectral mineral identification at the core scale. Mineral assemblages identified from hyperspectral data are consistent with those observed at the microscopic scale through petrographic analysis, and the spatial relationships among constituent mineral phases exhibit excellent correspondence with diagnostic spectral absorption characteristics. For example, sericite maintains persistently high modal abundance and exhibits continuous spatial distribution throughout all alteration zones as interpreted from spectral data, whereas chlorite and biotite mineral abundance decreases systematically with decreasing distance from orebody intervals [72]. These systematic compositional trends interpreted from hyperspectral data are unambiguously confirmed through detailed microscopic petrographic examination.

Overall, hyperspectral core scanning technology enables rapid, non-destructive, and spatially continuous characterization of mineral composition and alteration mineralogy, including along drill core intervals. The results obtained through this approach show good agreement with petrographic observations with traditional microscopic petrographic and

https://doi.org/10.3390/min16050526

geochemical analytical results, thereby demonstrating its applicability and reliability for mineralogical characterization. This methodology provides critical technical support for multiple aspects of economic geology research and mineral exploration, including (1) quantitative characterization of hydrothermal alteration zones in gold deposits; (2) systematic analysis of temporal and spatial evolution of mineralization-related alteration processes; and (3) delineation and prioritization of high-potential exploration target areas based on spectral-mineralogical alteration signatures.

While the current dataset provides reliable mineral identification and alteration mapping, there remains scope for further methodological refinement. For instance, standardized illumination protocols and surface preparation techniques could be explored in future studies to further improve data consistency and comparability. Such refinements would complement rather than supersede the present findings, potentially enabling more extensive spatial coverage in subsequent investigations.

7. Conclusions

Based on previous geological knowledge of alteration zonation in the Liba gold deposit, integrated with the hyperspectral core scanning and petrographic verification presented in this study, the main conclusions are as follows:

(1) Wall-rock alteration in the Liba gold deposit exhibits clear zonal differentiation, from an inner silicification–pyrite zone to a sericite–chlorite zone and an outer carbonate zone.
(2) Hyperspectral core scanning effectively delineates the spatial distribution of key alteration minerals, including sericite, chlorite, kaolinite, and calcite, with results consistent with petrographic observations.
(3) Hyperspectral data capture alteration zonation and mineral assemblage evolution and allow characterization of hydrothermal variation along drill cores.
(4) Hyperspectral core scanning is a practical and reliable method for identifying alteration zonation in orogenic gold deposits.
(5) Hyperspectral data can be used to delineate exploration targets and support mineral exploration studies.

Author Contributions: Conceptualization, Z.-H.L. and H.-J.T.; methodology, Z.-H.L. and H.-J.T.; software, Z.-H.L.; validation, Z.-H.L. and T.-T.H.; formal analysis, Z.-H.L.; investigation, Z.-H.L., H.-J.T. and J.-N.F.; resources, D.-F.X., T.Y. and K.-F.Q.; data curation, Z.-H.L., H.-J.T. and T.-T.H.; writing—original draft preparation, Z.-H.L.; writing—review and editing, Z.-H.L. and D.-F.X.; visualization, Z.-H.L.; supervision, D.-F.X.; project administration, K.-F.Q.; funding acquisition, K.-F.Q. All authors have read and agreed to the published version of the manuscript.

Funding: This research was funded by Deep Earth Probe and Mineral Resources Exploration-National Science and Technology Major Project (2025ZD1009600), Gansu Province Key Research and Development Plan—Industrial Projects (No. 24YFGA015), the projects of Ministry of Natural Resources, a new round of Mineral Exploration Breakthrough Strategic Action Science and Technology Support Project (grant ZKKJ202410), the Fundamental Research Funds for the Central Universities (grant 2652024008), the Fundamental Research Funds for the Central Universities of China (grant 2652025005), and the Frontiers Science Center for Deep-time Digital Earth (grant 2652023001).

Data Availability Statement: The data that support the findings of this study are available from the corresponding author upon reasonable request.

Acknowledgments: We thank the editor and anonymous reviewers for numerous helpful suggestions.

Conflicts of Interest: The authors declare no conflicts of interest.

https://doi.org/10.3390/min16050526

References

1. Zhang, J.-Y.; Qiu, K.-F.; Yin, R.; Long, Z.-Y.; Feng, Y.-C.; Yu, H.-C.; Gao, Z.-Y.; Deng, J. Lithospheric Mantle as a Metal Storage Reservoir for Orogenic Gold Deposits in Active Continental Margins: Evidence from Hg Isotopes. *Geology* **2024**, *52*, 423–428. [CrossRef]
2. Tan, H.J.; Qiu, K.F.; Liu, H.C.; Zhang, P.C.; Fu, J.N.; Wang, R.; Luo, J.X.; Shi, Y.M. Wall-Rock Alteration Characteristics of the Liba Orogenic Gold Deposit, West Qinling: Insights from GF-5B (Gaofen-5B) Satellite Hyperspectral Data Extraction. *Acta Petrol. Sin.* **2024**, *40*, 1784–1800.
3. Qiu, K.-F.; Yu, H.-C.; Deng, J.; McIntire, D.; Gou, Z.-Y.; Geng, J.-Z.; Chang, Z.-S.; Zhu, R.; Li, K.-N.; Goldfarb, R. The Giant Zaozigou Au-Sb Deposit in West Qinling, China: Magmatic- or Metamorphic-Hydrothermal Origin? *Miner. Depos.* **2020**, *55*, 345–362. [CrossRef]
4. He, D.-Y.; Qiu, K.-F.; Yu, H.-C.; Jowitt, S.M.; Zheng, X.; Mazumder, R.; Deng, J. Lithospheric Architecture and Evolution of the Qinling Orogen of Central China and Associated Controls on Metallogeny. *Earth-Sci. Rev.* **2025**, *264*, 105092. [CrossRef]
5. Groves, D.I. The Crustal Continuum Model for Late-Archaean Lode-Gold Deposits of the Yilgarn Block, Western Australia. *Miner. Depos.* **1993**, *28*, 366–374. [CrossRef]
6. Wang, Q.F.; Deng, J.; Zhao, H.S.; Yang, L.; Ma, Q.Y.; Li, H.J. Advances in Orogenic Gold Research: Implications for Orogenic Gold Metallogenesis in China. *Earth Sci.* **2019**, *44*.
7. Naleto, J.L.C.; Perrotta, M.M.; da Costa, F.G.; de Souza Filho, C.R. Point and Imaging Spectroscopy Investigations on the Pedra Branca Orogenic Gold Deposit, Troia Massif, Northeast Brazil: Implications for Mineral Exploration in Amphibolite Metamorphic-Grade Terrains. *Ore Geol. Rev.* **2019**, *107*, 283–309. [CrossRef]
8. Pu, F.C. Geological Characteristics and Orebody Occurrence Regularities of the Liba Gold Deposit, Li County, Gansu Province Geological Characteristics and Orebody Occurrence Regularities of the Liba Gold Deposit, Li County, Gansu Province. *Gansu Metall.* **2018**, *40*, 59–63, 68. [CrossRef]
9. Pan, Z.; Liu, J.; Ma, L.; Chen, F.; Zhu, G.; Qin, F.; Zhang, H.; Huang, J.; Li, Y.; Wang, J. Research on Hyperspectral Identification of Altered Minerals in Yemaquan West Gold Field, Xinjiang. *Sustainability* **2019**, *11*, 428. [CrossRef]
10. Wan, Y.; Fan, Y.; Jin, M. Application of Hyperspectral Remote Sensing for Supplementary Investigation of Polymetallic Deposits in Huaniushan Ore Region, Northwestern China. *Sci. Rep.* **2021**, *11*, 441–451. [CrossRef] [PubMed]
11. Fan, Y.; Wan, Y.; Wang, H.; Yang, X.; Liang, M.; Pan, C.; Zhang, S.; Wang, W.; Tan, F. Application of an Airborne Hyper-Spectral Survey System CASI/SASI in the Gold-Silver-Lead-Zinc Ore District of Huaniushan, Gansu, China. *Geol. Croat.* **2021**, *74*, 73–83. [CrossRef]
12. Ke, C.; Xiao, Y.; Fan, Y.; Luo, L.; Cai, Y.; Shen, X.; Wang, G.; Wu, J.; Zhuang, L.; Wang, J.; et al. A New Time-Stamped Glacier Inventory of High Mountain Asia Based on Deep Learning and Remote Sensing Big Data in 2020. *Sci. China Earth Sci.* **2026**, *69*, 236–252. [CrossRef]
13. Mao, Z.; Wang, M.; Chu, J.; Sun, J.; Liang, W.; Yu, H. Feature Extraction and Analysis of Reclaimed Vegetation in Ecological Restoration Area of Abandoned Mines Based on Hyperspectral Remote Sensing Images. *J. Arid Land* **2024**, *16*, 1409–1425. [CrossRef]
14. Corrado, F.; Putzolu, F.; Armstrong, R.N.; Mondillo, N.; Chirico, R.; Casarotto, B.; Massironi, M.; Fuller, D.; Ball, R.; Herrington, R.J. Application of Satellite and Proximal Hyperspectral Sensing to Target Lithium Mineralization in Volcano-Sedimentary Deposits: A Case Study from the McDermitt Caldera, USA. *Remote Sens. Environ.* **2025**, *323*, 114724. [CrossRef]
15. Pour, A.B.; Zoheir, B.; Pradhan, B.; Hashim, M. Editorial for the Special Issue: Multispectral and Hyperspectral Remote Sensing Data for Mineral Exploration and Environmental Monitoring of Mined Areas. *Remote Sens.* **2021**, *13*, 519. [CrossRef]
16. Bedini, E. The Use of Hyperspectral Remote Sensing for Mineral Exploration: A Review. *J. Hyperspectr. Remote Sens.* **2017**, *7*, 189–211. [CrossRef]
17. Mandende, H.; Ndou, C.; Mothupi, T. Hyperspectral Core Scanner: An Effective Mineral Mapping Tool for Apatite in the Upper Zone, Northern Limb, Bushveld Complex. *J. S. Afr. Inst. Min. Metall.* **2023**, *123*, 81–92. [CrossRef]
18. Géring, L.; Kirsch, M.; Thiele, S.; De Lima Ribeiro, A.; Gloaguen, R.; Gutzmer, J. Spectral Characterisation of Hydrothermal Alteration Associated with Sediment-Hosted Cu–Ag Mineralisation in the Central European Kupferschiefer. *Solid Earth* **2023**, *14*, 463–484. [CrossRef]
19. Guha, A. Mineral Exploration Using Hyperspectral Data. In *Hyperspectral Remote Sensing*; Elsevier: Amsterdam, The Netherlands, 2020; pp. 293–318.
20. Tuşa, L.; Khodadadzadeh, M.; Contreras, C.; Rafiezadeh Shahi, K.; Fuchs, M.; Gloaguen, R.; Gutzmer, J. Drill-Core Mineral Abundance Estimation Using Hyperspectral and High-Resolution Mineralogical Data. *Remote Sens.* **2020**, *12*, 1218. [CrossRef]
21. Contreras Acosta, I.C.; Khodadadzadeh, M.; Gloaguen, R. Resolution Enhancement for Drill-Core Hyperspectral Mineral Mapping. *Remote Sens.* **2021**, *13*, 2296. [CrossRef]
22. Tusa, L.; Andreani, L.; Khodadadzadeh, M.; Contreras, C.; Ivascanu, P.; Gloaguen, R.; Gutzmer, J. Mineral Mapping and Vein Detection in Hyperspectral Drill-Core Scans: Application to Porphyry-Type Mineralization. *Minerals* **2019**, *9*, 122. [CrossRef]

 https://doi.org/10.3390/min16050526

23. Lypaczewski, P.; Rivard, B.; Gaillard, N.; Perrouty, S.; Piette-Lauzière, N.; Bérubé, C.L.; Linnen, R.L. Using Hyperspectral Imaging to Vector towards Mineralization at the Canadian Malartic Gold Deposit, Québec, Canada. *Ore Geol. Rev.* **2019**, *111*, 102945. [CrossRef]
24. Arne, D.; House, E.; Pontual, S.; Huntington, J. Hyperspectral Interpretation of Selected Drill Cores from Orogenic Gold Deposits in Central Victoria, Australia. *Aust. J. Earth Sci.* **2016**, *63*, 1003–1025. [CrossRef]
25. Souza, S.R.C.D.; Ducart, D.F.; Botelho, N.F. Reflectance Spectroscopy Applied to Mapping the Hydrothermal Alteration at the Mamão Gold Deposit, Andorinhas Greenstone Belt, Carajás Mineral Province, Northern Brazil. *Braz. J. Geol.* **2024**, *54*, e20230048. [CrossRef]
26. Fu, M.; Dai, J.; Wang, X.; Hu, Z.; Ma, X.; Peng, B.; Wan, X. Characteristics of the Shortwave-Thermal Infrared Spectra of the Zhuxi Superlarge Tungsten Deposit in Jiangxi Province and Their Exploration Significance. *Ore Geol. Rev.* **2023**, *158*, 105535. [CrossRef]
27. Jiang, T.F.; Zhi, C.L.; An, M.G.; Han, Z.R.; Shang, Z.; Xie, L.L.; Tang, H.M. Zonation of Alteration Mineral Spectral Signatures Based on Hyperspectral Core Scanning and Its Application to Mineral Exploration: A Case Study of the Xishan Rare Earth Deposit, Shandong Province. *Shandong Land Resour.* **2023**, *39*, 110–118.
28. Yu, H.-C.; Qiu, K.-F.; Deng, J.; Zhu, R.; Mathieu, L.; Sai, S.-X.; Sha, W.-J. Exhuming and Preserving Epizonal Orogenic Au-Sb Deposits in Rapidly Uplifting Orogenic Settings. *Tectonics* **2022**, *41*, e2021TC007165. [CrossRef]
29. Qiu, K.-F.; Deng, J. Petrogenesis of Granitoids in the Dewulu Skarn Copper Deposit: Implications for the Evolution of the Paleotethys Ocean and Mineralization in Western Qinling, China. *Ore Geol. Rev.* **2017**, *90*, 1078–1098. [CrossRef]
30. Mao, J.; Qiu, Y.; Goldfarb, R.; Zhang, Z.; Garwin, S.; Fengshou, R. Geology, Distribution, and Classification of Gold Deposits in the Western Qinling Belt, Central China. *Miner. Depos.* **2002**, *37*, 352–377. [CrossRef]
31. Chang, L.L. Polyphase Deformation and Tectonic Evolution of the Northern Margin Faults of the West Qinling Orogen and Their Geodynamic Implications. Master's Thesis, Tianjin Chengjian University, Tianjin, China, 2023.
32. He, C.G. Metallogenic Characteristics and Genesis of Typical Gold Deposits in the Northern Sub-Belt of the West Qinling Orogen. Ph.D. Dissertation, China University of Geosciences (Wuhan), Wuhan, China, 2023.
33. Gansu Nonferrous Metals Geological Exploration Bureau Tianshui Mineral Exploration Institute. *Gansu Province Lixian Liba Gold Mine Exploration Report [Report]*; Longnan Zijin Mining Co., Ltd.: Tianshui, China, 2017.
34. Wang, Z.; Liu, J.; Carranza, E.J.M.; Xue, R.; Zhai, D.; Weng, G.; Zhao, Q.; Wang, L.; Wang, G.; Zhang, B. Constraints on the Metallogenesis of the Liba Gold Deposit, West Qinling Orogen, China: Evidence from Rutile U–Pb Dating, Pyrite Geochemistry and Sulfides in-Situ S–Pb Isotopes. *Ore Geol. Rev.* **2025**, *184*, 106741. [CrossRef]
35. Tang, Y.Q. Geological and Geochemical Characteristics and Mineral Assemblages of the Jinshan Gold Deposit, Li County, Gansu Province. *Gansu Metall.* **2004**, *2*, 7–9. [CrossRef]
36. Chen, Y.; Wang, Y.W.; Luo, J.X.; Li, D.D.; Wang, J.P.; Long, L.L.; Zhang, H.Q.; Geng, J. Characteristics and Genesis of Gold-Bearing Sulfides in the Liba Gold Deposit, West Qinling. *Acta Petrol. Sin.* **2025**, *41*, 1387–1405. [CrossRef]
37. Yang, S.-S.; Liu, J.-J.; Zhang, F.; Yuan, F.; Zhai, D.-G.; Li, Y.-J.; Liao, Y.; Cao, Q. Petrogenesis and Geodynamic Setting of the Triassic Granitoid Plutons in West Qinling, China: Insights from LA-ICP-MS Zircon U–Pb Ages, Lu–Hf Isotope Signatures and Geochemical Characteristics of the Zhongchuan Pluton. *Int. Geol. Rev.* **2017**, *59*, 1908–1928. [CrossRef]
38. Zhu, L.; Zhang, G.; Yang, T.; Wang, F.; Gong, H. Geochronology, Petrogenesis and Tectonic Implications of the Zhongchuan Granitic Pluton in the Western Qinling Metallogenic Belt, China. *Geol. J.* **2013**, *48*, 311–334. [CrossRef]
39. Ke, C.H.; Wang, X.X.; Yang, Y.; Tian, Y.F.; Li, J.B.; Nie, Z.R.; Lv, X.Q.; Wang, S.A.; Gong, M.Q. Petrogenesis of Dikes in the West Qinling Region and Their Relationship to Gold Mineralization: Constraints from Geochronology, Geochemistry, and Nd-Hf–S Isotopes of the Liba Gold Deposit. *Miner. Depos.* **2020**, *39*, 42–62. [CrossRef]
40. Langa, W.M.; Ndou, C.; Zieger, L.; Harris, P.; Wagner, N. Hyperspectral Imaging of Coal Core: A Focus on the Visible-near-Shortwave Infrared (VN-SWIR) Region. *Int. J. Coal Geol.* **2024**, *284*, 104456. [CrossRef]
41. Rotem, A.; Vidal, A.; Pfaff, K.; Tenorio, L.; Chung, M.; Tharalson, E.; Monecke, T. Interpretation of Hyperspectral Shortwave Infrared Core Scanning Data Using SEM-Based Automated Mineralogy: A Machine Learning Approach. *Geosciences* **2023**, *13*, 192. [CrossRef]
42. Thiele, S.T.; Kirsch, M.; Lorenz, S.; Saffi, H.; El Alami, S.; Contreras Acosta, I.C.; Madriz, Y.; Gloaguen, R. Maximising the Value of Hyperspectral Drill Core Scanning through Real-Time Processing and Analysis. *Front. Earth Sci.* **2024**, *12*, 1433662. [CrossRef]
43. Sinaice, B.B.; Owada, N.; Saadat, M.; Toriya, H.; Inagaki, F.; Bagai, Z.; Kawamura, Y. Coupling NCA Dimensionality Reduction with Machine Learning in Multispectral Rock Classification Problems. *Minerals* **2021**, *11*, 846. [CrossRef]
44. Kokaly, R.F.; Clark, R.N.; Swayze, G.A.; Livo, K.E.; Hoefen, T.M.; Pearson, N.C.; Wise, R.A.; Benzel, W.M.; Lowers, H.A.; Driscoll, R.L.; et al. *USGS Spectral Library Version 7*; U.S. Geological Survey Data Series; U.S. Geological Survey: Reston, VA, USA, 2017.
45. van der Meer, F.D.; van der Werff, H.M.A.; van Ruitenbeek, F.J.A.; Hecker, C.A.; Bakker, W.H.; Noomen, M.F.; van der Meijde, M.; Carranza, E.J.M.; de Smeth, J.B.; Woldai, T. Multi- and Hyperspectral Geologic Remote Sensing: A Review. *Int. J. Appl. Earth Obs. Geoinf.* **2012**, *14*, 112–128. [CrossRef]
46. Reniff, L. Transferring the 45/0 Spectral Reflectance Factor Scale. *Color Res. Appl.* **1994**, *19*, 332–340. [CrossRef]

 https://doi.org/10.3390/min16050526

47. Hubbard, B.E.; Gallegos, T.J.; Stengel, V.; Hoefen, T.M.; Kokaly, R.F.; Elliott, B. Hyperspectral (VNIR-SWIR) Analysis of Roll Front Uranium Host Rocks and Industrial Minerals from Karnes and Live Oak Counties, Texas Coastal Plain. *J. Geochem. Explor.* **2024**, *257*, 107370. [CrossRef]
48. Wei, J.; Ming, Y.; Jia, Q.; Yang, D. Simple Mineral Mapping Algorithm Based on Multitype Spectral Diagnostic Absorption Features: A Case Study at Cuprite, Nevada. *J. Appl. Remote Sens.* **2017**, *11*, 26015. [CrossRef]
49. Ni, L.; Xu, H.; Zhou, X. Mineral Identification and Mapping by Synthesis of Hyperspectral VNIR/SWIR and Multispectral TIR Remotely Sensed Data with Different Classifiers. *IEEE J. Sel. Top. Appl. Earth Obs. Remote Sens.* **2020**, *13*, 3155–3163. [CrossRef]
50. Moghadam, H.J.; Oskouei, M.M.; Nouri, T. Unmixing of Hyperspectral Data for Mineral Detection Using a Hybrid Method, Sar Chah-e Shur, Iran. *Arab. J. Geosci.* **2020**, *13*, 1041. [CrossRef]
51. Clark, R.N.; Roush, T.L. Reflectance Spectroscopy: Quantitative Analysis Techniques for Remote Sensing Applications. *J. Geophys. Res. Solid Earth* **1984**, *89*, 6329–6340. [CrossRef]
52. NV5 Geospatial. Continuum Removal. ENVI Documentation. Available online: https://www.nv5geospatialsoftware.com/docs/continuumremoval.html (accessed on 8 May 2026).
53. Baldridge, A.M.; Hook, S.J.; Grove, C.I.; Rivera, G. The ASTER Spectral Library Version 2.0. *Remote Sens. Environ.* **2009**, *113*, 711–715. [CrossRef]
54. Li, Z.H.; Ma, S.F. Applications of Infrared Spectroscopy in Mineral Exploration. *China Met. Bull.* **2024**, *3*, 149–151.
55. Yao, Y.; Li, Z.; Liu, J.; Fu, J.; Hou, T.; Zhang, Y. BIF-Hosted High-Grade Magnetite Iron Ore Targeting by Hyperspectral Wavelength Mapping of Chlorite: Case Study of Qidashan Iron Mine, Northeast China. *Geocarto Int.* **2023**, *38*, 2226109. [CrossRef]
56. Lypaczewski, P.; Rivard, B. Estimating the Mg# and AlVI Content of Biotite and Chlorite from Shortwave Infrared Reflectance Spectroscopy: Predictive Equations and Recommendations for Their Use. *Int. J. Appl. Earth Obs. Geoinf.* **2018**, *68*, 116–126. [CrossRef]
57. Santamaría-López, Á.; Suárez, M.; García-Romero, E. Detection Limits of Kaolinites and Some Common Minerals in Binary Mixtures by Short-Wave Infrared Spectroscopy. *Appl. Clay Sci.* **2024**, *250*, 107269. [CrossRef]
58. Tian, J.; Zhang, Y.; Cheng, J.; Sun, S.; Zhao, Y. Short Wavelength Infra-Red (SWIR) Characteristics of Hydrothermal Alteration Minerals in Skarn Deposits: Example from the Jiguanzui Cu–Au Deposit, Eastern China. *Ore Geol. Rev.* **2019**, *106*, 134–149. [CrossRef]
59. AusSpec International Ltd. *Spectral Interpretation Field Manual*, 3rd ed.; AusSpec International Ltd.: Sydney, Australia, 2008; Volume 1, pp. 59–110.
60. Hajaj, S.; El Harti, A.; Jellouli, A.; Beiranvand Pour, A.; Mnissar Himyari, S.; Hamzaoui, A.; Hashim, M. ASTER Data Processing and Fusion for Alteration Minerals and Silicification Detection: Implications for Cupriferous Mineralization Exploration in the Western Anti-Atlas, Morocco. *Artif. Intell. Geosci.* **2024**, *5*, 100077. [CrossRef]
61. Lovering, T.S. *Rock Alteration as a Guide to Ore—East Tintic District, Utah*; Society of Economic Geologists: Littleton, CO, USA, 1949.
62. Liu, Y.-H.; Mao, J.; Miggins, D.P.; Qiu, K.-F.; Hu, J.; Wang, L.; Xu, D.-M.; Goldfarb, R.J. 40Ar/39Ar Geochronology Constraints on Formation of the Tuwaishan Orogenic Gold Deposit, Hainan Island, China. *Ore Geol. Rev.* **2020**, *120*, 103438. [CrossRef]
63. Asadzadeh, S.; Koellner, N.; Chabrillat, S. Detecting Rare Earth Elements Using EnMAP Hyperspectral Satellite Data: A Case Study from Mountain Pass, California. *Sci. Rep.* **2024**, *14*, 20766. [CrossRef]
64. Parry, W.T.; Downey, L.M. Geochemistry of Hydrothermal Chlorite Replacing Igneous Biotite. *Clays Clay Miner.* **1982**, *30*, 81–90. [CrossRef]
65. Liu, Y.-H.; Mao, J.; Hu, J.; Ma, J.-D.; Wang, L.; Xu, D.-M.; Zhao, Z.-X.; Bishikwabo, G.K.; Gamboa-Herrera, J.A.; Fan, C.; et al. Element Migration and Hydrothermal Alteration of the Tuwaishan Orogenic Gold Deposit in West Hainan Island, South China. *Ore Geol. Rev.* **2025**, *180*, 106568. [CrossRef]
66. Eggleton, R.A.; Banfield, J.F. The Alteration of Granitic Biotite to Chlorite. *Am. Mineral.* **1985**, *70*, 902–910. [CrossRef]
67. Medina, C.M.; Ducart, D.F.; Passos, J.S.; De Oliveira, L.R. Exploration Vectoring from the White Mica Spectral Footprint in the Atypical Auriferous Lavra Velha Deposit, San Francisco Craton, Brazil. *Ore Geol. Rev.* **2021**, *139*, 104438. [CrossRef]
68. Groves, D.I.; Goldfarb, R.J.; Gebre-Mariam, M.; Hagemann, S.G.; Robert, F. Orogenic Gold Deposits: A Proposed Classification in the Context of Their Crustal Distribution and Relationship to Other Gold Deposit Types. *Ore Geol. Rev.* **1998**, *13*, 7–27. [CrossRef]
69. Clark, R.N. Spectroscopy of Rocks and Minerals, and Principles of Spectroscopy. In *Manual of Remote Sensing*; Rencz, A.N., Ed.; John Wiley & Sons: New York, NY, USA, 1999; pp. 3–58.
70. Campos, L.M.; Toledo, C.L.B.; Silva, A.M.; Ducart, D.F.; Dos Santos, B.A.; Campos, M.P.; Borges, C.C.A. The Hydrothermal Footprint of the Crixás Deposit: New Vectors for Orogenic Gold Exploration in Central Brazil. *Ore Geol. Rev.* **2022**, *146*, 104925. [CrossRef]

 https://doi.org/10.3390/min16050526

71. Clark, R.N.; King, T.V.V.; Klejwa, M.; Swayze, G.A.; Vergo, N. High Spectral Resolution Reflectance Spectroscopy of Minerals. *J. Geophys. Res. Solid Earth* **1990**, *95*, 12653–12680. [CrossRef]
72. Jianzhong, F.; Dongbo, W.; Xueming, W.; Yishan, Z.; Tiefeng, L. Magmatic Gold Mineralization in the Western Qinling Orogenic Belt: Geology and Metallogenesis of the Baguamiao, Liba and Xiaogouli Gold Deposits. *Acta Geol. Sin. Engl. Ed.* **2004**, *78*, 529–533. [CrossRef]

https://doi.org/10.3390/min16050526

MDPI AG
Grosspeteranlage 5
4052 Basel
Switzerland
Tel.: +41 61 683 77 34

Minerals Editorial Office
E-mail: minerals@mdpi.com
www.mdpi.com/journal/minerals

www.ingramcontent.com/pod-product-compliance
Lightning Source LLC
Chambersburg PA
CBHW040800250926
51850CB00007BA/1041
* 9 7 8 3 7 2 5 8 8 0 4 7 8 *